Spectroscopic Methods in Organic Chemistry

Ian Fleming • Dudley Williams

Spectroscopic Methods in Organic Chemistry

Seventh Edition

Ian Fleming
Department of Chemistry
University of Cambridge
Cambridge
UK

Dudley Williams
Department of Chemistry
University of Cambridge
Cambridge
UK

ISBN 978-3-030-18251-9 ISBN 978-3-030-18252-6 (eBook)
https://doi.org/10.1007/978-3-030-18252-6

This Springer imprint is published by the registered company Springer Nature Switzerland AG
The registered company address is: Gewerbestrasse 11, 6330 Cham, Switzerland

Preface

This book is the seventh edition of a well-established introductory guide to the interpretation of the mass, ultraviolet, infrared and nuclear magnetic resonance spectra of organic compounds. It is a textbook suitable for a course in the application of these techniques to structure determination and as a handbook for organic chemists to keep on their desks throughout their career.

These four spectroscopic methods are used routinely to determine the structure of organic compounds. Every organic chemist needs to be skilled in how to apply them and to know which method works for which problem. In outline, the mass spectrum gives the molecular formula, the ultraviolet spectrum identifies conjugated systems, the infrared or Raman spectrum identifies functional groups and the nuclear magnetic resonance spectra identify how the atoms are connected. One or more of these techniques are frequently enough to identify the complete chemical structure of an unknown compound. If they are not enough on their own, there are other methods that the organic chemist can turn to: X-ray diffraction, microwave absorption, electron spin resonance, atomic force spectroscopy and circular dichroism, among others. Powerful though they are, these techniques are all more specialised and less part of the everyday practice of most organic chemists.

In preparing this edition, I have moved the chapter on mass spectrometry to be the first, in order to reflect its importance in giving a molecular formula—a vitally important first step in a structure determination. Along with the following chapters on UV and IR spectra, it remains concise, to reflect the lesser importance of all these techniques, with the discussion of the theoretical background kept to a minimum, since application of spectroscopic methods is possible without a detailed command of the theory behind them.

Benefitting from my experience teaching the subject over the last 7 years, I have completely rewritten the chapter on NMR spectroscopy, dividing it into two to separate the discussion of 1D spectra from the 2D spectra. In addition, I have greatly expanded the text with explanations of the spin physics that the earlier edition entirely lacked. I have introduced each of the pulse sequences, and the vector description of the resultant magnetisa-

---

CRY page is updated for previous edition information.

tion, as each is needed in order to understand its effect on the spectra. It is nowhere near a rigorous account but simply provides the minimum level of understanding an organic chemist *using* NMR needs.

The augmented tables of data at the ends of the chapters on MS, IR and 1D-NMR remain an important resource for reference by any chemist practising organic synthesis. Finally, in Chap. 6, there are 11 worked examples illustrating the ways in which the four spectroscopic methods can be brought together to solve simple structural problems, and in Chap. 7, there are 34 problem sets with increasing difficulty for you to work through, many more with 2D NMR spectra than in the earlier editions.

I have been helped by Drs. Richard Horan, Chris Jones and Ed Anderson for their work on the sixth edition that has survived into this; Duncan Howe in Cambridge and Dr. Dean Olson and Lingyang Zhu in Urbana Champaign took the many new NMR spectra; my colleague Jeremy Sanders has given me much useful advice, both about this book and more generally; several colleagues (Chris Urch, Scott Denmark, Alain Krief, Mike Aldersley and Rob Britton) have been generous with samples or spectra that have found their way into the book; I am most grateful to them all and finally to Manuel Perez for the gift of yearly licences to use the Mestrenova program for processing NMR spectra; I used Photoshop and ChemDraw to prepare all the figures.

Cambridge, UK Ian Fleming

Contents

Mass Spectra

1

1.1 Introduction

A mass spectrometer is a device for producing and weighing ions from a compound for which we wish to obtain molecular weight and structural information. All mass spectrometers use three basic steps: molecules M are taken into the gas phase; ions, such as the cations $M^{\bullet+}$, MH^+ or MNa^+, are produced from them (unless the molecules are already charged); and the ions are separated according to their mass-to-charge ratios (m/ze). The value of z is normally one, and since e is a constant (the charge on one electron), m/z gives the mass of the ion. Some of the devices that are used to produce gas-phase ions put enough vibrational energy into the ions to cause them to fragment in various ways to produce new ions with smaller m/z ratios. Through this fragmentation, structural information can be obtained.

Because the molecular weight and the molecular formula of an unknown are usually the first pieces of information to be sought in the investigation of a chemical structure, mass spectrometry is often the first of the spectroscopic techniques to be called upon. For that reason this chapter comes first.

A mass spectrometer detects only the charged components (e.g. MH^+, and its associated fragments A^+, B^+, C^+, etc.), because only they are contained, accelerated or deflected by the electromagnetic or electrostatic fields used in the various analytical systems, and only they give an electric signal when they hit the collector plate. When the array of ions has been separated and recorded, the output is known as a mass spectrum. It is a record of the abundance of each ion reaching the detector (plotted vertically) against its m/z value (plotted horizontally). The mass spectrum is a result of a series of competing and consecutive unimolecular reactions, and what it looks like is determined by the chemical structure and reactivity of the sample molecules. It is not a spectro*scopic* method based on electromagnetic radiation, but since it complements information provided to the organic chemist by the UV, IR and NMR spectra, it is conveniently considered alongside them.

I. Fleming, D. Williams, *Spectroscopic Methods in Organic Chemistry*,
https://doi.org/10.1007/978-3-030-18252-6_1

Mass spectrometry is the most sensitive of all these methods. It can be carried out routinely with a few micrograms of sample, and in favourable cases even with picograms (10^{-12} g), making it especially important in solving problems where only a very small sample is available, as in the detection and analysis of such trace materials as pheromones, atmospheric pollutants, pesticide residues, and of drugs and their metabolites, especially in forensic science and in medical research. It is even possible to use mass spectrometry to analyse individual slices of tissue to detect the change from cancerous to non-cancerous as an operation is in progress.

There is a variety of instruments available for taking a mass spectrum—they differ in the methods by which molecules are taken into the gas phase, they differ by the way in which molecules are induced to produce ions, and they differ by the method used to analyse the ions. Perhaps most significantly, as far as this book is concerned, they also differ in the degree to which they induce fragmentation. In this chapter, we shall see how these instruments provide molecular weight and structural information for molecules with a relatively low molecular weight, say <1000 Da. These are the molecules that are the main concern of this book—natural products, for example, freshly isolated and with unknown structures, or the products of chemical reactions, where we want to know whether or not they have the structures we expected.

The methods commonly used to produce ions are covered first, followed by descriptions of the most common ways by which the ions may be separated according to their mass-to-charge ratios (m/ze). The bulk of the chapter then covers the appearance of mass spectra, and how they may be interpreted. One of the methods, electron impact, is described in more detail, because fragmentation induced by this method has for many years been an everyday workhorse in structure determination. A huge amount of information gathered by this method is stored in databases, which enable analysts rapidly to identify known compounds.

1.2 Ion Production

1.2.1 Electron Impact (EI)

This method of ion production is used in the analysis of relatively volatile organic molecules. These typically have molecular weights up to a maximum of around 400 Da, but, in cases like glucose with its numerous polar hydroxyl groups the volatility is too low, even though it has a much lower molecular weight. The sample is simply heated to evaporate it into the ionisation chamber (Fig. 1.1), which is kept at a very low pressure, typically $\leq 10^{-4}$ Nm^{-2} ($\leq 10^{-6}$ mmHg), to avoid collisions between molecules. Alternatively, and more commonly, the sample is placed on the ceramic tip of a probe which can be inserted into the ion source. When the low pressure has stabilised, the tip is heated to 200–300 °C to drive the molecules off the surface into the ionisation cham-

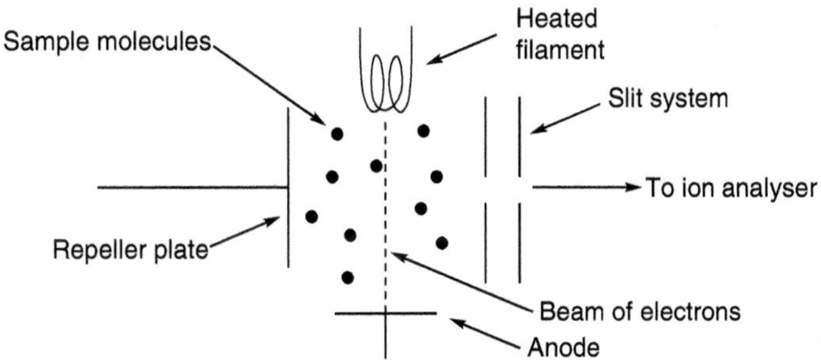

Fig. 1.1 Ionisation by electron impact (EI)

ber. Most organic molecules are stable at these temperatures provided that they do not collide with other molecules.

At the same time as the molecules enter the ionisation chamber, electrons are driven off a heated filament by attraction to an anode, usually through a potential difference of about 70 eV (1 eV ≈ 23 kcal mol^{-1} ≈ 96 kJ mol^{-1}). A 70 eV electron has enough energy to expel an electron from a molecule with which it collides; this expulsion of an electron corresponds to the ionisation potential (IP) of the molecule, which typically requires around 7–10 eV. Through this process, a radical-cation ($M^{\bullet+}$) is formed.

$$\mathrm{M} + \mathrm{e} \rightarrow \mathrm{M}^{\bullet+} + 2\mathrm{e}$$

A repeller plate at one end of the ionisation chamber is positively charged, and therefore repels the positively charged radical-cations through a slit system into the mass spectrometer, where they, and fragments derived from them, will be separated according to their m/z values.

Electron capture to give a radical-anion does not occur to a significant extent, because the bombarding electrons have too much translational energy to be captured. The 70 eV electron does not deposit all its energy into a molecule with which it interacts. In addition to the 7–10 eV required for ionisation of the sample, a further 0–6 eV is typically transferred as internal energy into the resulting ion. Since the strongest single bonds in organic molecules have strengths of about 4 eV, and many are much weaker, this is enough internal energy to lead to extensive fragmentation in most EI spectra.

1.2.2 Chemical Ionisation (CI)

In the numerous applications of mass spectrometry, frequently the single most important requirement is to obtain a molecular weight, and hence, if the mass measurement is precise enough, a molecular formula. Chemical ionisation (CI) cuts down the amount of fragmentation that is such a feature of EI spectra: it imparts less energy than EI into the molecule, and the molecular ion that is generated is intrinsically more stable because it is not a radical.

In CI, a reagent gas, usually a small hydrocarbon like methane or isobutane, or more commonly ammonia, is passed into the ion chamber at a pressure of about 10^2 Nm^{-2}. This gas is ionised by using electrons, produced from a hot filament as before, but with energies up to 300 eV, giving rise (in the case of methane) to $CH_4^{\bullet+}$. At the operating pressures of CI sources, this ion collides with its neutral counterparts, which are present in much higher concentration than the sample molecules. The main bimolecular reaction that occurs is, for methane as the carrier gas:

$$CH_4^{\bullet+} + CH_4 \rightarrow CH_5^{+} + CH_3^{\bullet}$$

If sample molecules are volatilised into this mixture of ions, they are protonated by the CH_5^+ ion, which is an exceptionally strong acid:

$$M + CH_5^{+} \rightarrow MH^{+} + CH_4$$

Thus, in positive-ion CI spectra, molecular weight information is obtained from protonation of sample molecules, and the observed *m/z* value is one unit higher than the molecular weight. When isobutane is used to generate CI spectra, the reagent ion which protonates the sample ($C_4H_{11}^+$) is still a strong acid, but less kinetic energy is passed to the sample being analysed and the degree of fragmentation is reduced. Similarly, because NH_4^+ binds a proton much more strongly than CH_5^+, less energy is released in the transfer of a proton to M from the former than the latter, and again the degree of fragmentation is less. Some molecules M are not basic enough to be protonated by NH_4^+, so in these cases CH_5^+ or $C_4H_{11}^+$ must be employed instead. Thus, ammonia CI producing MH^+ (and/or MNH_4^+) is an excellent technique for the determination of molecular weights of volatile and functionalised molecules, with little fragmentation occurring. In any CI mass spectrum, the ions have an even number of electrons, and are for that reason, as well as from having a lower internal energy, less susceptible to fragmentation than the ions produced in an EI source.

Although EI does not produce satisfactory negative-ion spectra, negative-ion CI works well for molecules with electron-accepting properties, such as trifluoroacetates, quinones and nitro compounds. This is because the collisions occurring in a CI source reduce the large initial kinetic energies of the bombarding electrons to lower values at which they can be captured to give a radical-anion. Alternatively, a reagent ion such as CH_3O^- may be generated in the CI source, and this can act as a Brønsted base, abstracting a proton from the sample molecule:

$$M + CH_3O^{-} \rightarrow (M - H)^{-} + CH_3OH$$

A schematic diagram of a CI source is given in Fig. 1.2. The sample may be introduced directly by lowering the upper probe, which can be a gas line or can incorporate a heated ceramic tip; or indirectly from the output of a GC instrument by raising the lower probe. The ion chamber has holes in it to allow an electron beam to be directed through it. The whole of the ion chamber can be withdrawn from the centre of the vacuum chamber, the

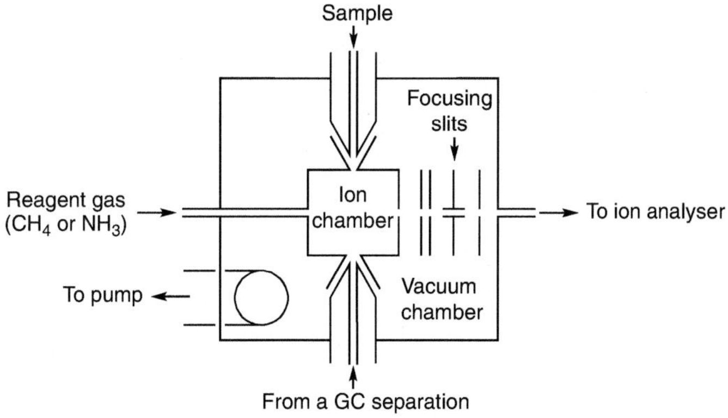

Fig. 1.2 Ionisation by chemical ionisation (CI)

Sample
Focusing slits
Reagent gas (CH_4 or NH_3)
Ion chamber
To ion analyser
To pump
Vacuum chamber
From a GC separation

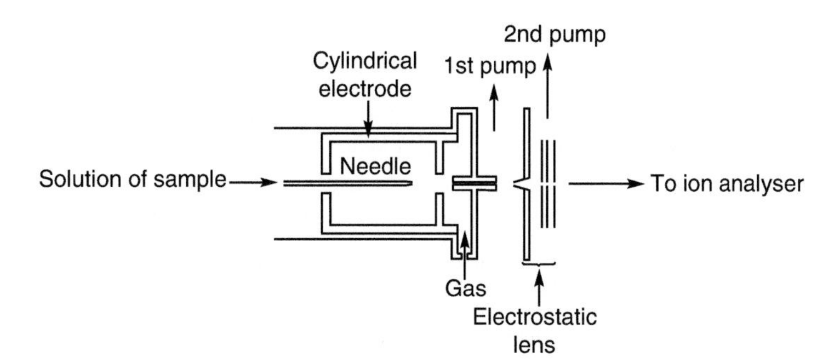

Fig. 1.3 Ionisation in an electrospray (ESI)

reagent gas supply can be cut off and the vacuum improved, allowing the same instrument to be used for EI spectra. In either CI or EI mode, the path to the analyser will be pumped to the usual low pressure to minimise further collisions.

1.2.3 Electrospray Ionisation (ESI)

In ESI, relatively involatile organic compounds are 'carried along' into the gas phase by the evaporation of a solution in the presence of a strong electric field. The 'electrospray' is the fine mist expelled from a capillary needle when a potential difference typically of 3–6 kV is applied between the end of the capillary (along which a solution flows, generally 1–10 μL/min) and a cylindrical electrode located 0.3–2 cm away (Fig. 1.3). The mist is at or near atmospheric pressure, and consists of highly charged liquid droplets. The charge on these droplets may be selected as positive or negative according to the sign of the voltage applied to the capillary. ESI is especially useful since it can be used to analyse directly the effluent from an HPLC column.

The use of a 'sheath' gas or 'nebulising' gas promotes efficient spraying of the solution of the sample from the capillary. Sample molecules dissolved in the spray are released from the droplets by evaporation of the solvent. This evaporation is accomplished by passing a drying gas across the spray before it enters another capillary and a vacuum chamber.

As the droplets are multiply charged, and reduced in size by evaporation, the rate of desolvation is increased by repulsive Coulombic forces. These forces eventually overcome the cohesive forces of the droplet, and an MH^+ ion free of solvent is produced. For organic acids (carboxylic acids, sulfate esters and phosphate esters), the electrode will be set to abstract rather than add a proton, and the ions produced will have a negative charge $(M - H^+)^-$. The charged particles are carried, by an appropriate electric field into an ion analyser. A typical droplet will contain a large number of sample molecules, and the separation of these, avoiding their aggregation, is promoted by charge-charge repulsion. Solvents used in ESI are typically water/methanol mixtures; and frequently, in the case of positive ion work, traces of a volatile organic acid such as formic acid.

1.2.4 Fast Ion Bombardment (FIB or LSIMS)

In FIB desorption, the energy is provided by a beam of ions (most commonly Cs^+) of high translational energy, typically several keV. In an older technique, which was called fast atom bombardment (FAB), a beam of atoms was used rather than ions, but atoms are rarely used today. (FIB spectra are still sometimes called FAB.) Typically, a solution of a few micrograms of the sample is dissolved in a few microlitres of glycerol [$HOCH_2CH(OH)CH_2OH$] acting as a matrix of low volatility. A schematic illustration of the FIB source is given in Fig. 1.4.

When the Cs^+ ions strike the solution of the sample, the sample is desorbed, often as an ion, by momentum transfer. It is for this reason that the name LSIMS is also used for this technique, since the **S**econdary **I**ons are desorbed from a **L**iquid matrix. Neutral molecules M will also be desorbed, but since the polar molecules being analysed by FIB usually have relatively acidic (e.g. $-CO_2H$) or basic (e.g. $-NH_2$) functional groups, the corresponding ions ($-CO_2^-$ or $-NH_3^+$) are also desorbed. The prevailing electric fields in the source then ensure that only the ions are efficiently transmitted into the analyser.

The beam only penetrates about 10 nm into the matrix. It is therefore helpful if the sample is marginally less hydrophilic than the matrix, so that the sample is concentrated near the surface. On the other hand, it should be in solution to avoid clumping. For both reasons, therefore, there is a need to have a range of matrices in order to find one that suits the sample. These include a thioglycerol-diglycerol mixture (1:1), tetragol

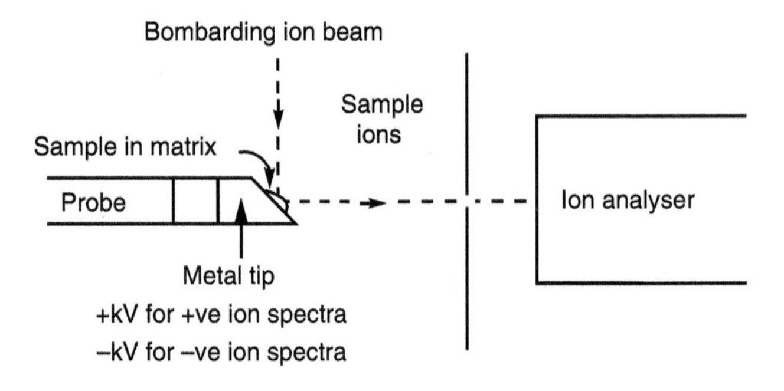

Fig. 1.4 Ionisation by fast ion bombardment (FIB)

$[HO(CH_2CH_2O)_4H]$ and teracol $[HO(CH_2CH_2CH_2CH_2O)_nH]$, which are successively less hydrophilic than glycerol itself.

It is not uncommon to observe $(M + Na)^+$ or $(M + K)^+$ ions in FIB spectra, from the presence of traces of the salts of these cations. Hence, it is also possible deliberately to create such ions by adding sodium or potassium salts to the matrix, where the presence of MH^+ and MNa^+ ions 22 mass units apart can help to identify the molecular ion.

1.2.5 Laser Desorption (LD) and Matrix-Assisted Laser Desorption (MALDI)

In this ionisation method, a large energy pulse is passed to the sample from a laser. The sample molecule is thereby induced to leave its solid or liquid environment within a time of the order of 10^{-12} s. This time is too short to achieve an equilibrium distribution of the energy. Thermal decomposition is thereby reduced, or avoided altogether, in spite of the large amount of energy used. Efficient and controllable energy transfer to the sample requires resonant absorption of the molecule at the laser wavelength, either in the UV, which can induce electronic excitation, or in the IR, which can excite vibrational states. Typically, the laser pulses are applied for times in the range 1–100 ns.

The version of the technique that is most often used, especially suitable for large peptides, proteins, oligonucleotides and oligosaccharides, is matrix-assisted laser desorption (MALDI). In this method, a low concentration of the sample is embedded either in a matrix (molar ratio ranging from 1:100 to 1:50000), which is selected to absorb strongly the laser light. A list of suitable matrix compounds is given in Table 1.1. In addition to these matrices, α-cyano-4-hydroxycinnamic acid (a UV absorbing matrix) is most commonly used in the analysis of peptides—an important field because of the importance of MS in the study of proteins, from which the peptides are often derived.

The energy absorbed by the matrix is transferred indirectly to the sample. The matrix is chosen to have solubility properties similar to those of the sample, in order that the sample molecules are properly dispersed. Higher molecular weight oligomeric 'clumps' are

Table 1.1 Matrices for MALDI mass spectra

Matrix	Form	Usable wavelengths
Nicotinic acid	Solid	266 nm, 2.94 μm, 10.6 μm
2,5-Dihydroxybenzoic acid	Solid	266 nm, 337 nm, 355 nm, 2.79 μm, 2.94 μm, 10.6 μm
Sinapinic acid	Solid	266 nm, 337 nm, 355 nm, 2.79 μm, 2.94 μm, 10.6 μm
Succinic acid	Solid	2.94 μm, 10.6 μm
Glycerol	Liquid	2.79 μm, 2.94 μm, 10.6 μm
Urea	Solid	2.79 μm, 2.94 μm, 10.6 μm
Tris buffer (pH 7.3)	Solid	2.79 μm, 2.94 μm, 10.6 μm

produced as 2 M^+, 3 M^+, and so on, but these are usually minor components of the spectrum if a well-matched matrix is chosen.

1.3 Ion Analysis

Once ions have been obtained in the gas phase, they are repelled into an ion analyser, in order to be separated according to their *m*/*ze* ratios.

1.3.1 Magnetic Analysers

The most simple method of analysis is to deflect the ions using a strong electrostatic or magnetic field, as in Fig. 1.5, in which the poles of the magnet lie above and below the plane of the paper. Within the magnetic sector, ions of larger mass are deflected less than ions of smaller mass, according to Eq. (1.1); where *B* is the strength of the magnetic field, *r* is the radius of the circular path in which the ion is travelling; and *V* is the potential used to accelerate the ions when they leave the source, determining the velocity of an ion when it enters the analyser.

$$\frac{m}{z} = \frac{B^2 r^2}{2V} \tag{1.1}$$

At specific values of *B* and *V*, paths of different radius *r* are followed for each value of *m*/*z*. By scanning the magnetic field *B*, Eq. (1.1) can be satisfied successively for ions of all *m*/*z* ratios, given the fixed values of *r* and *V*.

In older spectrometers, a high mass resolution was achieved by passing the ion beam through an electrostatic sector as well as a magnetic sector. In such a double-focusing mass spectrometer, ion masses can be measured with an accuracy of about 1 p.p.m., but these spectrometers are cumbersome, and largely unused today.

1.3.2 Time-Of–Flight (TOF) Analysers

TOF instruments are linear, bench-top devices, significantly smaller than the magnetic sector instruments. Additionally, TOF ion analysis is more sensitive because *all* the ions in

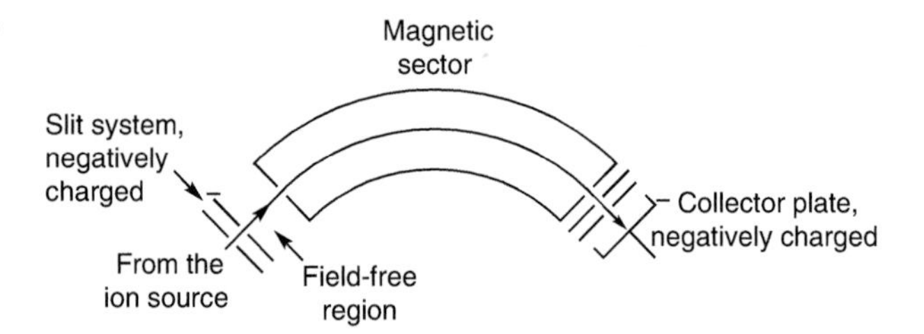

Fig. 1.5 The path followed by ions of a specific *m*/*z* value is indicated, for which the values of *V* and *B* are appropriate to let these ions pass through the collector slits

the sample are collected at the detector, instead of only those meeting the criterion of Eq. (1.1) at any instant. Ions with charge z are accelerated through a potential difference V, and thereby acquire a translational energy zeV. The relationship between the time of flight t and the distance d travelled through a field-free tube to reach the detector is given by Eq. (1.2).

$$t^2 = \frac{m}{z}\left(\frac{d^2}{2Ve}\right) \tag{1.2}$$

All those ions with a single charge possess a translational energy eV. In a given instrument, where the term in brackets in Eq. (1.2) is kept constant, the time taken for an ion of mass m to reach the detector is proportional to $\sqrt{m}$, with ions of largest mass having the lowest velocities and the longest time of flight over a given distance. Since mass analysis is achieved through the accurate measurement of time, the ions must be produced in pulses.

A voltage drop from the source to the slit A (Fig. 1.6) accelerates the ions entering the analyser, they pass into a field-free tube between plates A and B. The resolution of TOF instruments is improved by the use of a device known as a reflectron, which is more complicated than a simple mirror. Since it is common to have differences in arrival times between successive mass peaks of $<10^{-7}$ s, fast electronics are required to distinguish successive peaks.

1.3.3 Quadrupole Analysers

The arrangement of electrodes in a quadrupole mass filter is given in Fig. 1.7. A voltage U, typically 500–2000 V, and a radio-frequency potential V, oscillating from −3000 to +3000 V, are applied between opposite pairs of four parallel rods, which are 0.1–0.3 m long in most commercial instruments. Ions are injected along the z direction, along which

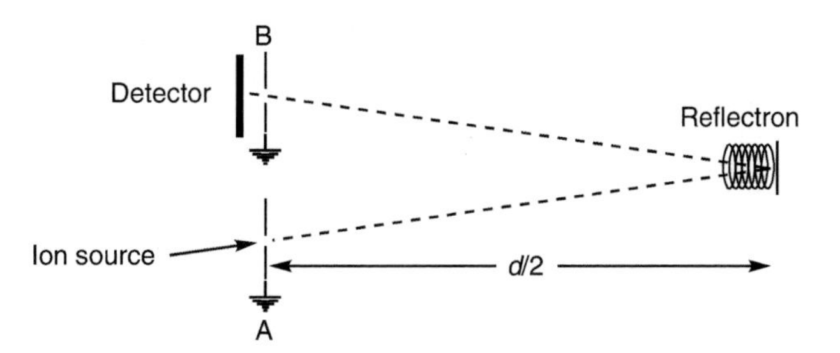

Fig. 1.6 Time-of-flight analyser (TOF)

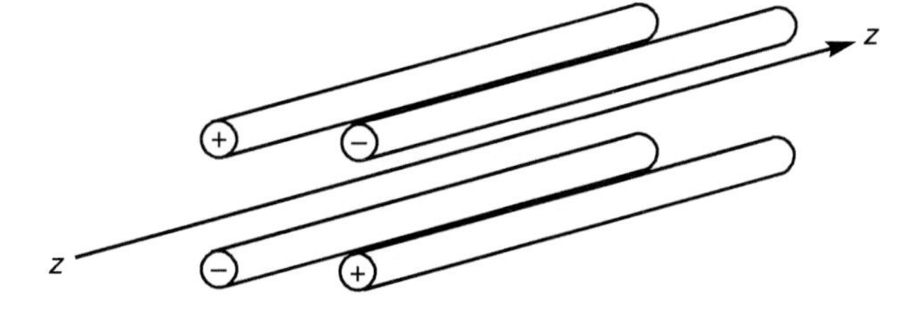

Fig. 1.7 Quadrupole analyser (QA)

they maintain a constant velocity. They travel with a wave pattern in the x and y directions (mutually orthogonal to each other and to z), controlled by the fluctuating potentials on the rods, so that under any given set of conditions ions of only one m/z value arrive at the detector, the others being captured by the rods. All ions are successively brought to the detector by varying the amplitude of U and V, while keeping the ratio U/V constant, or by varying the radio-frequency potential V.

Quadrupole mass spectrometers are relatively small, robust, and especially useful where GC and/or HPLC instruments are directly coupled to the mass spectrometer, when high resolution is not required, and where simplicity of operation and speed of analysis are important.

1.3.4 Ion Cyclotron Resonance (ICR) Analysers

In ICR, a pulse of ions is injected into a cell at low translational energy, where a uniform magnetic field B constrains them into a circular path perpendicular to the direction of the magnetic field. For a singly charged ion, the frequency ω_c (the number of turns round the circle each second) is given by Eq. (1.3).

$$\omega_c = \frac{eB}{m} \tag{1.3}$$

If an alternating electric field of frequency ω_l is applied normal to B, an ion will absorb energy when $\omega_c = \omega_l$. Thus, an ICR mass spectrum can be obtained by fixing B and scanning ω_l so that ions of different mass successively satisfy the equation. The absorption of energy by the ions at resonance is measured using an oscillator detector similar to that used in NMR instruments. Better still, a spread of frequencies may be generated by using a pulsed radio-frequency electric field, and the spectrum analysed by Fourier transformation (FT). For high sensitivity, the ions are not consumed, but kept resonating for times in the range of milliseconds–seconds, necessarily at low operating pressures, typically 10^{-6} Nm^{-2}.

FT-ICR has the advantage of high sensitivity and high resolving power—in a conventional 5 cm cubic cell, as few as ten ions can be detected. Since ion detection is non-destructive, signal detection can in principle continue for long periods, improving the signal-to-noise ratio. High mass resolution can be achieved typically ±1 p.p.m. for molecules of low molecular weight. Resolving power decreases linearly with increasing mass, a drop in resolution to $m/\Delta m$ = 10,000 can be expected for an ion of m/z 10,000.

1.3.5 Ion-Trap Analysers

Ion-trap mass spectrometers use electrodes or magnetic fields to trap gas-phase ions in a small volume. They have the advantages of being relatively compact and, through being able to trap and retain ions, of high sensitivity. The two most common types of ion traps used are the quadrupole ion trap and the Orbitrap.

In the quadrupole ion trap, all ions created over a given time period are trapped inside the analyser through appropriate voltages applied to three electrodes which surround the trapping region. The ions are then sequentially ejected (according to their *m/z* values) towards a conventional electron multiplier detector by gradually changing the potentials applied to the electrodes.

In an Orbitrap mass spectrometer, ions are trapped by an electrostatic attraction to a central electrode balanced by the centrifugal forces that act on the ions as they orbit round this central electrode. As in an FT-ICR mass spectrometer, the mass spectrum is obtained by Fourier transformation of the orbiting frequencies, and the resolving power is high.

In principle, any of the ionisation techniques may be combined with any of the methods of analysis, although only some of these combinations are commercially available. The analysers can also be used in tandem, one after the other (MS-MS), allowing an individual ion from the first analyser to be analysed separately in a second analyser, in order to identify fragmentation pathways. In between the two analysers, energy can be added to the ion by allowing it to collide with high-energy but chemically inert atoms such as argon or helium, or by irradiating it with a radio frequency pulse. This is called collision activated dissociation (CAD) or collision induced dissociation (CID). The high-energy ions are more likely to fragment to give information about their structure. This is especially useful with the *soft* ionisation techniques like CI and ESI, which initially give ions with little susceptibility to fragment.

1.4 Structural Information from EI Mass Spectra

1.4.1 The Features of an EI Spectrum

Figure 1.8 shows a representative mass spectrum illustrating the main features to be found in every mass spectrum. It is the mass spectrum of n-nonane, which proves to have a largely explicable fragmentation pattern.

The most intense peak, at *m/z* 43, is called the *base peak*. The peak at *m/z* 128 is called the *molecular ion* $M^{•+}$, with an intensity of 8% of the intensity of the base peak. There are several *fragment ions*, the intensities of which are similarly recorded as percentages of the height of the base peak. Fragmentation gives rise to a *pattern* of fragment ions like a fingerprint (compare the appearance of the mass spectra of the isomers in Figs. 1.8 and 1.10). A compound may be quickly identified by this pattern if it has already had its mass spectrum recorded. The mass spectrum of nonane would be recorded for publication thus, picking out only the more conspicuous peaks:

m/z (EI) 128 ($M^{•+}$, 8%), 99 (5%), 85 (28%), 71 (22%), 57 (68%), 41 (42%), 29 (37%) and 27 (31%).

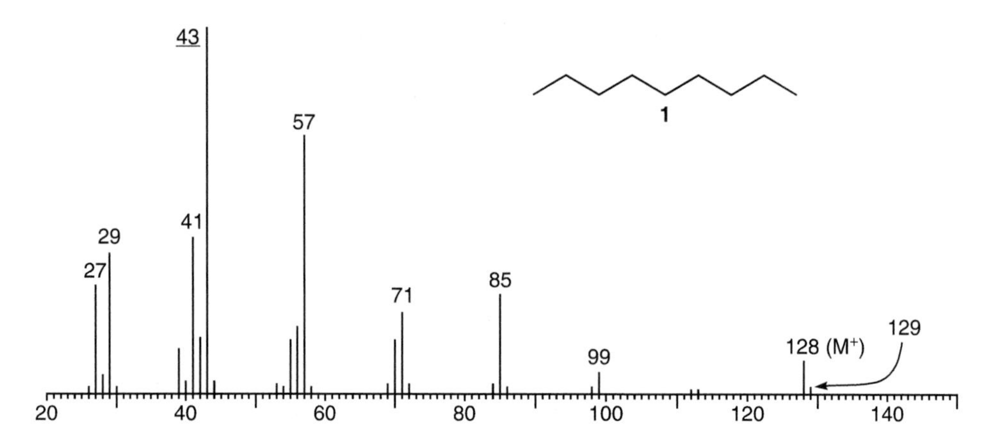

Fig. 1.8 EI mass spectrum of n-nonane

1.4.2 The Molecular Ion

We begin by looking at the molecular ion. If we did not know what this molecule was, the $M^{\bullet+}$ value of 128 could have been produced by molecules with formulae $C_{10}H_8$, C_9H_{20}, C_9H_6N, C_9H_4O, $C_8H_{18}N$, and other combinations with fewer and fewer carbon atoms. Ions with the same nominal integral mass possess different exact masses, because the individual isotopes do not have exactly integral masses. Based on the convention that the atomic weight of ^{12}C is 12.000, the other elements would have atomic masses: ^{1}H 1.008, ^{14}N 14.003 and ^{16}O 15.995 (other data in the first table at the end of this chapter). Using these numbers, the formulae $C_{10}H_8$, C_9H_{20}, C_9H_6N, C_9H_4O and $C_8H_{18}N$ would have accurate molecular weights of 128.0625, 128.1564, 128.0500, 128.0262 and 128.1439, respectively. A mass spectrometer set for high resolution (FTICR or Orbitrap) is accurate to about ±1 p.p.m. for ions with low *m*/*z* values, and is therefore easily able to distinguish between these values. Typically a measured value might be $M^{\bullet+} = 128.1568$, which would identify the sample unambiguously as one of the nonanes C_9H_{20}. Although the level of accuracy falls off with increase in mass, it is still possible to determine unambiguously the molecular formula of most organic compounds with molecular weights up to about 1000 Da.

This information is arguably the single most valuable use of the mass spectrum in everyday structure determination. Spectrometers set to determine high-resolution spectra are equipped with a computer that automatically generates acceptable formulae for the ions it detects, and the chemist can read the formulae directly on the computer's print-out. From the molecular formula it is easy to calculate the number of *double bond equivalents* (DBE), that is the total number of double bonds and rings in the molecule. If the molecule contains no elements other than C, H, N and O, then the number of DBE is given by Eq. (1.4).

$$\text{For } C_aH_bN_cO_d \quad \text{DBE} = \frac{(2a+2)-(b-c)}{2} \tag{1.4}$$

The $(2a + 2)$ term is the number of hydrogens in a saturated hydrocarbon having a carbon atoms. Relative to such a structure, every ring or double bond will mean that there are two fewer hydrogen atoms (cyclohexane is C_6H_{12}, and so is 1-hexene). Therefore, subtracting b, the actual number of hydrogens present, from $(2a + 2)$ and dividing by two gives the total number of double bonds and rings in the molecule. A halogen atom, being univalent, counts as a hydrogen atom. The term c comes in because any nitrogen atom in the molecule will add an extra hydrogen atom. The number of oxygen atoms does not affect the DBE. The knowledge of the number of double bonds and rings is a useful first step in assessing the complexity of the problem facing you, something to keep in your mind as you add information from the fragmentation pattern and, much more importantly, from the other spectroscopic data.

A high-resolution mass spectrum (HRMS) has largely replaced combustion analysis for determining molecular formulae, but it is wise to exercise caution before abandoning combustion analysis altogether. The exact mass measurement works whether the compound is pure or not, and it is easy to accept the existence of a peak with the expected exact mass as proving the formula, when all it does is prove that there was some of that compound present. Combustion analysis has different problems—it can give wrong answers when the compound is not pure, or when there is error in the analysis. It only suggests a possible empirical formula, and it is not capable of assessing the presence or measuring the amount of many elements now commonly found in organic structures.

1.4.3 Isotopic Abundances

The masses and natural abundances of some isotopes which are important in organic mass spectrometry are given in the first table at the end of this chapter. All singly charged ions in the mass spectra of compounds which contain carbon also give rise to a peak at one mass unit higher, because the natural abundance of ^{13}C is 1.1%. For an ion containing n carbon atoms, the abundance of the isotope peak is $n \times 1.1\%$ of the ^{12}C-containing peak. Thus, the molecular ion for nonane ($C_9H_{20}^{\bullet+}$) gives an isotope peak at 129, one mass unit higher than the molecular ion, with an approximate abundance of 10% (9 × 1.1) of the abundance at m/z 128. Obviously, larger molecules with a lot of carbon atoms give much more prominent ($M^{\bullet+}$+1) ions. In the case of small molecules, the probability of finding two ^{13}C atoms in an ion is low, and ($M^{\bullet+}$+2) peaks are accordingly of insignificant abundance. Conversely, $M^{\bullet+}$+2, and even $M^{\bullet+}$+3, $M^{\bullet+}$+4, etc. peaks do become important in very large molecules (see later for details). The ratio of the two peaks ($M^{\bullet+}$+1:$M^{\bullet+}$) gives a rough measure of the number of carbon atoms in the molecule.

While iodine and fluorine are monoisotopic, chlorine consists of ^{35}Cl and ^{37}Cl in a ratio of approximately 3:1, and bromine of ^{79}Br and ^{81}Br in a ratio of approximately 1:1. Molecular ions (or fragment ions) containing various numbers of chlorine or bromine

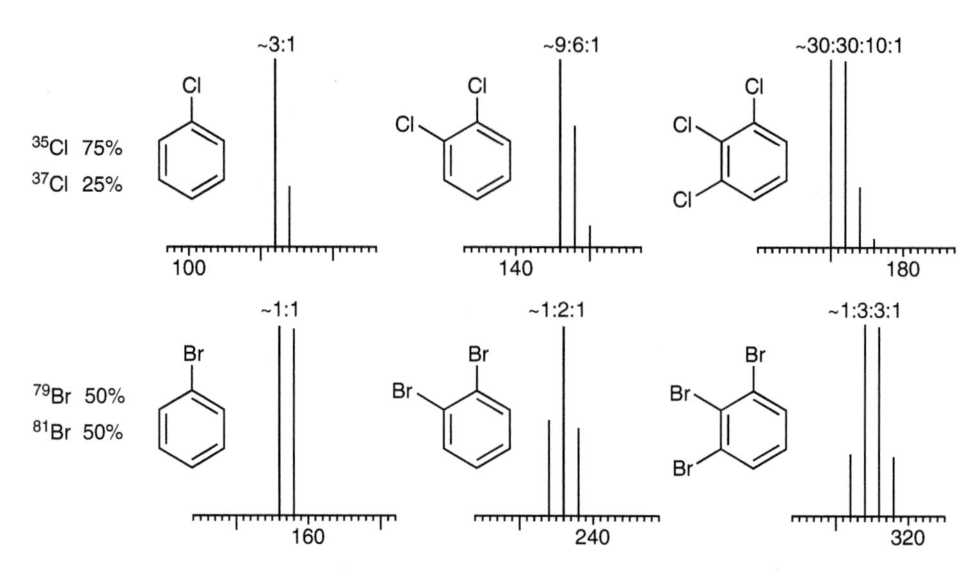

Fig. 1.9 Molecular ions in the mass spectra of halogen-containing compounds

atoms therefore give rise to the characteristic patterns shown in Fig. 1.9, with all peaks spaced 2 mass units apart. Isotope patterns to be expected from any combination of elements can readily be calculated, and they provide a useful test of ion composition in those cases where polyisotopic elements are involved.

1.4.4 Identifying the Molecular Ion

In EI spectra, when fragmentation is especially easy, the molecular ion does not always survive to be analysed, and it is often a problem to decide whether or not the peak at highest mass is a molecular ion. Figure 1.10 shows the EI mass spectrum of the nonane **2**,

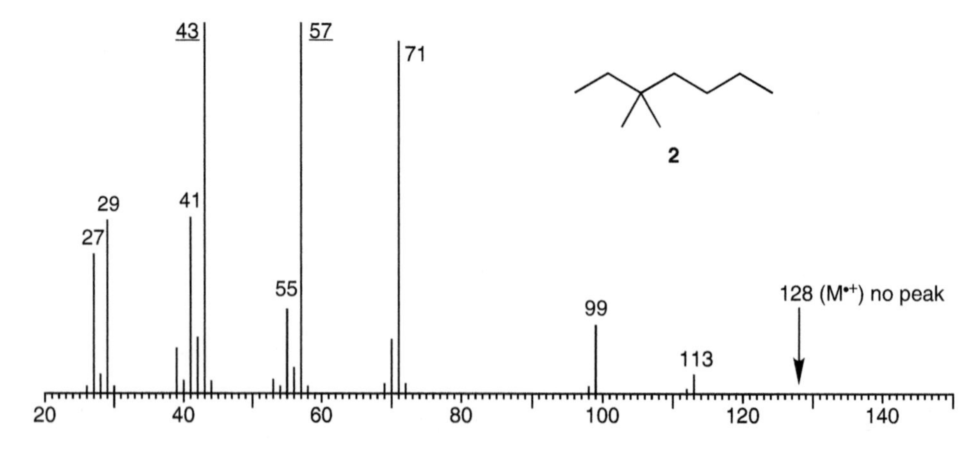

Fig. 1.10 EI mass spectrum of a highly branched nonane

which does not record a molecular ion. The peak at *m/z* 113 could not be the molecular ion *because it is at an odd number*. In this molecule, the only elements are C and H, and all neutral hydrocarbons have an even molecular weight. If the molecule has one or more C, H, O, Si or S atoms, the molecular ion will still be an even number. The same is true with one or more halogens replacing the same number of hydrogen atoms, since all the abundant isotopes of the halogens have odd atomic weights. But if there is a single nitrogen atom, the molecular ion will register as an odd number. Triethylamine is $C_6H_{15}N$, with a molecular weight of 101. More generally, if the neutral molecule has an odd number of nitrogen atoms, the molecular ion will be an odd number, and if it has an even number of nitrogen atoms, the molecular ion will be an even number.

Another useful criterion for identifying a molecular ion is to look at the gap below the ion you suspect is a molecular ion. If there are any peaks corresponding to the loss of 3–14 mass units from the highest recorded ion, that ion is probably *not* the molecular ion. Fragmentations corresponding to the loss of H_3 to H_{11} and of C, CH or CH_2 do not typically occur. Such ions, if they are present, must either come from impurities or be produced by fragmentations of an ion that is greater in mass than the highest recorded ion. Thus, the gap of 14 mass units between the ion at *m/z* 99 and the highest recorded ion at *m/z* 113 in Fig. 1.10 is not likely to represent the loss of a CH_2 group. Its presence is a further indication that the ion at *m/z* 113 is not a molecular ion. Conversely, the absence of any peaks between 113 and 128 in Fig. 1.8 is evidence that the peak at 128 is a molecular ion.

1.4.5 Fragmentation in EI Spectra

In the following sections, the spectra of some representative organic molecules are discussed according to the functional groups they contain (or, in the case of hydrocarbons, do not contain). Although there are some reliable fragmentation pathways induced in EI-MS, not all yield to a simple analysis. Nevertheless, there are some patterns among them that can be interpreted with confidence. If the quantity of material available is insufficient for the other spectroscopic methods, a mass spectrum will have to suffice, and a knowledge of what to expect can be invaluable. On the other hand it is important to remember that you should not expect always to be able to account for every peak in a mass spectrum, or in the worst case even for a majority of them. Structure determination using mass spectrometry is not an exact science.

Aliphatic hydrocarbons The molecular ion produced by EI has an unpaired electron (it is a radical-cation, $M^{•+}$). We can think of its structure as one in which an electron has been ejected from the highest occupied molecular orbital, since that will hold the most loosely bound electrons. In a hydrocarbon like n-nonane **1**, we can simplistically think of the electron as having been removed from one of the σ bonds. Thus, given sufficient vibrational energy, a bond like the one illustrated in the representation **3** can break with the single electron remaining in the bond moving to the right to give an ion **4** and a radical **5**. Equally plausible would be for it to move in the other direction to give the radical **6** and the ion **7**.

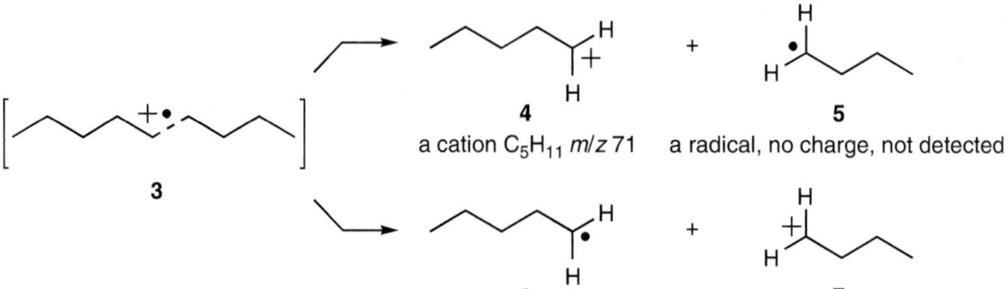

3

4 a cation C_5H_{11} *m/z* 71 + 5 a radical, no charge, not detected

6 a radical, no charge, not detected + 7 a cation C_4H_9 *m/z* 55

The ion **4** with a molecular formula of $C_5H_{11}^+$ gives rise to the peak at *m/z* 71 and the ion **7** with a molecular formula $C_4H_9^+$ gives rise to the peak at *m/z* 55. The radicals **5** and **6** with molecular formulae of C_4H_9• and C_5H_{11}• have no charge, they are not deflected, and they do not appear in the spectrum. In this way, several C_nH_{2n+1} cations can be generated, giving the ion series *m/z* 99, 85, 71, 57 and 43. The lower mass ions of this series may be formed, not only directly, but also by the loss of ethylene **9** from one of the ions with higher mass, as in the fragmentation of the pentyl cation **4** to give the propyl cation **8** in a retro-Friedel-Crafts reaction. Thus the ions with the lower *m/z* values are more abundant than the ions with higher *m/z* values.

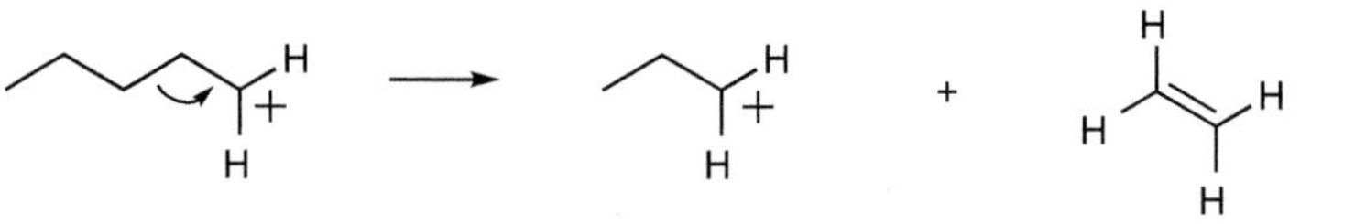

4 C_5H_{11} m/z 71 → 8 a cation C_3H_7 m/z 43 + 9 neutral, not detected

A useful way of conveying the structural features responsible for all the major ions is shown in the drawing **10**, where the wavy lines identify the bonds broken and the numbers (and the side on which they are placed) identify the *m/z* values of the cations produced. Note that the fragmentation of any of the C–H bonds is unfavourable, because hydrogen atoms are very high in energy, and the ion at *m/z* 113 is weak, because the methyl radical is not well stabilised.

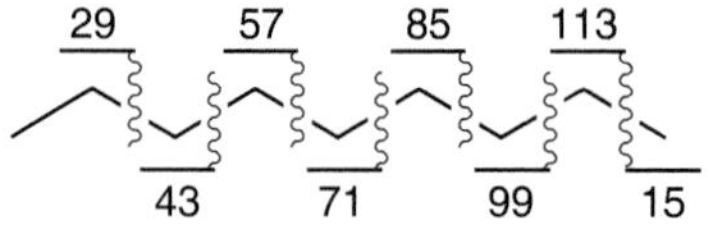

29 57 85 113

43 71 99 15

10

Ions of the general formula $C_nH_{2n-1}^+$ form a minor series of fragment ions at *m/z* 27, 41 and 55, two mass units below the more prominent ions for the ethyl, propyl and butyl cations at 29, 43 and 57. Their formation occurs by loss of a *saturated* hydrocarbon molecule or, less commonly, H_2 from ions of the $C_nH_{2n+1}^+$ series.

All the major fragmentations in this mass spectrum are nicely explained. They illustrate the two generalisations true of almost all EI mass spectra:

1. Radical cations fragment by loss of a radical to give a cation with an even number of electrons.
2. Once the unpaired electron has left the molecule, the fragmentation that takes place is normally by loss of a neutral molecule, rather than of a radical.

Thus molecular ions fragment in the ionisation chamber mainly by loss of a radical to give the cation A^+ in Fig. 1.11. Subsequently, the cation A^+ is most likely to break down with loss of a molecule having all its electrons paired, to give another cation B^+. More rarely, the molecular ion breaks down by loss of a molecule to give another radical cation $C^{\bullet+}$. (We can see an example in Fig. 1.8 in the formation of the weak fragments at *m/z* 98 and 84, which correspond to the loss of ethane and propane from the molecular ion.) When a radical cation does fragment to give another radical cation, further fragmentation will mainly take place by loss of a radical to give a new cation E^+.

The bonds breaking, and the extent of fragmentation in the ionisation chamber, are dependent upon the chemical structure; stable $M^{\bullet+}$ ions with no well stabilised fragments lead to abundant molecular ions in the mass spectrum. Conversely, unstable $M^{\bullet+}$ ions with well stabilised fragments can result in the absence of a molecular ion.

Let us look now at the mass spectrum of 3,3-dimethylheptane **2**, the isomer of n-nonane. The spectrum in Fig. 1.10 shows the same fragment ions as those for n-nonane in Fig. 1.8, but there are subtle differences of intensity giving distinctive fingerprints from each molecule, The not so subtle difference is the absence of the molecular ion at *m/z* 128. It is absent because there are now several energetically favourable fragmentation pathways, with the result that the molecular ion does not survive long enough to escape the ionisation chamber. One of the more subtle differences is the increase in the relative intensities of the peaks at *m/z* 99 and 71, produced by the fragmentations **11** and **12**. These ions carry a higher proportion of the ion current because they are tertiary, and therefore lower in energy

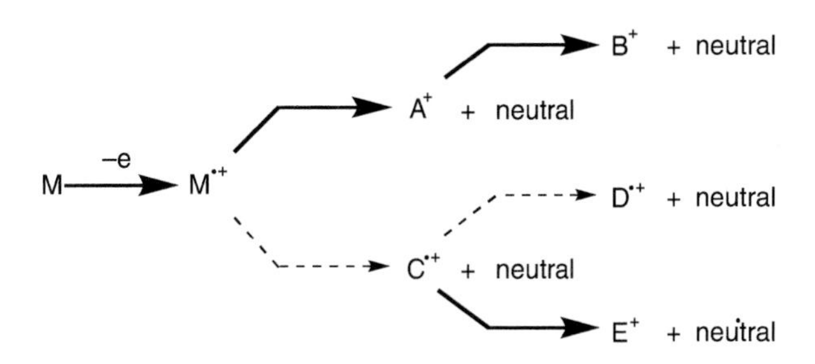

Fig. 1.11 The pathways of fragmentation

(better stabilised by alkyl substitution) than the primary cations produced in the spectrum of n-nonane. A second subtle difference is that the loss of the unstabilised methyl radical in the fragmentation **13** is more prominent, because the formation of the tertiary cation **14** compensates for the high energy of the accompanying methyl radical. The third subtle difference is that the ion at *m*/*z* 57 is also increased in intensity. In this case the fragmentation **12**, with the charge kept by the relatively unstabilised n-butyl cation, is compensated for by the formation of the well-stabilised tertiary pentyl radical.

99 71 57 –Me$^{\bullet}$ +

11 **12** **13** **14** *m*/*z* 113

From the differences between Figs. 1.8 and 1.10, we learn that *the preferred fragmentations are those where the sums of the energies of the fission products are at a minimum* (as is indeed demanded by theory). The second and third tables at the end of this chapter give the enthalpies of formation of common radicals and ions. These can be useful in identifying the preferred sites of fragmentation in radical cations.

The Control of Fragmentation by Functional Groups We now look at the ways in which aromatic rings, and some functional groups, influence fragmentation. In these molecules, the highest energy molecular orbital is frequently a non-bonding atomic orbital on the hetero-atom, or a π-bonded molecular orbital, and it is useful to consider $M^{\bullet+}$ to correspond to the species generated by loss of one of these pairs of electrons.

Benzylic Cleavage In aromatic compounds, the cleavage of a substituent leaving behind a benzyl, or substituted benzyl, cation, is frequently the most favourable pathway, because an aromatic ring can lower the energy of a radical or a cation conjugated to it. In the mass spectrum of toluene in Fig. 1.12, the loss of a hydrogen atom gives the benzyl

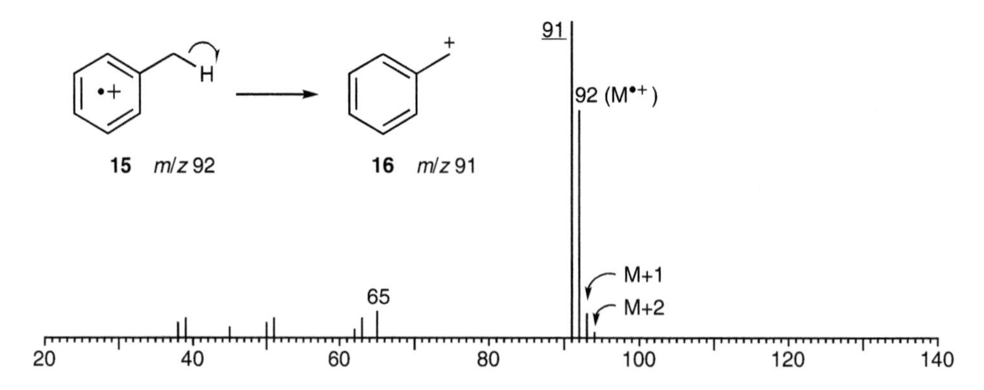

Fig. 1.12 EI mass spectrum of toluene

cation **16** as the base peak at *m*/*z* 91, in spite of the high energy of the naked hydrogen atom. Because the loss of a hydrogen atom is relatively unfavourable, the molecular ion is more abundant than usual, making the M + 1 and M + 2 ions easy to see. The benzyl cation **16** is probably the appropriate structure for *m*/*z* 91 when it is initially generated, but there is evidence that it can rearrange to the aromatic tropylium cation **17** before it fragments further. Such rearrangement is possible because the internal energy of the radical cation **15** and hence of the cation **16** must be very high to enable it to fragment.

Some further fragmentation of **17** occurs by the loss of a molecule of acetylene, probably to give the cyclopentadienyl cation **18**. In accord with the high-energy requirements for any fragmentation of **16**, there is no simple mechanistic pathway, yet, whatever the method of generating the benzyl cation, this sequence of events is always followed.

– H—≡—H

16 *m*/*z* 91 **17** *m*/*z* 91 **18** *m*/*z* 65

The stabilising effect of delocalisation makes the cleavage **15** of benzylic bonds easier than it would otherwise be, and this is helped by the relative ease with which an electron is removed from the π-bonds of the benzene ring in the first place. Cleavage of almost any benzylic substituent R is a common pathway, made even easier when the radical cleaved off is lower in energy than a hydrogen atom. Thus in the mass spectrum of ethylbenzene **19** (Fig. 1.13), the molecular ion, now at *m*/*z* 106, is less abundant than it is for toluene in Fig. 1.11, and the base peak at *m*/*z* 91 now dominates the spectrum, accompanied as usual by the smaller peak at *m*/*z* 65. The phenyl cation **20**, on the other hand, is not stabilised by delocalisation of the positive charge, and the fragment ion *m*/*z* 77 is of low abundance, although its presence hints that the molecule is a monosubstituted benzene.

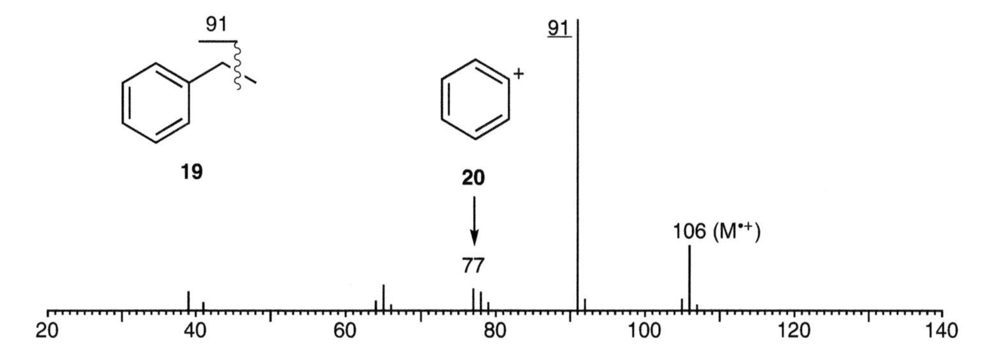

Fig. 1.13 EI mass spectrum of ethylbenzene

Cationic rearrangement is common in chemistry, and therefore also in the mass spectrometer, where it can take place counter-thermodynamically if it leads to structures from which a stable molecule can be easily lost. Thus, t-butylbenzene loses a methyl radical from $M^{•+}$, and then loses a molecule of ethylene. Probably the initially generated cumyl cation $PhCMe_2^+$ rearranges to a 1-phenyl-n-propyl cation $PhCH_2CH_2CH_2^+$, from which ethylene can be lost to give a benzyl cation.

In disubstituted benzenes, there is always formally a choice between the fragmentation of one or the other side chains. Since the energetically easier cleavage is always preferred, it is possible to deduce which fragmentation will occur by measuring the energy requirements for the fragmentation of each side chain when present alone (the second and third tables at the end of this chapter). Thus, *p*-cyano-t-butylbenzene loses a methyl radical, rather than suffering the loss of HCN. On the other hand, a bromine radical is lost exclusively from the molecular ion of *p*-bromoaniline, because the cleavages involving the NH_2 group require higher energies. *o*-Disubstituted benzenes are sometimes anomalous in that fragments made up from parts of *both* substituents can be lost. Thus, *o*-nitrotoluene fragments to give a benzisoxazoline cation by loss of an OH radical, with the hydrogen atom coming from the methyl group and the oxygen atom from the nitro group.

These features of the mass spectra of benzene compounds are also seen in other aromatic compounds and in heterocyclic aromatic compounds.

Alkenes In an alkene, the electron most easily displaced by electron bombardment will be one of the pair in the π bond, just as the electron ejected in aromatic compounds is initially from the aromatic π system. It might be hoped that this would lead to allyl cleavage, by analogy with the ubiquitous benzylic cleavage seen in aromatic systems. In the mass spectrum of 2-methyl-2-hexene **21** in Fig. 1.14 this type of fragmentation can be seen, it gives the peak at *m*/*z* 69 by loss of an ethyl radical, but it does not dominate the spectrum. Instead, the base peak appears to be from the loss of a methyl group, possibly by allylic cleavage from an allylically rearranged alkene **22**. Allylic rearrangements like this are not simple unimolecular organic reactions we can recognise—they are presumably the result of the high energy imparted to the molecular ion. Furthermore, the strong peaks at *m*/*z* 55 and 41, corresponding to a butenyl cation and an allyl cation, are not obviously derived from the starting alkene by any standard radical chemistry.

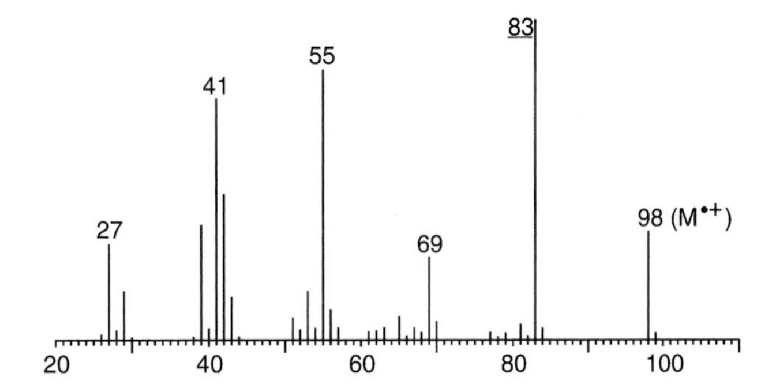

Fig. 1.14 EI mass spectrum of 2-methylpent-2-ene

When allylic cleavage is especially easy, as it is for the cyclohexene **23** in Fig. 1.15, it can dominate the mass spectrum—the loss of a methyl radical from the molecular ion gives a tertiary allylic cation. This mass spectrum also shows a fragment with an even molecular weight 82, an unusual feature but understandable, since it is produced by a retro-Diels-Alder reaction—it is one of the comparatively rare cases where a stable *molecule* is lost from the radical-cation, but it is quite common with cyclohexenes.

109

23•+ ⟶ Me• + *m/z* 109   **23**•+ ⟶ + *m/z* 82

Heteroatom-Based Cleavage Oxygen, sulfur, nitrogen and halogen atoms, often called heteroatoms, have one or more lone pairs of electrons. If they are present in the molecule, the lone pairs are less involved in bonding than are the other electrons, and are usually in

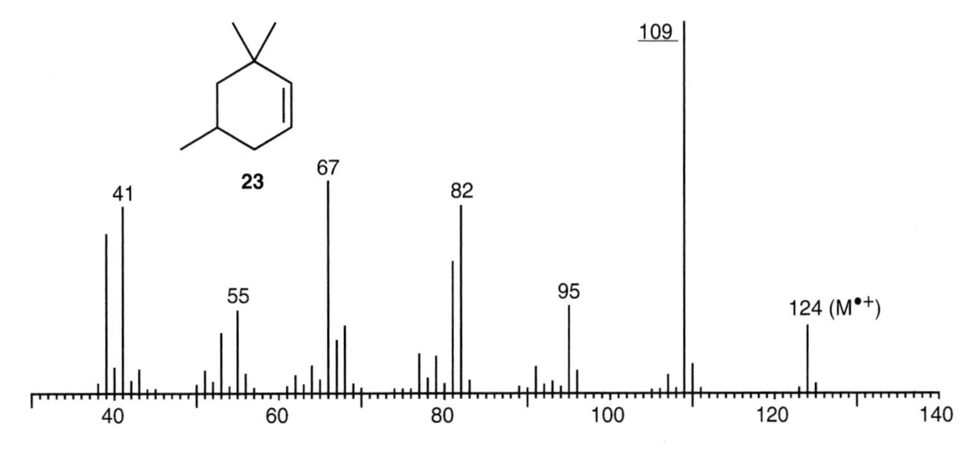

Fig. 1.15 EI mass spectrum of 3,3,5-trimethylcyclohexene

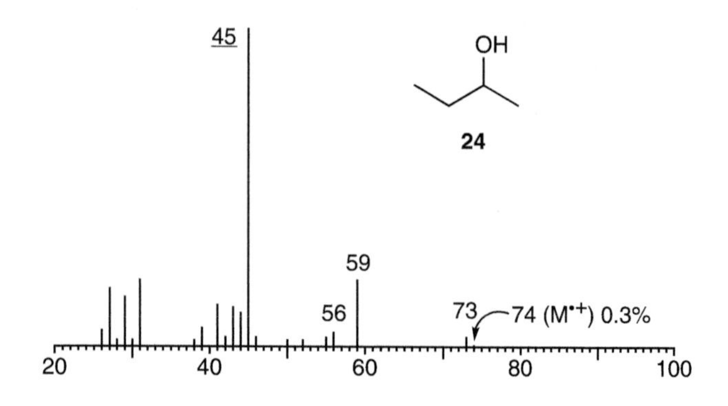

Fig. 1.16 EI mass spectrum of butan-2-ol

the highest-energy occupied molecular orbital. One of them is therefore more easily expelled than an electron in a σ or π bond. The radical cations produced are initially localised on the heteroatom, and fragmentation is therefore easy by cleavage of the β-bond. In contrast to alkenes, this leads to relatively well behaved patterns of fragmentation useful for structure determination. We can see how this works in the spectrum of 2-butanol **24** in Fig. 1.16.

The first-formed ion will be the radical-cation, with the odd electron localised in an orbital on the oxygen atom. This electron will pair with an electron from one of the three bonds β to the oxygen atom: the C–H bond, the bond between C-2 and C-1, and the bond between C-2 and C-3. The other electron in these bonds will be localised on a hydrogen atom, or a methyl radical, or an ethyl radical. Thus, the fragments carrying the current will be the protonated butanone **25**, the protonated propanal **26**, and the protonated acetaldehyde **27**, respectively, and the relative intensities of the peaks at *m/z* 73, 59 and 45 largely reflect the relative stabilities of the radicals that are lost.

OH 24 – e → •+OH – H• → 25 *m/z* 73

•+OH – Me• → 26 *m/z* 59

•+OH – Et• → 27 *m/z* 45

A characteristic peak from most alcohols is that from the loss of water (M – 18), which gives a peak at *m/z* 56 in this case. Note that this is another unusual fragment with an even molecular weight, and it occurs because of the thermodynamic stability of water. Tertiary alcohols almost never show a molecular ion, and usually show an abundant M – 18 ion, but primary alcohols are more apt to lose H_2 to give the aldehyde as the first fragmentation.

Ethers fragment in an entirely analogous manner, but one particular type of ether is especially useful. The acetal **28** in Fig. 1.17 fragments selectively to cleave off the groups on either side of the acetal function, because the cations left behind **29** and **30** are so well stabilised. As usual the larger radical is cleaved off more easily than the ethyl radical, making the ion **30** responsible for the exceptionally prominent base peak at *m*/*z* 101.

The fragmentation of simple ketones follows the same pattern—the electron ejected is one of those from one of the lone pairs on the oxygen, and the radicals expelled are those that are β to the oxygen. Thus, in the mass spectrum of 3-heptanone **31** in Fig. 1.18, the radicals that are cleaved off are the butyl and ethyl radicals.

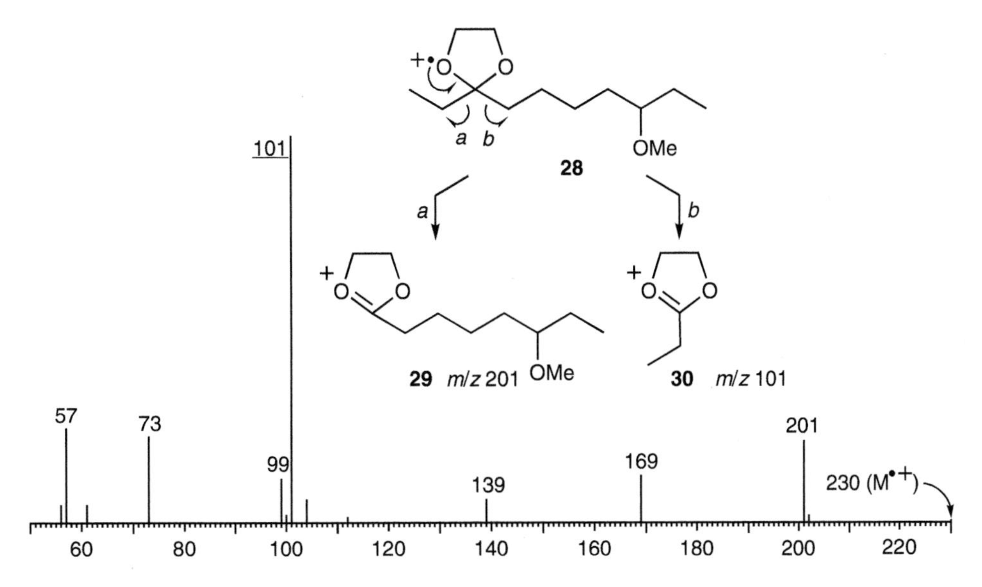

Fig. 1.17 EI mass spectrum of an ethylene acetal

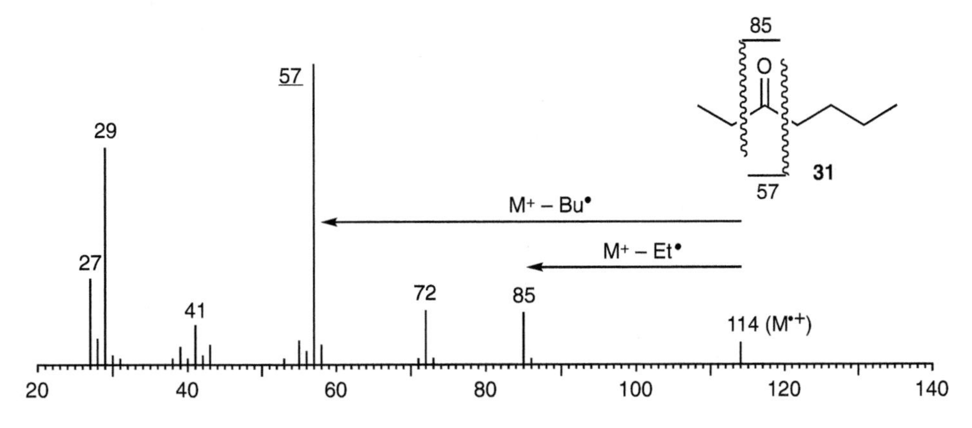

Fig. 1.18 EI mass spectrum of heptan-3-one

The nomenclature, however, is confusing because the bond that is cleaved is that between the carbonyl group and the carbon that is called the α carbon when we are naming ketones. As a result this cleavage is called α cleavage, even though the bonds cleaved in 3-heptanone **31** bear the same relationship to the oxygen atom as the bonds cleaved in an alcohol like 2-butanol **24**.

α α′
31
– e
– Et•
– Bu•
32 m/z 85
33 m/z 57

The peak at *m/z* 29 is the ethyl cation produced by loss of CO from the ion **33**. A similar loss of CO from the ion **32** gives a boost to the abundance of the ion at *m/z* 57, because the butyl cation and the propionyl cation **33** have the same integral mass. High resolution would reveal that the ion at *m/z* 57 is actually composed of two closely spaced peaks. The ion at *m/z* 41 is an allyl cation, produced by a fragmentation typical of hydrocarbons, as seen in the spectrum of nonane.

There is another abundant ion, at *m/z* 72, which is again a relatively unusual case of a fragment having an *even* molecular weight produced by the loss of a *molecule* from $M^{\bullet+}$. It corresponds to the radical-cation of the enol **35**, and arises by a McLafferty rearrangement **34**, in which propene **36** is lost by transfer of the γ-hydrogen atom from carbon to oxygen. For obvious reasons it is called β cleavage, since this is a pathway seen in ketones, for which this is the appropriate terminology.

34 → 35 (m/z 72) + 36

This fragmentation is useful, because it is diagnostic of a carbonyl group having a γ-hydrogen atom. The *m/z* value of the peak and the mass of the fragment lost can help to identify the substituents, if any, attached to the two halves of a ketone. Both the α-cleavage and the McLafferty rearrangement are reminiscent of the Norrish Type I and Type II fragmentations seen in the photolysis of ketones. Similar fragmentations are also seen with other carbonyl compounds, and summarised in a table at the end of this chapter listing the *m/z* values for the ions produced from the most simple member of each of the common carbonyl compounds.

Amines even more easily undergo the loss of one of the lone pair of electrons, followed by cleavage of the bonds β to the heteroatom. The amine **37** has three bonds to alkyl groups with this relationship, and they give rise to the most prominent peaks in the mass

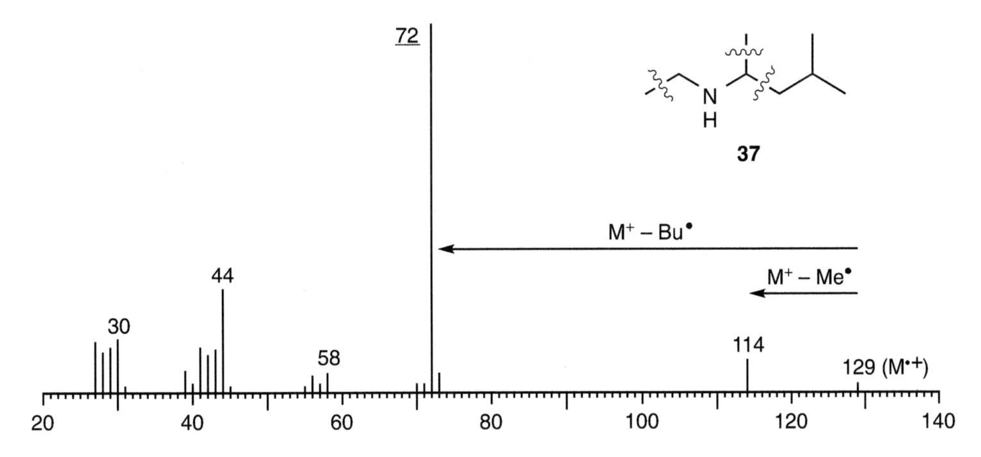

Fig. 1.19 EI mass spectrum of *N*-ethyl-4-methylpent-2-ylamine

spectrum shown in Fig. 1.19. This time, the molecular ion at *m/z* 129 is an odd number, and consequently all the major fragments are even-numbered, a characteristic pattern for a molecule containing an odd number of nitrogen atoms from which the major fragments still contain the nitrogen atoms.

There are two ways that a methyl radical can be lost, **38** and **40**, and one for an isobutyl radical to be lost **42**. The latter is the major pathway, because the larger radical is more stable than the methyl radical. Each of the cations **39** and **43**, can lose a molecule of an alkene, which accounts for the formation of the fragments at *m/z* 30 and 44. The loss of any of the three β-hydrogen atoms is, as usual, less favourable than the loss of the alkyl radicals, and there is no significantly abundant M – 1 peak.

38 —(– Me•)→ **39** *m/z* 114 —(– alkene)→ $CH_2{=}\overset{+}{N}H_2$ *m/z* 30

40 —(– Me•)→ **41** *m/z* 114

42 —(– Bu^i•)→ **43** *m/z* 72 —(– $CH_2{=}CH_2$)→ $H_2\overset{+}{N}{=}$ *m/z* 44

The halogens, although they do have lone-pairs, give rise to a different pattern of fragmentation—primarily because halonium cations are much less well stabilised than are iminium or oxonium ions, but also because the halogens form relatively low-energy radicals. Thus, in the spectrum of n-hexyl bromide **44** in Fig. 1.20, β-cleavage **45** giving

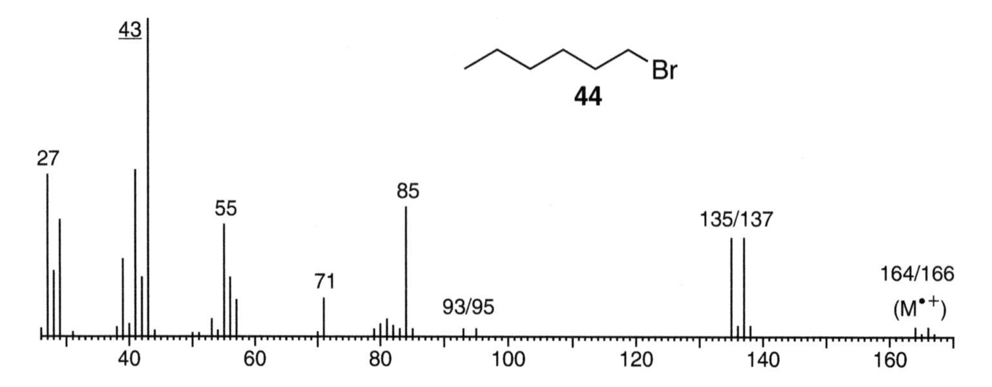

Fig. 1.20 EI mass spectrum of 1-bromohexane

the bromomethylene cation **46** at *m/z* 93/95 is a very minor pathway. The loss of the halogen atom **47** is a relatively major pathway giving the hexyl cation at *m/z* 85.

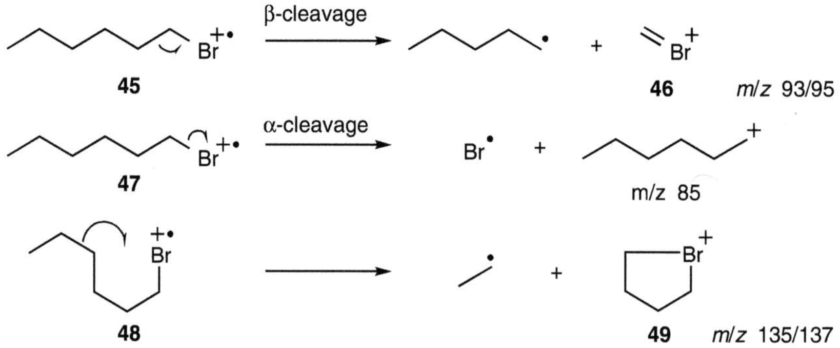

Charged fragments retaining the bromine atom are always recognisable, because they give rise to two peaks of equal intensity two mass units apart. One of these in this case is the cyclic bromonium cation **49** at *m/z* 135/137 from the loss of an ethyl radical by intramolecular displacement **48**. The strong ions at *m/z* 43 and *m/z* 41 are the propyl and allyl cations that are common fragments in n-alkyl chains, as we saw in Fig. 1.8.

This has been a superficial look at the myriad pathways found in mass spectra. Much more guidance is available in specialised texts, but it is not normally profitable to attempt to identify the origins of all the peaks in EI mass spectra. It is best, after extracting the vital molecular weight and the molecular formula, simply to glance at the spectrum to see if there are any of the informative and characteristic peaks of the easily recognised common fragments. Always bear in mind the useful generalisations that are affirmed in this section: the major fragmentation pathways from molecules with an even molecular weight take place by the loss of radicals to give charged fragments with odd molecular weights. Conversely, molecules with an odd molecular weight mainly give fragments with an even molecular weight. At the end of this chapter, several tables may help in identifying some of the more common fragments, both those characteristic of common functional groups, those broken off from the ions, and those carrying the charge.

1.5 Fragmentation in CI and FIB Spectra

1.5.1 Fragmentation in CI Spectra

Since a major piece of information from a mass spectrum is the molecular weight and the derived molecular formula, the absence of a molecular ion in a mass spectrum can be a serious problem. EI spectra quite often show no molecular ion, because fragmentation by radical pathways is so easy. Thus the EI spectrum of the common plasticizer dioctyl phthalate **50** in Fig. 1.21a has no peak for the molecular ion $M^{\bullet+}$, and so its molecular formula could not have been obtained in this way. The solution to this problem is to use one of the soft methods of ionisation like CI or ESI. The CI spectrum in Fig. 1.21b gives an abundant MH^+ peak at *m/z* 391.

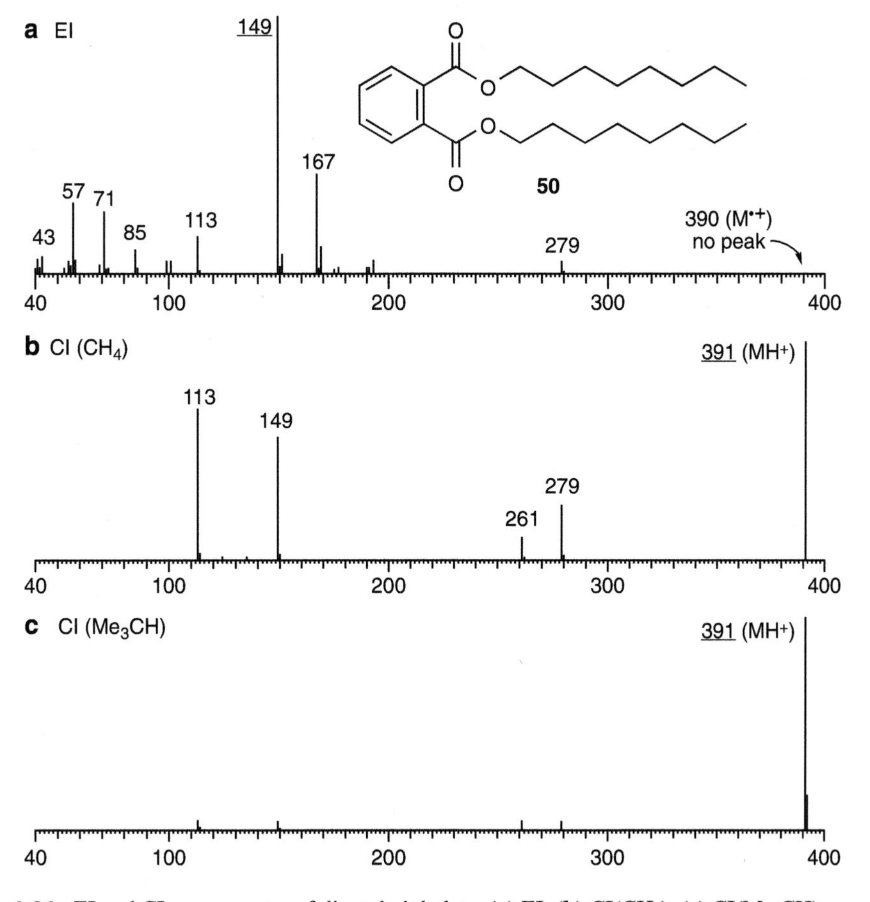

Fig. 1.21 EI and CI mass spectra of dioctyl phthalate. (**a**) EI, (**b**) CI(CH_4), (**c**) CI(Me_3CH)

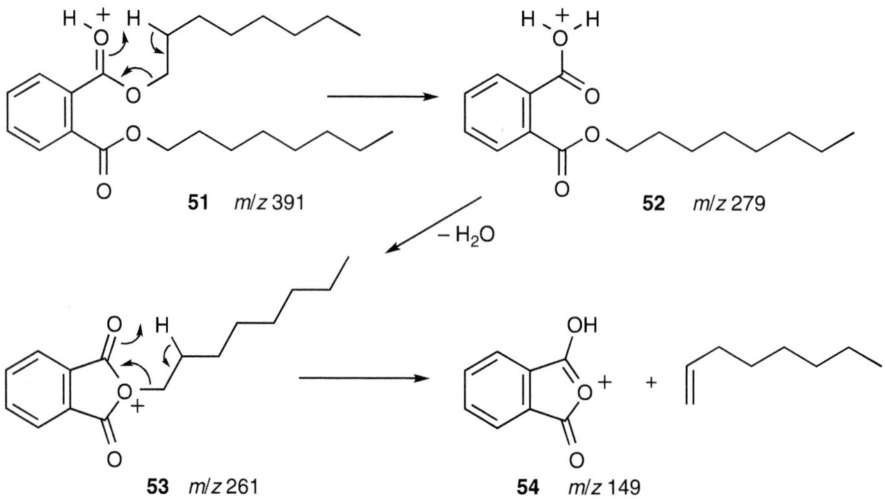

51 m/z 391 **52** m/z 279 $- H_2O$ **53** m/z 261 **54** m/z 149

The spectrum still shows some fragmentation, which takes place by pericyclic and ionic pathways, rather than by radical pathways. The protonated molecular ion **51** (the proton may be attached at other sites than the one drawn) undergoes a McLafferty-like rearrangement, which is recognisable as a retro-ene reaction, to give the ion **52**, and this in turn gives the phthalic anhydride derivatives **53** and then **54**. The ion at m/z 113 is the octyl cation $C_8H_{17}^+$, and so all the fragment ions are recognisable components of the structure. The key lesson is that soft ionisation is a powerful method for the determination of the molecular weights of volatile compounds because fragmentation is much reduced relative to that seen in EI spectra.

In some cases CI spectra still give too much fragmentation when methane is the carrier gas, because the CH_5^+ that is the proton source is highly energised. It is easily possible to minimise fragmentation by using a different carrier gas. Ammonia is commonly used, as are larger hydrocarbons. Thus isobutane as the CI gas transfers much less energy into the molecular ion, and largely suppresses fragmentation, as seen in the spectrum in Fig. 1.21c.

1.5.2 Fragmentation in FIB (LSMIS) Spectra

Both positive- and negative-ion FIB mass spectra can be obtained, although the latter method is often reserved for the analysis of compounds that form unusually stable anions (e.g. molecules containing sulfate groups). Molecular weights are normally available through the observation of abundant MH^+ ions in positive-ion spectra, and abundant $(M - H^+)^-$ ions in negative-ion spectra. FIB spectra frequently contain structurally useful fragment ions, but in the spectra of large molecules obtained from a matrix, MH^+ is usually the most abundant ion, making molecular weight determination easy by this method.

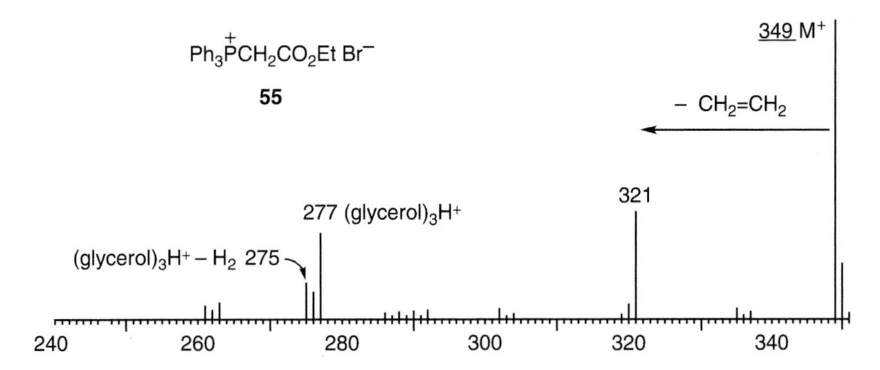

Fig. 1.22 FIB mass spectrum of the phosphonium salt **55**

FIB is used in the analysis of relatively involatile molecules up to molecular weights of *ca.* 4000 Da. Figure 1.22 shows the FIB spectrum of the phosphonium bromide salt **55**, where the phosphonium cation *(m/z* 349) evaporated directly, and did not need to be protonated. The molecular ion is the most abundant ion, and, since it is not a radical, the only significant fragmentation is the loss of the molecule ethylene. Hydrocarbons, and other highly hydrophobic molecules such as steroids with little functionality (low basicity), are not handled well by particle bombardment methods.

Peptides having molecular weights in the range 300–3000 Da fragment to give valuable sequence information. Cleavages occur on either side of the nitrogen atom of peptide bonds, and rarely anywhere else to a significant extent, quite different from the fragmentation seen in EI mass spectra. The fragment **56** is already charged, and appears only in positive-ion spectra. The fragments **57**, **58** and **59** are recorded in positive-ion spectra with an extra proton (H^+), and in negative-ion spectra with a missing proton.

56 **57** **58** **59**

The fragments **56**–**59** identify, residue by residue, the amino acids making up the peptide. Leu and Ile are isomers and have to be differentiated by amino acid analysis, but Lys and Gln, which have the same integral mass, can be differentiated by acetylation of the former. An example of peptide sequencing is illustrated by the negative ion spectrum of a

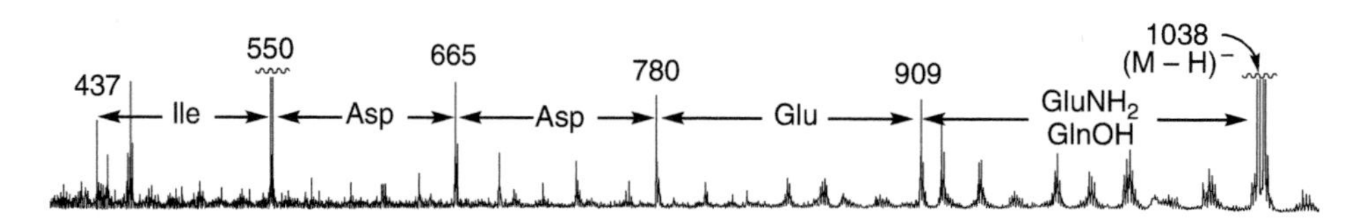

Fig. 1.23 FIB mass spectrum of the oligopeptide PhCO-Ala-Phe-Val-Ile-Asp-Asp-Glu-Gln

Table 1.2 Selective microscale reactions of some common functional groups

Functional group	Reagent	Product	Change in mass per functional group
RNH_2	Ac_2O/H_2O (30 min)[a]	RNHAc	+42
ROH	Ac_2O/pyridine (overnight)	ROAc	+42
RCO_2H	0.5% HCl/MeOH (overnight)	RCOOMe	+14
$RCONH_2$	$PhI(OCOCF_3)_2$	RNH_2	−28

[a]Reaction mixture buffered to pH ~ 8.5 with NH_4HCO_3

peptide toxin in Fig. 1.23, which gives the partial sequence X-Ile-Asp-Asp-Glu-Gln. In conjunction with the positive-ion spectrum and the molecular weight, the total sequence was found to be PhCO-Ala-Phe-Val-Ile-Asp-Asp-Glu-Gln.

In the interpretation of the FIB spectra of oligopeptides, one (or the computer) looks for a series of fragment ions that are successively separated by amino acid mass differences, greatly helped by performing an amino acid analysis to identify which amino acids are present. The atomic masses of the proteinogenic amino acids, which are used in this way in peptide sequencing, are given in a table at the end of this chapter. Multiply charged ions such as MH_2^{2+} are occasionally seen in FIB mass spectra, especially when there are two basic residues (as for example in arginine) in a peptide, but such ions are normally of low abundance.

Another aid in determining the structure of polar molecules with many functional groups is to carry out on a microgram scale a reliable reaction, characteristic of a given functional group, and measure the difference in molecular weight before and after reaction. The change in molecular weight, often measured using FIB, indicates the number of such functionalities. Thus the presence of serine, with its hydroxyl group and of glutamic acid with its carboxyl group can be revealed by one or another of these reactions. Table 1.2 lists some reactions reliably giving high yields.

1.6 Some Examples of Mass Spectrometry in Action

In general, structure determination is most powerfully carried out using NMR spectroscopy, with occasional input from MS, UV and IR spectra, as we shall see in the following chapters. There are however some situations, where MS alone is the method of choice. Here are six examples.

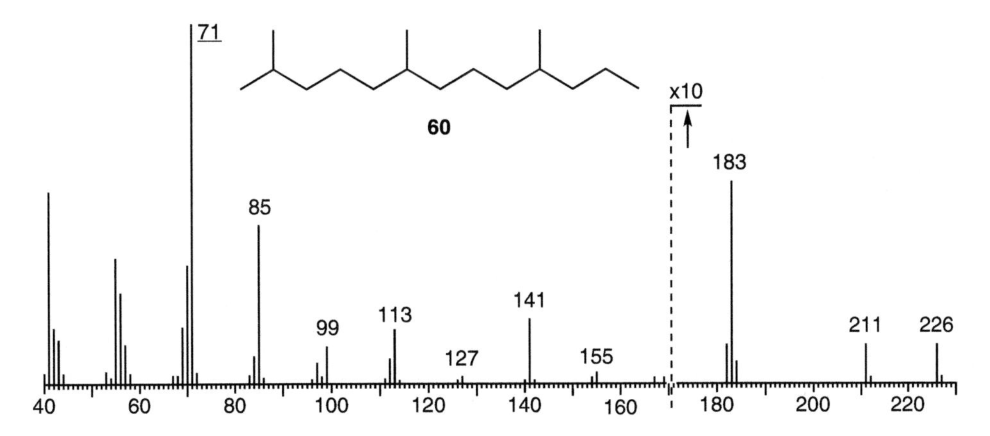

Fig. 1.24 EI mass spectrum of a component of San Joaquin oil

1.6.1 San Joaquin Oil

Mixtures of hydrocarbons are the main substances handled by the oil industry. They are readily separated by GC, and they can be identified quickly by the fingerprint pattern in their MS spectra, frequently using a mass spectrometer attached as the detector to a GC apparatus. Occasionally an oil is a new one, and the mass spectrum can provide the structure without recourse to anything more elaborate, because hydrocarbons are fairly well behaved in their fragmentation.

The oil from the San Joaquin field, for example, has the structure **60**. Its mass spectrum in Fig. 1.24 shows several fragments, almost all of which correspond to the breakage of the bonds at the branch points. The fragmentations summarised in the drawing **61** are those giving the secondary cations with expulsion of a primary radical. The fragmentations summarised in the drawing **62** are those giving the secondary radicals with expulsion of a primary cation. The primary cations can easily lose ethylene. Between them these fragmentations account for almost all the prominent peaks, and putting them together, with some help from the isoprene rule, indicates that the structure is **60**.

43 113 183 57 ← $-C_2H_4$ 85 127 ← $-C_2H_4$ 155

211 141 71 183 113

61 **62**

1.6.2 Oleic Acid

How might we find where the double bond is placed in an alkene like oleic acid **63**, given that we know that allylic cleavage in the EI mass spectra of alkenes is not reliable. In the case of oleic acid, its dimethylamide gives a mass spectrum with a peak **65** at *m/z* 87 from a McLafferty rearrangement **64**; this proves that the carbonyl group has a methylene group adjacent to it.

$CH_3(CH_2)_nCH{=}CH(CH_2)_mCO_2H$
63

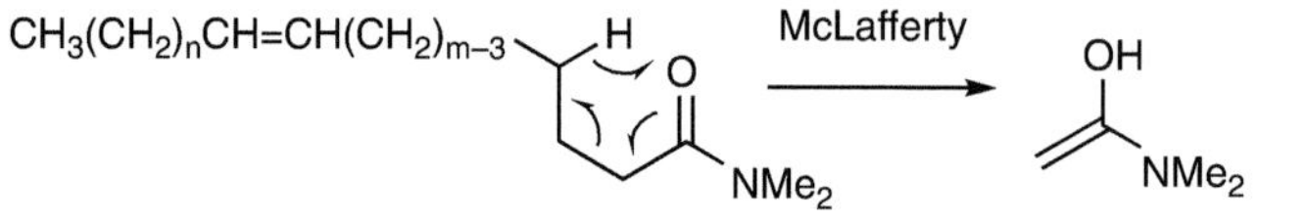

$CH_3(CH_2)_nCH{=}CH(CH_2)_{m-3}$ H O NMe$_2$ McLafferty → OH NMe$_2$

64 *m/z* 309 **65** *m/z* 87

To find where the double bond is, further back in the chain, we can use chemistry. Epoxidation and opening of the epoxide with dimethylamine in a protic solvent will reliably give a pair of amino alcohols. Amines are particularly well behaved in the mass spectrum, reliably showing cleavage β to the amino group (Fig. 1.19). In this case the most favourable fragmentations **66** and **68** will give the cations **67** and **69** and the oxygen-stabilised radicals. The former pair identify the values of both m and n to be 7. The structure of oleic acid is **70**.

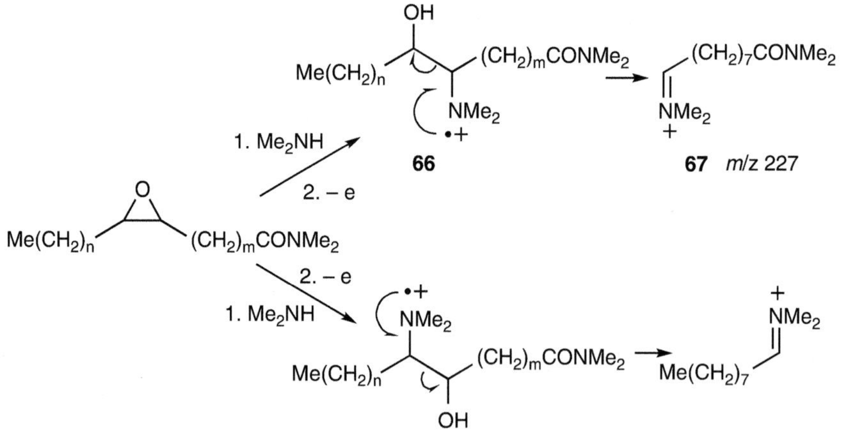

OH Me(CH$_2$)$_n$ (CH$_2$)$_m$CONMe$_2$ NMe$_2$ •+ → (CH$_2$)$_7$CONMe$_2$ NMe$_2$ +

1. Me$_2$NH 2. – e

66 **67** *m/z* 227

O Me(CH$_2$)$_n$ (CH$_2$)$_m$CONMe$_2$

2. – e 1. Me$_2$NH

•+ NMe$_2$ Me(CH$_2$)$_n$ (CH$_2$)$_m$CONMe$_2$ OH → + NMe$_2$ Me(CH$_2$)$_7$

68 **69** *m/z* 170

CO$_2$H

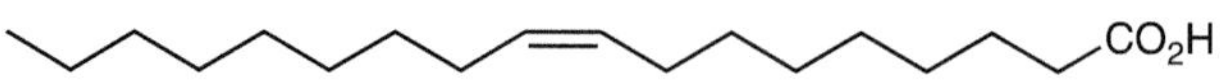

70

1.6.3 The Oviposition Pheromone

The female mosquito *Culex quinquefasciatus* lays her eggs in a raft, identifying where the raft is by detecting a pheromone given off by the eggs. The microgram amounts of pheromone available from the eggs precluded analysis by any other technique than mass spectrometry. GC-MS gave a fraction which was active, and the EI mass spectrum of this fraction is shown in Fig. 1.25.

Exact mass measurements allowed molecular formulae to be assigned to the peaks: 43: C_2H_3O; 99, $C_5H_7O_2$; 142, $C_7H_{10}O_3$; and 252, $C_{16}H_{28}O_3$. A soft CI spectrum using isobutane as the carrier gas identified a weak peak at *m*/*z* 313, which would fit a formula for MH^+ of $C_{18}H_{33}O_4$, suggesting that the peak at *m*/*z* 252 in Fig. 1.25 was not the molecular ion, but represented a product from the loss of acetic acid (M – 60) from the true molecular ion. The ions at *m*/*z* 43 and 99 are the two halves of the ion at 142. This suggested that in some way an acetyl fragment (with *m*/*z* 43) had been transferred from the acetate group to the fragment with an *m*/*z* value of 99. Thinking of reasonable structures for this fragment, the researchers considered a lactone, either six-membered or methyl-substituted five-membered. Furthermore, they considered that the acetyl group must have been situated close to this lactone ring in the parent molecule.

To reassure themselves that they had identified the molecular formula, they carried out some reliable chemical reactions: they hydrolysed the acetate with alkali, and trimethylsilylated the alcohol produced. The silyl ether gave them a mass spectrum with a distinct molecular ion (3%) at the expected value for a formula of $C_{19}H_{38}O_3Si$ of *m*/*z* 342. Further support came from reduction of the lactone alcohol to a triol with the appropriate molecular weight, indicating that the lactone hypothesis was correct. From all this evidence, they thought it likely that the sequence of reactions could be explained by the scheme in Fig. 1.26, and accordingly synthesised all four possible stereoisomers. The 5*R*,6*S*-stereoisomer **71** proved to be the active pheromone.

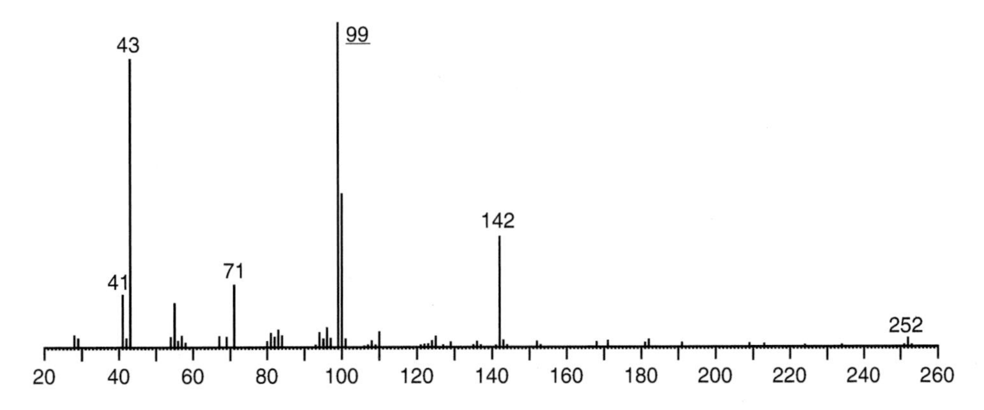

Fig. 1.25 EI mass spectrum of the oviposition pheromone

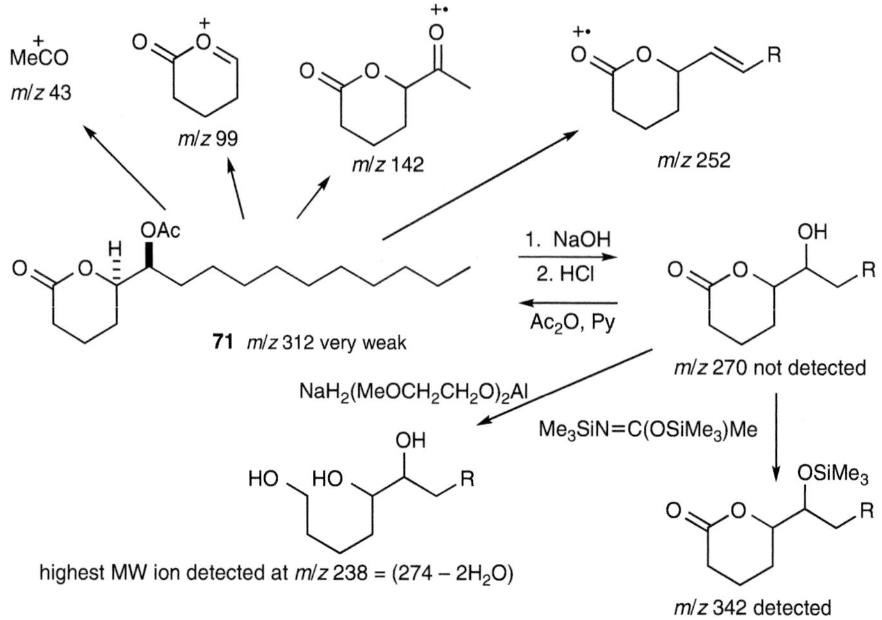

Fig. 1.26 Chemical reactions EI fragmentations of the pheromone **71** from *Culex quinquefasciatus*

1.6.4 Identifying Antibodies

The MALDI-TOF combination is a powerful tool to characterise biomolecules, even in the range 100,000–200,000 Da. The precision of the mass determination at these large masses may not be better than 10–100 Da, and the addition of a proton to the molecule in the production of the gas-phase ion is not something that can be tested. Hence, the species analysed are usually referred to as '$M^{+\cdot}$', even though it is likely that they actually correspond to MH^+. The spectra obtained from many laser shots can be accumulated in order to give an improved signal-to-noise ratio and extraordinary sensitivity. It is possible to detect 10^{-15} moles or less.

The power of the method is illustrated by the determination of the molecular weight of a monoclonal antibody (Fig. 1.27). Antibodies are moderately large proteins, produced by the immune system to bind to foreign substances that have appeared in the body. Normally, a given foreign substance (the antigen) causes the production of a mixture of such antibodies. However, if one particular cell of the activated set of cells in the immune system is cloned, then single (pure) antibodies can be produced. These *monoclonal antibodies* are of great importance in diagnostic medicine, since their binding to an antigen can be coupled to the generation of a colour response. The characterisation of monoclonal antibodies is therefore important. In the case illustrated in Fig. 1.27, all the prominent high molecular weight ions are either the molecular ion M^+ or are related to it, either as a simple multiple ($2M^+$), or as ions with more than one positive charge (M^{2+} and M^{3+}), or both ($2M^{3+}$ and

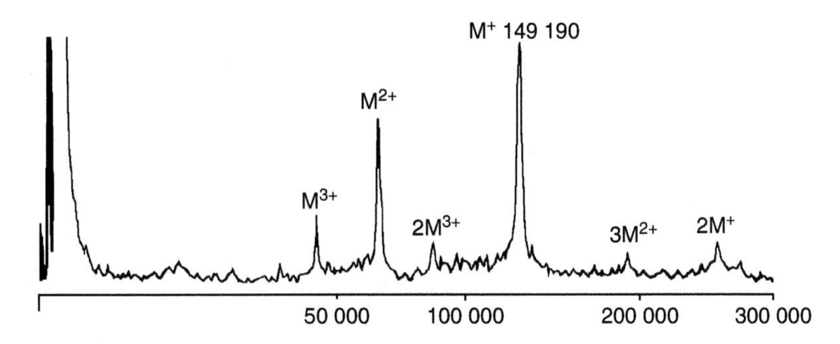

Fig. 1.27 The ions of a monoclonal antibody

$3M^{2+}$). They all contribute to the measurement of the molecular weight, for which the precision in this case is 149,190 ± 69 Da.

1.6.5 The ESI Spectra of Melittin and the Human Parathyroid Hormone

ESI typically produces multiply charged ions from biopolymers. Since the position of an ion in the mass spectrum is determined by the *m/z* ratio, multiple charges reduce the *m/z* value at which an ion of mass *m* appears in the spectrum, bringing the ions of large molecular weight to a low *m/z* value. It allows an ESI spectrum of a biopolymer to be measured by quadrupole analysers, even though these instruments are suitable for the determination of *m/z* values only up to about 4000. The principle is illustrated in the ESI spectra in Fig. 1.28.

On the left is the ESI mass spectrum of the peptide melittin **72** (a component of the sting venom of the honey bee, which is able to promote the bursting of cells by destabilising their membranes). Multiply charged ions characteristic of the molecular weight of 2846 Da occur even in the *m/z* region from 400 to 600. Similarly, in the spectrum on the right, the molecular weight of a fragment **73** from the human parathyroid hormone, made up of a sequence of 44 amino acids, is measured as 5064 Da. In representing the structures **72** and **73**, the one-letter codes for the amino acids are used (they can be found in a table at the end of this chapter). The molecular weights (M_r) in the figures are average values including all isotopes.

The width of the distribution of charged states is often about half of the mass of the charged state with the highest mass, as it is in Fig. 1.28. The highest charged state often correlates quite well with the number of relatively basic functionalities in the amino acids. The side chains $-(CH_2)_4NH_2$ (p*Ka* *ca.* 8) in lysine, $-(CH_2)_3NHC(=NH)NH_2$ (p*Ka* *ca.* 12) in arginine, and imidazole (pK_a *ca.* 6.5) in histidine, and the *N*-terminus of the peptide or protein, are relatively easily protonated to the corresponding positively charged ions.

Thus, melittin has a total of 5 Lys, Arg and His residues, and human parathyroid hormone (1–44) has 9. Adding an extra charge from the protonation of the *N*-terminus leads to an expectation of maximum charged states of 6 and 10, respectively, close to the observed values of 6 and 9.

Since the number of basic residues in a protein is usually roughly proportional to the molecular weight, much larger molecules than these peptides can still be studied. Charged states of 40–80 would not be unusual in a protein with a molecular weight as high as 80,000 Da, and this brings the *m/z* values of the ions within the range covered by quadrupole analysers. Since the molecular weight information is contained in a number of the prominent peaks, as in Fig. 1.28, there is an algorithm to calculate the molecular weight using the data from all of them—permitting a determination with relatively high confidence and precision. Table 1.3 gives some examples of molecular weights determined in this way, along with the calculated molecular weights based on the weighted average of all the isotopes that are present.

Determinations of molecular weights of proteins can be made with as little as 20 femtomoles (20×10^{-15} moles) of material. For a molecular weight of 10,000 Da, this corresponds to 200 picograms of material (<1000 of a microgram!).

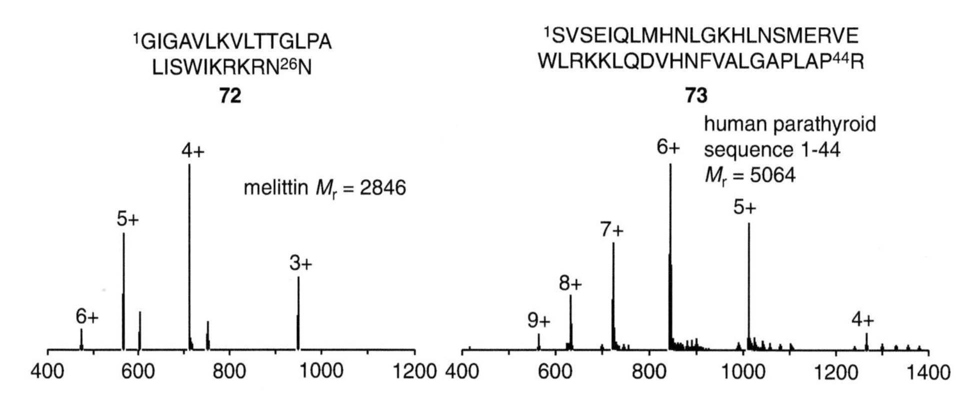

Fig. 1.28 ESI mass spectra of melittin **72** and the human parathyroid hormone **73**. Redrawn with permission from [1]

Table 1.3 Molecular weight determination of proteins by ESI-MS

Compound	M_r (measured)	M_r (calculated)	M_r error, %
Bovine insulin	5733.4	5733.6	–0.01
Ubiquitin	8562.6	8564.8	–0.02
Thioredoxin (*E. coli*)	11672.9	11673.4	–0.00
Bovine ribonuclease A	13681.3	13682.2	–0.01
Bovine α-lactalbumin	14173.3	14175.0	–0.01
Hen egg lysozyme	14304.6	14306.2	–0.01

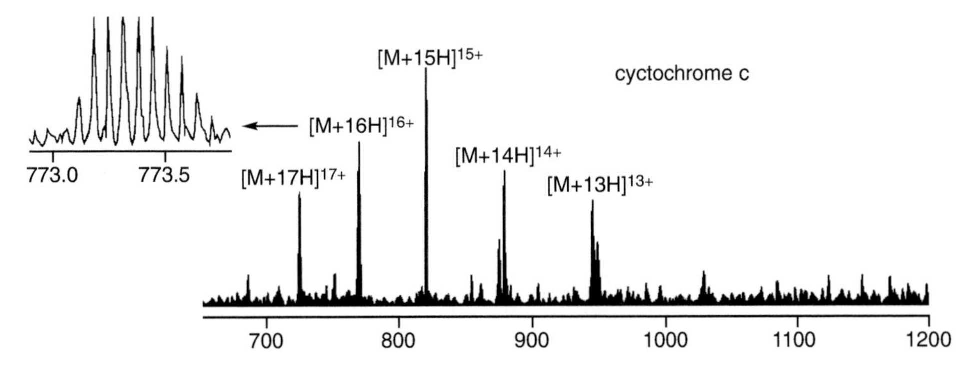

Fig. 1.29 ESI-FT-ICR mass spectrum of cyctochrome c. Redrawn with permission from [2]

1.6.6 ESI-FT-ICR and ESI-FT-Orbitrap Spectra

When an ESI source is used in combination with FT-ICR, the resolution is much better than that of the ESI-quadrupole combination. The spectrum of the protein cytochrome c is shown in Fig. 1.29, with the expected distribution of multiply charged ions.

If the peak from any one of these ions is expanded at high resolution, the peaks from the various combinations of ^{13}C isotopes are resolved, as shown in the expansion of the cluster of peaks just above *m/z* 773. For a protein of this size (just over 12,000 Da) with several hundred carbon atoms, the isotope peaks from species containing from 0 to 8 ^{13}C atoms are abundant. Since the carbon isotope peaks for a singly charged ion must be separated by 1.0034 Da, the number of these isotope peaks appearing within a single unit on the *m/z* scale defines the number of charges on the ion. The resolution of peaks separated by only 0.0625 units on the *m/z* scale at 773, from a molecule with an average molecular weight of 12,358 Da, emphasises the resolving power of ESI-FT-ICR. The acquisition of the mass spectra of proteins comparable to those of cytochrome c, is also possible using the Orbitrap mass analyser in conjunction with ESI. These instruments are robust and relatively easy to maintain.

1.7 Separation Coupled to Mass Spectrometry

1.7.1 GC/MS and LC/MS

The separation and detection of components from a mixture of organic compounds is readily achievable by gas chromatography (GC), or by high-performance liquid chromatography (HPLC). Mass spectrometry, because of its high sensitivity and fast scan speeds, is suited to the analysis of the small quantities of material eluted from these chromatographs, even though the components may be present in nanogram quantities and eluted over periods of only a few seconds. LC is readily coupled to the mass spectrometer with the LC eluate introduced directly into an ESI source, for example, with the mass spectrometer used as the detector, through measurement of the total ion current at any point in the chromatogram.

In LC-ESI-MS work, if the analyte contains groups that readily form ions (e.g. amino or carboxyl groups), and especially if the solvent includes a volatile buffer (such as formic acid or ammonium acetate) which can promote the formation of ions, the droplets in the mist are to some extent positively or negatively charged. Solvent molecules are removed by fast pumping, leaving MH^+ or $(M - H^+)^-$ ions in the gas phase, where they can be extracted from the chamber by the electric field, and injected into the mass analyser. Good mass spectra of amino acids, peptides, nucleotides, and antibiotics up to molecular weights of around 2000 Da can be obtained fairly routinely.

The majority of mass spectrometers are sold as integrated combinations of the separation and analysis components. Bench-top GC and LC/time-of-flight mass spectrometry systems offer sufficient mass accuracy to be used in the routine screening of the crude products of organic reactions and of complex biological samples. Although these instruments do not offer the highest mass resolution, they may be used to measure the 'monoisotopic mass' of ions to within 5 mDa (this is the mass of the isotopic peak which is composed only of the most abundant isotopes of those elements present). Thus, the monoisotopic mass can be calculated using the atomic masses of these most abundant isotopes. A further tool for identification is the degree to which the isotopic pattern of a detected compound agrees with the theoretical pattern of any analyte.

Because large amounts of data are collected from a GC/MS or LC/MS run, the system is usually coupled on-line to a computer-controlled data system. The output from the computer shows the usual trace from a GC or HPLC, with the peaks identified by code letters and numbers, from which the mass spectra can be called up.

For example, the methionyl human growth hormone is a protein, which was hydrolysed by digestion with the enzyme trypsin, which cleaves the protein at the *C*-terminal side of each of the basic residues lysine and arginine. The numerous peptides thus generated all contain two basic sites, namely the *C*-terminal Lys or Arg, and the *N*-terminus of the peptide. Figure 1.30 illustrates the HPLC trace obtained, and Fig. 1.31 show the ESI spectra obtained from two of the peaks, labelled T11 and T12.

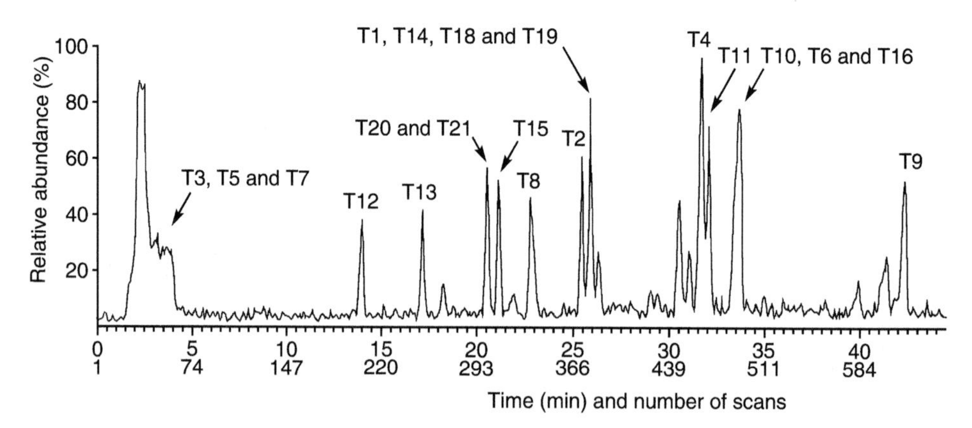

Fig. 1.30 HPLC trace after a trypsin digest of human growth hormone. Redrawn with permission from [3]

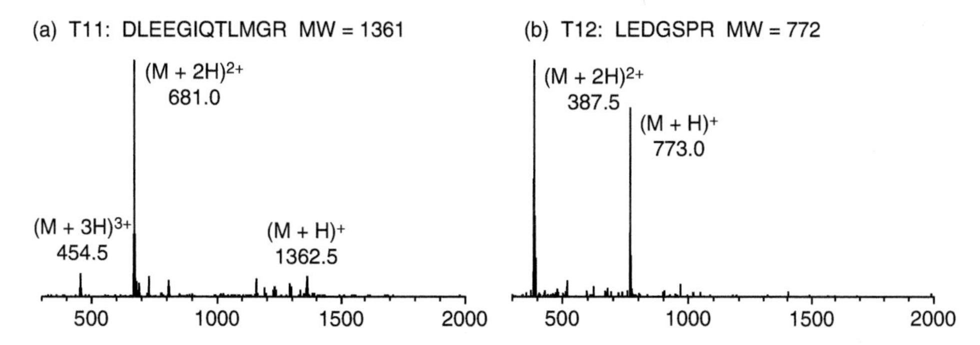

Fig. 1.31 ESI-MS spectra of two components of the trypsin digest of human growth hormone. (**a**) T11: DLEEGIQTLMGR MW = 1361, (**b**) T12: LEDGSPR MW = 772. Redrawn with permission from [3]

The mass spectra are relatively simple, containing molecular weight information only in MH^+ and MH_2^{2+} ions; they give the molecular weights of the two peptides DLEEGIQTLMGR and LEDGSPR, which are contiguous in the growth hormone sequence.

1.7.2 MS/MS

The output of one mass spectrometer can be directed to the input of a second. If a mixture of three compounds of molecular weights M_1, M_2 and M_3, for example, is ionised by one of the 'soft' ionisation techniques, CI, FIB or ESI, any of the standard methods of analysis can be set to pass only M_2H^+, say, on to the second mass spectrometer. If the second spectrometer has a collision chamber containing an inert gas, the M_2H^+ ions collide with an atom or molecule of the gas, and a small proportion of their translational (kinetic) energy is converted into vibrational energy. As a result, fragment ions are produced from M_2H^+, and the second analyser detects them. This gives a collisionally induced mass spectrum (CID-MS) of M_2H^+, and these spectra are similar to the mass spectra obtained when the energy is deposited by other means. Since all components of a mixture can be analysed in this way, all components of a mixture M_1, M_2 and M_3 (other than isomers) can in principle be separated, and a mass spectrum obtained for each one. The technique is commonly called MS/MS, since the molecular ion from an initial mass spectrum is selected and made to give a second mass spectrum.

For example, mixture analysis can be achieved by three quadrupole mass spectrometers connected in sequence. The first quadrupole is used to separate the molecular ions, the second as a collision chamber, and the third to separate the products of collision-induced decomposition. If collision-induced mass spectra are already available in a computer file, MS/MS can be a sensitive and rapid technique for identifying the components of a mixture, useful in medical and forensic applications.

The Orbitrap mass spectrometer is powerful in carrying out successive MS/MS experiments when structure determination is the goal. In the Orbitrap analyser, all ions except that of a selected molecular ion may be expelled, leaving the one ion to carry forward to CID. Its fragmentation may then be used to try to determine its structure. Such experiments, on whatever instrument they are carried out, are described as $(MS)^n$ experiments, where n is the number of successive CIDs carried out.

1.8 Interpreting the Spectrum of an Unknown

It is important to know whether the sample is pure or not. In the case of crystalline compounds, useful information can come from the sharpness, and constancy, of the melting point. More generally, purity can be assessed from thin-layer chromatography (TLC), GC and LC. Characteristic impurity peaks (summarised in a table at the end of this chapter) may be present in samples that have been extensively handled on thin-layer plates, in columns, or in greased apparatus. Since molecular ions of aliphatic compounds are frequently of low abundance in EI spectra, it is helpful in these cases to determine, in addition, a spectrum using a soft ionisation technique such as CI or ESI. For the identification of a peak in a mixture from a chromatographic separation, HPLC/MS or GC/MS, a search of a computer file of EI spectra of known compounds is a powerful method.

In the structure elucidation of an unknown compound, the molecular ion must be identified. Ensure that the supposed molecular ion is separated from peaks at lower masses by acceptable mass differences. Note whether the *m*/*z* value of the molecular ion is odd or even, and look for any characteristic isotope patterns. Once a molecular ion, $M^{\bullet+}$, MH^+ or $(M - H^+)^-$, has been identified with some conviction, decide whether it is appropriate to determine the molecular formula of the compound by a high-resolution measurement. Be aware that such a measurement will be definitive at low molecular weights but will be consistent with increasing numbers of combinations of elements at higher molecular weights.

Once you know the molecular formula, calculate the number of double bond equivalents (DBE, Eq. (1.4)), in order to have useful information to hold in your mind as you advance through the analysis. Use any available UV, IR and NMR spectra and chemical knowledge to determine a suggested partial or total structure. At this stage, return to the EI mass spectrum to see if the observed fragmentation pattern is consistent with the proposal or limits the range of possibilities, bearing in mind that single-bond cleavages correspond to the energetically more favourable pathways discussed in outline above.

The recognition of structural units from *m*/*z* values is of limited value, even for molecules of molecular weights <200, but where the nature of the functionality is limited and/or known, it is probably worth while just to check the observed *m*/*z* values for the various fragments with those given in the penultimate table at the end of this chapter. It is often more helpful to search for the neutrals that are lost from $M^{\bullet+}$, rather than just identify the fragment ion itself. The final table lists a number of the common fragments that are easily lost from the molecular ion. Remember that the real power of mass spectrometry is in the determination of a molecular formula, in its sensitivity, and (in the case of a previously isolated compound) the possibility of structure identification through pattern recognition.

1.9 Internet

The Internet is an evolving system, with links and protocols changing frequently. The following information is inevitably incomplete and may no longer apply, but it gives you a guide to what you can expect. Some websites require particular operating systems and may only work with a limited range of browsers, many require payment, and some require you to register and to download programs before you can use them.

There are vast resources on the Internet, for MS data. Several of these (CDS, Bio-Rad, ACD, Sigma-Aldrich, ChemGate) cover most or all of the spectroscopic methods.

A website listing databases MS is: http://www.lohninger.com/spectroscopy/dball.html

The Japanese Spectral Database for Organic Compounds (SDBS) at: http://www.aist.go.jp/RIODB/SDBS/cgi-bin/cre_index.cgi has free access to MS data.

The Wiley-VCH website gives access to the SpecInfo data: http://www3.interscience.wiley.com/cgi-bin/mrwhome/109609148/HOME and to ChemGate, which has a large collection of mass spectra: http://chemgate.emolecules.com

The Sadtler website is: http://www.bio-rad.com follow the leads to Sadtler, KnowItAll and MS for 198,000 mass spectra.

http://www.acdlabs.com/products/spec_lab/exp_spectra/ms/ takes you to the ACD website. In its entry-level configuration, ACD/MS Manager can process MS, tandem mass spectra (MS/MS, MS/MS/MS, MS^n) and MS combined with separation techniques (LC/MS, LC/MS/MS, LC/DAD, CE/MS, GC/MS).

1.10 Further Reading

Development of the Subject

- Beynon JH, Saunders RA, Williams AE (1968) The mass spectra of organic molecules. Elsevier, London
- Biemann K (1962) Mass spectrometry. McGraw-Hill, New York
- Budzikiewicz H, Djerassi C, Williams DH (1964) Structure elucidation of natural products by mass spectrometry, vols I and II. Holden-Day, San Francisco
- Budzikiewicz H, Djerassi C, Williams DH (1967) Mass spectra of organic compounds. Holden-Day, San Francisco
- Chapman JR (1978) Computers in mass spectrometry. Academic Press

- Howe I, Williams DH, Bowen RD (1981) Mass spectrometry—principles and applications. McGraw-Hill, New York
- McFadden WH (1973) Techniques of combined GC/MS. Wiley, New York
- Millard BJ (1978) Quantitative mass spectrometry. Heyden, London
- Rose ME, Johnstone RAW (1982) Mass spectrometry for chemists and biochemists. Cambridge University Press, Cambridge

Later Textbooks

- Gross JH (2003) Mass spectrometry. Springer, Heidelberg
- Herbert C, Johnstone R (2002) Mass spectrometry basics. CRC Press, Boca Raton
- de Hoffmann E, Stroobant V (2002) Mass spectrometry, 2nd edn. Wiley, Chichester
- McMaster MC (1998) GC/MS, a practical user's guide. Wiley, New York
- McLafferty FW, Turecek F (1993) Interpretation of mass spectra, 4th edn. University Science Books, Sausalito
- Nibbering N (ed) (2004) The encyclopedia of mass spectrometry, vol. 4: fundamentals of and applications to organic (and organometallic) compounds. Elsevier. This is the most relevant volume of a 10-volume set to structure determination in everyday organic chemistry. The other volumes cover more specialised aspects of the technique
- Smith RM (2004) Understanding mass spectra, 2nd edn. Wiley, New York

Chromatography and Mass Spectrometry

- Ardrey RE (2003) Liquid chromatography—mass spectrometry, an introduction. Wiley, New York.
- McMaster MC (2005) LC/MS, a practical user's guide. Wiley, New York.
- Niessen WMA (2006) Liquid chromatography-mass spectrometry, 3rd edn. CRC Press, Boca Raton

Data

- McLafferty FW, Stauffer DB (1989) Wiley/NBS registry of mass spectral data, 7 vols. Wiley, New York
- McLafferty FW, Stauffer DB (1991) Important peak index of the registry of mass spectral data, 3 vols. Wiley, New York
- (1991) The eight peak index of mass spectra, 3 vols, 4th edn. RSC, Cambridge
- Bruno TJ, Svoronos PDN (2006) CRC handbook of fundamental spectroscopic correlation charts. CRC Press, Boca Raton

1.11 Problems

Mass spectra alone are rarely used to identify structures, but for practice at interpreting mass spectra here are a few simple examples, showing some of the more characteristic peaks that it is useful to be familiar with. Each of the problems includes a useful piece of information, which it would be artificial to withold: the number of *different* carbon atoms, readily derived from the ^{13}C-NMR spectrum (Chap. 4).

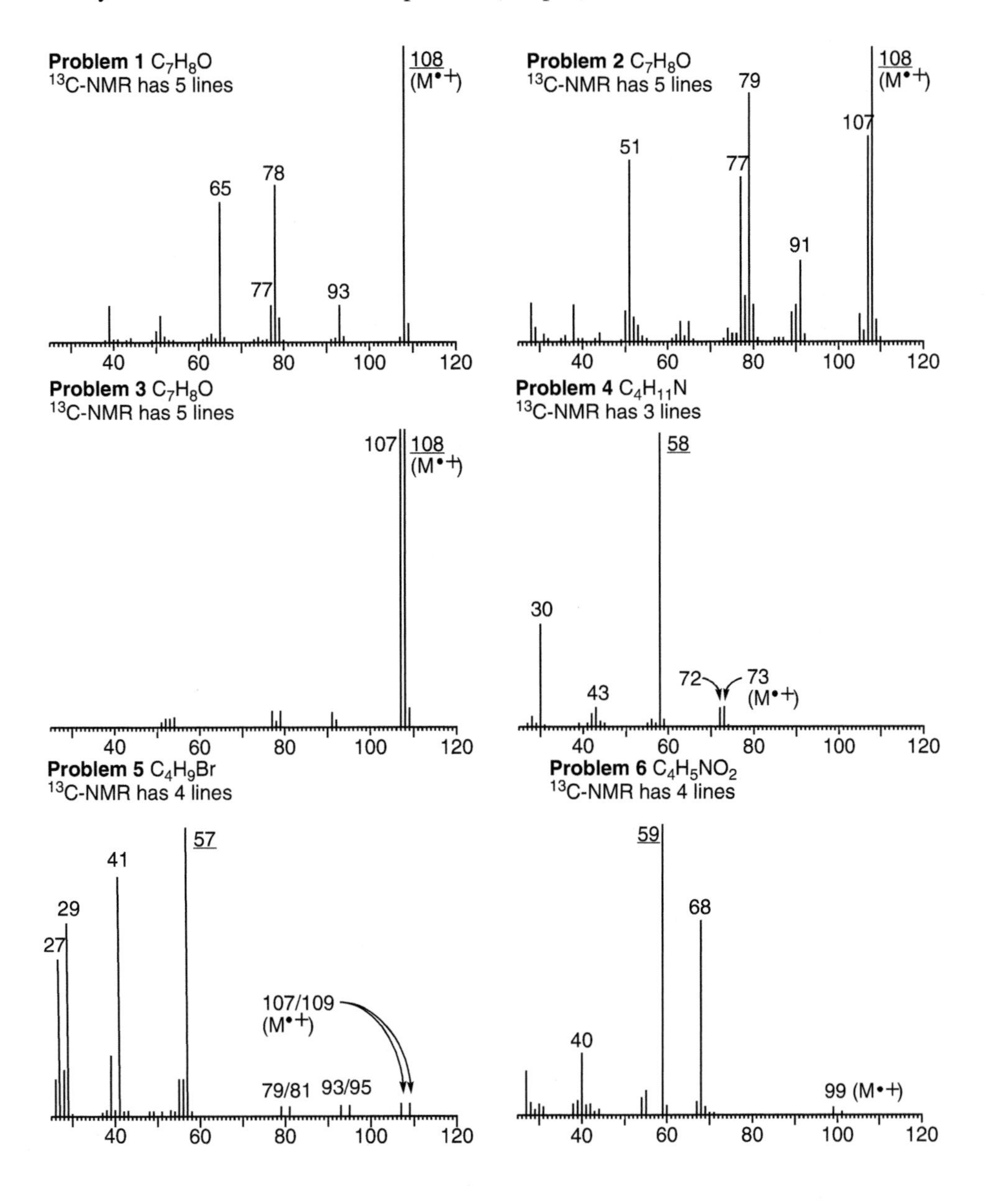

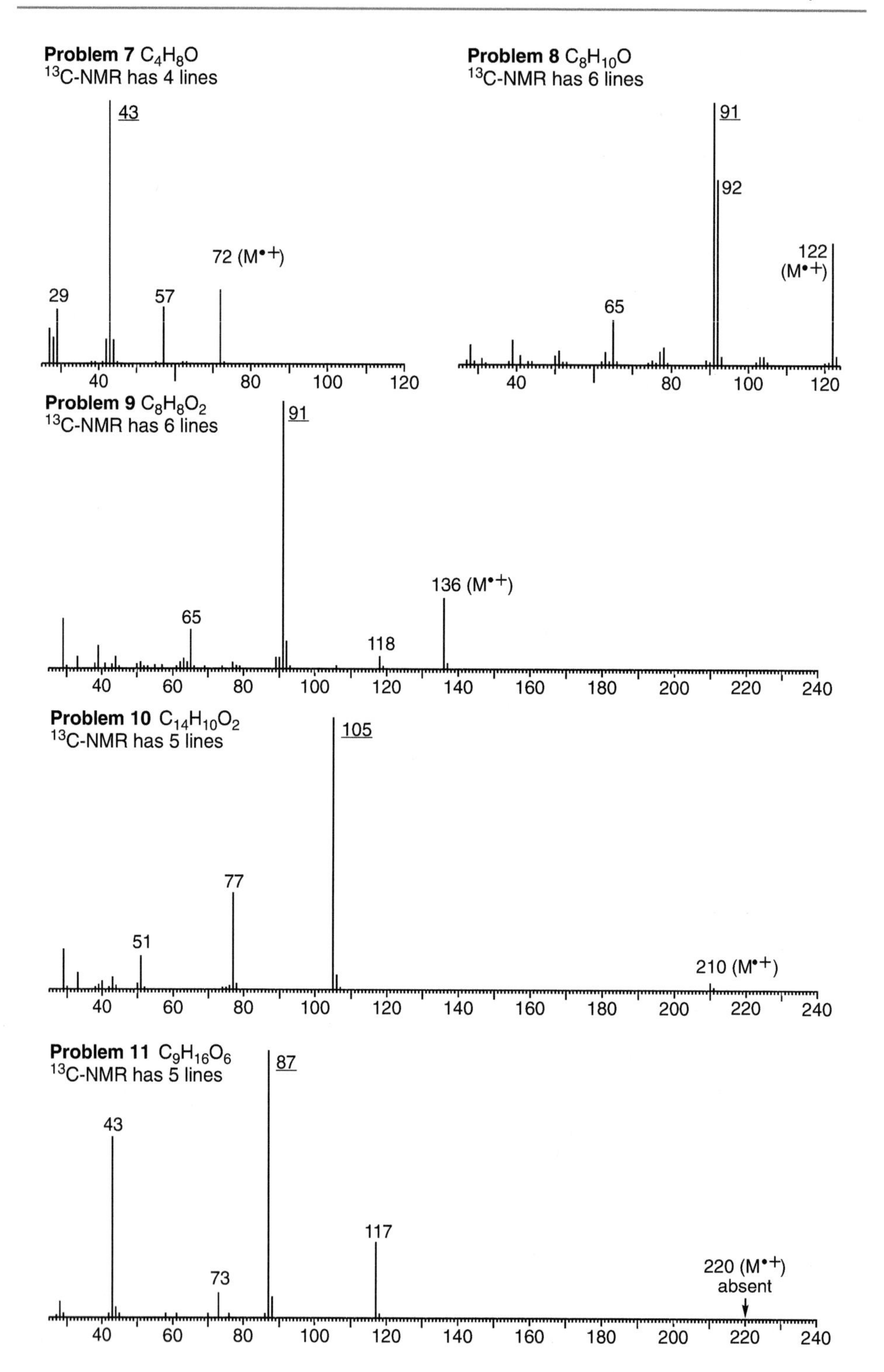

Problem 7 C_4H_8O
^{13}C-NMR has 4 lines
43
29
57
72 (M•+)
40
80
100
120
Problem 8 $C_8H_{10}O$
^{13}C-NMR has 6 lines
91
92
65
122 (M•+)
40
80
100
120
Problem 9 $C_8H_8O_2$
^{13}C-NMR has 6 lines
91
65
118
136 (M•+)
40
60
80
100
120
140
160
180
200
220
240
Problem 10 $C_{14}H_{10}O_2$
^{13}C-NMR has 5 lines
105
77
51
210 (M•+)
40
60
80
100
120
140
160
180
200
220
240
Problem 11 $C_9H_{16}O_6$
^{13}C-NMR has 5 lines
87
43
117
73
220 (M•+)
absent
40
60
80
100
120
140
160
180
200
220
240

1.12 Tables of Data

Table 1.4 Atomic weights and approximate natural abundance of some common isotopes

Isotope	Atomic weight (^{12}C = 12.000000)	Natural abundance (%)
^{1}H	1.007825	99.985
^{2}H	2.014102	0.015
^{12}C	12.000000	98.9
^{13}C	13.003354	1.1
^{14}N	14.003074	99.64
^{15}N	15.000108	0.36
^{16}O	15.994915	99.8
^{17}O	16.999133	0.04
^{18}O	17.999160	0.2
^{19}F	18.998405	100
^{28}Si	27.976927	92.2
^{29}Si	28.976491	4.7
^{30}Si	29.973761	3.1
^{31}P	30.973763	100
^{32}S	31.972074	95.0
^{33}S	32.971461	0.76
^{34}S	33.967865	4.2
^{35}Cl	34.968855	75.8
^{37}Cl	36.965896	24.2
^{79}Br	78.918348	50.5
^{81}Br	80.916344	49.5
^{127}I	126.904352	100

Table 1.5 ΔH_f (kJ mol^{-1}) of some ions

Ion	ΔH_f	Ion	ΔH_f
H^+	1530	Me_2C^+–CH=CH_2	770
Me^+	1090	Ph^+	1140
Et^+	920	$PhCH_2^+$	890
n-Bu^+	840	EtC^+=O	600
$EtCH^+Me$	770	PhC^+=O	730
Me_3C^+	690	MeCH=OH^+	600
CH_2=CH^+	1110	MeO^+=CH_2	640
CH_2=CH–CH_2^+	950	MeCH=NH_2^+	650

Note: (1) Formation of small ions such as H^+ and CH_3^+ is unfavourable. (2) Vinyl and phenyl cations have high ΔH_f values. (3) The ease of formation of alkyl cations is tertiary > secondary > primary. (4) Delocalised cations, and acylium, oxonium, and imminium ions are stabilised in the gas phase, just as they are in solution

Table 1.6 ΔH_f (kJ mol^{-1}) of some radicals

Radical	ΔH_f	Radical	ΔH_f
H·	218	PhCH$_2$·	188
Me·	142	MeC·=O	−23
Et·	108	HO·	39
n-Pr·	87	MeO·	−4
Me$_2$CH·	74	H$_2$N·	172
CH$_2$=CH·	250	Cl·	122
CH$_2$=CH–CH$_2$·	170	Br·	112
Ph·	300	I·	107

Note the relative instability of vinyl radicals and the increased stability of radicals with increasing substitution

Table 1.7 Molecular ion abundances in relation to molecular structure

Strong	*Medium*	*Weak or absent*
Aromatic hydrocarbons ArH	Conjugated alkenes	Long-chain aliphatics
ArF	Ar≀Br and Ar≀I	Branched alkanes
ArCl	ArCO≀R	Tertiary aliphatics
ArCN	ArCH$_2$≀R	Tertiary aliphatic bromides
ArNH$_2$	ArCH$_2$≀Cl	Tertiary aliphatic iodides

Table 1.8 Order of ease of fragmentation of some C_6H_5X compounds in EI spectra

X	Neutral fragments lost from $M^{\bullet+}$	X	Neutral fragments lost from $M^{\bullet+}$
COMe	Me	OH	CO
CMe$_3$	Me	Me	H
CHMe$_2$	Me	Br	Br
CO$_2$Me	OMe	NO$_2$	NO$_2$ & NO
NMe$_2$	H	NH$_2$	HCN
CHO	H	Cl	Cl
Et	Me	CN	HCN
OMe	CH$_2$O & Me	F	C$_2$H$_2$ & HF
I	I	H	C$_2$H$_2$

Table 1.9 Primary single-bond cleavage processes associated with some common functional groups in an approximate order of ease[a]

Functional group	*Fragmentation*
Amine	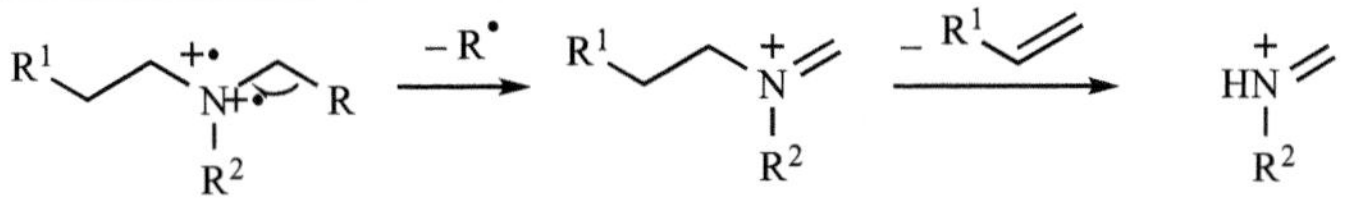 $R^1CH_2CH_2N^{+\bullet}(R^2)CH_2R \xrightarrow{-R^\bullet} R^1CH_2CH_2N^+(R^2)=CH_2 \xrightarrow{-R^1CH=CH_2} HN^+(R^2)=CH_2$
Acetal	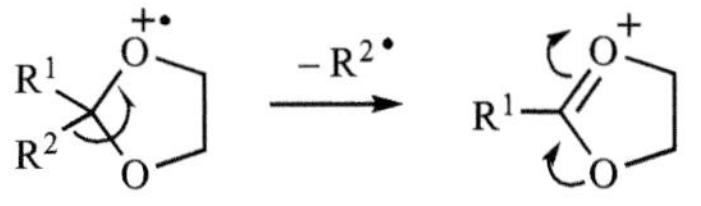 $R^1R^2C(OCH_2CH_2O)^{+\bullet} \xrightarrow{-R^{2\bullet}} R^1C^+(OCH_2CH_2O)$ Exceptionally favourable because of stabilised cation
Iodide	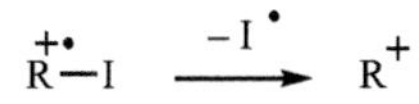 $R\text{—}I^{+\bullet} \xrightarrow{-I^\bullet} R^+$
Ether (X = O) Thioether (X = S)	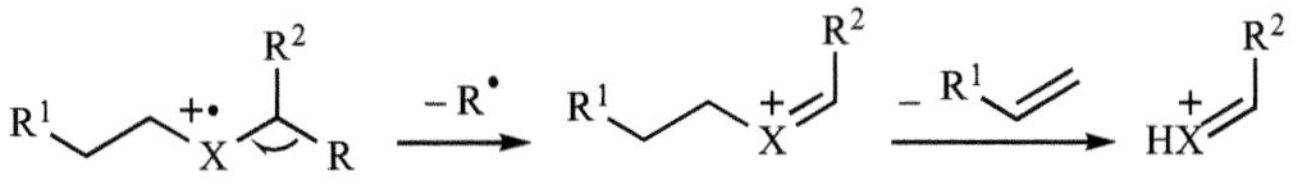 $R^1CH_2CH_2X^{+\bullet}CH(R^2)R \xrightarrow{-R^\bullet} R^1CH_2CH_2X^+=CHR^2 \xrightarrow{-R^1CH=CH_2} HX^+=CHR^2$
Ketone	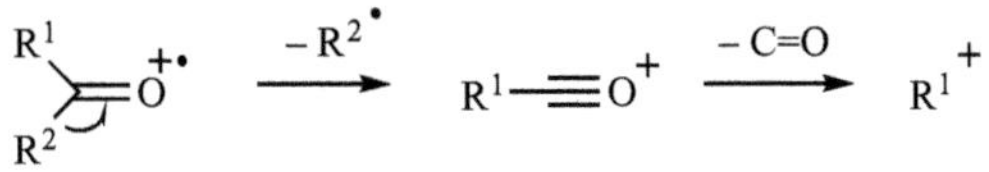 $R^1R^2C=O^{+\bullet} \xrightarrow{-R^{2\bullet}} R^1\text{—}C\equiv O^+ \xrightarrow{-C=O} R^{1+}$
Alcohol (X = O) Thiol (X = S)	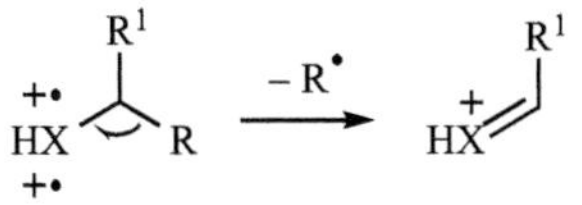 $HX^{+\bullet}CH(R^1)R \xrightarrow{-R^\bullet} HX^+=CHR^1$
Bromide	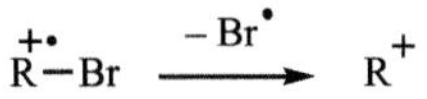 $R\text{—}Br^{+\bullet} \xrightarrow{-Br^\bullet} R^+$
Ester	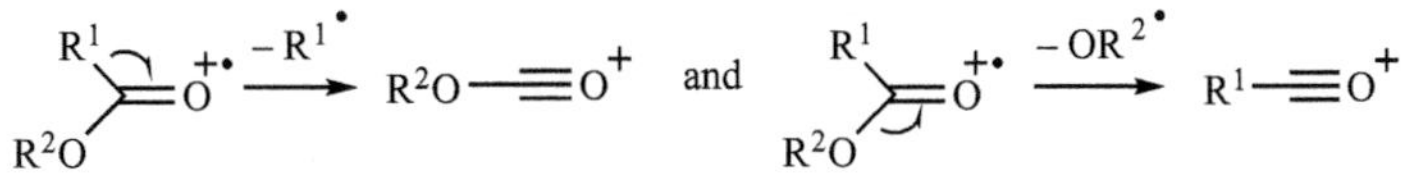 $R^1C(OR^2)=O^{+\bullet} \xrightarrow{-R^{1\bullet}} R^2O\text{—}C\equiv O^+$ and $R^1C(OR^2)=O^{+\bullet} \xrightarrow{-OR^{2\bullet}} R^1\text{—}C\equiv O^+$

[a]In polyfunctional aliphatic molecules, cleavages associated with functional groups higher up the table are preferred over those cleavages associated with lower entries

Table 1.10 Useful ion series

Functional group	Simplest ion type	Ion series (m/z)
Amine	CH_2=NH+ m/z 30	30, 44, 58, 72, 86, 100…
Ether and alcohol	CH_2=OH+ m/z 31	31, 45, 59, 73, 87, 101…
Ketone	MeC≡O+ m/z 43	43, 57, 71, 85, 99, 113…
Aliphatic hydrocarbon	C_2H_5+ m/z 29	29, 43, 57, 71, 85, 99, 113…

Table 1.11 *m/z* Values of some McLafferty rearrangement ions found in the mass spectra of carbonyl compounds

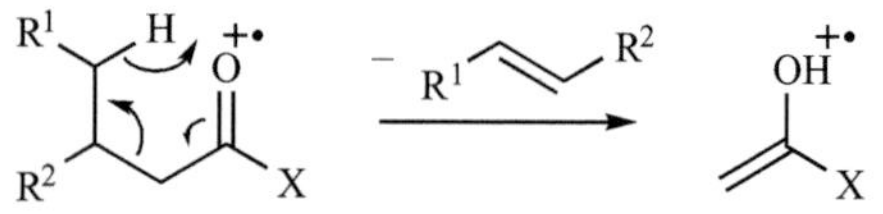

Compound	*X*	*m/z*
Aldehyde	H	44
Ketone (methyl)	Me	58
Ketone (ethyl)	Et	72
Acid	OH	60
Ester (methyl)	OMe	74
Ester (ethyl)	OEt	88
Amide	NH2	59

Table 1.12 Some common impurity peaks

m/z Values	Cause
149, 167, 279	Plasticizers (phthalic acid derivatives)
129, 185, 259,329	Plasticizer (tri-n-butyl acetyl citrate)
133, 207, 281, 355, 429	Silicone grease
99, 155, 211	Plasticizer (tributyl phosphate)

Table 1.13 Integral masses of amino acid residues –NHC(R)CO– with the one-letter code in parenthesis after the three-letter code

Gly(G)	57	Ser(S)	87	Gln(Q)	128	His(H)	137
Ala(A)	71	Thr(T)	101	Phe(F)	147	Pro(P)	97
Val(V)	99	Asp(D)	115	Tyr(Y)	163	Met(M)	131
Leu(L)	113	Asn(N)	114	Trp(W)	186	Arg(R)	156
Ile(I)	113	Glu(E)	129	Lys(K)	128	Cys(C)	103

Based on C = 12.0, H = 1.0, N = 14.0 and O = 16.0. When the first amino acid is lost from the carboxyl terminus of a peptide to give an ion of type **56**, the mass lost from $(M - H)^-$ or MH^+ is one mass unit greater than the values given in the table

Table 1.14 Some common losses from molecular ions

Ion	Groups commonly associated with the mass lost	Possible inference
M – 1	H	
M – 2	H_2	
M – 14		Homologue?
M – 15	CH_3	
M – 16	O	$ArNO_2$. R_3N–O, R_2SO
M – 16	NH_2	$ArSO_2NH_2$, $-CONH_2$
M – 17	OH	
M – 17	NH_3	
M – 18	H_2O	Alcohol, aldehyde, ketone, etc.
M – 19	F	Fluoride
M – 20	HF	Fluoride

(continued)

Table 1.14 (continued)

Ion	Groups commonly associated with the mass lost	Possible inference
M – 26	C_2H_2	Aromatic hydrocarbon
M – 27	HCN	Aromatic nitrile, nitrogen heterocycle
M – 28	CO	Quinone
M – 28	C_2H_4	Aromatic ethyl ether, ethyl ester, n-propyl ketone
M – 29	CHO	
M – 29	C_2H_5	Ethyl ketone, n-Pr–Ar
M – 30	C_2H_6	
M – 30	CH_2O	Aromatic methyl ether
M – 30	NO	Ar–NO_2
M – 31	OCH_3	Methyl ester
M – 32	CH_3OH	Methyl ester
M – 32	S	
M – 33	$H_2O + CH_3$	
M – 33	SH	Thiol
M – 34	H_2S	Thiol
M – 41	C_3H_5	Propyl ester
M – 42	CH_2CO	Methyl ketone, aromatic acetate, $ArNHCOCH_3$,
M – 42	C_3H_6	n- or i-butyl ketone, aromatic propyl ether, n-Bu–Ar
M – 43	C_3H_7	Propyl ketone, n-Pr–Ar
M – 43	CH_3CO	Methyl ketone
M – 44	CO_2	Ester (skeleton rearrangement, anhydride)
M – 44	C_3H_8	
M – 45	CO_2H	Carboxylic acid
M – 45	OC_2H_5	Ethyl ester
M – 46	C_2H_5OH	Ethyl ester
M – 46	NO_2	Ar–NO2
M – 48	SO	Aromatic sulfoxide
M – 55	C_4H_7	Butyl ester
M – 56	C_4H_8	n- or i-C_5H_{11}–Ar, n- or i-Bu–OAr, pentyl ketone
M – 57	C_4H_9	Butyl ketone
M – 57	C_2H_5CO	Ethyl ketone
M – 58	C_4H_{10}	
M – 60	CH_3CO_2H	Acetate

Table 1.15 Masses and some possible compositions of common fragment ions

m/z	*Groups commonly associated with the mass*	*Possible inference*
15	CH_3^+	
18	H_2O^+	
26	$C_2H_2^+$	
27	$C_2H_3^+$	
28	C_2H_3+CO^+, $C_2H_4^+$, N_2^+	
29	CHO^+, $C_2H_5^+$	
30	$CH_2{=}NH_2^+$	Some primary amines
31	$CH_2{=}OH^+$	Some primary alcohols
36/38 (3:1)	HCl^+	
39	$C_3H_3^+$	
40	Argon$^+$, $C_3H_4^+$	
41	$C_3H_5^+$	
42	$C_2H_2O^+$, $C_3H_6^+$	
43	CH_3CO^+	CH_3CO—X
44	$C_2H_6N^+$	Some aliphatic amines
44	$O{=}C{=}NH_2^+$	Primary amides
44	CO_2^+, $C_3H_8^+$	
44	$CH_2{=}CH(OH)^+$	Some aldehydes
45	$CH_2{=}O^+CH_3$, $CH_3CH{=}OH^+$	Some ethers and alcohols
47	$CH_2{=}SH^+$	Aliphatic thiol
49/51 (3:1)	CH_2Cl^+	
50	$C_4H_2^+$	Aromatic compound
51	$C_4H_3^+$	C_6H_5—X
55	$C_4H_7^+$	
56	$C_4H_8^+$	
57	$C_4H_9^+$	C_4H_9—X
57	$C_2H_5CO^+$	Ethyl ketone, propionate ester
58	$CH_2{=}C(OH)CH_3^+$	Some methyl ketones, some dialkyl ketones
58	$C_3H_8N^+$	Some aliphatic amines
59	$CO_2CH_3^+$	Methyl ester
59	$CH_2{=}C(OH)NH_2^+$	Some primary amides
59	$C_2H_5CH{=}OH^+$	$C_2H_5CH(OH)$—X
59	$CH_2{=}O^+{-}C_2H_5$ and isomers	Some ethers
60	$CH_2{=}C(OH)OH^+$	Some carboxylic acids
61	$CH_3CO(OH_2)^+$	$CH_3CO_2C_nH_{2n+1}$ (n>1)
61	$CH_2CH_2SH^+$	Aliphatic thiol
66	$H_2S_2^+$	Dialkyl disulfide
60	$CH_2{=}C(OH)OH^+$	Some carboxylic acids
61	$CH_3CO(OH_2)^+$	$CH_3CO_2C_nH_{2n+1}$ (n>1)
61	$CH_2CH_2SH^+$	Aliphatic thiol
66	$H_2S_2^+$	Dialkyl disulfide
68	$CH_2CH_2CH_2CN^+$	
69	CF_3^+, $C_5H_9^+$	
70	$C_5H_{10}^+$	

Table 1.15 (continued)

m/z	*Groups commonly associated with the mass*	*Possible inference*
71	$C_5H_{11}^+$	C_5H_{11}—X
71	$C_3H_7CO^+$	Propyl ketone, butyrate ester
72	$CH_2{=}C(OH)C_2H_5^+$	Some ethyl alkyl ketones
72	$C_3H_7CH{=}NH_2^+$ and isomers	Some amines
73	$C_4H_9O^+$	
73	$CO_2C_2H_5^+$	Ethyl ester
73	$(CH_3)_3Si^+$	$(CH_3)_3Si$—X
74	$CH_2{=}C(OH)OCH_3^+$	Some methyl esters
75	$C_2H_5CO(OH_2)^+$	$C_2H_5CO_2C_nH_{2n+1}$ (n>1)
75	$(CH_3)_2Si{=}OH^+$	$(CH_3)_3SiO$—X
76	$C_6H_4^+$	C_6H_5—X, X—C_6H_4—Y
77	$C_6H_5^+$	
78	$C_6H_6^+$	
79	$C_6H_7^+$	
79/81 (1:1)	Br^+	
80/82 (1:1)	HBr^+	
80	$C_5H_6N^+$	N—CH_2—X (pyrrole); X—CH_2-pyrrole (N–H)
81	$C_5H_5O^+$	X—CH_2-furan
83/85/87 (9:6:1)	$HCCl_3^+$	$CHCl_3$
85	$C_6H_{13}^+$	C_6H_{13}—X
85	$C_4H_9CO^+$	C_4H_9CO—X
85	dihydropyranyl oxonium ion (O+)	2-X-tetrahydropyran
85	γ-butyrolactone oxonium ion (O+)	5-X-γ-butyrolactone
86	$CH_2{=}CHC(OH)C_3H_7^+$	Some propyl alkyl ketones
86	$C_4H_9CH{=}NH_2$ & isomers	Some amines
87	$CH_2{=}CHC({=}OH^+)OCH_3$	X—$CH_2CH_2CO_2CH_3$
91	$C_7H_7^+$	$C_6H_5CH_2$—X
92	$C_7H_8^+$	$C_6H_5CH_2$—alkyl

Table 1.15 (continued)

m/z	*Groups commonly associated with the mass*	*Possible inference*
92	$C_7H_8^+$	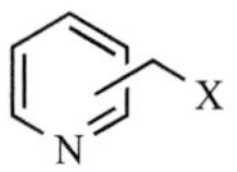
91/93 (3:1)	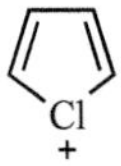	n-Alkyl chloride
93/95 (1:1)	CH_2Br^+	R—CH_2Br
	$C_6H_6O^+$	C_6H_7 O—alkyl (alkyl> CH_3
94	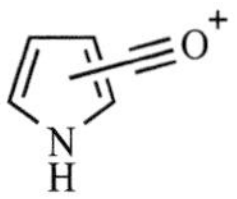	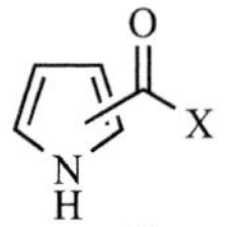
95	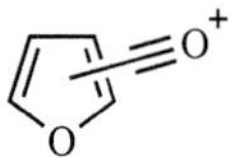	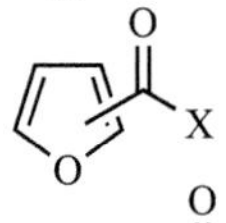
95	$C_6H_7O^+$	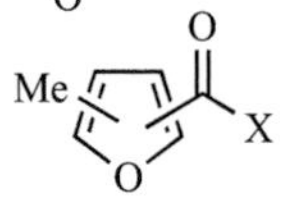
97	$C_5H_5S^+$	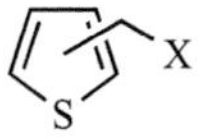
97	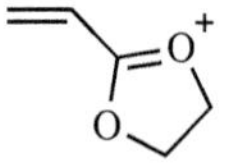	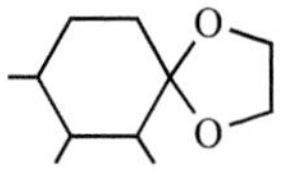
99	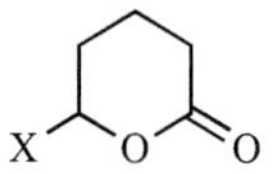	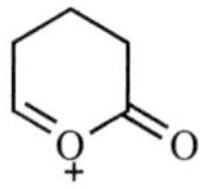
105	$C_6H_5CO^+$	C_6H_5CO—X
105	$C_8H_9^+$	$CH_3C_6H_4CH_2$—X
106	$C_7H_8N^+$	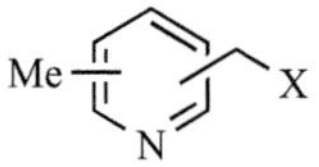
107	$C_7H_7O^+$	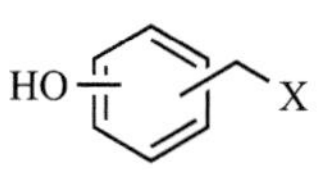
107/109 (1:1)	$C_2H_4Br^+$	
111	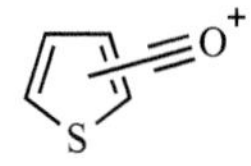	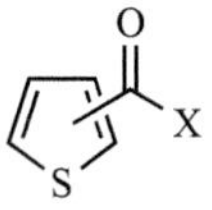

Table 1.15 (continued)

m/z	*Groups commonly associated with the mass*	*Possible inference*
121	$C_8H_9O^+$	
122	$C_6H_5CO_2H^+$	Alkyl benzoates
123	$C_6H_5CO_2H_2^+$	Alkyl benzoates
127	I^+	
128	HI^+	
135/137 (1:1)		n-Alkyl bromide (>pentyl)
130	$C_9H_8N^+$	
141	CH_3I^+	
147	$(CH_3)_3Si{=}O^-Si(CH_3)_3^+$	
149		Dialkyl phthalates
160	$C_{10}H_{10}NO^+$	
190	$C_{11}H_{12}NO_2^+$	

References

1. Smith RD, Loo JA, Edmonds CG, Baringa CJ (1990). Anal Chem 62:882–899
2. Loo JA, Quinn JP, Ryu SI, Henry KD, Senko MW, McLafferty FW (1992). Proc Natl Acad Sci U S A 89:286–289
3. Covey TR, Huang EC, Henion JD (1991). Anal Chem 63:1193–1200

2 Ultraviolet and Visible Spectra

2.1 Introduction

The ultraviolet and visible spectra of organic compounds are associated with transitions between electronic energy levels in which an electron from a low-energy orbital in the ground state is promoted into a higher-energy orbital. Normally, the transition occurs from a filled to a formerly empty orbital (Fig. 2.1) to create a singlet excited state. The wavelength of the absorption is a measure of the separation of the energies of the ground and excited *states*, related to but not the same as the energy separation E in Fig. 2.1 of the *energy levels* of the orbitals concerned.

Energy is related to wavelength by Eq. (2.1).

$$E\left(kJ\text{mol}^{-1}\right) = \frac{1.19\times10^{5}}{\lambda\left(\text{nm}\right)} \tag{2.1}$$

Thus, 297 nm, for example, is equivalent to 400 kJ (~96 kcal)—enough energy to initiate many interesting reactions; compounds should not, therefore, be left in the ultraviolet beam any longer than is necessary.

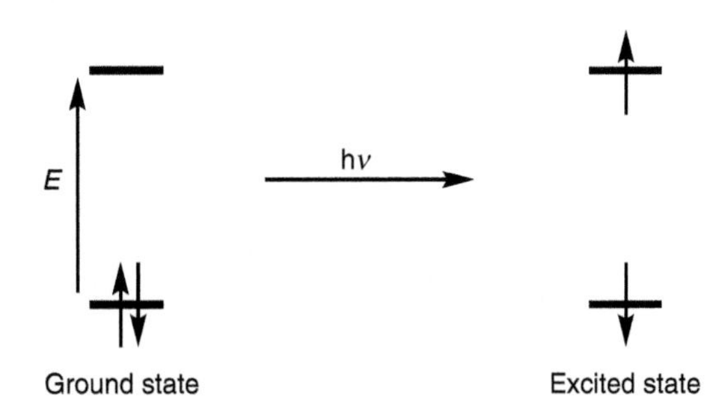

Fig. 2.1 Electronic transitions between occupied and unoccupied molecular orbitals

I. Fleming, D. Williams, *Spectroscopic Methods in Organic Chemistry*,
https://doi.org/10.1007/978-3-030-18252-6_2

2.2 Chromophores

The word *chromophore* is used to describe the system containing the electrons responsible for the absorption in question. Chromophores leading to the shortest wavelength absorption, in other words the highest energy separation, are found when electrons in σ bonds are excited, giving rise to absorption in the 120–150 nm (1 nm = 10^{-7} cm = 10 Å = 1 mμ) range, corresponding to the transition x in Fig. 2.2. Isolated double bonds like that in ethene give rise to a strong absorption maximum at 162 nm, corresponding to the transition y in Fig. 2.2. Since the air is full of σ and π bonds, it strongly absorbs UV light below 200 nm, and this range is known as the vacuum ultraviolet, since air must be excluded from the instrument in order to detect the absorption. The absorption at these short wavelengths is difficult to measure and of little use in structure determination.

Above 200 nm, however, excitation of electrons from conjugated π-orbitals, gives rise to readily measured and informative spectra. When two double bonds are conjugated, the energy level of the highest occupied molecular orbital ψ_2 (the HOMO) is raised in energy relative to the π orbital of the isolated double bonds, and that of the lowest unoccupied molecular orbital ψ_3* (the LUMO) is lowered relative to π*. The transition from ψ_2 to ψ_3* is now associated with the even smaller value z in Fig. 2.2. This transition appears in the spectrum of butadiene as a strong, easily detected, and easily measured maximum at 217 nm. The same principle governs the energy levels when unlike chromophores, for example those of an α,β-unsaturated ketone, are conjugated together. Thus, methyl vinyl ketone has an absorption maximum at 225 nm, while neither a carbonyl group nor an isolated C=C double bond has a strong maximum above 200 nm. Since the longest wavelength absorption is usually that caused by promotion of an electron from the HOMO to the LUMO, it measures how far apart in energy those important orbitals are.

If yet another π bond is brought into conjugation, the separation of the HOMO and LUMO is further reduced, and absorption occurs at a longer wavelength, with hexatriene absorbing at 267 nm. Each successive addition of a double bond reduces the energy gap, and moves the longest wavelength maximum further towards the visible. The long conjugated polyene lycopene, with 11 conjugated double bonds, has its longest wavelength absorption maximum at 504 nm (ε 158,000) with a tail reaching far into the visible region (absorbing the light in the blue to the orange range). Lycopene is responsible for the red colour of tomatoes. The most important point to be made is that, in general:

> The longer the conjugated system, the longer the wavelength of the absorption maximum.

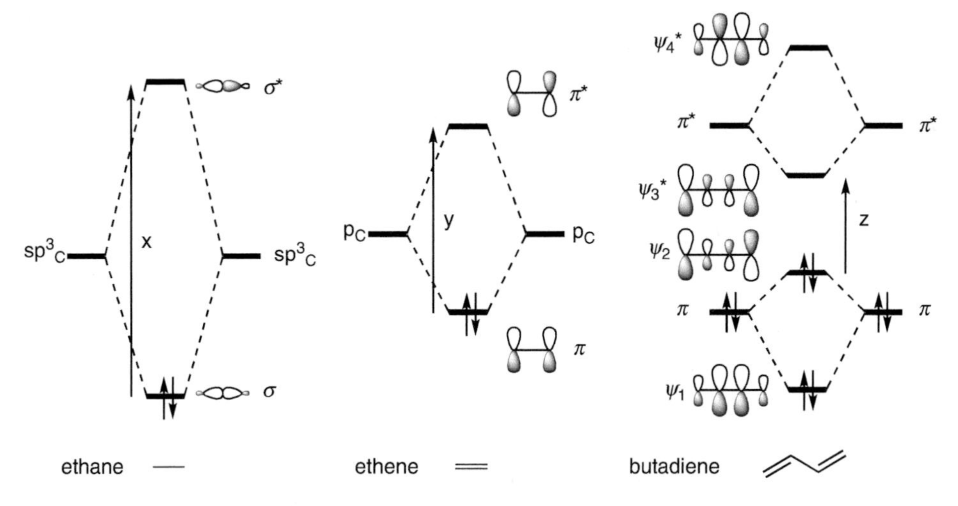

Fig. 2.2 The molecular orbitals of single, double and conjugated double bonds

2.3 The Absorption Laws

Two empirical laws have been formulated about the absorption intensity. Lambert's law states that the fraction of the incident light absorbed is independent of the intensity of the source. Beer's law states that the absorption is proportional to the number of absorbing molecules. From these laws, the remaining variables give the Eq. (2.2).

$$\log_{10}\frac{I_0}{I} = \varepsilon.l.c \tag{2.2}$$

I_0 and I are the intensities of the incident and transmitted light respectively, l is the path length of the absorbing solution in centimetres, and c is the concentration in moles per litre. $\text{Log}_{10}(I_0/I)$ is called the *absorbance* or *optical density*; ε is known as the *molar extinction coefficient* and has units of 1000 $cm^2\ mol^{-1}$ but the units are, by convention, not normally expressed.

2.4 Measurement of the Spectrum

The ultraviolet or visible spectrum is usually taken using a dilute solution. An appropriate quantity of the compound (often about 1 mg when the compound has a molecular weight of 100–400) is weighed accurately, dissolved in the solvent of choice, and made up to, for instance, 100 mL. Common solvents are 95% ethanol (commercial absolute ethanol contains residual benzene, which absorbs in the ultraviolet), hexane and cyclohexane. A portion of this solution is transferred to a silica cell 1 cm from front to back internally (the value l in Eq. (2.2)), and the pure solvent is transferred into an accurately matched cell. Two equal beams of ultraviolet or visible light are passed, one through the solution of the sample, and one through the pure solvent. The intensities of the transmitted beams are then compared over the whole wavelength range of the instrument. The spectrum is plotted automatically on most instruments as a $\log_{10}(I_0/I)$ ordinate and λ abscissa and might look something like Fig. 2.3, which is the spectrum of styrene **1** (molecular weight 104) as a solution of 0.535 mg in 100 mL of hexane and a path length of 1 cm.

For publication and comparisons the optical density is converted to an ε versus λ or log ε versus λ plot using Eq. (2.2). The unit of λ is almost always nm. The intensity of a transition is better measured by the area under an absorption peak (when plotted as ε against frequency), but for convenience, and because of the difficulty of dealing with overlapping bands, ε_{max}, the maximum intensity of the absorption, is adopted in everyday use. Spectra are quoted, therefore, in terms of λ_{max}, the wavelength at the maximum of the absorption peak read directly off the plot like that in Fig. 2.3, where it is 250 nm, and the ε value of 14,700 calculated from the value of $\log_{10}(I_0/I)$, which is 0.756 on the plot in Fig. 2.3. Typically, this spectrum would be reported thus:

λ_{max} 250 (hexane) (ε 14 700) and 282 (ε 750) nm

with the second of the peaks, barely resolved in Fig. 2.3, explained in Sect. 2.15.

2.5 Vibrational Fine Structure

The excitation of electrons is accompanied by changes in the vibrational and rotational quantum numbers so that what would otherwise be an absorption line becomes a broad peak containing all the vibrational and rotational fine structure. Because of interactions of solute with solvent molecules this is usually not resolved, and a smooth curve is observed like that illustrated in Fig. 2.3. In the vapour phase, in non-polar solvents, and with certain peaks (e.g. benzene with the 260 nm band), vibrational fine structure is sometimes resolved.

2.6 Selection Rules and Intensity

The irradiation of organic compounds does not always give rise to excitation of electrons from any filled orbital to any unfilled orbital, because there are *selection rules* based on symmetry governing which transitions are allowed. The intensity of the absorption is therefore a function of the 'allowedness', or otherwise, of the electronic transition and of the

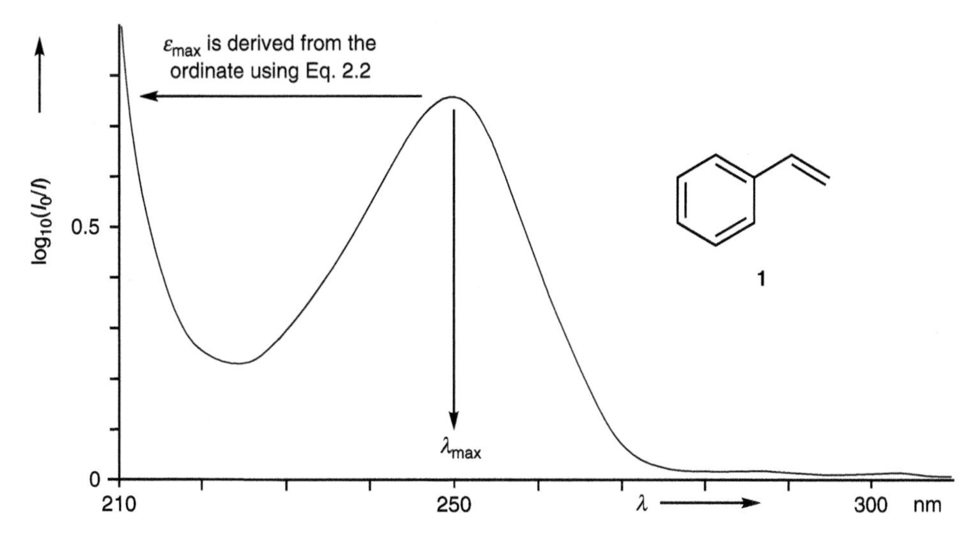

Fig. 2.3 Ultraviolet spectrum of styrene

target area able to capture the light. In practice, a chromophore with two double bonds conjugated together gives rise to absorption by a fully allowed transition with ε values of about 10,000, while many forbidden transitions (which do nevertheless occur) have ε values below 1000. The important point is that, in general:

The longer the conjugated system, the more intense the absorption.

The selection rules are a function of the symmetry and multiplicity of the ground state and excited state orbitals concerned, and whether the molecular orbitals of each overlap. A theoretical picture is given in the books by Jaffe and Orchin and by Murrell, listed in the section on further reading, but a simple knowledge of which of the commonly encountered transitions are allowed and which are forbidden is adequate for anyone using UV spectra simply to determine organic structures or to follow reaction kinetics. Thus, the important promotion of an electron from the HOMO of a linear conjugated system to the LUMO of the same system is allowed, and always leads to intense absorption. In contrast, two important forbidden transitions are the $n \rightarrow \pi^*$ band near 300 nm of ketones, with ε values of the order of 10–100, and the benzene 260 nm band, and its equivalent in more complicated systems, with ε values from 100 upwards. 'Forbidden' transitions like these, with ε_{max} typically <1000, are observed because the symmetry in the lowest vibrational state, which makes absorption strictly forbidden, is broken by molecular vibrations or by the presence of unsymmetrical substitution.

2.7 Solvent Effects

$\pi \rightarrow \pi^*$ The Frank-Condon principle states that during an electronic transition atoms do not move. Electrons, however, including those of solvent molecules, may reorganise. Most transitions result in an excited state more polar than the ground state; the dipole-dipole interactions with solvent molecules will, therefore, lower the energy of the excited state more than that of the ground state. Thus, it is usually observed that ethanol solutions give longer wavelength maxima than do hexane solutions. In other words, there is a small red shift of the order of 10–20 nm in going from hexane as solvent to ethanol.

$n \rightarrow \pi^*$ The weak transition of the oxygen lone pair in ketones—the $n \rightarrow \pi^*$ transition—shows a solvent effect in the opposite direction. The solvent effect is now a result of the lesser extent to which solvents can hydrogen bond to the carbonyl group in the excited state than in the ground state. In hexane solution, for example, the absorption maximum of acetone is at 278 nm ($\varepsilon = 15$), whereas in aqueous solution the maximum is at 264.5 nm. The shift in this direction is known as a blue shift.

2.8 Searching for a Chromophore

There is no easy rule or set procedure for identifying a chromophore—too many factors affect the spectrum and the range of structures that can be found is too great. The examination of a spectrum with particular regard for the following points is the first step to be taken.

- *The complexity and the extent to which the spectrum encroaches on the visible region.* A spectrum with many strong bands stretching into the visible shows the presence of a long conjugated or a polycyclic aromatic chromophore. A compound giving a spectrum with one band (or only a few bands) below about 300 nm probably contains only two or three conjugated units.
- *The intensity of the bands, particularly the principal maximum and the longest wavelength maximum.* This observation can be informative, and indicates why it is important to derive ε values from the optical density. Simple conjugated chromophores such as dienes and α,β-unsaturated ketones have ε values of 10,000–20,000. The longer simple conjugated systems have principal maxima (usually also the longest wavelength maxima) with correspondingly higher ε values. Low intensity absorption bands in the 270–350 nm region, on the other hand, with ε values of 10–100, are the result of the $n \rightarrow \pi*$ transition of ketones. In between these extremes, the existence of absorption bands with ε values of 1000–10,000 almost always shows the presence of an aromatic system. Many unsubstituted aromatic systems show bands with intensities of this order of magnitude, the absorption being the result of a transition with a low transition probability, low because the symmetry of the ground and excited states make the transition

forbidden. When the aromatic nucleus is substituted with groups that can extend the chromophore and break the symmetry, strong bands with ε values above 10,000 appear, but bands with ε values below 10,000 are often still present.

- *Confidence in the purity of the sample.* It is always possible that weak bands are caused by small amounts of intensely absorbing impurities. Before any confidence can be put on an absorption with a low ε value, one must be sure of the purity of the sample.

Having made these observations, one should search for a model system which contains the chromophore and therefore gives a similar spectrum to that which is being examined. This may be difficult in rare cases; but so many spectra are now known, and the changes caused by substitution so well documented, that the task can be a simple one. The first tool which an organic chemist requires is a general knowledge of the simple chromophores and the changes which structural variations make in the absorption pattern. The remaining task, that of searching through the literature, is greatly facilitated by the existence of indexes and compilations. The major collection of data is [1]. This most valuable collection has been prepared by a complete search of the major journals from 1945 until 1989, but has been discontinued since then. The compounds are indexed by their empirical formulae, and λ_{max} and $\log_{10} \varepsilon$ values are quoted together with literature references.

The search for a chromophore is likely to be assisted by the other and more powerful physical methods described in this book. The UV spectrum will mainly help to decide on the likely degree to which the functional groups are conjugated. The range of structures in which a search must be made can be narrowed, for example, to aromatic compounds on the strength of infrared or ^{1}H NMR aromatic C–H absorptions. Similarly the presence of an α,β-unsaturated ketone may be inferred from the C=O stretching vibration observed in the infrared spectrum, the presence of a low-field carbon resonance in the ^{13}C NMR spectrum and then confirmed by an appropriate ultraviolet spectrum. One area where the UV spectrum can be especially important at an early stage is in the assignment of a structure to a natural product. These compounds, isolated from a natural source, have no history to help in the structure determination, in contrast to the products of a reaction between two known chemicals. The positive identification of a likely chromophore in a natural product can help to identify to which class the natural product belongs.

2.9 Definitions

The following words and symbols are commonly used:

- *Red shift* or *bathochromic effect.* A shift of an absorption maximum towards longer wavelength. It may be produced by a change of medium or by the presence of an auxochrome.
- *Auxochrome.* A substituent on a chromophore which leads to a red shift. For example, the conjugation of the lone pair on the nitrogen atom of an enamine shifts the absorption maximum from the isolated double bond value of 190 nm to about 230 nm. The

nitrogen substituent is the auxochrome. An auxochrome, then, extends a chromophore to give a new chromophore.

- *Blue shift* or *hypsochromic effect.* A shift towards shorter wavelength. This may be caused by a change of medium and also by such phenomena as the removal of conjugation. For example, the conjugation of the lone pair of electrons on the nitrogen atom of aniline with the π bond system of the benzene ring is removed on protonation. Aniline absorbs at 230 nm (ε 8600), but in acid solution the main peak is almost identical with that of benzene, being now at 203 nm (ε 7500). A blue shift has occurred on protonation.
- *Hypochromic effect.* An effect leading to decreased absorption intensity.
- *Hyperchromic effect.* An effect leading to increased absorption intensity.
- λ_{max}. The wavelength of an absorption maximum.
- ε. The extinction coefficient defined by Eq. (2.2).
- $E^{1\%}_{1cm}$. Absorption [$\log_{10}(I_0/I)$] of a 1% solution in a cell with a 1 cm path length. This is used in place of ε when the molecular weight of a compound is not known, or when a mixture is being examined, so that Eq. (2.2) cannot be used to define the intensity of the absorption.
- *Isosbestic point.* A point common to all curves produced in the spectra of a compound taken at several pH values—the one point where the absorption intensity does not change as the pH changes.

2.10 Conjugated Dienes

The energy levels of butadiene have been illustrated in Fig. 2.2. The transition z gives rise to strong absorption at 217 nm (ε 21,000). Alkyl substitution extends the chromophore, in the sense that there is a small interaction (hyperconjugation) between the σ-bonded electrons of the alkyl groups and the π bond system. The result is a small red shift with alkyl substitution, just as there is a red shift (though a relatively large one) in going from an isolated double bond to a conjugated diene or to an enamine.

The effect of alkyl substitution, in dienes at least, is approximately additive, and a few rules suffice to predict the position of absorption in open chain dienes and dienes in six-membered rings. Open chain dienes exist normally in the energetically preferred s-*trans* conformation, while homoannular dienes must be in the s-*cis* conformation. These conformations are illustrated in the part structures **2** (heteroannular diene) and **3** (homoannular diene). The s-*cis* arrangement, as in the diene **3**, leads to longer wavelength absorption than does the s-*trans* arrangement in the diene **2**. Also, because of the shorter distance between the ends of the chromophore, s-*cis* dienes give maxima of lower intensity (ε ~10,000) than the maxima of s-*trans* dienes (ε ~20,000).

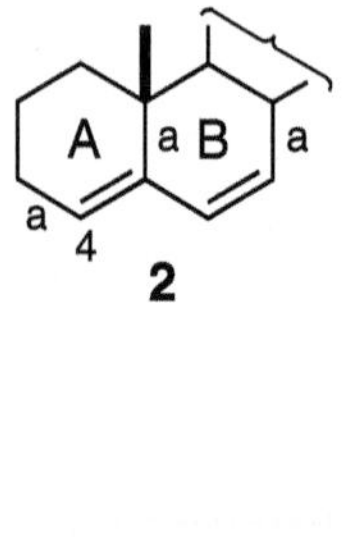

A a B a
a 4
2

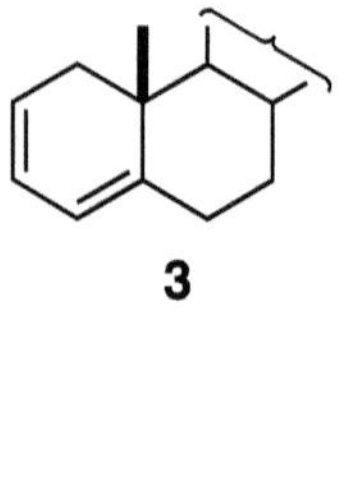

3

Table 2.1 Rules for diene and triene absorption

Value assigned to parent s-*trans* diene (like **2**)		214 nm
Value assigned to parent s-*cis* diene (like **3**)		253 nm
Increment for:		
(a) Each alkyl substituent or ring residue		5 nm
(b) The exocyclic nature of any double bond		5 nm
(c) A double-bond extension		30 nm
(d) Auxochrome:		
–OAcyl		0 nm
–OAlkyl		6 nm
–SAlkyl		30 nm
–Cl, –Br		5 nm
–NAlkyl$_2$		60 nm
λ_{calc}	Total	

Reprinted with permission from [2]

The rules for predicting the absorption of open chain and six-membered ring dienes were first made by Woodward in 1941, and were a breakthrough in showing that physical methods could be used in the details of structure determination. Since that time they have been modified by Fieser and by Scott as a result of experience with a larger number of dienes and trienes. The modified rules are given in Table 2.1.

For example, the diene **2** would be calculated to have a maximum at 234 nm by the following addition:

Parent value	214 nm
Three-ring residues (marked *a*) 3 × 5 =	15 nm
One exocyclic double bond (the Δ^4 bond exocyclic to ring B)	5 nm
Total	234 nm

A typical value observed for a steroid with this part structure is 235 nm (ε 19,000).

By similar calculation, the diene **3** would be expected to have a maximum at 273 nm, and steroids like this typically have maxima at 275 nm. Though ethanol is the usual solvent, change of solvent has little effect.

There are a large number of exceptions to the rules, when special factors are present. Distortion of the chromophore may lead to red or blue shifts, depending on the nature of the distortion.

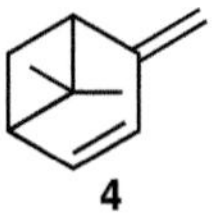

4 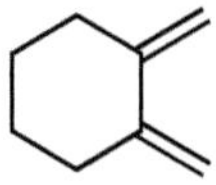 **5** 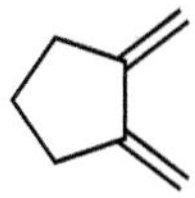 **6**

Thus, the strained molecule verbenene **4** has a maximum at 245.5 nm, whereas the usual calculation gives a value of 229 nm. The diene **5** might be expected to have a maximum at 273 nm, but distortion of the chromophore, presumably out of planarity with consequent loss of conjugation, causes the maximum to be as low as 220 nm with a similar loss in intensity (ε 5500). The diene **6**, in which coplanarity of the diene is more likely, gives a maximum at 248 nm (ε 15,800), but it still does not agree with the expected value. Change of ring size in the case of simple homoannular dienes also leads to departures from the predicted value of 263 nm as follows: cyclopentadiene, 238.5 nm (ε 3400); cycloheptadiene, 248 nm (ε 7500); while cyclohexadiene is close at 256 nm (ε 8000). The lesson, an important one, is that when the ultraviolet spectrum of an unknown compound is to be compared with that of a model compound, then the choice of model must be a careful one. Allowance must be made for the likely shape of the molecule and for any unusual strain. Some general comments on the effect of steric hindrance to coplanarity are given in Sect. 2.19.

2.11 Polyenes and Poly-Ynes

As the number of double bonds in conjugation increases, the wavelength of maximum absorption encroaches on the visible region. A number of subsidiary bands also appear and the intensity increases. Table 2.2 gives examples of the longest wavelength maxima of some simple conjugated polyenes showing these trends.

The appearance of the spectra of some of these simple polyenes is illustrated in Fig. 2.4, which shows vividly the two main lessons enshrined in the boxes used for emphasis in Sects. 2.4 and 2.6 about the wavelength and intensity as a function of the length of the conjugated system. In addition, in each spectrum, shorter wavelength absorption maxima are visible. They are the result of other transitions than just the HOMO-LUMO transition that is responsible for the longest wavelength maximum. The longer the conjugated system the more transitions become possible, and the pattern of the maxima is characteristic of the polyene concerned. It can be used as a kind of fingerprint, but in using UV spectra for structure determination we largely concentrate on the longest wavelength maximum, and the hint it gives us about the length of the conjugated system.

Several attempts, both empirical and theoretical, have been made to correlate quantitatively the principal or longest wavelength maximum with chain length. Increasing values of λ_{max} are found for increasing length in a conjugated polyene. As the chain length increases, the value of λ_{max} increases, but by less for each addition of a double bond. Quantitatively the correlation is less satisfactory. Hückel calculations for simple polyenes based on the 'electron in the box' give a good linear correlation between the energy of the transition and the separation of the energy levels in the β units of simple Hückel theory, but extrapolation does not lead back to zero on both axes, showing that the absolute values for the energy are not correct.

Table 2.2 Longest wavelength maxima of some simple polyenes

	trans-Me(CH=CH)$_n$Me		*trans*-Ph(CH=CH)$_n$Ph	
n	λ_{max} (nm)	ε	λ_{max} (nm)	ε
3	274.5	30,000	358	75,000
4	310	76,500	384	86,000
5	342	122,000	403	94,000
6	380	146,500	420	113,000
7	401	–	435	135,000
8	411	–	–	

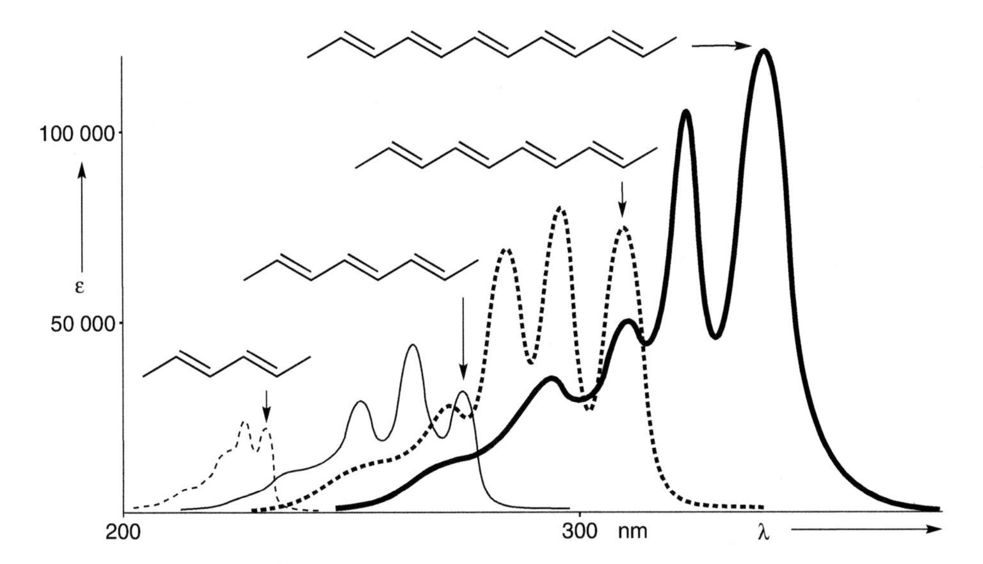

Fig. 2.4 Ultraviolet spectra of polyenes. Replotted with data from [3]

In a long-chain polyene, change from a *trans* to a *cis* configuration at one or more double bonds lowers both the wavelength and the intensity of the absorption maximum as a result of steric problems in attaining coplanarity.

The ultraviolet spectra of natural polyeneynes illustrate how a family of natural products can be detected and identified. A distinctive feature in the UV spectrum, when more than two triple bonds are conjugated, is a series of low-intensity 'forbidden' bands at regular intervals, together with high-intensity bands at longer intervals. The characteristic spiky appearance of the spectra has been helpful in screening crude plant extracts for acetylenic compounds.

In a representative application in structure determination, Fig. 2.5 shows the UV spectrum of dehydromatricaria ester **8**. In this compound, the longer wavelength bands are like a fingerprint, characteristic of an enetriyne chromophore. The structure of a

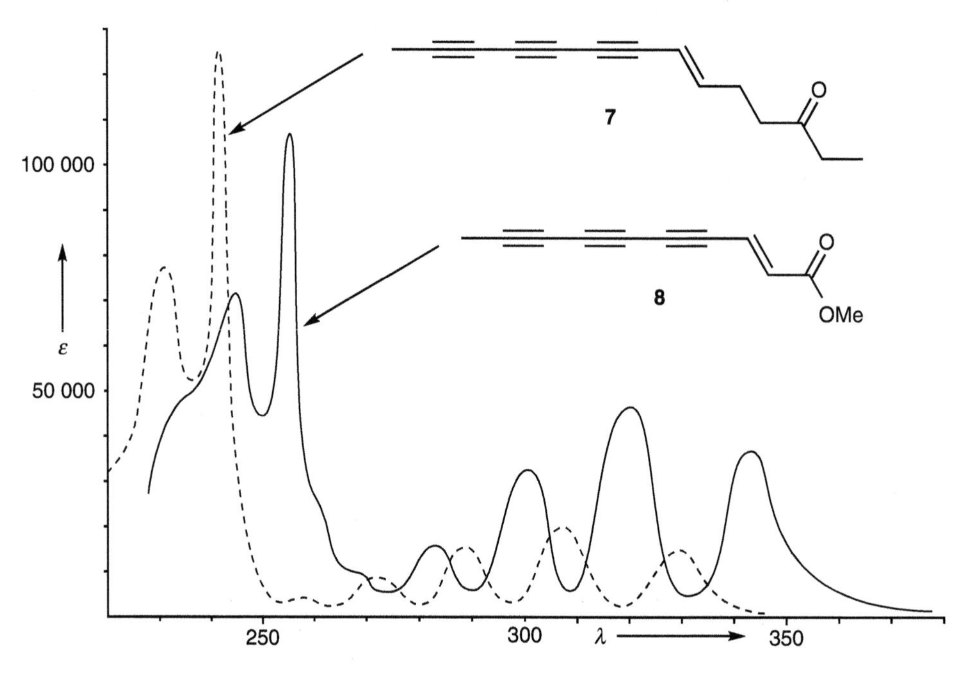

Fig. 2.5 Ultraviolet spectra of two related natural products. Replotted with data from [4, 5]

natural product **7** was assigned with the help of its UV spectrum, which showed a similar pattern, with each of the peaks shifted to shorter wavelength. Thus, the UV spectrum indicated that the structure might be an enetriyne without the extra conjugation of the ester carbonyl group. When the particular enetriyne **7** was synthesised, it proved to be identical to the natural product. This is an example of the way in which an organic chemist deals with the comparison of UV spectra: the enetriyne chromophore of the ester **8** is present in the ketone **7**, and the latter therefore continues to show similar features to those of the former, with a blue shift because of the relatively shorter conjugated system.

2.12 Ketones and Aldehydes; $\pi \rightarrow \pi*$ Transitions

As with dienes, Woodward formulated a set of rules for predicting the UV absorption of α,β-unsaturated ketones and aldehydes in ethanol. These rules, subsequently modified by Fieser and by Scott, are given in Table 2.3.

Table 2.3 Rules for α,β-unsaturated ketone and aldehyde absorption in ethanol

δ β γ α O

ε values are usually above 10,000 and increase with the length of the conjugated system

Value assigned to parent α,β-unsaturated six-ring or acyclic ketone	215 nm
Value assigned to parent α,β-unsaturated five-ring ketone	202 nm
Value assigned to parent α,β-unsaturated aldehyde	207 nm
Increments for	
(a) A double bond extending the conjugation	30 nm
(b) Each alkyl group or ring residue	
α	10 nm
β	12 nm
γ and higher	18 nm
(c) Lone-pair auxochromes	
(i) –OH	
α	35 nm
β	30 nm
γ	50 nm
(ii) –OAc	
α, β, γ	6 nm
(iii) –OMe	
α	35 nm
β	30 nm
γ	17 nm
δ	31 nm
(iv) –SAlk	
β	85 nm
(v) –Cl	
α	15 nm
β	12 nm
(vi) –Br	
α	25 nm
β	30 nm
(vii) –NR_2	
β	95 nm
(d) The exocyclic nature of any double bond	5 nm
(e) Homodiene component	39 nm
$\lambda_{calc(EtOH)}$	Total

Reprinted with permission from [2]

For example, mesityl oxide (Me_2C=CHCOMe) may be calculated to have λ_{max} at 215 + (2 × 12) = 239 nm. The observed value is 237 nm (ε 12,600). A more complicated example, the trienone chromophore of **9**, would be calculated to have a maximum at 349 nm by the following addition.

9

Parent value	215 nm
β-Alkyl substituent (marked *a*)	12 nm
ω-Alkyl substituent (marked *b*)	18 nm
2 × Extended conjugation	60 nm
Homoannular diene component	39 nm
Exocyclic double bond (the α, β double bond exocyclic to ring A)	5 nm
Total	349 nm

The observed values of λ_{max} for a trienone with the substructure **9** are 230 nm (ε 18,000), 278 nm (ε 3720) and 348 nm (ε 11,000). As was the case with simple polyenes, the long chromophore present in this example gives rise to several peaks, with the longest wavelength peak in good agreement with the prediction.

An important general principle is illustrated by the calculation for the cross-conjugated trienone **10**. In this case the main chromophore is the linear dienone portion, with the Δ^5-double bond cross-conjugated to it. Ignoring the Δ^5-double bond, the calculation, along the lines above, gives a value of 324 nm. Observed values are 256 nm and 327 nm. The former might be from the Δ^5–7-one system (λ_{calc} = 244 nm), but a positive identification of this sort in a complicated system is unjustified.

10 **11**

Certain special changes in structure, as noted in the case of dienes in Sect. 2.10, also lead to departures from the rules given above. The effect of the five-membered ring in cyclopentenones is accommodated in the rules; but when the carbonyl group is in a five-membered ring and the double bond is exocyclic to the five-membered ring, a parent value of about 215 nm holds. Another special case, verbenone **11**, would be calculated to have a maximum at 239 nm but actually has a maximum at 253 nm, an increment for strain of 14 nm, close to the increment for the corresponding diene **4**.

Table 2.4 Solvent corrections for α,β-unsaturated ketones

Solvent	Correction (nm)	Solvent	Correction (nm)
Water	−8	Dioxan	+5
Ethanol	0	Ether	+7
Methanol	0	Hexane	+11
Chloroform	+1	Cyclohexane	+11

Reprinted with permission from [2]

For λ_{calc} in other solvents than ethanol, a solvent correction from Table 2.4 must be subtracted from the above value, because the spectra are affected significantly by the solvent as a result of the change in polarity on excitation.

2.13 Ketones and Aldehydes; n→π* Transitions

Saturated ketones and aldehydes show a weak forbidden band, in the 275–295 nm range (ε ~20), from excitation of an oxygen lone-pair electron (n = p_O) into the antibonding π* orbital of the carbonyl group, as shown on the left of Fig. 2.6. This transition is forbidden because the orbitals do not overlap.

Aldehydes and the more heavily substituted ketones absorb at the long wavelength end of this range. Electronegative substituents on the α-carbon atoms increase (when oriented like an axial substituent on C-2 in a cyclohexanone) or decrease (when equatorial) the wavelength. When the carbonyl group is directly attached to an electronegative element X—as in an ester, an acid, or an amide—the π* orbital is slightly raised in energy because the substituent, having a lone pair, is a π-donor. The n (p_O) level of the lone pair, on the other hand, is lowered because

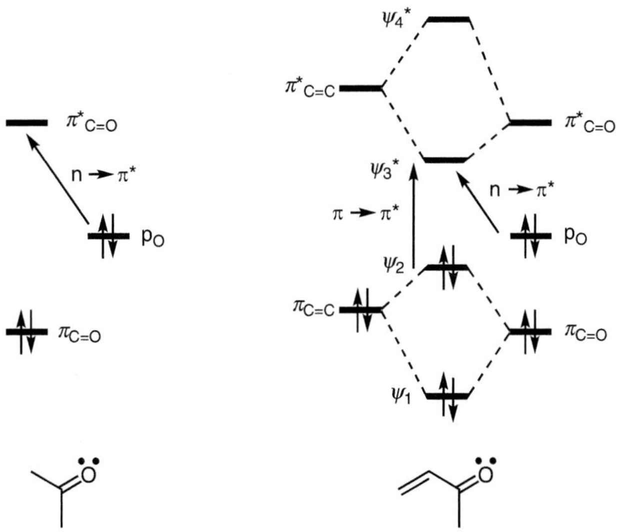

Fig. 2.6 Electronic transitions from lone pairs to unoccupied π-orbitals

it is conjugated to the C–X bond, which is a σ-withdrawing group. The result is that the n→π* transition of these compounds is shifted to shorter wavelength into the relatively inaccessible 200–215 nm range. The presence, therefore, of a weak band in the 275–295 nm region is positive identification of a ketone or aldehyde carbonyl group (nitro groups show a similar band and, of course, impurities must be absent). In contrast, if the carbonyl group is directly attached to an electropositive substituent—as in an acylsilane where the silyl group is σ-donating and π-withdrawing—the π* orbital is lowered in energy, the n orbital is raised in energy and the n→π* transition is shifted to longer wavelength, close to 370 nm for saturated acylsilanes and 420 nm for aryl or α,β-unsaturated acylsilanes. These compounds are yellow and green, respectively, as a consequence of the tail of these absorptions reaching into the visible.

α,β-Unsaturated ketones show slightly stronger n→π* absorption (ε ~100) in the 300–350 nm range, since the ψ_3^* orbital is lowered in energy by conjugation relative to the π* level of a simple carbonyl group, but the n level of the lone pair is largely unaltered, as shown on the right in Fig. 2.6. The precise position of these bands is not predictable from the extent of alkylation, but is a regular function of the conformation of γ-substituents, substituents oriented so that they overlap with the π-system shifting the absorption to longer wavelength, since they extend the conjugation. The position and intensity of n→π* bands are also influenced by transannular interactions (see Sect. 2.18) and by solvent (see Sect. 2.8).

An acyl group conjugated to a carbonyl group lowers the energy of the π* orbital even more than conjugation with a simple π bond or with a silyl group. The n→π* transitions of α-diketones in the diketo form give rise to two bands, one in the usual region near 290 nm (ε ~30) and a second (ε ~10–30) that stretches into the visible in the 340–440 nm region, and gives rise to the yellow colour of some of these compounds. (See also quinones in Sect. 2.16, quinones being α-, or vinylogous α-, diketones.)

2.14 α,β-Unsaturated Acids, Esters, Nitriles and Amides

α,β-Unsaturated acids and esters follow a trend similar to that of ketones but at slightly shorter wavelength and without the n→π* band. The rules for alkyl substitution, summarised by Nielsen, are given in Table 2.5. The change in going from acid to ester is usually not more than 2 nm.

α,β-Unsaturated nitriles absorb at wavelengths slightly lower than the corresponding acids, and α,β-unsaturated amides lower still, usually near 200 nm (ε ~8000). α,β-Unsaturated lactams have an additional band at 240–250 nm (ε ~1000).

2.15 Aromatic Compounds

Benzene absorbs at 184 (ε 60,000), 203.5 (ε 7400) and 254 (ε 204) nm in hexane solution; it is illustrated by the dashed line in Fig. 2.7. The latter band, also called the *B*-band, shows vibrational fine structure. Although a 'forbidden' band, it owes its appearance to the loss

Table 2.5 Rules for α,β-unsaturated acids' and esters' absorption (ε values are usually above 10,000)

β-Monosubstituted		208 nm
αβ- or ββ-Disubstituted		217 nm
αββ-Trisubstituted		225 nm
Increment for		
(a) A double bond extending the conjugation		30 nm
(b) The exocyclic nature of any double bond		5 nm
(c) When the double bond is endocyclic in a five-or seven-membered ring		5 nm
λ_{calc}	Total	

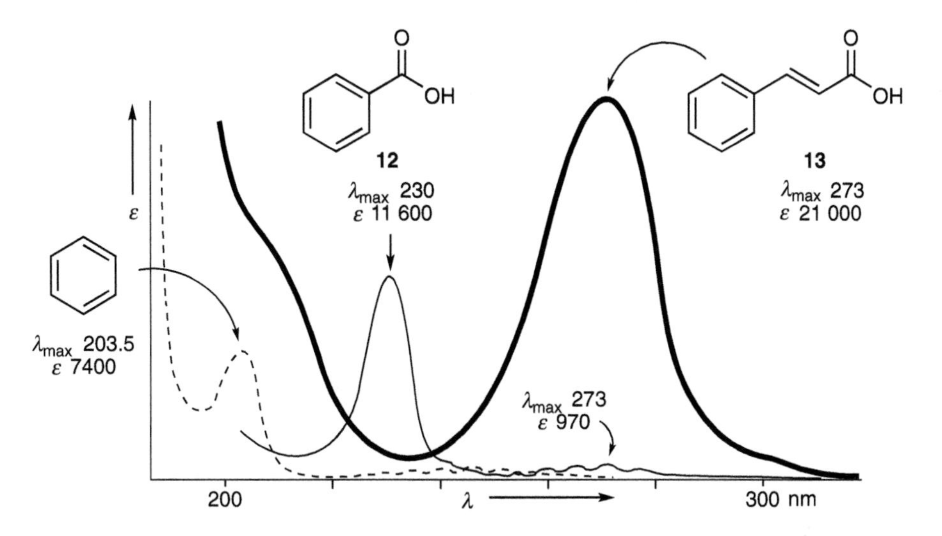

Fig. 2.7 Ultraviolet spectra of aromatic compounds

of symmetry caused by molecular vibrations; indeed, the 0→0 transition (the transition between the ground state vibrational energy level of the electronic ground state to the ground state vibrational energy level of the electronic excited state) is not observed.

When the aromatic ring is substituted by alkyl groups, for example, or is an aza analogue such as pyridine, the symmetry is lowered and the 0→0 transition is then observed, although the spectrum is little changed otherwise. The presence of fine structure resembling that shown in Fig. 2.7 is characteristic of the simpler aromatic molecules. When the benzene ring has a lone-pair or π-bonded substituent, in other words an auxochrome, the chromophore is extended. Table 2.6 gives the wavelength of absorption maxima in the spectra of a range of monosubstituted benzenes, showing how, as usual, the wavelength and intensity of the absorption peaks increase with an increase in the extent of the chromophore.

As more conjugation is added to the benzene ring, the band originally at 203.5 nm (also called the *K*-band) effectively 'moves' to longer wavelength, and moves 'faster' than the *B*-band, which was originally at 254 nm, eventually overtaking it. This can be seen in the

Table 2.6 Absorption maxima of substituted benzene rings Ph–R

R	λ_{max} nm (ε) (solvent H_2O or MeOH)					
–H	203.5	(7400)	254	(204)		
$–NH_3^+$	203	(7500)	254	(160)		
–Me	206.5	(7000)	261	(225)		
–I	207	(7000)	257	(700)		
–Cl	209.5	(7400)	263.5	(190)		
–Br	210	(7900)	261	(192)		
–OH	210.5	(6200)	270	(1450)		
–OMe	217	(6400)	269	(1480)		
$–SO_2NH_2$	217.5	(9700)	264.5	(740)		
–CN	224	(13,000)	271	(1000)		
$–CO_2^-$	224	(8700)	268	(560)		
$–CO_2H$	230	(11,600)	273	(970)		
$–NH_2$	230	(8600)	280	(1430)		
$–O^-$	235	(9400)	287	(2600)		
–NHAc	238	(10,500)				
–COMe	245.5	(9800)				
$–CH=CH_2$	248	(14,000)	282	(750)	291	(500)
–CHO	249.5	(11,400)				
–Ph	251.5	(18,300)				
–OPh	255	(11,000)	272	(2000)	278	(1800)
$–NO_2$	268.5	(7800)				
$–CH=CHCO_2H$	273	(21,000)				
–CH=CHPh	295.5	(29,000)				

Most values taken with permission from [6]

two other spectra recorded on Fig. 2.7: benzoic acid **12** shows the *K*-band at 230 nm with the *B*-band still clearly visible at 273 nm; but with the longer chromophore of cinnamic acid **13** the *K*-band has moved to 273 nm and the *B*-band is almost completely submerged. In the latter case, we can see how the even stronger band, originally at 184 nm in benzene itself, has also moved, but has still not reached the accessible region. It is responsible for what is called *end absorption*, that is the short-wavelength side of an absorption peak, the maximum of which is below the range of the instrument.

14

λ_{max} 375 nm (ε 16 000)

15

λ_{max} 260 nm (ε 13 000)

In disubstituted benzenes, two situations are important. When electronically complementary groups, such as amino and nitro, are situated *para* to each other as in *p*-nitroaniline

14, there is a pronounced red shift in the main absorption band, compared to the effect of either substituent separately, caused by the extension of the chromophore from the electron-donating group to the electron withdrawing group through the benzene ring, as symbolised by the curly arrows. Alternatively, when two groups are situated *ortho* or *meta* to each other, or when the *para* disposed groups are not complementary, as in *p*-dinitrobenzene **15**, then the observed spectrum is usually close to that of the separate, non-interacting, chromophores. These principles are illustrated by the examples in Table 2.7. The values in this table should be compared with each other and with the values for the single substituents separately given in Table 2.6.

In particular it should be noted that those compounds with non-complementary substituents, or with an *ortho* or *meta* substitution pattern, actually have a band (though a much weaker one) at longer wavelength than the compounds with interacting *para*-disubstituted substituents. This fact is not in accord with the simple resonance picture; neither is the similarity of the *ortho* to the *meta* disubstituted cases. This is another case in which molecular orbital theory (too complicated to be introduced here, but dealt with in Murrel's book) gives a better picture.

In the case of disubstituted benzene rings in which the electron-donating group is complemented by an electron-withdrawing carbonyl group, some quantitative assessments may be made. These apply to the compounds RC_6H_4COX in which X is alkyl, H, OH, or OAlkyl, and refer to the strongest band in the accessible region, which is often the only measured band in the highly conjugated *para*-disubstituted systems. The calculation is based on a parent value with increments for each substituent. Polysubstituted benzene rings should be treated with caution, particularly when the substitution might lead to steric hindrance preventing coplanarity of the carbonyl group and the ring. Table 2.8 gives the rules for this calculation. In the absence of steric hindrance to coplanarity, the calculated values are usually within 5 nm of the observed values.

Table 2.7 Absorption maxima of disubstituted benzene rings

$R^1–C_6H_4–R^2$								
R^1	R^2		λ_{max} (EtOH) nm (ε)					
–OH	–OH	*o*	214	(6000)	278	(2630)		
–OMe	–CHO	*o*	253	(11,000)	319	(4000)		
$–NH_2$	$–NO_2$	*o*	229	(16,000)	275	(5000)	405	(6000)
–OH	–OH	*m*	277	(2200)				
–OMe	–CHO	*m*	252	(8300)	314	(2800)		
$–NH_2$	$–NO_2$	*m*	235	(16,000)	373	(1500)		
–Ph	–Ph	*m*	251	(44,000)				
–OH	–OH	*p*	225	(5100)	293	(2700)		
–OMe	–CHO	*p*	277	(14,800)				
$–NH_2$	$–NO_2$	*p*	229	(5000)	375	(16,000)		
–Ph	–Ph	*p*	280	(25,000)				

Table 2.8 Rules for the principal band of substituted benzene derivatives RC_6H_4COX

Parent chromophore: *X*			λ_{max} (EtOH) nm		
Alkyl or ring residue			246		
H			250		
OH or Oalkyl			230		
Increment (nm) for each substituent:			Increment (nm) for each substituent:		
R	*o, m or p*	Increment	R	*o, m or p*	Increment
Alkyl or ring residue	*o, m*	3	Br	*o, m*	2
	p	10		*p*	15
OH, OMe, Oalkyl	*o, m*	7	NH_2	*o, m*	13
	p	25		*p*	58
O^-	*o*	11	NHAc	*o, m*	20
	m	20		*p*	45
	p	78	NHMe	*p*	73
Cl	*o, m*	0	NMe_2	*o, m*	20
	p	10		*p*	85

Reprinted with permission from [2]

6-Methoxytetralone **16** provides an example:

MeO O **16**

Parent value	246 nm
ortho alkyl	3 nm
para MeO	25 nm
λ_{calc}	274 nm

The maximum is actually at 276 nm (ε 16,500).

Polycyclic aromatic hydrocarbons have many more energy levels between which electronic transitions can take place. Their spectra are usually complicated, and for that reason are useful as fingerprints. When they only have relatively poorly conjugating substituents, such as alkyl and acetoxy groups, the spectra are similar in the shape and position of the absorption peaks to the unsubstituted hydrocarbons. The degradation products of natural products often contain polycyclic nuclei which can be identified in this way as, for example, a phenanthrene or a perylene. The spectra of a typical series, naphthalene **17**, anthracene **18**, and naphthacene **19**, are illustrated in Fig. 2.8; this figure uses a logarithmic ordinate in order to encompass the range of intensities.

In general heteroaromatic compounds resemble the spectra of the corresponding hydrocarbons, but only in the crudest way. The heteroatom, whether like that in a pyrrole or that in a pyridine, leads to pronounced effects analogous to those of substituents, and depend on the electron-donating (in pyrroles) or withdrawing (in pyridine) effect of the heteroatom and its orientation. The effects of these factors are predictable in a qualitative

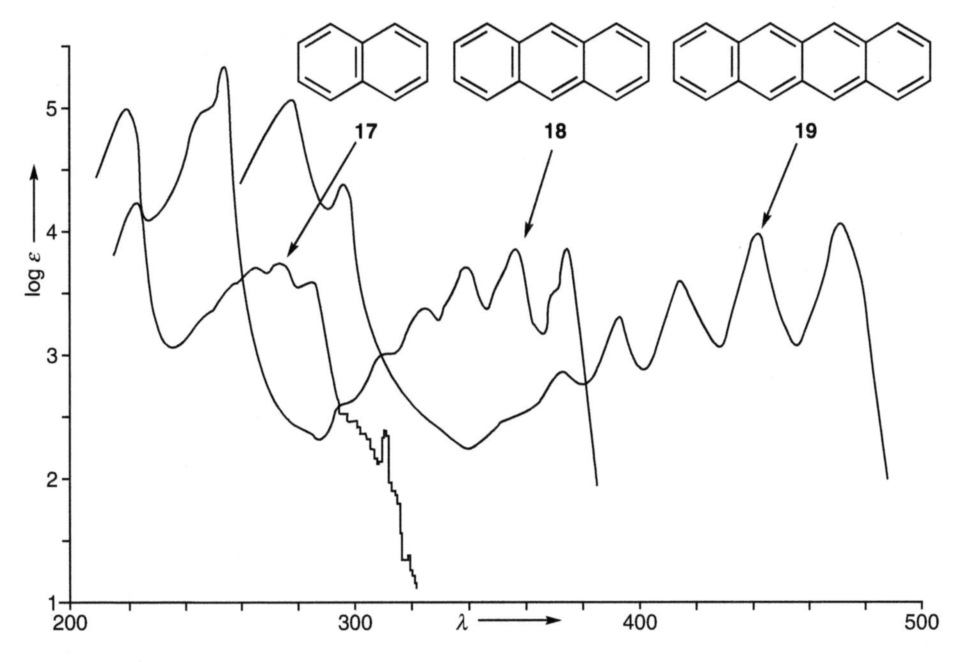

Fig. 2.8 Ultraviolet spectra of fused aromatic systems. Most values taken with permission from [6]

way using the same sorts of criteria as before. For example, the spectrum of a simple pyrrole **20** resembles the spectrum of benzene. In the substituted pyrrole **21**, the conjugation from the nitrogen lone pair through the pyrrole ring to the carbonyl group (arrows) increases the length of the chromophore and leads to substantially longer wavelength absorption. The linear conjugation to the carbonyl group in the 2-position in the pyrrole **21** is longer, and leads to longer wavelength absorption, than the cross-conjugation present in the 3-substituted pyrrole **22**.

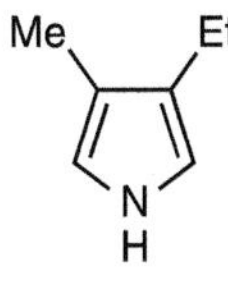

20	**21**	**22**
λ_{max} (EtOH)	λ_{max} (EtOH)	λ_{max} (EtOH)
203 nm (ε 5670)	262 nm (ε 12 000)	245 nm (ε 4800)

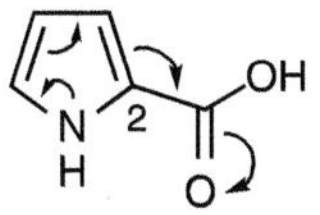

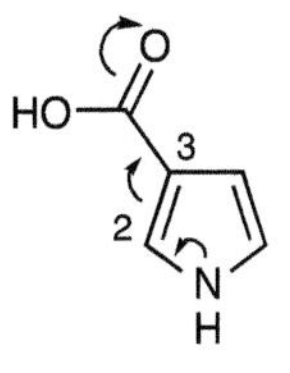

The data under the structures **23–30** illustrate the absorption of some common aromatic heterocyclic systems, including the four nucleoside bases **27–30**. Since some of these compounds are acidic and basic, the pH makes a difference to the spectra; an extra proton or one proton fewer changes the conjugated system, and hence the UV spectrum. The changes in absorption maxima with change of pH are also useful diagnostically, since they serve in some systems to identify the pattern of substitution.

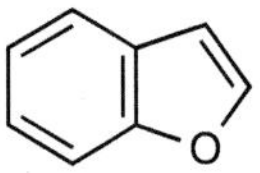

23

λ_{max} ($CHCl_3$)
245 nm (ε 12 000)
275 nm (ε 2800)
282 nm (ε 3020)

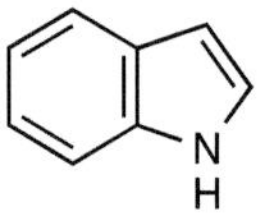

24

λ_{max} (cyclohexane)
220 nm (ε 26 000)
262 nm (ε 6310)
280 nm (ε 5620)
288 nm (ε 4170)

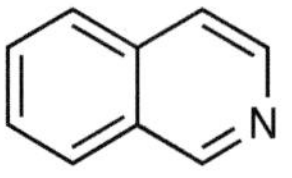

25

λ_{max} ($CHCl_3$)
218 nm (ε 79 000)
266 nm (ε 3900)
305 nm (ε 2000)
318 nm (ε 3000)

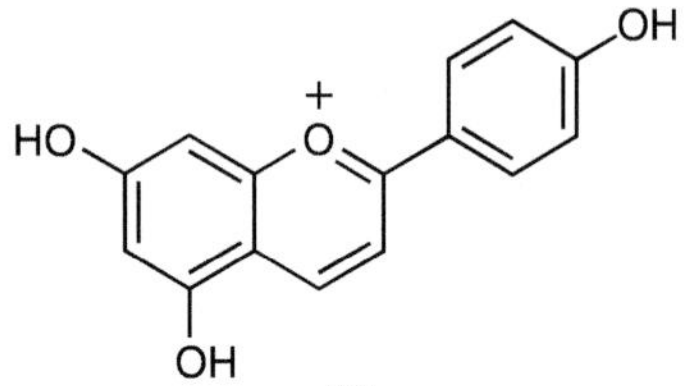

26

λ_{max} (MeOH) 520 nm

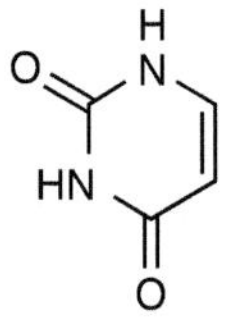

27

λ_{max} (H_2O)

pH 4	259.5 nm
pH 7	260 nm (ε 11 000)
pH 9.5	261 nm

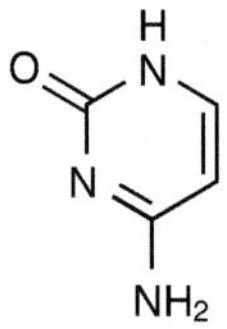

28

λ_{max} (H_2O)

pH 1	210 nm (ε 9700) 276 nm (ε 110 000)
pH 5	269 nm (ε 6650)
pH 7	267 nm (ε 6130)
pH 12	272 nm (ε 5630)

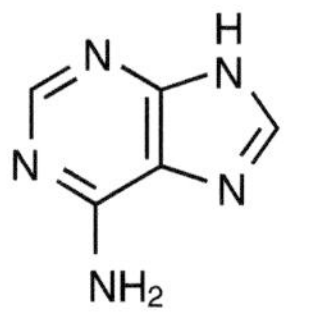

29

λ_{max} (H_2O)

pH 2	262 nm
pH 7	260 nm (ε 13 500)
pH 12	267 nm

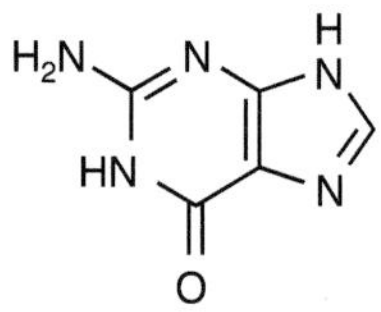

30

λ_{max} (H_2O)

pH 1	248 nm 271 nm
pH 6	246 nm (ε 10 000) 275 nm (ε 7800)
pH 11	245 nm 273 nm

Ultraviolet spectroscopy can be used to identify which is the stable tautomer of this type of aromatic compound. For example, evidence that the equilibrium between 2-hydroxypyridine **31** and pyrid-2-one **32** lies far to the right, comes from a comparison of its UV spectrum with those of 2-methoxypyridine **33** and *N*-methylpyrid-2-one **34**.

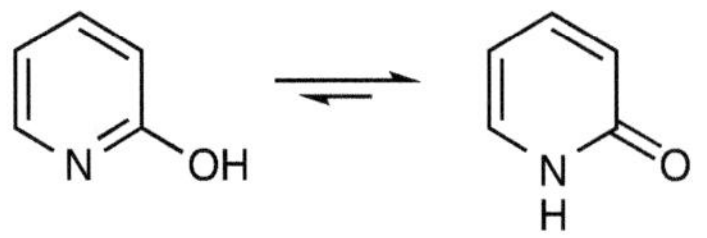

31

32
λ_{max} 224 nm (ε 7230)
293 nm (ε 5900)

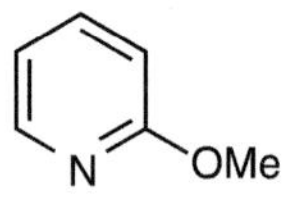

33
λ_{max} <205 nm (ε >5300)
269 nm (ε 3230)

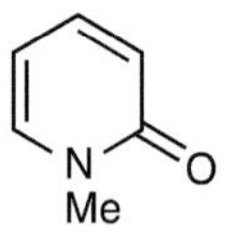

34
λ_{max} 226 nm (*e* 6100)
297 nm (*e* 5700)

2.16 Quinones

The quinones **35–38** are a representative series of these usually coloured compounds. The colour of the simpler members is from the weak n → π* transition, similar to that of α-diketones (Sect. 2.13).

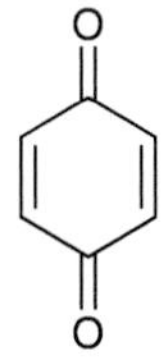

35
λ_{max} (hexane)
242 nm (ε 24 000)
281 nm (ε 400)
434 nm (ε 20)

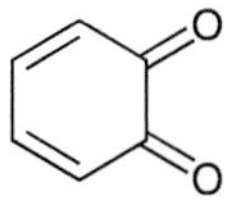

36
λ_{max} (EtOH)
276 nm (ε 2000)
387 nm (ε 800)

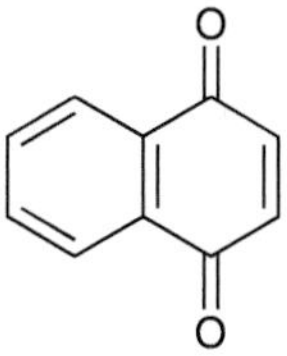

37
λ_{max} (hexane)
241 nm (ε 20 000)
246 nm (ε 23 500)
251 nm (ε 19 000)
256 nm (ε 13 000)
330 nm (ε 2750)

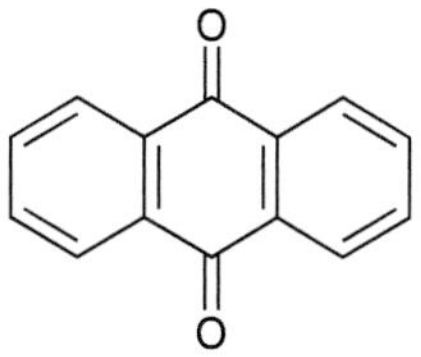

38
λ_{max} (EtOH)
243.5 nm (ε 33 000)
252.5 nm (ε 51 500)
263 nm (ε 20 000)
272 nm (ε 20 000)
325 nm (ε 5600)
405 nm (ε 90)

2.17 Corroles, Chlorins and Porphyrins

Figure 2.9 shows the visible spectra of representative members of each of the three main classes of pyrrole pigments: hydrogenobyrinic acid **39**, with the chromophore of vitamin B_{12}, chlorophyll **40**, and protoporphyrin IX **41**. The long conjugated systems, with the chromophores emphasised in black, give rise to an exceptionally strong and sharp band, the shoulder of which can be seen on the left in each spectrum in Fig. 2.9. In chlorins and porphyrins this band is called the Soret band, and it occurs near 400 nm (ε 100,000). Changes in the chromophore in each class can often be recognised by changes in the position and relative intensity of the four or more weaker, but still strong, bands found in the visible region. These spectra illustrate the general principle that the longer conjugated systems lead to more intense absorption at longer wavelength, and they also reveal how the detailed pattern changes diagnostically with the presence or absence of extra double bonds. The chlorophyll spectrum, in particular, shows the strong absorption at the blue and the red end of the spectrum, leaving the green colour with which we are so familiar. Another conjugated macrocyclic aromatic system, [18]-annulene, shows a similar intense band at 369 nm (ε 303,000).

The pyrrole pigments are mentioned here to stress the importance and usefulness of ultraviolet and visible spectroscopy in the study of groups of compounds possessing a

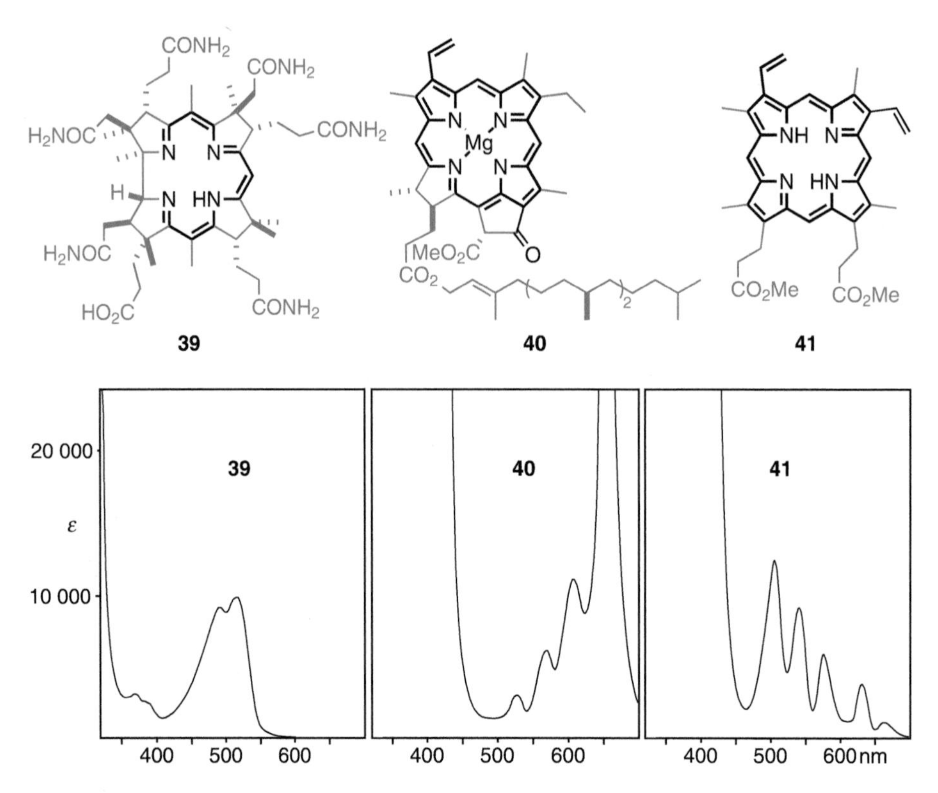

Fig. 2.9 Ultraviolet and visible spectra of representative pyrrole pigments

long, complicated chromophore. The large number of model systems available makes it relatively easy to recognise a chromophore, which the other spectroscopic methods do not probe. For example, the oxidations involved in the biodegradation of chlorophyll and haem interrupt the conjugated system in the middle, and are immediately picked up by the very dramatic changes in the visible absorption spectra.

2.18 Non-conjugated Interacting Chromophores

Non-conjugated systems usually have little effect on each other; diphenyl methane has a spectrum similar to that of toluene; the cross-conjugation of the trienone **10** was successfully ignored when calculating the expected absorption maximum; and even diphenyl ether is not very different from anisole, because the two phenyl rings are not conjugated to each other. However, conjugation through space is possible when an auxochrome is held close enough to a conjugated system without being directly conjugated to it.

Thus, the βγ-unsaturated ketone norbornenone **42** shows $n \rightarrow \pi^*$ and $\pi \rightarrow \pi^*$ transitions shifted towards the blue relative to the absorptions of the isolated components. There is evidently some interaction across space raising the HOMO and lowering the LUMO, but not as effectively as in αβ-unsaturated ketones. Also, the cumulated double bonds in allenes **43** and ketenes **44**, although not conjugated, cause some ultraviolet light to be absorbed weakly in the accessible region.

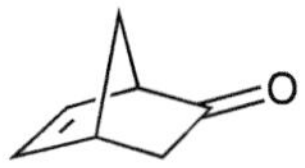

	42	**43**	**44**
λ_{max}	210 nm (ε 3000) 305 nm (ε 290)	170 nm (ε 4000) 227 nm (ε 630)	227 nm (ε 360) 375 nm (ε 20)

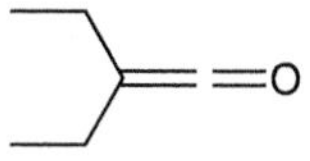

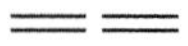

2.19 The Effect of Steric Hindrance to Coplanarity

Steric hindrance preventing full coplanarity in a conjugated system, as in *cis*-stilbene **45**, interferes with the effect conjugation has on the energy of the HOMO and the LUMO. In a poorly conjugated system, the HOMO is not raised in energy to the same extent, and the LUMO is not lowered in energy to the same extent. As a result the gap (z in Fig. 2.2) is larger. The result for the longest wavelength absorption band in *cis*-stilbene **45** is a shorter wavelength and less intense maximum than for *trans*-stilbene **46**.

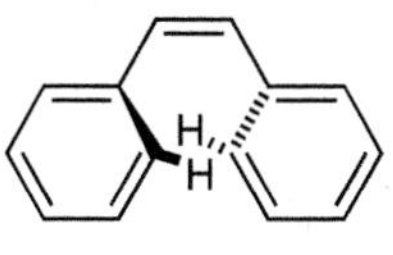

45

λ_{max} (EtOH) 224 nm (ε 24 400)
280 nm (ε 10 500)

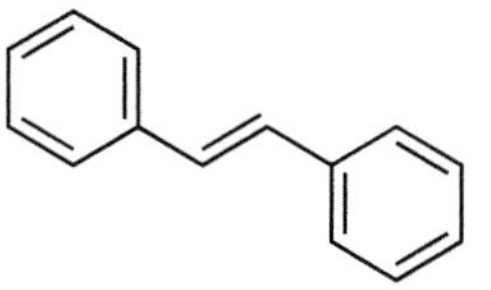

46

228 nm (ε 16 400)
295.5 nm (ε 29 000)

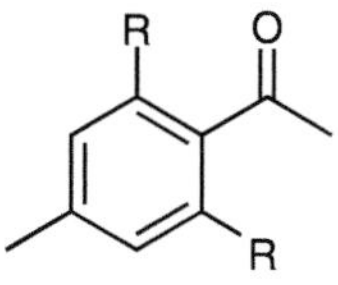

47

R = Me
λ_{max} 242 nm (ε 3200)
R = H
λ_{max} 252 nm (ε 15 000)

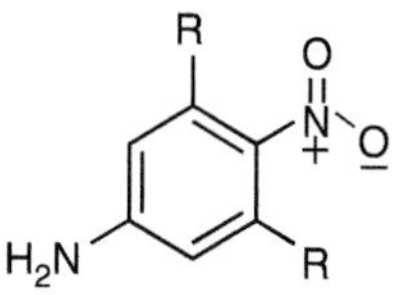

48

R = Me
λ_{max} 385 nm (ε 4840)
R = H
λ_{max} 375 nm (ε 16 000)

Similarly, the *ortho* substituents in 2,4,6-trimethylacetophenone **47** (R = Me) prevent the carbonyl group from lying coplanar with the benzene ring; this ketone has weaker absorption at shorter wavelength than *p*-methylacetophenone **47** (R = H). On the other hand, the absorption maximum of 3,5-dimethyl-*p*-nitroaniline **48** (R = Me) shows the usual reduction in intensity but this time a red shift relative to that of the parent compound *p*-nitroaniline **48** (R = H). It is possible that the former absorption is from a different transition to that monitored in the latter case.

The dilactone **49** produced from shellolic acid misleadingly showed no maximum in the accessible ultraviolet region, but hydrolysis gave a product which showed the expected absorption for an αβ-unsaturated acid **50**. The steric constraints in the polycyclic structure had prevented effective conjugation between the double bond and the carbonyl group, but the release of this constraint allowed the two π bonds to overlap.

$\xrightarrow{H_2O}$

49
no strong absorption >210 nm

50
λ_{max} 227 nm (ε 5500)

This observation provides an opportunity to stress that changes between the ultraviolet spectrum of a starting material and a product make it one of the easiest tools to use for following the kinetics of a chemical reaction, and that ultraviolet spectroscopy is possibly used more for this purpose than in structure determination. Nevertheless, the immediate and highly sensitive detection of conjugated systems is still a powerful application for this the oldest of the spectroscopic methods.

2.20 Internet

The Internet is a continuously evolving system, with links and protocols changing frequently. The following information is inevitably incomplete and may no longer apply, but it gives you a guide to what you can expect. Some websites require particular operating systems and may only work with a limited range of browsers, some require payment, and some require you to register and to download programs before you can use them.

Ultraviolet spectroscopy is not as well served on the Internet as the other spectroscopic methods. The set of books, *Organic Electronic Spectral Data*, is still the best source for UV and visible spectroscopic data.

There is a database of 1600 compounds with UV data on the NIST website belonging to the United States Secretary of Commerce: http://webbook.nist.gov/chemistry/name-ser.html

Type in the name of the compound you want, check the box for UV/Vis spectrum, and click on Search, and if the ultraviolet spectrum is available it will show it to you.

ACD (Advanced Chemistry Development) Spectroscopy sell proprietary software called ACD/SpecManager that handles all four spectroscopic methods, as well as other analytical tools: http://www.acdlabs.com/products/spec_lab/exp_spectra/

It is able to process and store the output of the instruments that take spectra, and can be used to catalogue, share and present your own data. It also gives access to a few free databases for UV spectra.

Wiley-VCH keep an up-to-date website on their spectroscopic books and provide links. The URL giving access to information about spectroscopy, including UV, is:

http://www.spectroscopynow.com/Spy/basehtml/SpyH/1,1181,7-4-773-0-773-directories%2D%2D0,00.html

2.21 Further Readings

Data

- (1960–96) Organic electronic spectral data, vols. 1–31. Wiley, New York
- (1979) Sadtler handbook of ultraviolet spectra. Sadtler Research Laboratories
- Kirschenbaum DM (ed) (1972) Atlas of protein spectra in the ultraviolet and visible region. Plenum, New York

Textbooks

- Jaffe HH, Orchin M (1962) Theory and applications of ultraviolet spectroscopy. Wiley, New York
- Murrell JN (1963) The theory of the electronic spectra of organic molecules. Methuen, London
- Barrow GR (1964) Introduction to molecular spectroscopy. McGraw-Hill, New York
- Brittain EFH, George WO, Wells CHJ (1970) Introduction to molecular spectroscopy. Academic, London
- Scott AI (1964) Interpretation of the ultraviolet spectra of natural products. Pergamon Press, Oxford
- Mason SF (1963) Chapter 7. The electronic absorption spectra of heterocyclic compounds. In: Physical methods in heterocyclic chemistry, vol II. Academic, New York
- Rao CNR (1975) Ultraviolet and visible spectroscopy, 3rd edn. Butterworths, London
- Thomas MJK (1996) Ultraviolet and visible spectroscopy, 2nd edn. Wiley, Chichester

2.22 Problems

1. The reaction of cyclohexanone with isopropenyl Grignard followed by treating the product with acid gave a hydrocarbon, C_9H_{14} which was incorrectly formulated as the structure **51**. Suggest an alternative structure in better agreement with its observed UV spectrum: λ_{max} 242 nm.

BrMg OH H_2SO_4

51

2. The UV spectra for the three compounds C_7H_8O in Problems 1–3 in Ch. 1, are given below. Assign which UV spectrum **A-C** is derived from which isomer in Problems 1–3.

   A: λ_{max} 219 (ε 6900), 265sh (ε 1200), 271 (ε 1800) and 278 (ε 1500) nm, unaffected by adding sodium hydroxide.

   B: λ_{max} 210 (ε 7600) and 252 (ε 250) nm unaffected by adding sodium hydroxide.

   C: λ_{max} 228 (ε 5800) and 279 (ε 1900) nm changed to λ_{max} 295 (ε 2630) nm on adding sodium hydroxide

3. The five isomers **52–56** of formula $C_8H_8O_2$ can be distinguished by each of the four spectroscopic methods. The ^{13}C-NMR spectra show that they all have six different carbon atoms. The UV spectra are the least decisive but it is possible to be reasonably confident of all of the assignments. The UV spectra are printed below, labelled with the letters **D-H**. Assign the letters **D-H** to the numbers **52–56**.

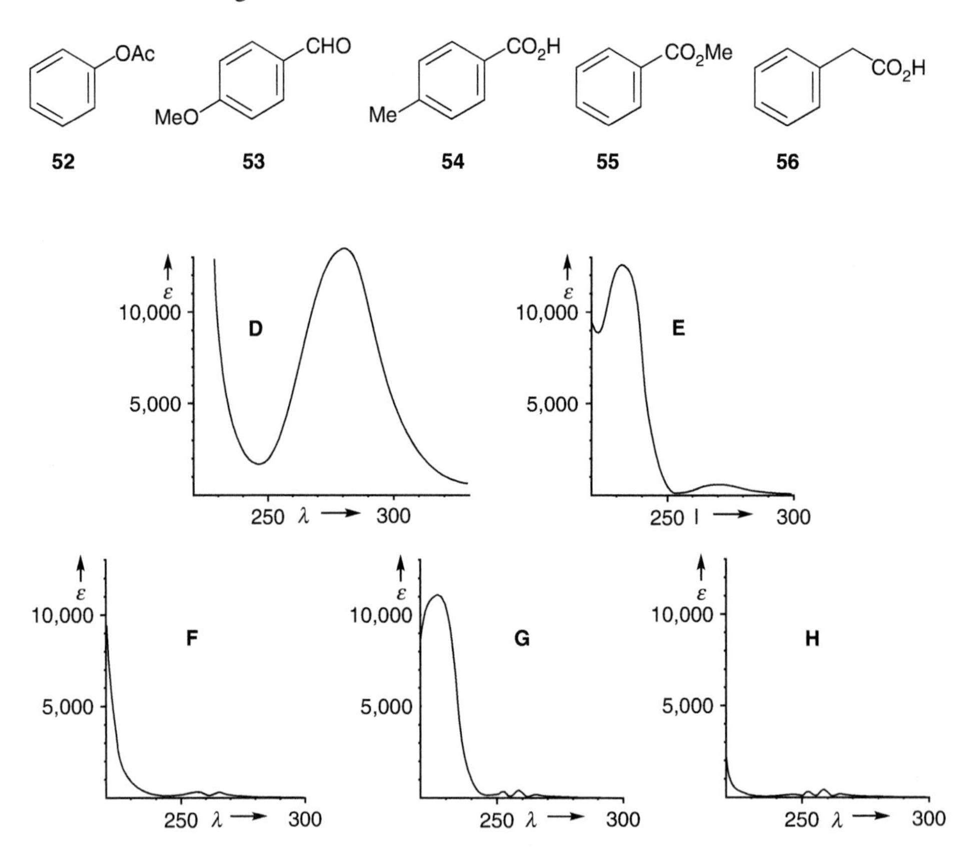

4. The cyclopropyl ketone **57** on treatment with base gave an isomer to which the stereostructure **58** was first assigned. Later, Ranganathan and Shechter realised that this could not possibly be true, because, unlike the starting material, the isomer was coloured bright yellow. Suggest a better structure for the product, which must have more conjugation in order to be yellow.

57 58

1. KOH, MeOH
2. HCl, cold

5. The intensely black solution of heptafulvalene **59** in dichloromethane was rendered colourless by treatment with one equivalent of dichlorocarbene. On adding water, the aqueous layer became deep blue, with λ_{max} at 600 nm. Suggest structures for the intermediate **I** and the product **J**, compatible with these dramatic changes in electronic absorption.

59 :CHCl I $C_{15}H_{13}Cl$ H_2O J $C_{15}H_{13}^+$

References

1. Kamlet MJ (ed) (1960–1996) Organic electronic spectral data. Wiley, New York
2. Scott AI (1964) Interpretation of the ultraviolet spectra of natural products. Pergamon Press, Oxford
3. Nayler P, Whiting MC (1955). J Chem Soc 1955:3037–3047
4. Sörensen JS, Bruun T, Holme D, Sörensen NA (1954). Acta Chem Scand 8:26–33
5. Bohlmann F, Mannhardt H-J, Viehe H-G (1955). Chem Ber 88:361–370
6. Jaffe HH, Orchin M (1962) Theory and applications of ultraviolet spectroscopy. Wiley, New York

Infrared and Raman Spectra

3

3.1 Introduction

The infrared spectra of organic compounds are associated with transitions between vibrational energy levels. Molecular vibrations may be detected and measured either in an infrared spectrum or indirectly in a Raman spectrum. The most useful vibrations, from the point of view of the organic chemist, occur in the narrower range of 3.5–16 μm (1 μm = 10^{-6} m). The position of an absorption band in the spectrum may be expressed in microns (μm), but standard practice uses a frequency scale in the form of wavenumbers, which are the reciprocals of the wavelength, cm^{-1}. The useful range of the infrared for an organic chemist is between 4000 cm^{-1} at the high-frequency end and 600 cm^{-1} at the low-frequency end.

Many functional groups have vibration frequencies, characteristic of that functional group, within well-defined regions of this range; these are summarised in Charts 3.1, 3.2, 3.3, and 3.4 at the end of this chapter, with more detail in the tables of data that follow. Because many functional groups can be identified by their characteristic vibration frequencies, the infrared spectrum is the simplest, most rapid, and often most reliable means for identifying the functional groups.

Equations (3.1) and (3.2), which are derived from the model of a mass *m* vibrating at a frequency ν on the end of a spring, are useful in understanding the range of values of the vibrational frequencies of various kinds of bonds.

$$\nu = \sqrt{\frac{k}{m}} \tag{3.1}$$

$$\frac{1}{m} = \frac{1}{m_1} + \frac{1}{m_2} \tag{3.2}$$

I. Fleming, D. Williams, *Spectroscopic Methods in Organic Chemistry*,
https://doi.org/10.1007/978-3-030-18252-6_3

The value k is a measure of the strength of the spring and m, m_1 and m_2 the masses of the objects at each end. The value of m in Eq. (3.1) is determined by the relationship in Eq. (3.2). If one of the masses (say, m_1) is much larger than the other, $1/m_1$ is correspondingly smaller, and the relevant mass m for Eq. (3.1) is close to that of m_2—making it analogous to the case where one end of the 'spring' is fixed. This is the situation when one of the elements is hydrogen and the other element is any of the common elements C, N, O, P or S. Other things being equal: (1) C–H bonds will have higher stretching frequencies than C–C bonds, which in turn are likely to be higher than C–halogen bonds; (2) O–H bonds will have higher stretching frequencies than O–D bonds; and (3), since k increases with increasing bond order, the relative stretching frequencies of carbon-carbon bonds lie in the order: C≡C > C=C > C–C.

3.2 Preparation of Samples and Examination in an Infrared Spectrometer

Older spectrometers used a source of infrared light which had been split into two beams of equal intensity. Only one of these was passed through the sample, and the difference in intensities of the two beams was then plotted as a function of wavenumber. Using this technology, a scan typically took about 10 min as each wavelength was examined. Most spectrometers in use today use a Fourier transform method, and the spectra are called Fourier transform infrared (FT-IR) spectra. These are taken on a spectrometer with the features simplified in Fig. 3.1.

A source of infrared light emitting radiation throughout the whole frequency range of the instrument, typically 4600–400 cm^{-1} is again divided into two beams of equal intensity. One beam, reflected from the fixed mirror, is passed through the sample typically placed onto a diamond on the outside of the instrument with a reflecting surface above it. The other beam is reflected from a moving mirror causing the beam to traverses a range of path lengths before passing through the sample. Recombination of the two beams produces an interference pattern. The sum of the interference patterns in the time domain is known as an interferogram, which looks nothing like a spectrum, but which contains information about all the frequencies absorbed by the sample. Fourier transformation of the interferogram, using a computer built into the instrument, converts it into a plot of absorption against wavenumber in the frequency domain. There is more about Fourier transformation in Sect. 3.11 on Raman spectra and in Sect. 4.2 on NMR spectra.

There are several advantages to FT-IR over the older method, and few disadvantages. Because it is not necessary to scan each wavenumber successively, the whole spectrum is measured in a few seconds, making it easier to use to extract the spectrum of compounds produced only for a short period in the outflow of a chromatograph. Because it is not dependent upon a slit and a prism or grating, the resolution is higher. FT-IR is better for examining small samples, because several scans can be added together.

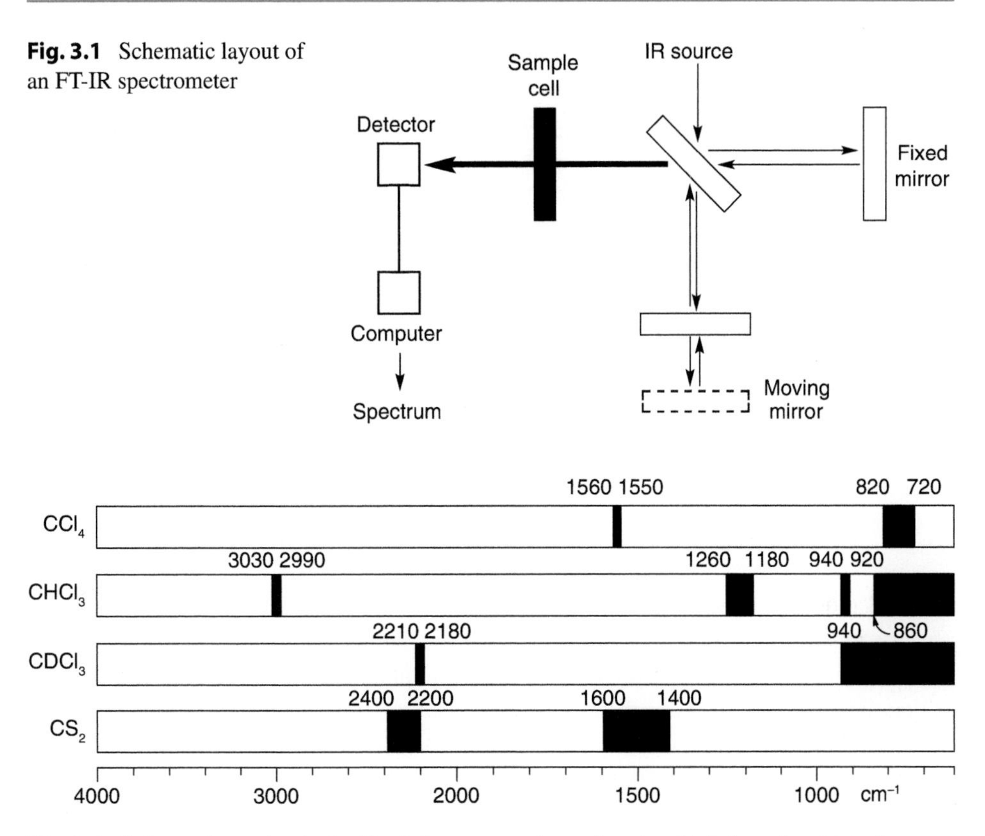

Fig. 3.1 Schematic layout of an FT-IR spectrometer

Fig. 3.2 Solvent absorption in the IR; dark areas are the regions in which the solvent interferes

Finally, the digital form in which the data are handled in the computer allows for adjustment and refinement by subtracting the background absorption of the medium in which the spectrum was taken, by subtracting the unavoidable signal from the carbon dioxide in the air, or by subtracting the spectrum of a known impurity from that of a mixture to reveal the spectrum of the pure component. However, the way in which infrared spectra are taken does not significantly affect their appearance. The older spectra and FT-IR spectra look very similar, and older spectra in the literature are still valuable for comparison.

A drop of a liquid (the sample or a solution) is simply placed on the diamond on the surface of the spectrometer. Common solvents are carbon tetrachloride, chloroform, deuterochloroform and carbon disulfide, which between them have transparency over the whole of the range (Fig. 3.2).

Solids, 1 mg is plenty, may be deposited like the liquids described above, by evaporation from solution dropped onto the diamond, or directly as a powder, with a sacrifice, usually small, from scattering off the solid surface. Older spectrometers used sodium chloride plates, between which the sample, liquid or solid, was squeezed, and through which the beam was passed. A hydrocarbon mulling agent (Nujol, Kaydol) was used to reduce light scattering from solids. Alternatively, the solid was ground with potassium bromide and the mixture pressed into a transparent disc. You will come across spectra taken in all these ways in the older literature.

Because of intermolecular interactions, band positions in solid state spectra are often different from those of the corresponding solution spectra. This is particularly true of those functional groups which take part in hydrogen bonding. The number of resolved lines is usually greater in solid state spectra, so that comparison of the spectra of, for example, synthetic and natural samples in order to determine identity is best done in the solid state. This is only true, of course, when the same crystalline modification is in use; racemic, synthetic material, for example, should be compared with enantiomerically pure, natural material in solution.

3.3 Selection Rules

Infrared radiation is only absorbed by the sample when the oscillating dipole moment (from a molecular vibration) interacts with the oscillating electric vector of the infrared beam. A simple rule for deciding if this interaction (and hence absorption of light) occurs is that the dipole moment at one extreme of a vibration must be different from the dipole moment at the other extreme of the vibration. In the Raman effect a corresponding interaction occurs only when the polarisability of the molecule changes in the course of the vibration. The different selection rules lead to the different applications of infrared and Raman spectra.

The most important consequence of these selection rules is that in a molecule with a centre of symmetry those vibrations symmetrical about the centre of symmetry are inactive in the infrared but active in the Raman (see Sect. 3.11); those vibrations which are not centrosymmetric are usually active in the infrared but inactive in the Raman. This is doubly useful, for it means that the two types of spectra are complementary. Furthermore, the more easily obtained, the infrared, is the more useful, because most functional groups are not centrosymmetric. The symmetry properties of a molecule in a solid can be different from those of an isolated molecule. This can lead to the appearance of infrared absorption bands in a solid state spectrum which would be forbidden in solution.

3.4 The Infrared Spectrum

A complex molecule has many vibrational modes which involve the whole molecule. To a good approximation, however, many of these molecular vibrations are largely associated with the vibrations of individual bonds and are called localised vibrations. These localised vibrations are useful for the identification of functional groups, especially the stretching vibrations of O–H and N–H single bonds and all kinds of triple and double bonds, almost all of which occur with frequencies greater than 1500 cm^{-1}. The stretching vibrations of other single bonds, most bending vibrations and the soggier vibrations of the molecule as a whole give rise to a series of absorption bands at lower energy, below 1500 cm^{-1}, the positions of which are characteristic of that molecule. The net result is a region above 1500 cm^{-1} showing absorption bands assignable to a number of functional groups, and a region containing many bands, characteristic of the compound in question and no other compound, below 1500 cm^{-1}. For obvious reasons, this is called the fingerprint region.

Figure 3.3 shows a representative infrared spectrum, that of cortisone acetate **1**. It shows the strong absorption from the stretching vibrations above 1500 cm^{-1} demonstrating the presence of each of the functional groups: the O–H group, three different C=O groups and the weaker absorption of the C=C double bond, as well as displaying a characteristic fingerprint below 1500 cm^{-1}. By convention absorbance is plotted downwards, opposite to the convention for ultraviolet spectra, but the maxima are still called peaks or bands. Rotational fine structure is smoothed out, and the intensity is frequently not recorded. When intensity is recorded, it is usually expressed subjectively as strong (s), medium (m) or weak (w). To obtain a high-quality spectrum, the quantity of substance is adjusted so that the strongest peaks absorb something close to 90% of the light. The scale on the abscissa is linear in frequency, but most instruments change the scale, either at 2200 cm^{-1} or at 2000 cm^{-1} to double the scale at the low-frequency end. The ordinate is linear in percent transmittance, with 100% at the top and 0% at the bottom.

The regions in which the different functional groups absorb are summarised below Fig. 3.3. The stretching vibrations of single bonds to hydrogen give rise to the absorption at the high-frequency end of the spectrum as a result of the low mass of the hydrogen atom, making it easy to detect the presence of O–H and N–H groups. Since most organic compounds have C–H bonds, their absorption close to 3000 cm^{-1} is rarely useful, although C–H bonds attached to double and triple bonds can be usefully identified. Thereafter, the order of stretching frequencies follows the order: triple bonds at higher frequency than double bonds and double bonds higher than single bonds—on the whole the greater the strength of the bond between two similar atoms the higher the frequency of the vibration. Bending vibrations are of lower frequency and usually appear in the fingerprint region below 1500 cm^{-1}, but one exception is the N–H bending vibration, which can appear in the 1600–1500 cm^{-1} region.

Although many absorption bands are effectively associated with the vibrations of individual bonds, many vibrations are coupled vibrations of two or more components of the whole molecule. Whether localised or not, stretching vibrations are given the symbol ν, and the various bending vibrations are given the symbol δ. Coupled vibrations may be subdivided into asymmetric and symmetric stretching, and the various bending modes into scissoring, rocking, wagging and twisting, as defined for a methylene group in Fig. 3.4. A coupled asymmetric and symmetric stretching pair is found with many other groups, like

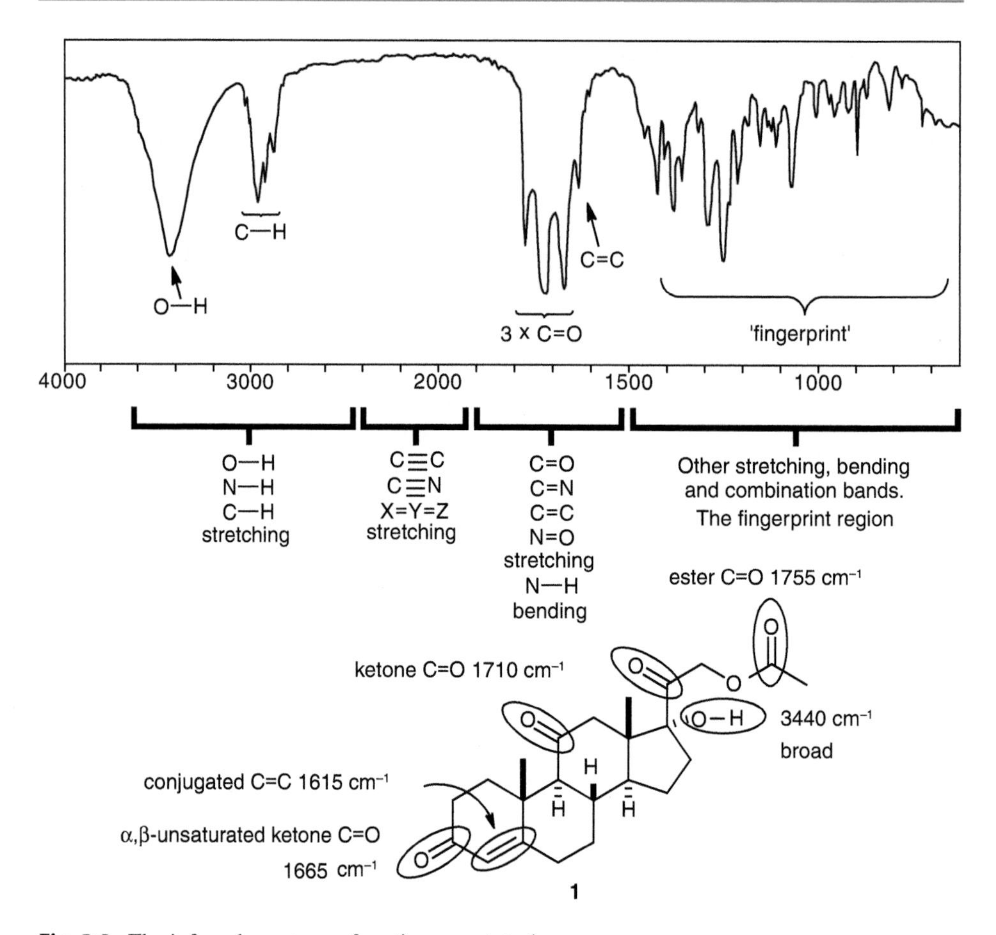

Fig. 3.3 The infrared spectrum of cortisone acetate **1**

primary amines, carboxylic anhydrides, carboxylate ions and nitro groups, each of which has two equal bonds close together.

3.5 The Use of the Tables of Characteristic Group Frequencies

Reference charts and tables of data are collected together at the end of this chapter for ready reference. Each of the three frequency ranges above 1500 cm^{-1} shown in Fig. 3.3 is expanded to give more detail in Charts 3.1, 3.2, 3.3, and 3.4 in Sec. 3.15. These charts summarise the narrower ranges within which each of the functional groups absorbs. The absorption bands which are found in the fingerprint region and which are assignable to functional groups are occasionally useful, either because they are sometimes strong bands in otherwise featureless regions or because their absence may rule out incorrect structures, but such identifications should be regarded as helpful rather than as definitive, since there

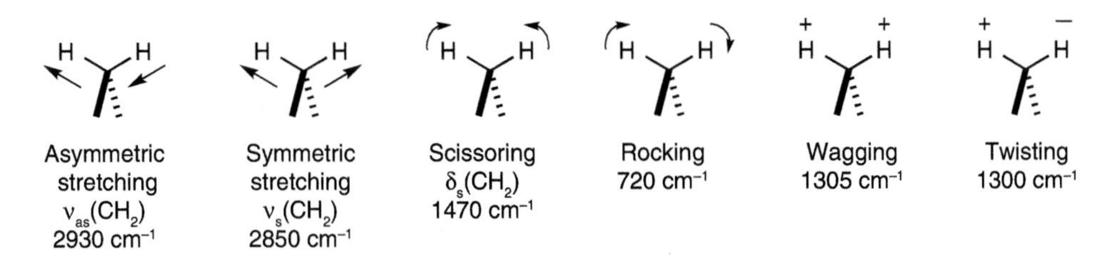

Fig. 3.4 Coupled vibrations of a methylene group

are usually many bands in this area. Tables of detailed information can be found in Sec. 3.16 at the end of this chapter, arranged by functional groups roughly in descending order of their stretching frequencies.

One could deal with the spectrum of an unknown as follows. Examine each of the three main regions of the spectrum above the fingerprint region, as identified on Fig. 3.3; at this stage certain combinations of structures can be ruled out—the absence of O–H or C=O, for example—and some tentative conclusions reached. Where there is still ambiguity—which kind of carbonyl group, for example—the tables corresponding to those groups that might be present should be consulted. It is important to be sure that the bands under consideration are of the appropriate intensity for the structure suspected. A *weak* signal in the carbonyl region, for example, is not good evidence that a carbonyl group is present, since carbonyl stretching is always strong—it is more likely to be an overtone or to have been produced by an impurity.

The text following this section amplifies some of the detail for each of the main functional groups, and shows the appearance, sometimes characteristic, of several of the bands. Cross-reference to the tables at the end is inevitable and will need to be frequent.

3.6 Stretching Frequencies of Single Bonds to Hydrogen

C–H Bonds C–H bonds do not take part in hydrogen bonding and so their position is little affected by the state of measurement or their chemical environment. C–C vibrations, which absorb in the fingerprint region, are generally weak and not practically useful. Since many organic molecules possess saturated C–H bonds, their absorption bands, stretching in the 3000–2800 cm^{-1} region and bending in the fingerprint region, are of little diagnostic value, but a few special structural features in saturated C–H groupings do give rise to characteristic absorption bands. Thus, methyl and methylene groups usually show two sharp bands from the symmetric and asymmetric stretching (Fig. 3.4), which can sometimes be picked out, but the general appearance of the accumulation of all the saturated C–H stretching vibrations often leads to broader and not fully resolved bands like those illustrated in many of the spectra above and below. The absence of saturated C–H absorption

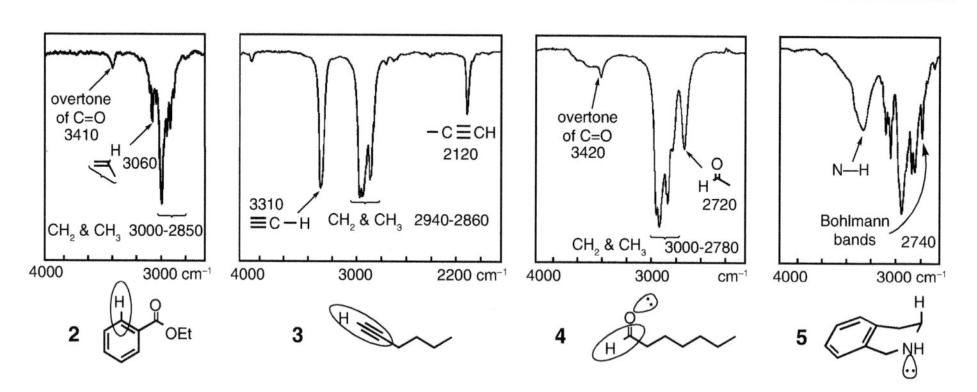

Fig. 3.5 Some characteristic C–H absorptions in the infrared

in a spectrum is, of course, diagnostic evidence for the absence of such a part structure, as in the spectrum of benzonitrile **14** (Fig. 3.8). Unsaturated and aromatic C–H stretching frequencies can be distinguished from the saturated C–H absorption, since they occur a little above 3000 cm^{-1} and are relatively weak, as in the spectrum of ethyl benzoate **2** (Fig. 3.5) and benzonitrile **14** (Fig. 3.8). Terminal acetylenes give rise to a characteristic strong, sharp line close to 3300 cm^{-1} from ≡C–H stretching, as in the spectrum of hexyne **3** (Fig. 3.5). A C–H bond antiperiplanar to a lone pair on oxygen or nitrogen is weakened, and the stretching frequency is lowered. Thus, the C–H bond of aldehydes gives rise to a relatively sharp band close to 2760 cm^{-1}, as seen in the spectrum of heptanal **4** (Fig. 3.5). Similarly, ethers and amines show low-frequency bands in the region 2850–2700 cm^{-1} when the lone pair is antiperiplanar to a C–H bond, as it is in six-membered rings like that of tetrahydroisoquinoline **5**, where the peak or peaks in this range are known as Bohlmann bands.

O–H Bonds The value of the O–H stretching frequency has been used for many years as a test for, and measure of, the strength of hydrogen bonds. The stronger the hydrogen bond the longer the O–H bond, the lower the vibration frequency, and the broader and the more intense the absorption band. O–H bonds not involved in hydrogen bonding have a sharp band in the 3650–3590 cm^{-1} range, typically observed for samples in the vapour phase, in very dilute solution, or when such factors as steric hindrance limit hydrogen bonding. Frequently solution phase spectra show two bands, one sharp at high frequency for the OH groups not involved in H-bonding and another broader and at somewhat lower frequency for those that are involved in H-bonding, as seen in the somewhat hindered alcohol **6** (Fig. 3.6). Pure liquids, solids, and many solutions, on the other hand, show only the broad strong band in the 3600–3200 cm^{-1} range, because of exchange and because of the different degrees of hydrogen bonding present within the sample, as seen in the spectrum of the primary alcohol **7** (Fig. 3.6) and oleol **26** (Fig. 3.18).

Weak intramolecular hydrogen bonds, like those in 1,2-diols for example, show a sharp band in the range 3570–3450 cm^{-1}, the precise position being a measure of the

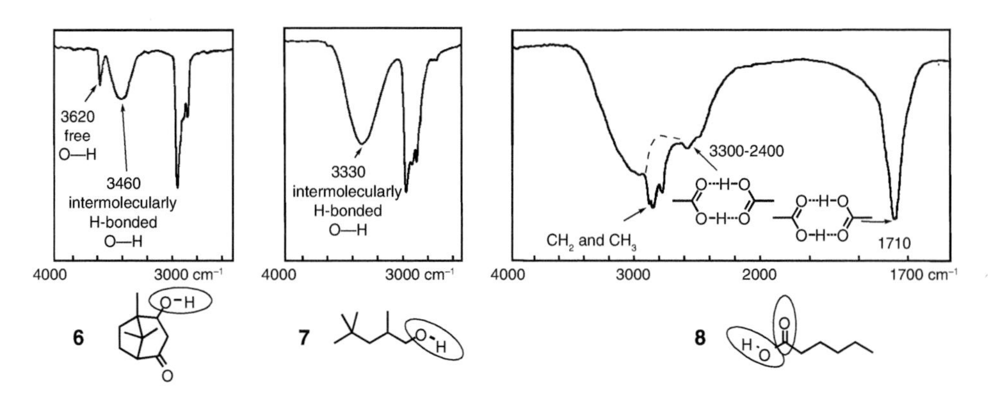

Fig. 3.6 Some characteristic O–H absorptions in the infrared

strength of the hydrogen bond. Strong intramolecular hydrogen bonds usually give rise to broad and strong absorption in the 3200–2500 cm^{-1} range. When a carbonyl group is the hydrogen bond acceptor, its characteristic stretching frequency is lowered, as seen for example in the dimeric association of most carboxylic acids. A broad absorption in the 3200–2500 cm^{-1} range, usually seen under and surrounding the C–H absorption, accompanied by carbonyl absorption in the 1710–1650 cm^{-1} region, is highly characteristic of carboxylic acids, like hexanoic acid **8** (Fig. 3.6), or of their vinylogues, β-dicarbonyl compounds in the enol form. Distinctions can be made among the various hydrogen-bonding possibilities by testing the effect of dilution: intramolecular hydrogen bonds are unaffected, while intermolecular hydrogen bonds are broken, leading to an increase in—or the appearance of—free O–H absorption. Spectra taken of samples in the solid state almost always show only the broad strong band in the range 3400–3200 cm^{-1}. Replacing an H with a D atom leads to absorption at a frequency 0.73 times lower. Note that the weak bands in the O–H region of the spectra of the ester **2** and the aldehyde **4** are clearly not O–H stretching, because they are too weak. S–H bond stretching is significantly lower in frequency, typically appearing as a weak and slightly broadened band near 2600 cm^{-1}.

N–H Bonds The stretching frequencies of the N–H bonds of amines are typically in the range 3500–3300 cm^{-1}. They are less intense than those of O–H bonds but can easily be confused with hydrogen-bonded O–H stretching. Because an N–H has a weaker tendency to form a hydrogen bond, its absorption is often sharper, and can be very sharp, as in the N–H band shown by the indole N–H from tryptophan **9** (Fig. 3.7).

Even in very dilute solutions N–H bonds never give rise to absorption as high in frequency as the free O–H near 3600 cm^{-1}. Primary amines and amides like the amide **10** (Fig. 3.7) give rise to two bands, typically one at 3350 and the other at 3250 cm^{-1}, because there are two stretching vibrations, one symmetrical and the other unsymmetrical, similar to those of a methylene group (Fig. 3.4). Secondary amines like morpholine **11** (and tetrahydroisoquinoline **5**) only have the one band. Secondary amides in an s-*trans* configu-

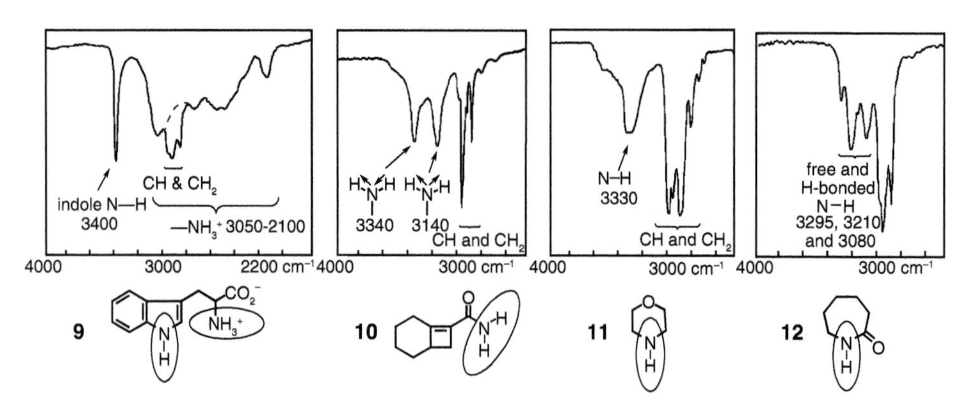

Fig. 3.7 Some characteristic N-H absorptions in the infrared

ration have only one band, but lactams, in which the amide group is in an s-*cis* configuration, often show several bands from various hydrogen-bonded associations, especially in the solid state, as in the spectrum of caprolactam **12** (Fig. 3.7). Amine salts and the zwitterions of amino acids give rise to several N–H stretching bands on the low-frequency side of any C–H absorption, and sometimes reaching as low as 2000 cm^{-1}, as in the spectrum of tryptophan **9** (Fig. 3.7).

3.7 Stretching Frequencies of Triple and Cumulated Double Bonds

Terminal acetylenes absorb in the narrow range 2140–2100 cm^{-1}, as in the spectrum of hexyne **3** in Fig. 3.5. Internal acetylenes absorb in the range 2300–2150 cm^{-1}, with conjugated triple bonds and enynes at the lower end of the range. Remembering Eqs. (3.1) and (3.2), we can understand how an internal acetylene will behave as though both carbons have somewhat larger masses than will the carbon of an acetylenic C–H group. The consequence is that internal acetylenes absorb at higher frequency, but the absorptions are often weak, because of the small change in dipole moment on stretching. *A symmetrical acetylene shows no triple-bond stretch at all.* Conjugation with carbonyl groups usually has little effect on the position of C≡C absorption, but substantially increases the intensity. When more than one acetylenic linkage is present, and sometimes when there is only one, there can be more absorption bands in this region than there are triple bonds to account for them.

Nitriles absorb in the range 2260–2200 cm^{-1}, but are often weak, as in the spectrum of the nitrile **13**, for example, which was originally assigned the wrong structure because the weak nitrile absorption had been overlooked. Cyanohydrins are notorious in this respect, sometimes showing no nitrile stretching band at all. As usual, conjugation lowers the frequency and increases the intensity, as in the spectrum of benzonitrile **14**. Isonitriles, nitrile oxides, diazonium salts and thiocyanates such as **15** absorb strongly in the region 2300–2100 cm^{-1} (Fig. 3.8).

The stretching vibrations of the two double bonds in cumulated double-bonded systems X=Y=Z, such as those found in carbon dioxide, isocyanates, isothiocyanates, diazoalkanes, ketenes and allenes, are coupled, with an unsymmetrical and a symmetrical pair. The former gives rise to a strong band in the range 2350–1930 cm^{-1} and the latter to a band in the fingerprint region, except for symmetrical systems, for which it is IR-forbidden. Carbon dioxide has a sharp band at 2349 cm^{-1}.

Allenes such as dimethylallene **17** give a sharp absorption in the characteristically narrow range 1960–1930 cm^{-1}. The other cumulated double-bond systems come in between, with isocyanates and isothiocyanates, such as cyclohexylisothiocyanate **16**, giving rise to a broad band in contrast to the narrow band seen for thiocyanates **15** and allenes **17** (Fig. 3.9).

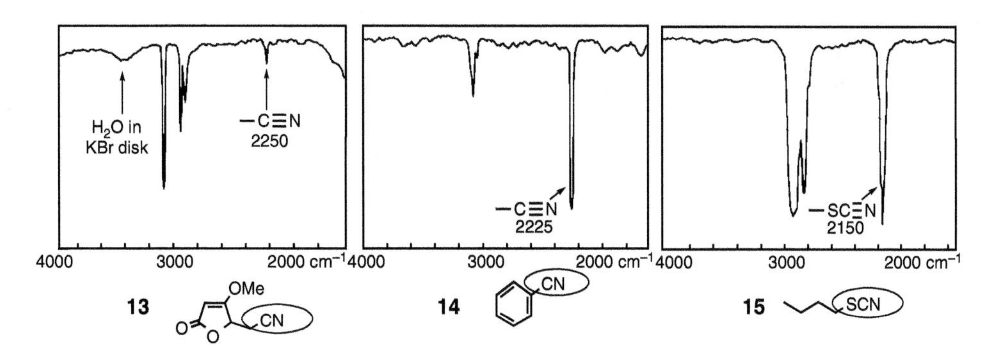

Fig. 3.8 Unusually weak saturated and strong conjugated nitrile absorption

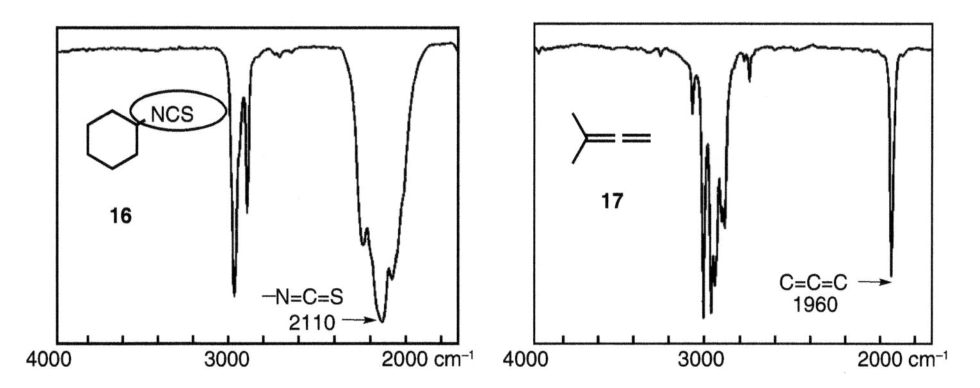

Fig. 3.9 Characteristic absorptions of an isothiocyanate and an allene

3.8 Stretching Frequencies in the Double-Bond Region

C=O Double Bonds Identifying which of the several kinds of carbonyl group is present in a molecule is one of the most important uses of infrared spectroscopy. Carbonyl bands are always strong, with carboxylic acids generally stronger than esters, and esters stronger than ketones or aldehydes. Amide absorption is usually similar in intensity to that of ketones but is subject to greater variations.

The precise position of carbonyl absorption is governed by the electronic structure and the extent to which the carbonyl group is involved in hydrogen bonding. The general trends of structural variation on the position of C=O stretching frequencies are summarised in Fig. 3.10. It is helpful to use the stretching vibration of a saturated ketone at 1710 cm^{-1} as a reference point, and to compare the consequence of all the variations in structure with that number.

1. The more electronegative the group X in the system R–C(=O)–X, the higher is the frequency, except that this trend from the inductive effect of X is offset by the effect in the π system of any lone pairs on X. Thus, the inductive effect raises the frequency of the absorption. The C=O stretching frequency falls in the order: acid fluorides, chlorides, bromides, esters and amides, as the electronegativity of the attached atom falls. In the opposite direction, an electropositive substituent like a silyl group lowers the frequency. However, ketones have their C=O stretching frequencies between those of comparable esters and amides, and so the electronegativity of X is not the only factor.
2. The overlap of the lone pair of electrons on X with the C=O bond, illustrated by the curly arrows in the amide in Fig. 3.10, reduces the double-bond character of the C=O bond, while increasing the double-bond character of the C–N bond. In molecular orbital terms, the total π bonding is increased by the overlap, lowering the π energy, but the extra bonding is shared over three atoms, instead of being isolated on the two atoms of the C=O group. This overlap will have most effect when the lone pair is relatively high in energy. The less electronegative X is, the higher is the energy of its lone-pair orbitals, and the more effective the overlap in reducing the π-bonding in the C=O bond. The net result is to lower the frequency of the C=O stretching vibration in amides below that of ketones, even though N is more electronegative than C. On the other hand, the higher electronegativity of O, coupled with the less-effective π-overlap of its lone pairs, keeps the stretching

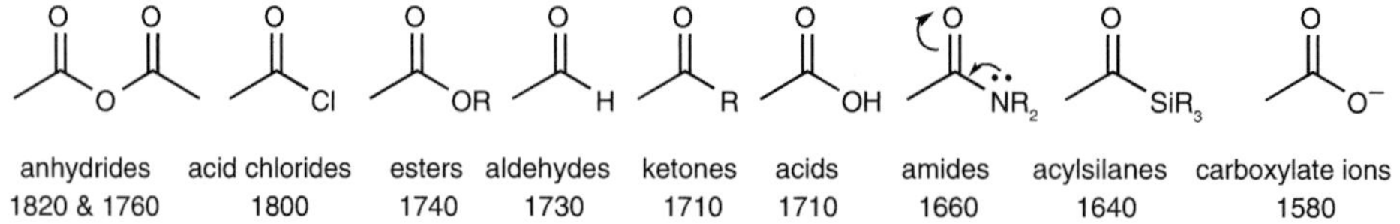

anhydrides	acid chlorides	esters	aldehydes	ketones	acids	amides	acylsilanes	carboxylate ions
O, O	Cl	OR	H	R	OH	NR_2	SiR_3	O^-
1820 & 1760	1800	1740	1730	1710	1710	1660	1640	1580

Fig. 3.10 Representative stretching frequencies (in cm^{-1}) of C=O groups

frequency of esters above that of ketones. Similarly, the hyperconjugative overlap of the neighbouring C–H (or C–C) bonds in a ketone reduces the C=O double-bond character, and lowers its stretching frequency relative to that of an aldehyde. On the other hand, if the oxygen lone pair in an ester overlaps with another double bond, it is less effective in lowering the frequency of the carbonyl vibration, with the result that vinyl and phenyl esters absorb at higher frequency than alkyl esters (see vinyl acetate **21** in Fig. 3.13), and carboxylic anhydrides even more so. Anhydrides also show two bands rather than one, because the two carbonyl groups have an unsymmetrical and a symmetrical vibration, with the former pushed a little above that of a comparable acid chloride and the latter rather more below it.

3. Hydrogen bonding to a carbonyl group causes a shift to lower frequency of 40–60 cm^{-1}. Acids, amides with an N–H, enolised β-dicarbonyl systems, and *o*-hydroxy- and *o*-aminophenyl carbonyl compounds show this effect, as illustrated by the carboxylic acid's coming in the list in Fig. 3.10 between the ester and the amide, instead of being the same as the ester. All carbonyl compounds tend to give slightly lower values for the carbonyl stretching frequency in the solid state compared with the value for dilute solutions.
4. Ring strain in cyclic compounds causes a relatively large shift to higher frequency (Fig. 3.11). This phenomenon provides a remarkably reliable test of ring size, distinguishing clearly between four-, five- and larger-membered ring ketones, lactones and lactams. Six-ring and larger ketones, lactones and lactams, show the normal frequency found for the open-chain compounds.
5. α,β-Unsaturation causes a lowering of frequency of 15–40 cm^{-1} because overlap of the π molecular orbitals over four atoms, while lowering the energy overall, reduces the double-bond character of the C=O bond itself. The unsaturated ketone group in cortisone acetate **1** can be seen as having the lowest frequency of the three carbonyl peaks in Fig. 3.3, with the saturated ketone just above it and the ester just above that. The same effect can be seen in the spectrum of ethyl crotonate **20** in Fig. 3.13. The effect of α,β-unsaturation is noticeably less in amides, where little shift is observed, and that sometimes even to higher frequency.

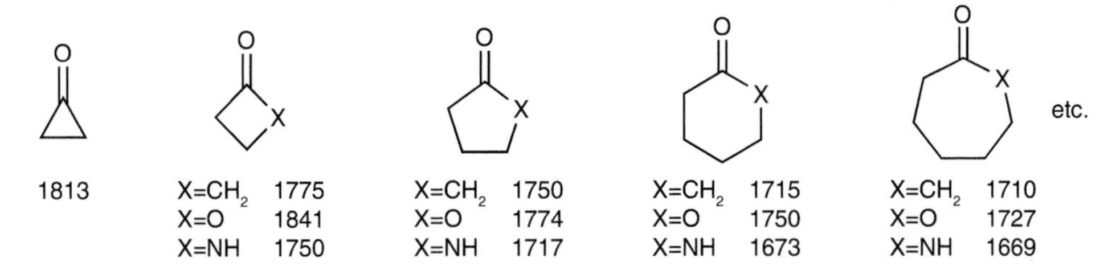

Fig. 3.11 The effect of ring size on C=O stretching frequencies (in cm^{-1})

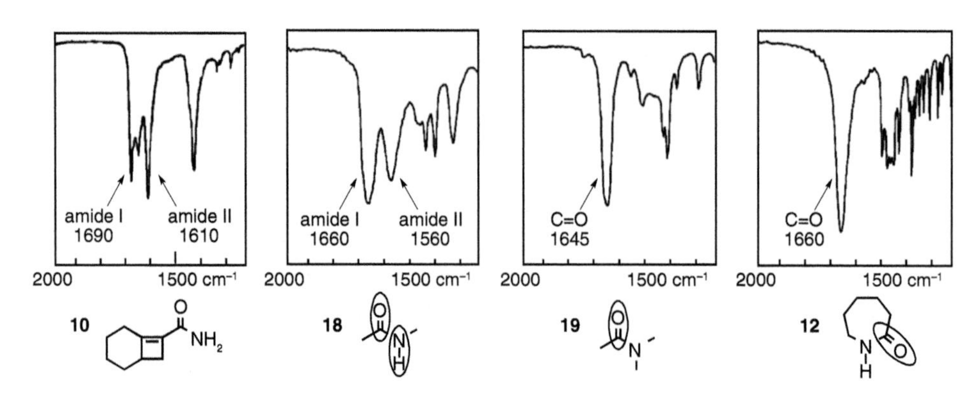

Fig. 3.12 Amide C=O stretching vibrations

6. Where more than one of the structural influences on a particular carbonyl group is operating, the net effect is usually close to additive, with α,β-unsaturation and ring strain having opposite effects.
7. Amides are special in showing greater variation in the number and position of the bands in the carbonyl region. In particular, they show extra bands in addition to the localised stretching of the C=O bond. Thus, primary and secondary amides, which also have an N–H bond, show at least two bands. The one at higher frequency is called the amide I band, and the other, at lower frequency is called the amide II band. The spectra of the primary amide **10** and of *N*-methylacetamide **18** in Fig. 3.12 show the amide I and II bands, whereas the spectra of *N*,*N*-dimethylacetamide **19** and caprolactam **12** show a single peak. Both bands are affected by hydrogen bonding and are therefore significantly different when the spectra are taken of solutions or of the solid amide.

For the full range of all types of carbonyl absorption, see the tables at the end of this chapter.

C=N Double Bonds Imines, oximes and other C=N double bonds absorb in the range 1690–1630 cm^{-1}. They are weaker than carbonyl absorption, and are difficult to identify because they absorb in the same region as C=C double bonds.

C=C Double Bond Unconjugated alkenes absorb in the range 1680–1620 cm^{-1}. The more substituted C=C double bonds absorb at the high-frequency end of the range, the less substituted at the low-frequency end. The intensity is generally less than that for the C=O stretching of carbonyl groups. The absorption may be very weak when the double bond is more or less symmetrically substituted, and absent when the double bond is symmetrically substituted. Oleol **29** has a *cis* double bond near the middle of a long chain. Since it is locally almost symmetrically substituted, its stretching is only just detectable, too weakly to be diagnostic in the infrared spectrum (Fig. 3.19).

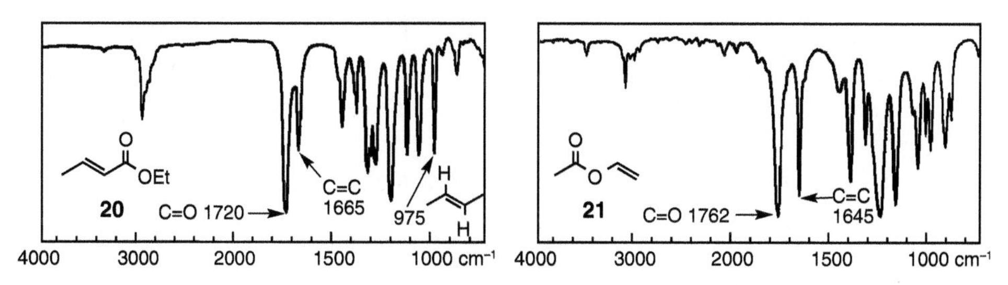

Fig. 3.13 Conjugated C=C and C=O stretching vibrations

Conjugation lowers the stretching frequency of C=C double bonds only a little, as seen in the spectrum of ethyl crotonate **20** in Fig. 3.13, where it gives rise to the moderately strong peak at 1665 cm^{-1}, which is nevertheless weaker than the carbonyl peak at 1720 cm^{-1}. Vinyl esters like vinyl acetate **21**, where the C=C double bond is conjugated to an oxygen lone pair, also give rise to a strong band at the low end of the C=C range. The C=C stretching vibration can also be seen as the small peak at 1615 cm^{-1} on the low-frequency side of the three carbonyl peaks in Fig. 3.3, and as the small peak at 1640 cm^{-1} between the amide I and amide II bands of the amide **10** in Fig. 3.12.

If the double bond is exocyclic to a ring the frequency rises as the ring size decreases. A double bond within a ring shows the opposite trend: the frequency falls as the ring size decreases.

N=O Double Bonds Nitro groups have asymmetrical and symmetrical N=O stretching vibrations, the former a strong band in the range 1570–1540 cm^{-1} and the latter a strong band in the range 1390–1340 cm^{-1} in the fingerprint region. They can be seen easily as the strongest peaks when they are conjugated with the benzene ring as they are in *o*-nitrobenzyl alcohol **22**, but they are weaker, although still evident, when they are not conjugated, as in ω-nitroacetophenone **23** in Fig. 3.14. The relative weakness of the bands when they are not

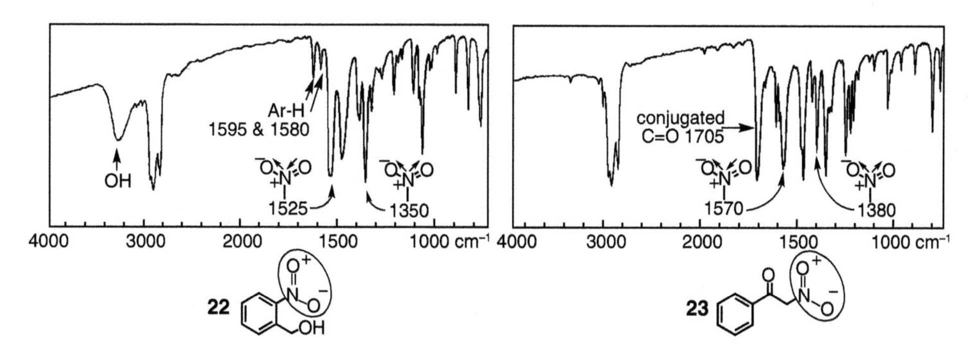

Fig. 3.14 Asymmetrical and symmetrical stretching absorption of nitro groups

conjugated makes it uncertain in **23** whether the symmetric stretching is the band at 1380 or the stronger band at 1330 cm^{-1}. If the assignment in Fig. 3.14 is correct, conjugation with the benzene ring has lowered the N=O stretching bands in the nitrobenzene **22** by 45 and 30 cm^{-1}, as well as increasing their intensity. Nitrates, nitramines, nitrites and nitroso compounds also absorb in the 1650–1400 cm^{-1} region.

3.9 Characteristic Vibrations of Aromatic Rings

There are several features in infrared spectra that combine to give evidence of the presence of an aromatic ring. The C–H stretch in aromatic rings can almost always be detected just above 3000 cm^{-1} as a weak peak or peaks, but this does not distinguish them from the corresponding alkene signals. More characteristic are the two or three bands in the 1600–1500 cm^{-1} region shown by most six-membered aromatic rings such as benzenes, polycyclic aromatic rings and pyridines. Typically, a benzene ring conjugated to a double bond has three bands like those marked in the spectrum of ethyl benzoate **2** in Fig. 3.15, usually close to 1600, 1580 and 1500 cm^{-1}. The two bands at higher frequency are usually the more intense when the benzene ring is further conjugated, but they are usually weak when the benzene ring is not conjugated, as seen in the infrared spectrum of the aromatic ring in tetrahydroisoquinoline **5** in Fig. 3.15. The relative intensities vary in other less predictable ways, and sometimes the two higher frequency bands appear as one.

There are bands in the fingerprint region in the 1225–950 cm^{-1} range which are of little use in providing structural information. The shape and number of the two to six overtone and combination bands which appear in the 2000–1600 cm^{-1} region is a function of the substitution pattern of the benzene ring, but the use of this region for this purpose has been overtaken by NMR spectroscopy, which does the job much better. A fourth group of strong bands below 900 cm^{-1} produced by the out-of-plane C–H bending vibrations is affected by the number of adjacent hydrogen atoms on the ring. It too can be used, but rarely is nowadays, to identify the substitution pattern.

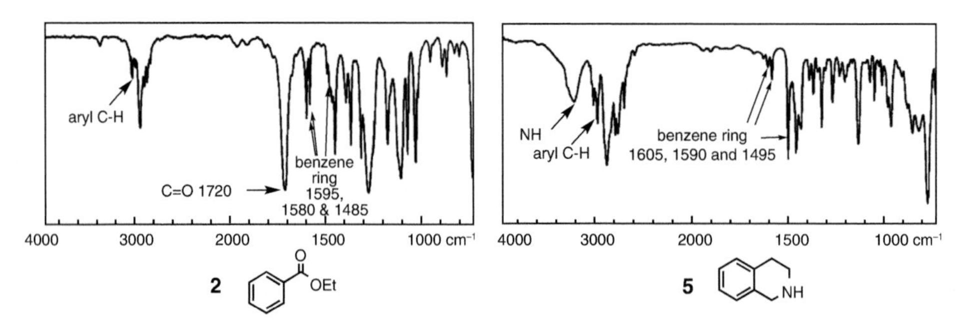

Fig. 3.15 Characteristic vibrations of aromatic rings

3.10 Groups Absorbing in the Fingerprint Region

Strong bands in the fingerprint region arise from the stretching vibrations of a few other doubly bonded functional groups like sulfonyl, thiocarbonyl and phosphoryl, but they are easily confused with the stretching vibrations of single bonds like C–O and C–halogen, which are always strong, and the bending vibrations of C–H bonds. They are rarely diagnostically useful, unless their absence is informative. The last few tables at the end of this chapter give an indication where some of these absorptions commonly occur, and include some of the less common and more specialised functional groups, like those of boron, silicon, phosphorus and sulfur.

A few of these bands are noteworthy. The methine C–H bending and methyl CH_3 and methylene CH_2 symmetrical bending vibrations give rise in many aliphatic compounds to two broad bands close to 1450 and 1380 cm^{-1}, which can clearly be seen in the spectrum of oleol **29** in Fig. 3.19. The out-of-plane vibration of *trans* –CH=CH– double bonds is one of the more usefully diagnostic bending vibrations. It occurs in a narrow range 970–960 cm^{-1}, or at slightly higher frequency if conjugated, and it is always strong. In contrast, the corresponding vibration of the *cis* isomer is of lower intensity and at lower frequency, typically in the range 730–675 cm^{-1}. The band at 975 cm^{-1} in the fingerprint of ethyl *trans*-crotonate **20** (Fig. 3.13) clearly shows that such a feature may be present; if there were no band there, it would be diagnostic of the *absence* of this feature, as in the spectrum of the *cis*-alkene oleol **26** in Fig. 3.18. The N–H bending vibration of primary and secondary amides can appear just above the fingerprint region, and may be responsible for what is called the amide-II band.

3.11 Raman Spectra

Raman spectra are generally taken on instruments using laser sources, and the quantity of material needed is of the order of a few mg. The sample may be a gas, liquid or solid, and may be in solution if the solvent does not interfere with the signals of interest. The sample is irradiated with a laser beam of monochromatic light, and the scattered light is examined using photoelectric detection. The sample does not even need to be taken out of its bottle or ampoule if the container is transparent to the wavelength used. Most of the scattered light has the frequency of the laser beam (the Rayleigh line) produced by scattering without any interaction with molecular vibrations. It also has frequencies lower and higher than that of the laser beam caused by scattering of the light coupled with vibrational excitation or decay, respectively. The lines of lower frequency (the Stokes lines) are the stronger, because organic samples are usually in low vibrational states, and are excited into higher vibrational states, with a loss of energy in the scattered beam. The difference in frequency between the Rayleigh line and the Stokes line is the frequency of the corresponding vibration, and the sum of these frequencies is the Raman spectrum.

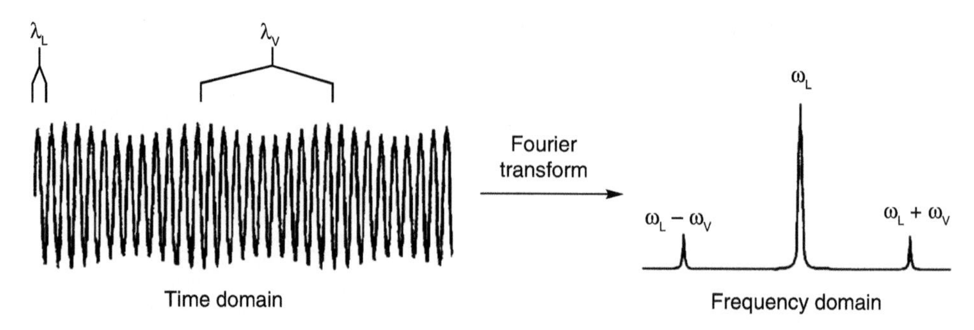

Fig. 3.16 Time and frequency domains for a scattered beam with two frequencies in it

The scattered light is a wave with all the frequencies in it, both that of the incident light and the many frequencies slightly different from it as a result of the subtraction and addition of the vibrational frequencies that have modulated it. The wave is described as being in the *time domain*, because it would be plotted with time as the abscissa and amplitude as the ordinate. Fourier transformation converts the data in the time domain into data in the *frequency domain*, plotted with frequency on the abscissa.

In the simplified case in Fig. 3.16 with only two frequencies, the two wavelengths λ_L and λ_V, having large and small amplitudes, respectively, would be plotted after Fourier transformation as the frequencies $\omega_L - \omega_V$ (the Stokes line), ω_L (the Rayleigh line) and $\omega_L + \omega_V$ (the anti-Stokes line).

Raman spectroscopy is used by organic chemists for the detection and analysis of a few structural features that are inactive in the infrared. Infrared vibrations interact with the incident light when the *dipole moment* changes from one extreme of the vibration to the other extreme. As a result infrared spectroscopy is still much more frequently used than Raman spectroscopy, because most functional groups are polar, and have a substantial change in dipole moment on stretching.

The scattered beam in Raman spectra is affected when the *polarisability* changes from one extreme of the vibration to the other extreme. Although both methods of detecting molecular vibrations have bands in common, there is a high level of complementarity between the two techniques. We can see this by comparing the infrared and Raman spectra of *p*-cresol and *trans*-3-hexenoic acid in Fig. 3.17. The Raman spectrum is the lower in each. It is plotted with absorption upwards as Raman spectra usually are.

p-Cresol **24** has a strong O–H stretching band in the infrared spectrum that is essentially absent in the Raman. On the other hand the Raman spectrum shows an identifiable and strong aromatic C–H stretch at 3065 cm^{-1}, which is either absent or lost under the OH signal in the infrared. The aromatic bands in cresol also show a difference between the infrared and the Raman spectra: the band at 1505 cm^{-1} is much stronger in the infrared and the band near 1600 cm^{-1} is strong in the Raman.

Similarly, the infrared spectrum of *trans*-3-hexenoic acid **25** shows the broad, strongly hydrogen-bonded OH stretching band between 3500 and 2300 cm^{-1}, characteristic of carboxylic acids, and a strong C=O stretching band at 1712 cm^{-1}. These are both polar

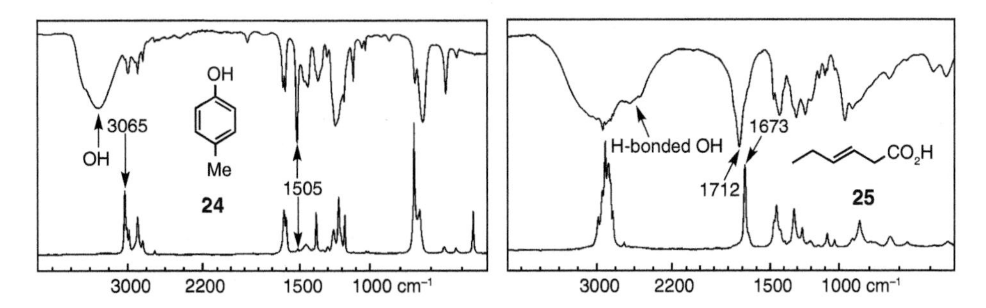

Fig. 3.17 Infrared (above) and Raman (below) spectra of *p*-cresol and *trans*-3-hexenoic acid

groups which regularly give strong infrared bands. In contrast they are both nearly absent in the Raman spectrum, which reveals instead the isolated C=C double bond stretching band at 1673 cm^{-1} that was, at best, a shoulder on the low-frequency side of the carbonyl absorption in the infrared. Carbonyl groups are not always as weak in Raman spectra as in this case. Finding the C=C or C≡C stretching band, from a symmetrical or nearly symmetrical double or triple bond, is the most useful kind of structural feature that Raman spectra reveal.

In a truly centrosymmetric compound like 3-hexyne **26** in Fig. 3.18, the stretching vibration has no effect on the dipole moment, and the infrared spectrum shows no absorption at all in the triple bond region 2300–2150 cm^{-1}. In contrast, the Raman spectrum has strong lines at 2275 and 2260 cm^{-1}, with more than one band because the stretching, although largely localised in the one band, is combined with other molecular vibrations. Equally, the symmetric stretching of the N=N bond in diethyl azodicarboxylate **27** in Fig. 3.18 is inactive in the infrared, but can be seen in the Raman spectrum as a sharp line at 1455 cm^{-1}. A peak is actually present at 1450 cm^{-1} in the infrared, close to the N=N frequency, but it cannot be N=N stretching. At the same time the carbonyl group at 1780 cm^{-1}, strong in the infrared, is much weaker in the Raman.

When the triple bond is technically unsymmetrical but the two substituents are alike, then the triple bond stretching band, although not forbidden, is weak. In Fig. 3.19 the

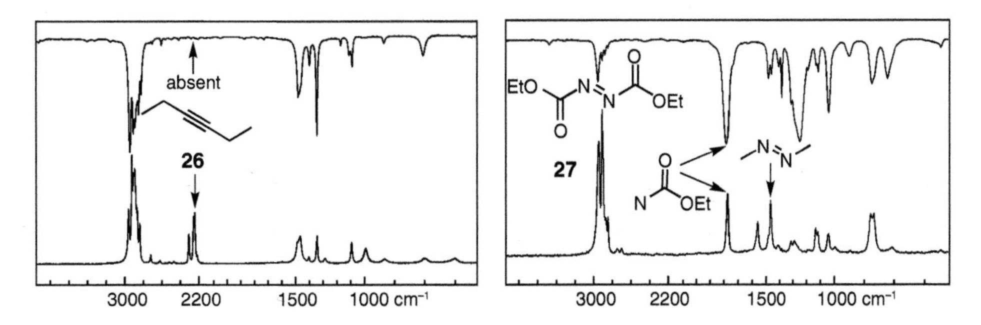

Fig. 3.18 Infrared and Raman spectra of centrosymmetric molecules: 3-hexyne and diethyl azodicarboxylate

infrared and Raman spectra of the nearly symmetrical acetylene 2-hexyne **28** shows infrared absorption at 2240 cm^{-1} at the triple bond frequency, but it is far too weak to identify for us the presence of a triple bond. It is clearly brought out in the Raman spectrum. Similarly, with the nearly symmetrical C=C double bond in oleol **29** in Fig. 3.19, the C=C stretching band at 1660 cm^{-1} is just discernible in the infrared spectrum, but in the Raman spectrum it is prominent.

The triple bond of a nitrile, even though it is polar, sometimes gives such weak absorption in the infrared that it can be overlooked, as in the cyanohydrin **30** of acetone. It is unmistakable in the Raman spectrum as the sharp peak at 2240 cm^{-1}. In both oleol in Fig. 3.19 and the cyanohydrin **30** in Fig. 3.20, the strong O–H peaks in the infrared are absent in the Raman spectra. Similarly, the strong N–H stretching in the infrared spectrum of tetrahydroisoquinoline **5** in Fig. 3.15, repeated in Fig. 3.20, is weak in the corresponding Raman spectrum, which also shows the usual differences in intensity of the three aromatic stretching bands in the 1500–1600 cm^{-1} region. A conspicuously sharp band near 1000 cm^{-1} in the Raman spectra of many aromatic compounds is a breathing vibration, in and out, that is always inactive in the infrared. In benzene, at 995 cm^{-1}, it is by far the strongest peak.

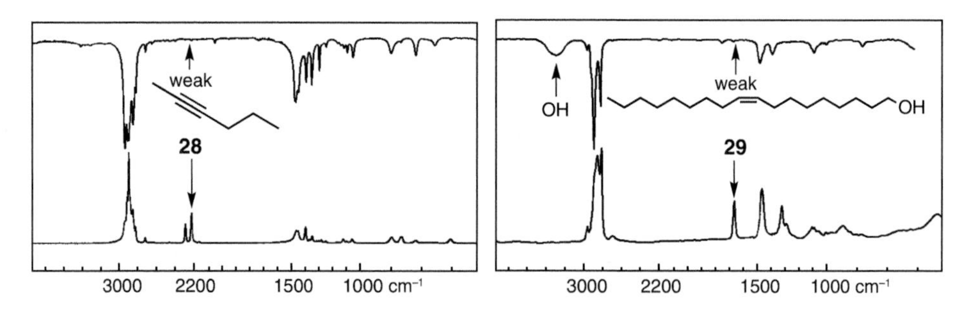

Fig. 3.19 Infrared and Raman spectra of 2-hexyne and oleol

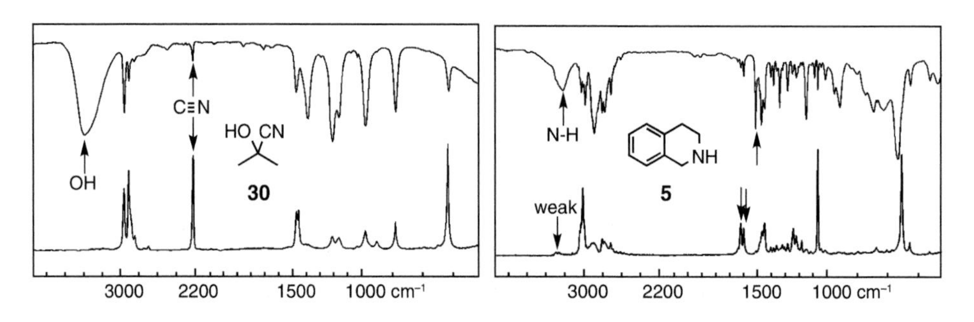

Fig. 3.20 Infrared and Raman spectra of acetone cyanohydrin and tetrahydroisoquinoline

3.12 Internet

The Internet is a continuously evolving system, with links and protocols changing frequently. The following information is inevitably incomplete and may no longer apply, but it gives you a guide to what you can expect. Some websites require particular operating systems and may only work with a limited range of browsers, some require payment, and some require you to register and to download programs before you can use them.

Infrared spectroscopy is better served on the Internet than ultraviolet spectroscopy.

The Chemical Database Service is available free to UK academic institutions, and it includes the SpecInfo system from Chemical Concepts, covering IR, NMR and mass spectra. To register go to: http://cds.dl.ac.uk.

Once you have an ID and Password, go to http://cds.dl.ac.uk/specsurf. To learn what is there and how to use the system, click on the online demo link or on the Illustrated Guide link. The service has 21,000 IR spectra. To look for the IR spectrum of a compound you are interested in, go to: http://cds.dl.ac.uk/specsurf.

Click on Start SpecSurf. When the window loads, pull down Edit-Structure to go to a new window in which you draw the structure. Pull down File-Transfer to drop the structure back into the earlier window, and pull down Search-Structure, which will create a list at the bottom of the window of any available spectra for the compound you have drawn. These may be NMR spectra, mass spectra or IR spectra, and unfortunately are not identified as such. The item at the top of the list, usually a ^{1}H-NMR spectrum, will be displayed in the lower spectrum window. Click on each member of the list until the spectrum that appears is an IR spectrum, or until one of them creates a sub-list on the right-hand side, with an itemised list of the kinds of spectra within it. Then click on the IR example, and it will appear in the spectrum window. Hover the cursor over a peak and its frequency in wavenumbers will be displayed in purple just below the spectrum.

The Sadtler database administered by Bio-Rad Laboratories has over 220,000 IR spectra of pure organic and commercial compounds. For information, go to: http://www.bio-rad.com/ and follow the leads to Sadtler, KnowItAll, and infrared.

Acros provide free access to IR spectra of compounds in their catalogue at: www.acros.be/ Enter the name or draw the structure of the compound you are interested in, and click on the IR tab.

A website listing databases for IR, NMR and MS is: http://www.lohninger.com/spectroscopy/dball.html.

The Japanese Spectral Database for Organic Compounds (SDBS) has free access to IR, Raman, ^{1}H- and ^{13}C-NMR and MS data at: http://www.aist.go.jp/RIODB/SDBS/cgi-bin/cre_index.cgi.

There is a database of >5000 compounds with gas phase infrared data on the NIST website belonging to the United States Secretary of Commerce: http://webbook.nist.gov/chemistry/name-ser.html Type in the name of the compound you want, check the box for IR spectrum, and click on Search, and if the infrared spectrum is available it will show it to you.

Sigma-Aldrich has a library of >60,000 FTIR spectra, access to which requires payment; for information go to: http://www.sigmaaldrich.com/Area_of_Interest/Equip____Supplies_Home/Spectral_Viewer/FT_IR_Library.html (that is 4 underline symbols between Equip and Supplies).

ACD (Advanced Chemistry Development) Spectroscopy sell proprietary software called ACD/SpecManager that handles all four spectroscopic methods, as well as other analytical tools: http://www.acdlabs.com/products/spec_lab/exp_spectra/.

It is able to process and store the output of the instruments that take spectra, and can be used to catalogue, share and present your own data. It also gives access to free databases, and to prediction and analysis tools—assigning peaks in the infrared to functional groups, for example:

Wiley-VCH keep an up-to-date website on spectroscopic books and links. The URL for infrared spectroscopy is: http://www.spectroscopynow.com/Spy/basehtml/SpyH/1,1181,3-0-0-0-0-home-0-0,00.html and they also provide a link to the SpecInfo databases: http://www3.interscience.wiley.com/cgi-in/mrwhome/109609148/HOME and to ChemGate, which has a collection of 700,000 IR, NMR and mass spectra: http://chemgate.emolecules.com.

There are expensive collections for industrial chemists through IR Industrial Organic Chemicals Vols. 1 and 2, BASF Software, January 2006.

3.13 Further Reading

DATA

- The Aldrich library of FT-IR spectra. Aldrich Chemical Company, Milwaukee
- Dolphin D, Wick A (1977) Tabulation of infrared spectral data. Wiley, New York
- Liu-Vlen D, Colthup NB, Fately WG, Grasselli JG (1991) Handbook of IR and Raman frequencies of organic molecules. Academic Press, San Diego
- Pachler KGR, Matlock F, Gremlich H-U (1988) Merck FT-IR atlas. VCH
- Pretsch E, Bühlmann P, Badertscher M (2009) Structure determination of organic compounds tables of spectral data, 4th edn. Springer, Berlin
- Sadtler handbook of infrared grating spectra. Heyden, London
- Schrader B (1989) Raman/infrared atlas of organic compounds, 2nd edn. VCH
- Bruno TJ, Svoronos PDN (2006) CRC handbook of fundamental spectroscopic correlation charts. CRC Press, Boca Raton

TEXTBOOKS

- Bellamy LJ (1975/1980) The infrared spectra of complex molecules, vols. 1 and 2. Chapman & Hall, London
- Nakanishi K (1977) Infrared absorption spectroscopy, 2nd edn. Holden-Day, San Francisco
- Griffiths PR, de Haseth JA (1986) Fourier transform infrared spectroscopy. Wiley, New York
- Roeges NPG (1998) A guide to the complete interpretation of infrared spectra of organic structures. Wiley, Chichester
- Günzler H, Gremlich H-U (2002) IR spectroscopy. Wiley-VCH, Weinheim
- Hendra P, Jones C, Wames G (1991) Fourier transform Raman spectroscopy. Ellis Horwood, Chichester
- Johnston SF (1992) FT-IR. Ellis Horwood, London
- Smith BC (1996) Fundamentals of Fourier transform infrared spectroscopy. CRC Press, Boca Raton
- Smith BC (1999) Infrared spectral interpretation. CRC Press, Boca Raton
- McCreery RL (2000) Raman spectroscopy for chemical analysis. Wiley-VCH, New York
- Stuart BH (2002) Infrared spectroscopy. Wiley, New York
- Socrates G (2001) Infrared and Raman characteristic group frequencies. Wiley, Chichester
- Smith E, Dent G (2004) Modern Raman spectroscopy: a practical approach. Wiley, New York
- Diem M (2015) Modern vibrational spectroscopy and micro-spectroscopy. Theory, instrumentation and biomedical applications. Wiley, New York

THEORETICAL TREATMENTS

- Herzberg G (1945) Infrared and Raman spectra of polyatomic molecules. Van Nostrand, Princeton
- Barrow GR (1964) Introduction to molecular spectroscopy. McGraw-Hill, New York
- Schrader B (ed) (1995) Infrared and Raman spectroscopy. Wiley-VCH, Weinheim
- Chalmers JM, Griffiths PR (eds) (2001) Handbook of vibrational spectroscopy, 5 vols. Wiley, Chichester. www.wileyeurope.com/vibspec
- Long DA (2001) The Raman effect. Wiley-VCH, New York

3.14 Problems

1. The five carbonyl compounds **52-56** from Question 3 in Chap. 2 gave the five infrared spectra labelled **a-e**, not respectively. Assign the structures **52-56** to the spectra **a-e**. Not all the significant peaks have been labelled, but critical carbonyl values are given,

since the scale of these pictures makes it difficult to measure them accurately enough here.

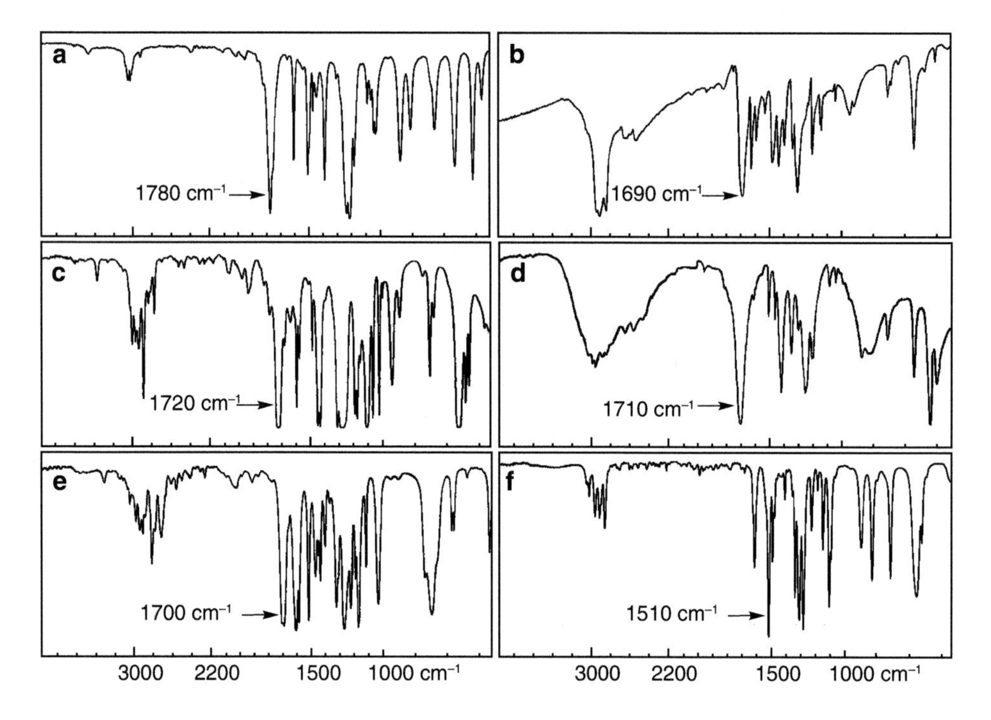

2. A sixth compound, an isomer of the five carbonyl compounds $C_8H_8O_2$, gave the infrared spectrum labelled **f**. Where they had six distinct carbon atoms each, the sixth compound only had four distinct carbon atoms. Suggest a structure for this compound.
3. There is only one structure that is compatible with the chemical formula and the infrared spectrum for each of these four compounds. What are they?

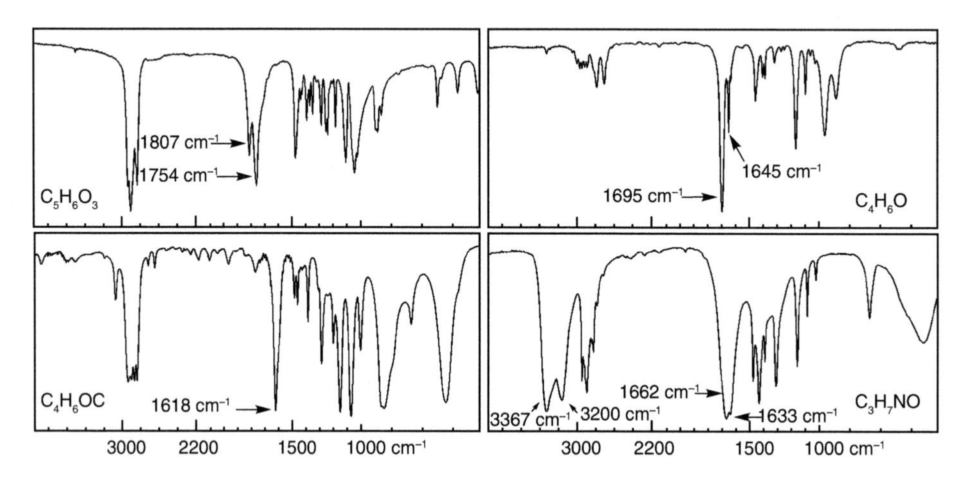

4. There are only two or three structures that are compatible with the chemical formula and the infrared spectrum for each of these compounds. What are they?

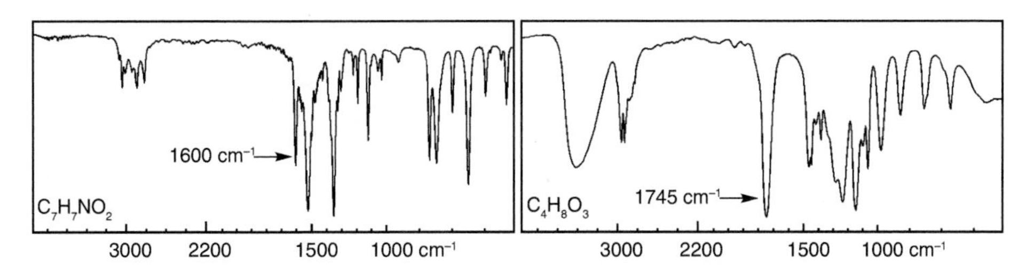

3.15 Correlation Charts

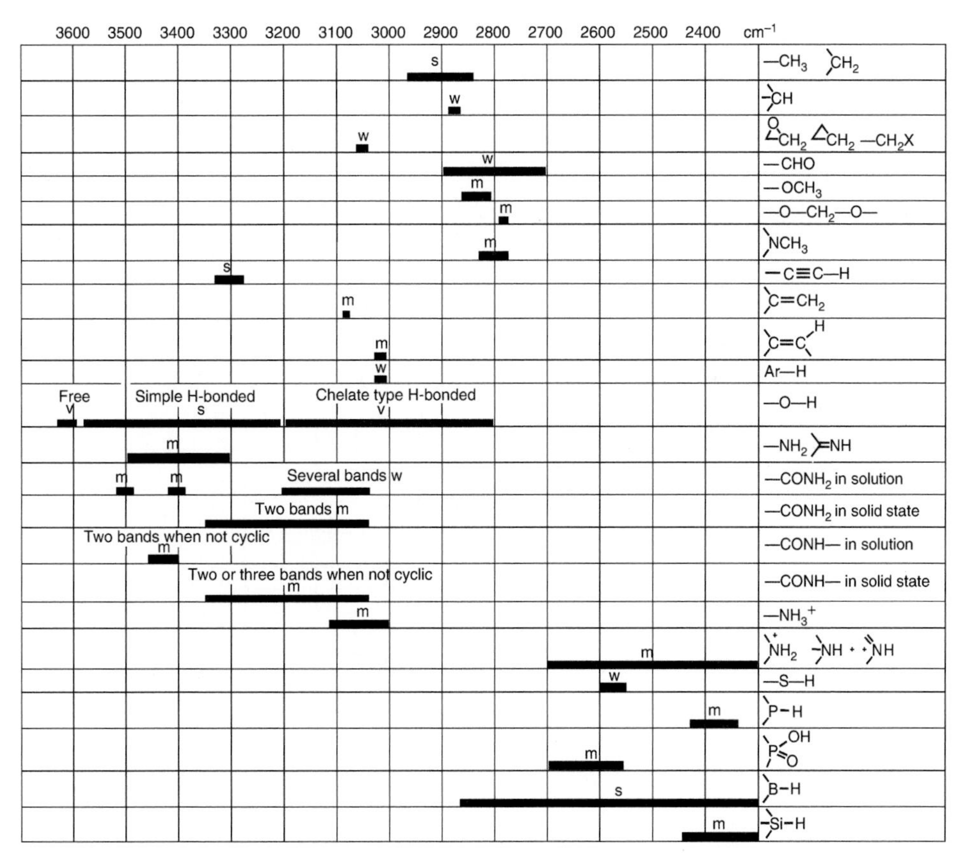

Chart 3.1 Stretching frequencies of single bonds to hydrogen (for more detail see the tables in Sect. 3.16)

Chart 3.2 Stretching frequencies of triple and cumulated double bonds (for more detail see the tables in Sect. 3.16.)

2400 2300 2200 2100 2000 1900 cm^{-1}

O=C=O
$-C\equiv N^{+}-O^{-}$
—N=C=O
$-C\equiv C-$
$-C\equiv N$
$-N_2^+$
$>N-C\equiv N$
$-S-C\equiv N$
$-N_3$
—N=C=N—
>C=C=O
$-C\equiv CH$
$-N^{+}\equiv C^{-}$
—N=C=S
$-COCHN_2^+$
$>C=N_2$
>C=C=N<
>C=C=C<

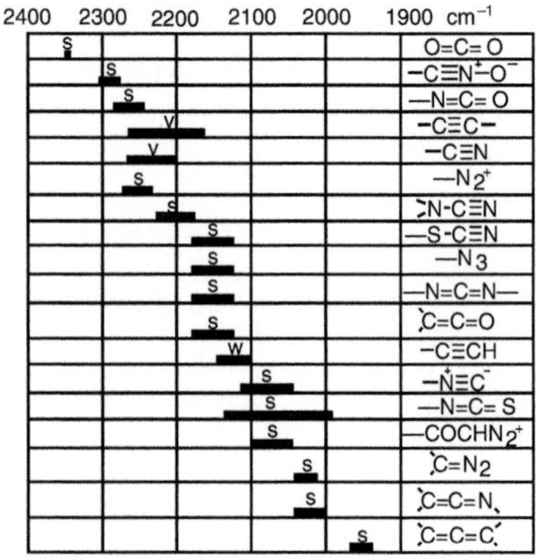

Chart 3.3 Stretching frequencies of double bonds other than carbonyl groups (for more detail see the tables in Sect. 3.16)

1700 1600 1500 1400 cm^{-1}

>C=N<
>C=N< α,β-unsaturated
>C=N< conjugated, cyclic
—N=N—
O^{-} + R N=N
$>C=C<^{NR_2}$ $>C=C<^{OR}$
>C=C< isolated
>C=C< aryl-conjugated
Dienes, trienes etc.
>C=C< carbonyl-conjugated
One or two bands — Benzenes, pyridines etc.
O—N=O (two bands)
$O-NO_2$
$N-NO_2$
C—N=O
$C-NO_2$
N—N=O
—CS—NH—
$-NH_2$
>NH
$-N_2^+$

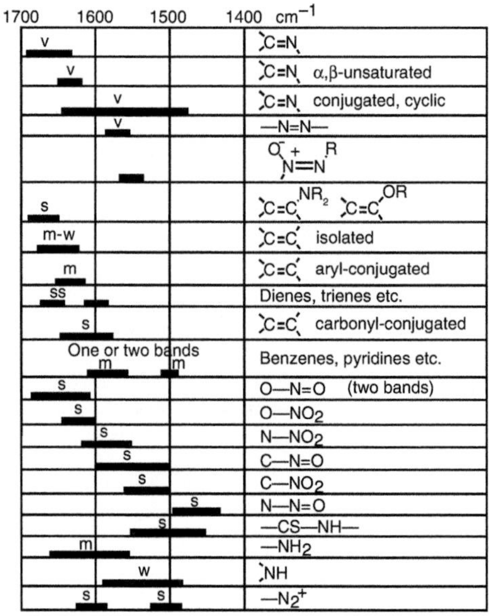

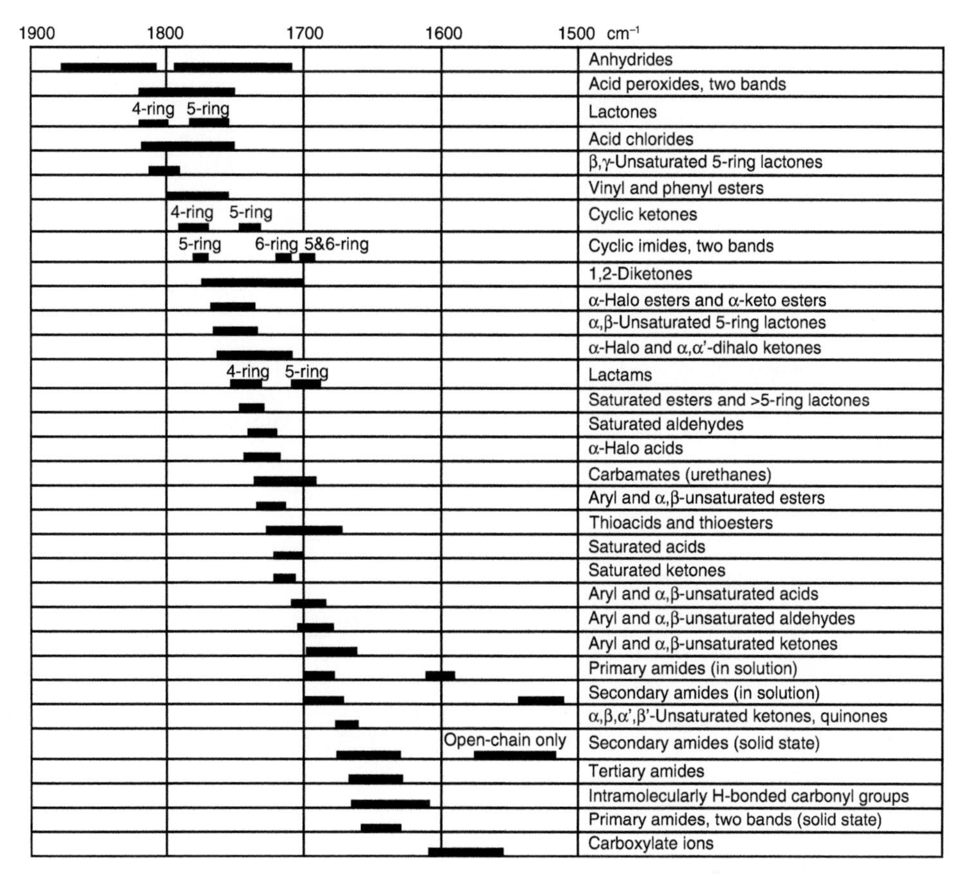

Chart 3.4 Stretching frequencies of carbonyl groups; all bands are strong (for more detail see the tables in Sect. 3.16)

3.16 Tables of Data

Table 3.1 C–H stretching vibrations

Group	Band	Remarks
C≡C–H	~3300 (s)	Sharp
$>C=CH_2$	3095–3075 (m)	Sometimes obscured by the stronger bands of saturated C–H groups, which occur below 3000 cm^{-1}
$>C=C(H)-$	3040–3010 (m)	
Aryl-H	3040–3010 (m)	Often obscured

(continued)

Table 3.1 (continued)

Group	Band	Remarks
Cyclopropane C–H Epoxide C–H	~3050 (w)	
$-CH_2-$halogen		
$-CO-CH_3$	3100–2900 (w)	Often very weak
Unfunctionalised C–H	2960–2850 (s)	Usually 2 or 3 bands
$>CH_2$ and $-CH_3$		
$\overset{\diagdown}{\underset{\diagup}{-C}}-H$	2890–2880 (w)	
$-CHO$	2900–2700 (w)	Usually 2 bands, one near 2720 cm^{-1}
$-OCH_3$	2850–2810 (m)	
$>CH_2$ and $>N-CH_2-$	2820–2700 (m)	Sometimes called Bohlmann bands
$-OCH_2O-$	2790–2770 (m)	

Table 3.2 C–H bending vibrations

Group	Band	Remarks
$>CH_2$ and $-CH_3$	1470–1430 (m)	Asymmetrical deformations
$-C(CH_3)_3$	1395–1385 (m) and 1365 (s)	
$>CH_2$ and $-CH_3$	1390–1370 (m)	Symmetrical deformations
$-OCOC-H$	1385–1365 (s)	The high intensity of these bands often dominates this region of the spectrum
$>C(CH_3)_2$	~1380 (m)	A roughly symmetrical doublet
$-COCH_3$	1360–1355 (s)	
$\overset{\diagdown}{\underset{H}{C}}=\overset{H}{\underset{H}{C}}$	995–985 (s) and 940–900 (s)	
$\overset{\diagdown}{\underset{H}{C}}=\overset{H}{\underset{\diagdown}{C}}$	970–960 (s)	C–H out-of-plane deformation; conjugation shifts the band towards 990 cm^{-1}
$\overset{\diagdown}{\underset{\diagup}{C}}=\overset{H}{\underset{H}{C}}$	895–885 (s)	
$\overset{\diagdown}{\underset{\diagup}{C}}=\overset{H}{\underset{\diagdown}{C}}$	840–790 (m)	
$\overset{H}{\underset{\diagup}{C}}=\overset{H}{\underset{\diagdown}{C}}$	730–675 (m)	
$>CH_2$	~720 (w)	Rocking

Table 3.3 O–H stretching and bending vibrations

Group	Band	Remarks
Water in dilute solution	3710	
Water of crystallisation in solids	3600–3100 (w)	Usually accompanied by a weak band at 1640–1615 cm^{-1}; residual water in KBr discs shows a broad band at 3450 cm^{-1}
Free O–H	3650–3590 (v)	Sharp
H-bonded O–H	3600–3200 (s)	Often broad but may be sharp for some intramolecular single-bridge H-bonds; the stronger the H-bond the lower the frequency
Intramolecularly H-bonded O–H of the chelate type and as found in carboxylic acids	3200–2500 (s)	Broad; the stronger the H-bond the lower the frequency; sometimes so broad as to be overlooked
O–H	1410–1260 (s)	O–H bending
\ –C–OH /	1150–1040 (s)	C–O stretching

Table 3.4 N–H stretching and bending vibrations

Group	Band	Remarks
–N(H)(H) \N–H / =N–H	3500–3300 (m)	Primary amines show two bands, ν_{asymm} and ν_{symm}; secondary amines absorb weakly, but pyrrole and indole N–H is sharp
–CONH–	3460–3400 (m)	Two bands lowered in the solid state and when involved in H-bonding; only one band with lactams
	3100–3070 (w)	A weak extra band with H-bonded and solid state samples
Amino acids $–NH_3^+$	3130–3030 (m)	Values for solid state; broad bands also (but not always) near 2500 and 2000 cm^{-1}
Amine salts $–NH_3^+$	~3000 (m)	
\ + NH2 / , –NH + (with \ /), \\+ NH /	2700–3250 (m)	Values for the solid state; broad because of the presence of overtone bands, etc.
–N(H)(H)	1650–1560 (m)	N–H bending
\ N–H /	1580–1490 (w)	Often too weak to be observed
$–NH_3^+$	1600 (s) and 1500 (s)	Secondary amine salts have the 1600 cm^{-1} band

Table 3.5 Miscellaneous R–H stretching vibrations

Group	Band	Remarks
\ P(=O) / OH	2700–2560 (m)	Associated O–H
\ B–H /	2640–2200 (s)	

(continued)

Table 3.5 (continued)

Group	Band	Remarks
–S–H	2600–2550 (w)	Weaker than O–H and less affected by H-bonding
\P–H/	2440–2350 (m)	Sharp
–Si–H (with \ and /)	2360–2150 (s) and 890–860	Sensitive to the electro-negativity of substituents
SiH_2 (H \ / Si / \ H)	~2135 (s) and 890–860	
R–D	1/1.37 times the corresponding R–H frequency	Useful when assigning R–H bands; deuteration leads to a recognisable shift to lower frequency

Table 3.6 Stretching frequencies of triple bonds and cumulated double bonds

Group	Band	Remarks
Carbon dioxide O=C=O	2349 (s)	ν_{asymm}; appears in many spectra because of inequalities in path length; ν_{symm} is IR inactive but Raman active
Nitrile oxides $–C≡N^+–O^-$	2305–2280 (m)	
Isocyanates –N=C=O	2275–2250 (s)	Very intense; position little affected by conjugation
Internal acetylenes –C≡C–	2300–2150 (v)	Strong and at low end of range when conjugated; weak or absent for nearly symmetrical substitution; strong in Raman
R_3Si –C≡CH	2040	
Nitriles –C≡N	2260–2200 (v)	Strong when conjugated; conjugated at the low end of the range; occasionally very weak; some cyanohydrins do not absorb in this region; strong in Raman
Isonitriles $–N^+≡C^-$	2180–2120 (s)	
Diazonium salts $–N^+≡N$	~2260	
Thiocyanates –S–C≡N	2175–2140 (s)	
Azides $–N=N^+=N^-$	2160–2120 (s)	
Carbodiimides –N=C=N–	2155–2130 (s)	Very intense; conjugation leads it to split into two bands of different intensity
Ketenes >C=C=O	2155–2130 (s)	Very intense
Terminal acetylenes –C≡C–H	2140–2100 (w)	C≡C stretching ($\nu_{C–H}$ at 3300 cm^{-1})
Isothiocyanates –N=C=S	2140–1990 (s)	Broad and very intense
Diazoketones $–CO•CH=N^+=N^-$	2100–2050 (s)	
Diazoalkanes $>C=N^+=N^-$	2050–2010 (s)	
Ketenimines >C=C=N\	2050–2000 (s)	Very intense
Allenes >C=C=C<	1950–1930 (s)	Two bands when terminal allenes or when bonded to carbonyl and similar groups

Table 3.7 Stretching frequencies of C=O groups (all bands listed are strong)

Group	Band	Remarks
Acid anhydrides –CO–O–CO–		
Saturated	1850–1800 and 1790–1740	Two bands usually separated by about 60 cm^{-1}; the higher-frequency band is more intense in acyclic anhydrides and the lower frequency band is more intense in cyclic anhydrides
Aryl and α,β-unsaturated	1830–1780 and 1770–1710	
Saturated five-ring	1870–1820 and 1800–1750	
All classes of anhydride	1300–1050	One or two bands; C–O stretching
Acid chlorides –COC1		
Saturated	1815–1790	COF higher, COBr and COI lower
Aryl and α,β-unsaturated	1790–1750	
Acid peroxides –CO–O–O–CO–		
Saturated	1820–1810 and 1800–1780	
Aryl and α,β-unsaturated	1805–1780 and 1785–1755	
Esters and lactones –CO–O–		
Saturated	1750–1735	
Aryl and α,β-unsaturated	1730–1715	
Aryl and vinyl esters C=C–O–CO–	1800–1750	The C=C stretch is also shifted to higher frequency
Esters with electronegative α-substituents	1770–1745	e.g. 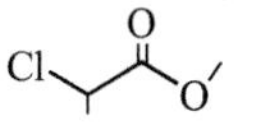
α-Keto esters	1755–1740	
Six-ring and larger lactones	Similar values to the corresponding open-chain esters	
Five-ring lactones	1780–1760	
α,β-Unsaturated five-ring lactones	1770–1740	When there is an α-C–H present, there are two bands, the relative intensity depending upon solvent
β,γ-Unsaturated five-ring lactones	~1800	
Four-ring lactones	~1820	
β-Keto ester in H-bonding enol form	~1650	Chelate-type H-bond causes shift to lower frequency than normal ester; the C=C is usually near 1630 (s) cm^{-1}
Aldehydes –CHO	Values below are for solution spectra; lowered by 10–20 cm^{-1} in liquid film or solid state and raised in the gas phase. See also Table 3.1 for C–H	
Saturated	1740–1720	
Aryl	1715–1695	*o*-Hydroxy or amino groups shift this range to 1655–1625 cm^{-1} because of intramolecular H-bonding
α-Chloro or bromo	1765–1695	

(continued)

Table 3.7 (continued)

Group	Band	Remarks
α,β-Unsaturated	1705–1680	
α, β, γ δ-Unsaturated	1680–1660	
β-Keto aldehyde in enol form	1670–1645	Lowering caused by intramolecular H-bonding
Ketones $>C=O$	Values below are for solution spectra; lowered by 10–20 cm^{-1} in liquid film or solid state and raised in the gas phase	
Saturated	1725–1705	α-Branching lowers the frequency
Aryl	1700–1680	
α,β-Unsaturated	1685–1665	Often two bands
α,β,α,β-Unsaturated and diaryl	1670–1660	
Cyclopropyl	1705–1685	
Six-ring and larger	Similar values to the corresponding open-chain ketones	
Five-ring	1750–1740	α,β-Unsaturation, etc. has a similar effect on these values as on those of open -chain ketones
Four-ring	~1780	
α-Chloro or bromo	1745–1725	Affected by conformation; highest values when halogens are in the simple plane as C=O group; α-F has larger effect; α-I has no effect
α,α-Dichloro or dibromo	1765–1745	
1,2-Diketones s-*trans* (i.e. open-chain α-diketones)	1730–1710	Antisymmetric stretching of both C=O groups; the symmetrical stretch is inactive in IR and active in Raman
1,2-Diketones s-*cis* six-ring	1760 and 1730	In diketo form
1,2-Diketones s-*cis* five-ring	1775 and 1760	In diketo form
1,3-Diketones enol form	1650 and 1615	Lowered by H-bonding and C=C conjugation
o-Hydroxy- or *o*-amino-aryl ketones	1655–1635	Lowering caused by intramolecular H-bonding
Diazoketones	1645–1615	
Quinones	1690–1660	C=C usually near 1600 (s) cm^{-1}
Extended quinones	1655–1635	
Tropone	1650	Near 1600 cm^{-1} in tropolones
Carboxylic acids R–CO_2H		
All types	3000–2500	O–H stretching; a charateristic group of bands from H-bonding in the dimer
Saturated	1725–1700	Monomer near 1760 cm^{-1}, rare; both free monomer and H-bonded dimer can be seen in dil. solution spectra; ether solvents give one band ~1730 cm^{-1}
α,β-Unsaturated	1715–1690	
Aryl	1700–1680	

Table 3.7 (continued)

Group	Band	Remarks
α-Chloro or bromo	1740–1720	
Carboxylate ions —CO_2^-	1610–1530 and 1420–1300	For the special features of amino acids, see text
Amides and lactams (O=C–N<)	See Table 3.4 for N–H bands	
Primary –$CONH_2$	~1690 and ~1600	Solution state amide I and II
	~1650 and ~1640	Solid state amide I and II; sometimes overlap; amide I is generally more intense than amide II
Secondary –CONH–	1700–1670 and 1550–1510	Solution state amide I and II; amide II not seen in lactams
	1680–1630 and 1570–1515	Solid state amide I and II; amide II not seen in lactams; amide I is generally more intense than amide II
Tertiary –$CONR_2$	1670–1630	Solid and solution spectra are little different
Five-ring lactams	~1700	Shifted to higher frequency when the N atom is in a bridged system in which overlap of the N lone pair with the C=O π-bond is diminished
Four-ring lactams (β-lactams)	~1745	
–CO–N–C=C		Shifted by +15 cm^{-1} from the corresponding amide or lactam
C=C–CO–N		Shifted by up to +15 cm^{-1} from the corresponding amide or lactam, unusual for the effect of α,β-unsaturation
Imides –CO–N–CO–		
Six-ring	~1710 and ~1700	α,β-Unsaturated shifted +15 cm^{-1}
Five-ring	~1770 and ~1700	
Ureas N–CO–N		
N–CO–N		
–NHCONH–	~1660	
Six-ring	~1640	
Five-ring	~1720	
Carbamates (= urethanes)		
O–CO–N	1740–1690	Also shows an amide II band when un- or mono-substituted on N
S–CO–N	1700–1670	
Thioesters and acids		
–CO–SH	~1720	Shifted −25 cm^{-1} when aryl or α,β-unsaturated
–CO–S–alkyl	~1690	
–CO–S–aryl	~1710	

(continued)

Table 3.7 (continued)

Group	Band	Remarks
Carbonates		
–O–CO–Cl	~1780	
–O–CO–O–	~1740	
Ar–O–CO–O–Ar	~1785	
Five-ring	~1820	
–S–CO–S–	~1645	
Ar–S–CO–S–Ar	~1715	
Acylsilanes –CO–SiR_3		
Saturated	~1640	
α,β-Unsaturated	~1590	

Table 3.8 C=N; Imines, oximes, etc

Group	Band	Remarks
>=N–H	3400–3300 (m)	N–H stretching
>=N–	1690–1640 (v)	Difficult to identify because of large variations in intensity and the closeness of C=C stretching bands; oximes usually have weak absorptions
α,β-Unsaturated	1600–1630 (v)	
Conjugated cyclic systems	1660–1480 (v)	

Table 3.9 N=N; Azo compounds

Group	Band	Remarks
–N=N–	1500–1400 (w)	Weak or inactive in IR, sometimes seen in Raman
$N=N^{+}-O^{-}$	1480–1450 and 1335–1315	Asymmetric and symmetric stretching

Table 3.10 C=C; Alkenes and arenes

Group	Band	Remarks
>=<	See Table 3.2 for =C–H bands	
Unconjugated	1680–1620 (v)	May be very weak if more or less symmetrically substituted; strong in Raman
Conjugated with aromatic ring	~1625 (v)	More intense than unconjugated C=C
Dienes, trienes etc.	1650 (s) and 1600 (s)	Lower-frequency band usually more intense and may obscure the higher-frequency band
α,β-Unsaturated carbonyl compounds	1640–1590 (s)	Usually weaker than the C=O band
Enol esters, enol ethers and enamines	1690–1650 (s)	
Aromatic rings	~1600 (m)	
	~1580 (m)	Stronger when the ring is further conjugated
	~1500 (m)	Usually the strongest of the three peaks in IR but weakest in Raman

Table 3.11 Nitro and nitroso groups, etc

Group	Band	Remarks
Nitro compounds C–NO_2	1570–1540 (s) and 1390–1340 (s)	Asymmetric and symmetric N=O stretching; lowered by ~30 cm^{-1} when conjugated
Nitrates O–NO_2	1650–1600 (s) and 1270–1250 (s)	
Nitramines N–NO_2	1630–1550 (s) and 1300–1250 (s)	
Nitroso compounds C–N=O		
Saturated	1585–1540 (s)	
Aryl	1510–1490 (s)	
Nitrites O–N=O	1680–1650 (s)	s-*trans* conformation
	1625–1610 (s)	s-*cis* conformation
N-Nitroso compounds N–N=O	1500–1430 (s)	
N-Oxides		
Aromatic	1300–1200 (s)	Very strong bands
Aliphatic	970–950 (s)	
Nitrate ions NO_3^-	1410–1340 and 860–800	

Table 3.12 Ethers

Group	Band	Remarks
–C–O–C	1150–1070 (s)	C–O stretching
C–O–C–	1275–1200 (s) and 1075–1020 (s)	
C–O–CH_3	2850–2810 (m)	C–H stretching
Epoxides C–C (O)	~1250, ~900 and ~800	

Table 3.13 Boron compounds

Group	Band	Remarks
B–H	2640–2200 (s)	
B–O	1380–1310 (vs)	
B–N	1550–1330 (vs)	
B–C	1240–620 (s)	

Table 3.14 Silicon compounds

Group	Band	Remarks
$-\overset{\backslash}{\underset{/}{Si}}-H$	2360–2150 (s) and 890–860	Sensitive to the electronegativity of the substituents
$\backslash\!/\,Si\,/\!\backslash$ H, H (SiH₂)	~2135 (s) and 890–860	
–SiMe$_n$	1275–1245 (s)	Typically sharp at 1260 cm^{-1}
	~840	n = 3
	~855	n = 2
	~765	n = 1
Si–OH	3690 (s)	Free O–H
	3400–3200 (s)	H-bonded O–H
Si–OR	1110–1000	
R_3Si–O–SiR_3	1080–1040 (s)	
Si–C≡C–	~2040	
Si–CH=CH_2	1600 and 1410	
	~1010 and ~960	
Si–Ph	1600–1590	
	1430	Sharp
	1130–1110 (s)	Split into two if Ph_2
	1030 (w) and 1000 (w)	
Si–F	1030–820	SiF_3 and SiF_2 show 2 bands

Table 3.15 Phosphorus compounds

Group	Band	Remarks
P–H	2440–2350 (s)	Sharp
P–Ph	1440 (s)	Sharp
P–O–alkyl	1050–1030 (s)	
P–O–aryl	1240–1190 (s)	
P=O	1300–1250 (s)	
P=S	750–580	
P–O–P	970–910	Broad
>P(=O)OH	2700–2560	H-bonded O–H
	1240–1180	P=O stretching
P–F	1110–760	

Table 3.16 Sulfur compounds

Group	Band	Remarks
S–H	2600–2550 (w)	Weaker than O–H and less affected by H-bonding
>=S	1200–1050 (s)	
C(=S)–N<	~3400 (m)	N–H stretching, when present; lowered in solid state by H-bonding
	1550–1450 (s) and 1300–1100 (s)	Amide-II and amide I, respectively
–O–CS–O	~1225	
–O–CS–N	~1170	
N–CS–N	1340–1130	
–S–CS–N	~1050	
–S–CS–S–	~1070	
>S=O	1060–1040 (s)	
>S(=O)=O	1350–1310 (s)	
	1160–1120 (s)	
$-SO_2-N$	1350–1330 (s) and 1180–1160 (s)	
$-SO_2-O-$	1420–1330 (s) and 1200–1145 (s)	
$-SO_2-Cl$	1410–1375 and 1205–1170	
–S–F	815–755	

Table 3.17 Halogen compounds

Group	Band	Remarks
C–F	1400–1000 (s)	Sharp
	780–680	Weaker bands
C–Cl	800–600 (s)	
C–Br	750–500 (s)	C–H stretching
C–I	~500	

Table 3.18 Inorganic ions (all bands are strong)

Group	Band	Remarks
NH_4^+	3300–3030	
CN^-, ^-SCN, ^-OCN	2200–2000	
CO_3^{2-}	1450–1410	
SO_4^{2-}	1130–1080	
NO_3^-	1380–1350	
NO_2^-	1250–1230	
PO_4^{2-}	1100–1000	

1D Nuclear Magnetic Resonance Spectra

4

4.1 Nuclear Spin and Resonance

Some atomic nuclei have a nuclear spin (I), and the presence of a spin makes these nuclei behave like bar magnets. In the presence of an applied magnetic field the nuclear magnets can orient themselves in $2I + 1$ ways. Those nuclei with an odd mass number have nuclear spins of 1/2, or 3/2, or 5/2, …, etc. (given in the first table at the end of this chapter). In the applications of NMR spectroscopy in organic chemistry, the five most important nuclei are ^{1}H and ^{13}C, followed by ^{19}F, ^{29}Si and ^{31}P, all of which have spins of 1/2. These nuclei, therefore, can take up one of only two orientations, a low-energy orientation aligned with the applied field and a high-energy orientation opposed to the applied field. The difference in energy between these two orientations is given by Eq. (4.1), the Boltzmann distribution by Eq. (4.2), and the frequency ν in Hz corresponding to the difference in energy by Eq. (4.3):

$$\Delta E = \frac{h\gamma B_0}{2\pi} \tag{4.1}$$

$$\frac{N_\beta}{N_\alpha} = e^{-\frac{\Delta E}{kT}} \tag{4.2}$$

$$\nu = \frac{\gamma B_0}{2\pi} \tag{4.3}$$

where γ is the magnetogyric ratio (also called the gyromagnetic ratio), which is a measure of the relative strength of the nuclear magnet, and is different for each element and for

I. Fleming, D. Williams, *Spectroscopic Methods in Organic Chemistry*,
https://doi.org/10.1007/978-3-030-18252-6_4

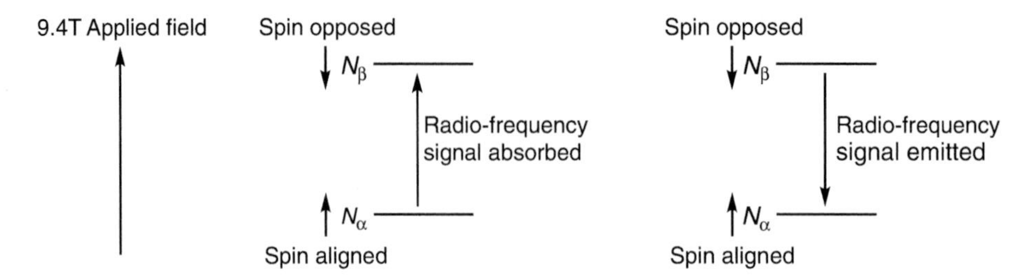

Fig. 4.1 Energy levels for aligned and opposed nuclear spins

each isotope of each element, B_0 is the strength of the applied magnetic field, N_α is the number of nuclei in the low-energy state, and N_β the number in the high-energy state. A radio-frequency signal applied to the system changes the Boltzmann distribution when the radio frequency matches the frequency ν. It is called the resonance frequency or Larmor frequency for that particular nucleus. The effect is to promote nuclei from the low-energy N_α level to the high-energy level N_β (Fig. 4.1).

The Larmor frequency is therefore dependent upon the applied field strength and the nature of the nucleus in question. The frequencies at which nuclei come into resonance at a field strength of 9.4 T (94 kG) are given in in the first table at the end of this chapter, but for now it is enough to know that at this field strength the common magnetic nuclei come into resonance each in a narrow range close to: 400 MHz for ^{1}H, 376.3 MHz for ^{19}F, 161.9 MHz for ^{31}P, 100.6 MHz for ^{13}C and 79.4 MHz for ^{29}Si. Because of the widespread use of proton NMR spectroscopy, it is usual to refer to an instrument with this field strength as a 400 MHz instrument. The common nuclei ^{12}C and ^{16}O having $I = 0$ are inactive. A few other common nuclei have spins, of which ^{2}H and ^{14}N ($I = 1$) are perhaps the most important. When present in organic molecules, they affect ^{1}H and ^{13}C NMR spectra, but it is comparatively unusual to study the NMR spectra of these nuclei themselves in the ordinary course of a structure determination.

The higher the field strength, B_0, the greater the difference between N_α and N_β (Eqs. (4.1) and (4.2)), and so the higher the field strength, the greater the difference in the populations in the N_α and N_β levels, and the more sensitive the instrument. For protons the difference between N_α and N_β, determined by the Boltzmann distribution (Eq. (4.2)), is very small—in a 400 MHz instrument at 300 K, it is only 6 in 10^5. NMR spectroscopy is therefore a relatively insensitive technique compared with UV, IR and mass spectrometry. It is even less sensitive for those nuclei with low magnetogyric ratios, made worse when the magnetically active nucleus is not the most abundant isotope (Table 4.1).

If it were possible to scan all these nuclei in one experiment (it is not), and if each of the elements were to be present equally, the spectrum might look something like Fig. 4.2, in which intensity is plotted upwards. All the different protons, for example, would come into resonance in a very narrow band (4000 Hz wide) at 400 MHz, and all the different ^{19}F signals close to 376 MHz, with the intensity of the signals from these two elements on a different scale from the others. Thus all the different ^{13}C signals in the narrow band

Table 4.1 Representative data for common nuclei in NMR spectra in a 9.4T instrument

	^{1}H	^{19}F	^{31}P	^{13}C	^{29}Si
Larmor frequency (MHz)	400	376	162	100	79
Natural abundance (%)	100	100	100	1.11	4.7
Relative sensitivity	1.00	0.83	0.07	0.164	0.0078
Relative sensitivity × natural abundance	1.00	0.83	0.07	0.00018	0.00037

(20,000 Hz wide) close to 100 MHz would be exceptionally weak because of the combination of the inherent insensitivity of the nucleus and its low natural abundance.

4.2 Taking a Spectrum

The following discussion is about NMR in the liquid state. Solid state NMR is different in a number of ways. It is not discussed in this book, because it is not much used by organic chemists for structure determination.

In order to take an NMR spectrum in the liquid state, the sample is dissolved in a solvent, preferably one that does not itself give rise to signals in the same range as the sample in the NMR spectrum. The most commonly used solvent is $CDCl_3$, but more polar solvents like d_8-THF, CD_3CN, $(CD_3)_2CO$, d_6-DMSO [$(CD_3)_2SO$], CD_3OD and D_2O are often used, especially for polar compounds. The solvent itself provides a signal that is used by the computer to calibrate the spectrum, replacing the older system where an internal standard (usually tetramethylsilane) was added for this purpose. Other solvents, like d_8-THF, d_6-benzene, d_8-toluene and d_5-pyridine are used, the choice of solvent being largely determined by the solubility of the compound under investigation and the extent to which the signals of individual nuclei are resolved in any one solvent. The solution should be free both of paramagnetic and insoluble impurities, and it should not be viscous, or resolution will suffer. Routinely a good ^{1}H NMR spectrum might use 5–10 mg, and a good ^{13}C spectrum might use 30 mg, but smaller quantities can still give good spectra with more effort.

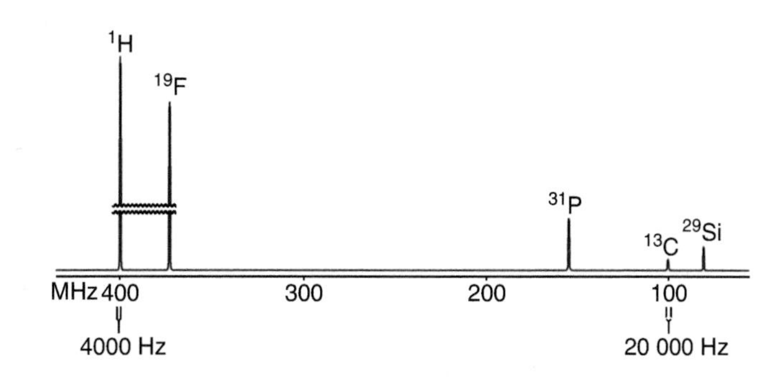

Fig. 4.2 Imaginary NMR spectrum with several different nuclei

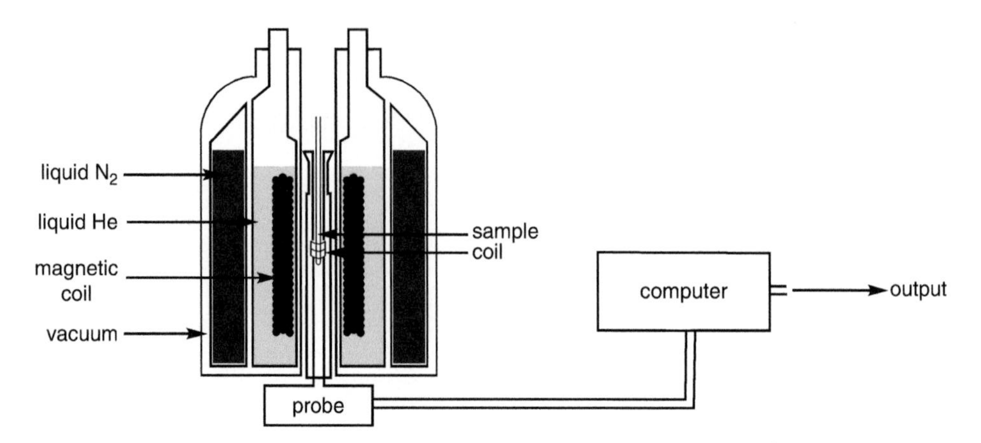

Fig. 4.3 Schematic cross-section of an NMR spectrometer

Until the early 1970s, spectra were measured at 100 MHz or less using a permanent magnet, and a continuous radio wave (CW). The magnetic field was varied to scan the range of frequencies to be detected, plotting the spectrum directly as it was being taken, the whole process taking a few minutes. This method has now gone out of use, in favour of high-field spectrometers (Fig. 4.3) with supercooled electromagnets. The solution of the sample in a precision-ground tube is lowered into the centre of a magnetic field, in which the probe sits. The magnet is tuned to give the highest possible level of homogeneity, generally about 1 in 10^9, using shimming coils which only need adjusting occasionally. The tube may be spun (at about 30 r.p.s.) about its vertical axis to improve the effective homogeneity even further, but this is less necessary today. Integral to the probe are coils, which fit immediately around the sample, and which provide pulses at the Larmor frequency for the nucleus under study. The same or other coils pick up the electrical current given back from the sample as the net magnetisation induced in the sample by the promotion of the nuclear spins from the N_α level to the N_β level oscillates in the applied magnetic field. An attached computer carries out a Fourier transformation (FT) of the digitised signal to give the spectrum.

When the sample is placed in the spectrometer the nuclear magnets settle into the equilibrium distribution, with more in the N_α level than in N_β. This gives the sample a bulk magnetisation represented by the vector M_0 pointing along the axis of the applied field B_0 (Fig. 4.4, left), which is drawn as the z axis. The RF pulse applied along the x axis tips the magnetisation through an angle θ given by $\theta = \gamma B_1 t_p$, in which B_1 is the magnetic field strength corresponding to the RF frequency. The vector representing the bulk magnetisation precesses around the z axis, just as a spinning top precesses in a gravitational field if the top is tipped over.

The angle θ is usually expressed in radians, so that if the strength of the RF pulse (B_1) and the time (t_p) are chosen to make $\theta = 90°$, the pulse is called a $\pi/2$ pulse. The value of $B_1 t_p$ appropriate for a $\pi/2$ pulse can be measured using a program called *pw90*.

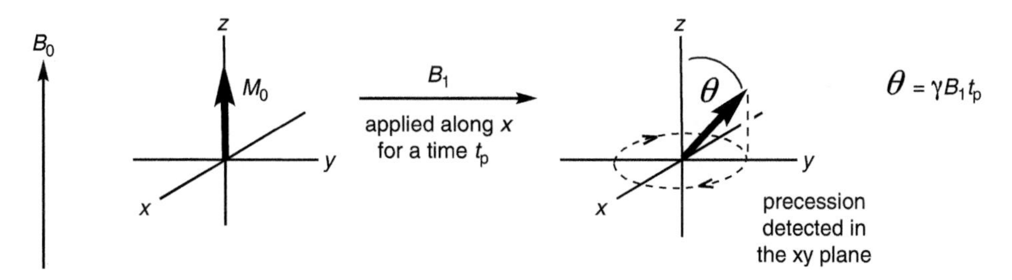

Fig. 4.4 Vector diagram for the basic pulse sequence

Successive increments of t_p are applied using a narrow range of frequencies from the middle of the spectrum, and spectra plotted. Each successive spectrum increases in intensity as the value of B_1t_p increases the angle θ, and increases the projection of the vector onto the xy plane. The intensity of the spectrum reaches a maximum at the optimal value for a $\pi/2$ pulse, and then decreases in intensity as the angle θ moves beyond 90°. It reaches zero intensity with a perfect π pulse, because the vector has no component in the xy plane at that point. It then increases in intensity again, but upside down, as it passes between a π and $3\pi/2$ pulse, and returns towards zero as it approaches a 2π pulse. One quarter of the easily detected value of B_1t_p for the 2π pulse is a good measure of the optimal value for a $\pi/2$ pulse. It is rarely necessary for taking routine spectra to know this parameter, but if careful tuning of all the variables is needed for a critical spectrum, it can be useful to know.

There are benefits and penalties from using a strong pulse and a short time (called a hard pulse), and equally from using a weak pulse and a long time (called a soft pulse). A hard pulse stimulates a wide bandwidth of frequencies to cover the whole range, but there are difficulties in the electronics from crowding a lot of power into a short time. A soft pulse makes it possible to select a particular frequency, but it leads to problems with phasing, since the vectors have more time to precess away from their starting point on the y axis while the pulse is still going on. Compromises have to be made, which depend upon the type of spectrum and the nucleus under study. A ^{19}F spectrum, for example, needs an especially hard pulse to cover the wide range of resonance frequencies (chemical shifts) found for fluorine atoms. Alternatively, a selective irradiation experiment uses a soft pulse to stimulate a narrow range of frequencies, followed by a hard pulse with which to collect the spectrum.

However it is achieved, a $\pi/2$ pulse equalises the number of nuclei in the N_β and the N_α levels. A pulse (B_1) that is twice as strong, or a time (t_p) that is twice as long, will be a π pulse, which tips the magnetisation vector to align perfectly opposed to the field, inverting the Boltzmann distribution, and leaving no component of magnetisation in the xy plane. Except when θ equals 0 or π, the vector will have a component of the magnetisation in the xy plane, whether it is a full $\pi/2$ pulse or not, and it is this projection of the vector onto the xy plane that is detected in the xy plane. Receiver coils on the x and y axes independently detect the fluctuating magnetic field in the *xy* plane produced by the precessing vectors. After the π/2 pulse has been applied, the detected signals from all the different nuclei start along +*y* (zero signal on the *x* axis), precess to the *x* axis (positive signal), then to −*y* (zero signal), and so on, allowing a receiver coil on the *x* axis to detect the decay of the magnetisation in the *xy* plane as a fluctuating magnetic field with the form of many superimposed sine curves, one from each frequency in the spectrum. A signal is simultaneously collected by a receiver coil on the *y* axis in the form of many superimposed cosine curves. These waves contain all the different resonance frequencies for each of the nuclei in the sample in their slightly different environments. This is the basis of what is called *quadrature detection*, and why we need it is discussed in Sect. 4.14.

After the sample has been lowered into the magnetic field, there is a delay t_r of a few seconds while the nuclei come to magnetic and thermal equilibrium (Fig. 4.5). The pulse time t_p is typically a few microseconds, following which there has to be a small delay of a few more microseconds while the powerful electronic signals from the hard pulse die away. Coils then pick up the frequencies from the precessing magnetic vectors cutting through the *x* and *y* axes. These signals, decaying over a few seconds, are called the Free Induction Decay (FID). Leaving out the initial wait t_r, the pulse sequence can be repeated n times, and the FIDs added up. The signal intensity increases linearly, whereas the noise increases by a factor $\sqrt{n}$, improving the signal-to-noise ratio. There are limits to how many such pulses are useful, since the signal-to-noise ratio only improves by a factor of $\sqrt{n}$. A ^{1}H NMR spectrum with 5 mg of sample would routinely use 16 pulses in a minute or two, and a good ^{13}C spectrum with 30 mg might use 256 pulses over 15 min, but these numbers are dropping as spectrometers improve. With a large number of pulses, it is possible to obtain ^{1}H spectra from less than 0.1 mg and ^{13}C spectra from 1 mg, if the molecular weight is not more than a few hundred.

Ideally the FID will have dropped to zero amplitude before the next pulse. If it doesn't, the signal intensity will give a wrong measure in the integration. In practice, it is usual to tip the magnetisation through a smaller angle than π/2, both because it is difficult to get it

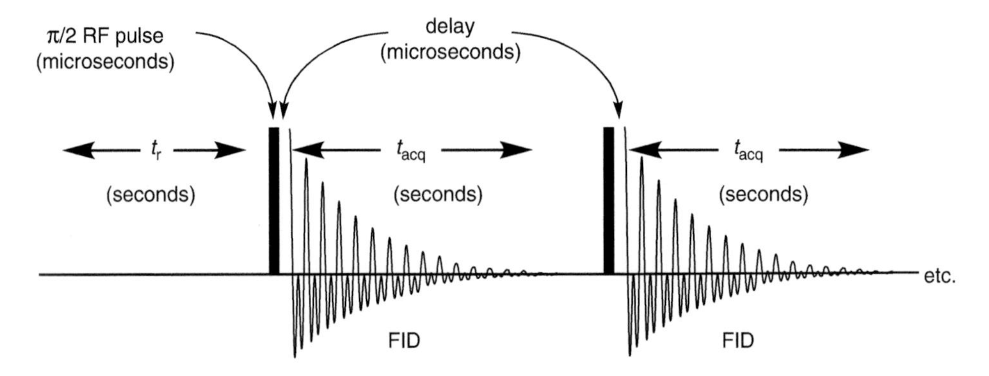

Fig. 4.5 The basic pulse sequence

exactly right, and because the relaxation from 90° back to 0° would take too long to give us time to put in a lot of pulses and still collect a complete FID. Compromises have to be made. Although the pulse may not be a full π/2 pulse, we can treat it in most of the discussion from now on as though it were π/2, because the vector always has a component in the *xy* plane, which is the only place where anything is detected.

In order to process the sum of all the FIDs, it must be digitised, and will need enough data points to define the frequency. The period between data points is called the dwell time (DW). The Nyquist theorem says that the sampling rate must be at least twice the signal frequency if the wave is to be reported with its true frequency. Given that the Larmor frequencies are hundreds of millions of Hz, this rate of sampling would give more data points than computers could process in a reasonable time. Instead a single Larmor frequency is subtracted from the original FID, leaving only the frequencies that were modulating it, rather as a radio signal is processed to send to the speakers only the sound frequencies modulating it. Furthermore, since we know from experience the normal range of frequencies we are likely to find in our spectra, we only need to digitise a limited range, called the spectral window (SW). Typically, for a 400 MHz spectrometer, this is spread over 20,000 Hz for ^{13}C and over 4000 Hz for protons (Fig. 4.2), numbers which can be digitised and processed efficiently if, following the Nyquist theorem, we set the dwell time to be the reciprocal of twice the spectral window. Should we choose too narrow a spectral window, the signals outside it would actually appear in the spectrum, but in the wrong place and severely distorted. They are said to be folded back, and can be corrected for by taking the spectrum again, choosing a different dwell time, thereby extending the spectral window. Because signals from outside the spectral window appear in the spectrum, some noise is also folded back. This is countered in the spectrometer using frequency filters that suppress signals from outside the range, but unavoidably the filters do not have a sharp cutoff at the edge of the SW.

In order for the computer to be able to store and process all the data points, it is also important that the intensity of the FID falls within a manageable range. If it does not, the early part of the FID will be reported as weaker than it actually is; it is described as being clipped, and the signal after Fourier transformation is distorted. The computer is usually set to check for overload of this kind, and to respond accordingly. Worse still, unavoidable voltages in the electronics would be added as constant numbers to the intensity of the FID, giving rise to an intense signal at δ 0, and overloading the computer. This problem is solved by *phase cycling*, routinely carried out with FIDs collected on both the x axis and on the y axis, using the 4-step sequence, called quadrature detection, described at the end of Sect. 4.14.

The first step in processing the digitised FID is Fourier transformation (FT), which, as we saw in an earlier chapter, takes all the frequencies in the time domain, pictured as a complex wave, to produce a picture in the frequency domain as plots of frequency against intensity. We can regard it as a black box, and read no more, but the essence of the technique is described below.

To have some understanding of how FT works, we shall look only at the FID collected from the coil on the *y* axis. The FT program creates a series of trial frequencies, expressed as cosine waves, with data points at the same intervals as the sample wave. The sample wave and each trial wave are multiplied together, and expressed as a new wave. If the trial frequency is present in the FID, the multiplication will lead to a wave predominantly above the mid-line and with little or no intensity below. Integration gives a positive value, giving a measure of the intensity of the signal at the trial frequency.

Figure 4.6 shows at the top left the effect of multiplying together two cosine waves lasting for 1 s of the same frequency, 10 Hz: the resultant wave below them is all above the horizontal axis, and integration of the area under it would give a positive value. Figure 4.6 at the top right shows the effect of multiplying together two cosine waves of different frequencies, the same one of 10 Hz and a trial frequency of 15 Hz: the resultant wave below them has the same area under the curve above the horizontal axis (positive) as below (negative), giving a zero integration. This would be true no matter what different frequency we chose, as we can see in Fig. 4.6 at the bottom left with a trial frequency only 10% larger than the frequency being tested. The result of multiplying the two cosine curves is shown below the 10 and 11 Hz waves, and again the total area above and below the horizontal axis is equal, the only difference being that it is necessary to wait longer for the equality to be established. Finally, in Fig. 4.6 at the bottom right, is a composite wave that contains frequencies of both 10 and 15 Hz. Multiplying this wave with a trial cosine curve of 10 Hz leads to the wave below, almost all of which is above the horizontal axis, showing that the sample wave included a 10 Hz component.

The FT plots the area from integration of the new wave against the trial frequency to produce a spectrum in the frequency domain. With a real spectrum there is a large number of trial frequencies, and a correspondingly large number of calculations to be carried out, but the computer does it seemingly instantaneously, producing a line in the frequency domain when there is a frequency present in the FID, and no line when it is not present.

In practice the frequency domain does not show *lines* corresponding to each frequency. The observed signals are more or less sharply pointed Lorentzian curves. To understand one reason why FT does not give a single line corresponding to each frequency, consider again what happens when the trial frequency is close to that of the signal, as shown at the bottom left in Fig. 4.6. In the early part of the time in which the wave is collected, the multiplied wave is mostly above the horizontal axis, because the two waves stay close to being in phase for a few cycles before they begin to get out of phase, and their product remains positive. As we wait a little longer, the two waves pass from being more or less in phase to being out of phase, whereupon the product is negative. Eventually, persevering long enough with steady waves like those in Fig. 4.6, the areas above and below the line cancel each other out. In practice, however, the FID is a decaying signal, and not a steady wave—the early part of the FID is more intense than the later part. As a result the trial wave and the wave near in frequency to it in the FID would not cancel, even though the trial frequency is not precisely present, because the early positive integration will have

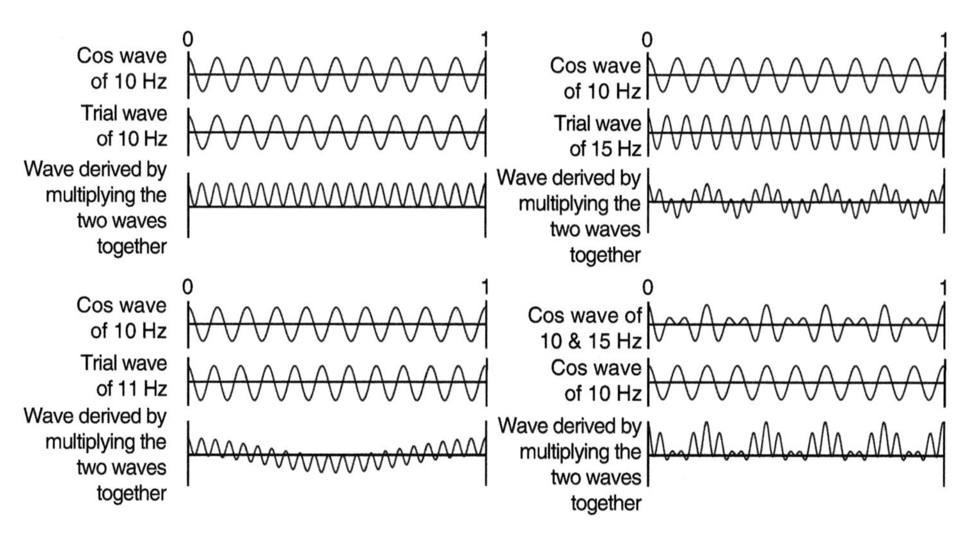

Fig. 4.6 Trial waves used in a Fourier transformation

been emphasised. Instead the FT gives a signal of weak intensity at a trial frequency close to that of the true frequency—the closer it is the more intense—inherently leading to a Lorentzian distribution in place of a sharp line.

Let us suppose that we are recording the idealised proton NMR spectrum of a greatly simplified molecule that contains only two protons which experience different magnetic environments. Let us suppose that the two protons differ in resonance frequency by 10 and 15 Hz from an arbitrary reference frequency, to which we assign the value zero. The signal which differs by 10 Hz from the reference frequency oscillates from positive to negative (and back to positive) 10 times per second. As it does so, it decays exponentially as relaxation gradually allows the nuclear magnets to return to their equilibrium distribution. The signal is therefore an exponentially decaying wave of frequency 10 Hz (Fig. 4.7 top left). Similarly, the signal arising from the resonance which differs from the reference frequency by 15 Hz oscillates 15 times per second. Signals from two protons that are in different magnetic environments will decay at different rates, and in this example the latter signal has been shown arbitrarily as relaxing more slowly than the former (Fig. 4.7 middle left). If the two frequencies are collected simultaneously, then the two decaying waves combine (Fig. 4.7 bottom left), with both frequencies still present, but with the later part of the wave looking more and more like the 15 Hz wave.

The Fourier transformation would multiply this wave by waves with a succession of frequencies. Only when it was multiplied by waves with frequencies of 10 and 15 Hz would the area under the upper part of the resultant wave be greater than the area above the lower part of the wave. Accordingly, absorption peaks would appear at these frequencies and little or none at the other frequencies. The spectrum can then be plotted with absorption (upwards) against frequency (increasing towards the left, as in UV and IR spectroscopy) (Fig. 4.7 right). The rate at which the signal decays only alters the relative intensity

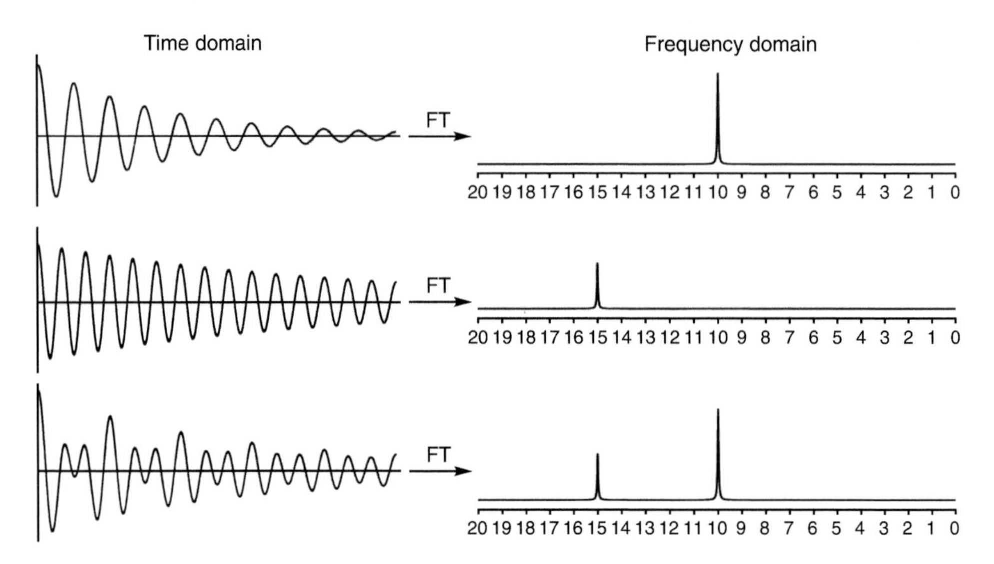

Fig. 4.7 Extracting frequencies in the time domain to create the frequency domain

of the two lines because the signal at 15 Hz had not reached zero before the next pulse. The decaying signals in real NMR spectra are of course much more complicated—spectra typically containing many peaks.

With the instrument controlled from a computer keyboard, the pulses are transmitted, the FIDs are accumulated, and stored. The FID is then processed using a computer loaded with one or more of several programs such as Topspin, Mestrenova, ACD/NMR-Processor, and others. The digitised signal is tweaked in a number of ways, of which the following are the best known. Zero-filling adds an equal number of data points to the end of the FID before processing. This adds no information but doubles the number of data points used to plot the signal (Fig. 4.8b), slightly improving the appearance of each signal. There are two tweaks called apodization: exponential and Gaussian. The first amplifies selectively the early part of the FID (Fig. 4.8d), improving the signal-to-noise ratio, since noise is constant throughout the FID, but most of the signal is in the early part. The second selectively amplifies the early-middle part of the FID (Fig. 4.8e), where the best signal quality lies, improving resolution and changing the Lorentzian curve to a Gaussian.

Fourier transformation is more complicated than the simple description above, since the trial frequency must be perfectly in phase to produce the results in Fig. 4.6. If it were to be 90° out of phase the result would be a zero signal instead of a positive signal, and if it were to be 180° out of phase the result would be a negative signal. The full mathematics of FT solves this problem, taking the data points from the sine curves from the phase cycling as real components, and the data points from the cosine curves as imaginary components, producing two solutions for each data point, one real and one imaginary. The

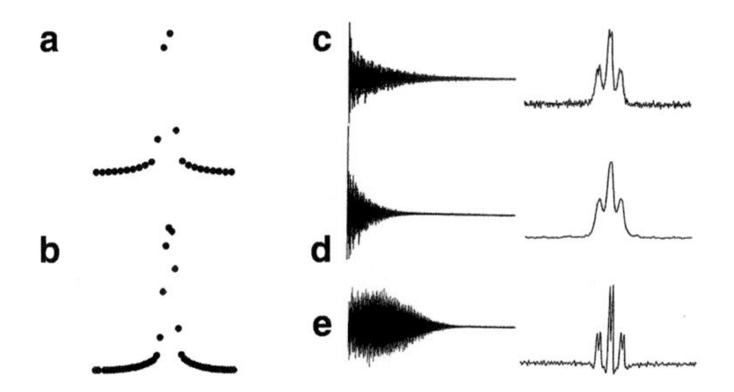

Fig. 4.8 The common tweaks that improve the NMR signal after FT. (**a**) A peak defined by data points without zero filling. (**b**) The same peak defined by twice as many data points after zero filling. (**c**) Raw FID. (**d**) Exponential multiplication. (**e**) Gaussian multiplication

peaks produced have very different appearances (Fig. 4.9). The real solution produces a plot that is called the absorption mode, and looks like the usual NMR signal; the imaginary solution gives a plot called the dispersion mode, which changes sign at the mid point. The former is relatively narrow close to the base line, and in 1D spectra it is normal to use only this output from the FT, and reject the dispersion mode signal. It is possible to use both at the same time, by squaring them and taking the square root, before adding them together. In 2D NMR spectra (Chap. 5), where the signals are all much weaker than in the usual 1D spectra, this manipulation is normal, and has little disadvantage, because 2D spectra are plotted as contour diagrams, with few contours sampled near the base of the peaks.

The number of data points used to define these peaks must be an integral power of 2 for the usual FT protocol; if we use 2^{15} data points (actually 32,768, but conventionally referred to as 32K) in the 4000 Hz spectral width of a typical 400 MHz proton spectrum, half of them will be in the real and half in the imaginary set, which means that the separation of the data points will be 0.244 Hz. As a result we cannot expect to resolve signals that are separated by 0.5 Hz or less. We need to remember that the frequencies reported on the spectrum to four places of decimals are meaningless beyond the first place, and that coupling constants (Sect. 4.5) should be reported only to the first place of decimals.

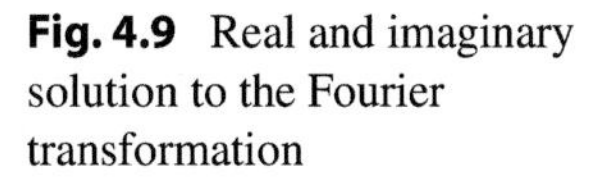

Fig. 4.9 Real and imaginary solution to the Fourier transformation

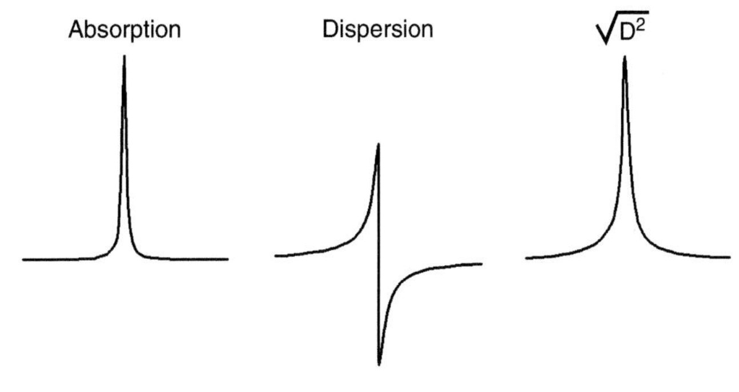

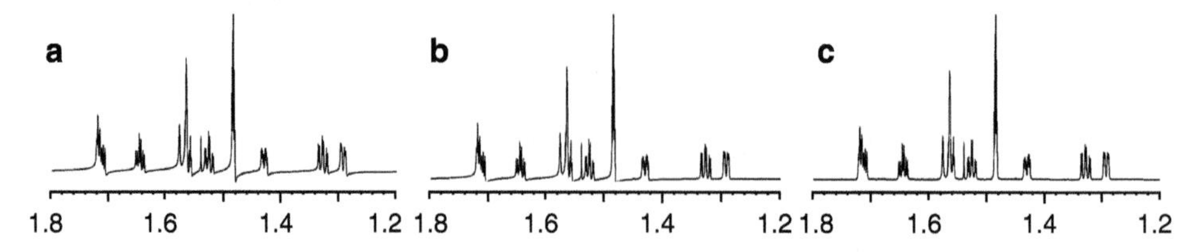

Fig. 4.10 Linear and non-linear phase corrections

In practice, most NMR spectra as first produced from the FT are not seen cleanly as absorption mode signals. The finite time (t_p) of the B_1 pulse together with the necessary delay of several microseconds before the FID is collected allow the signals to get linearly out of phase, because the higher frequency vectors, having precessed further than the lower frequency vectors, are further out of phase than the lower frequency signals. Any phase error between the transmitter signal and the receiver signal will give rise to a non-linear phase error evenly across the whole spectrum. Further internal features within the molecules also allow some individual signals to get out of phase with others in a non-linear way. The result of all these accumulated phase errors is that some dispersion mode signals get mixed in with the absorption mode signals in the output of the FT. Figure 4.10a shows the appearance of a ^{1}H spectrum with uncorrected out-of-phase signals. Figure 4.10b shows the result of a zero-order phase correction evenly across the whole spectrum, improving the less distorted peaks at the low-frequency end (lower frequencies are on the right), and Fig. 4.10c shows the result of a first-order phase correction applied selectively in proportion to frequency, bringing the more distorted, high-frequency signals on the left into line. These corrections are automatic in the processing programmes, but they are available for fine tuning by the operator.

4.3 The Chemical Shift

The range of frequencies found in any one spectrum is a relatively narrow band around the fundamental frequency for the specified nucleus at the field strength of the instrument. Thus, in ^{13}C spectra—taken, for example, on a 400 MHz instrument—the range of frequencies is a narrow segment of about 20,000 Hz in the neighbourhood of the resonance frequency for ^{13}C of 100.56 MHz. Within this range each of the different ^{13}C atoms in an organic compound resonates at its Larmor frequency, the precise magnitude of which is determined not only by the applied field, B_o, but also by minute differences in the magnetic environment experienced by each nucleus. These minute differences are caused largely by the variation in electron population in the neighbourhood of each nucleus, with the result that each chemically distinct carbon atom in a structure, when it happens to be a ^{13}C, will come into resonance at a slightly different frequency from all the others. The electrons affect the microenvironment, because electrons have a high magnetogyric ratio, and their movement creates a substantial local magnetic field. Similarly, in ^{1}H spectra taken on a

400 MHz instrument, a narrower range—about 4000 Hz in the neighbourhood of the resonance frequency of 400 MHz—is enough to bring most protons into resonance.

In practice it is inconvenient to characterise the ^{13}C and ^{1}H peaks by assigning to them their absolute frequency, all very close to 100.56 or 400 MHz: the numbers are cumbersome and much too difficult to measure accurately. Furthermore, they change from instrument to instrument, and even from day to day, as the applied field changes. It is convenient instead to measure the difference of the frequency (ν_s) of the peak from some internal standard (both measured in Hz) and to divide this by the operating frequency in MHz to obtain a field-independent number in a convenient range. Because the spread of frequencies is caused by the different chemical (and hence magnetic) environments, the signals are described as having a *chemical shift* from some standard frequency. The internal standard used for ^{1}H and ^{13}C spectra is tetramethylsilane (TMS), and the chemical shift scale δ is then defined by Eq. (4.4).

$$\delta = \frac{\nu_s(\text{Hz}) - \nu_{\text{TMS}}(\text{Hz})}{\text{operating frequency}(\text{MHz})} \tag{4.4}$$

The chemical shift δ, which measures the position of the signal, will now be the same whatever instrument it is measured on, whether operating at 200 MHz, 400 MHz or 600 MHz. It has no units and is expressed as fractions of the applied field in parts per million (p.p.m.). In a 400 MHz spectrometer, each unit of δ would be 400 Hz on the proton scale and 100 Hz on the carbon scale. Tetramethylsilane is chosen as the reference point because it is inert, volatile, non-toxic, and cheap, and it has only one signal (^{13}C or ^{1}H), which comes into resonance conveniently at the low-frequency extreme of the frequencies found for most carbon and hydrogen atoms in organic structures. By definition it has a δ value of 0 in ^{13}C and ^{1}H spectra.

The scale on Fig. 4.11 shows the common range of δ values within which most ^{13}C nuclei and most protons come into resonance. They are written from right to left, a convention that denotes frequencies as higher and positive on the left and lower on the right. Unfortunately, the left and right sides are hardly ever referred to as being at the high- and low-frequency end of the spectrum. Instead, for historical reasons, they are almost invariably referred to by the value of the applied field which corresponds to this relative fre-

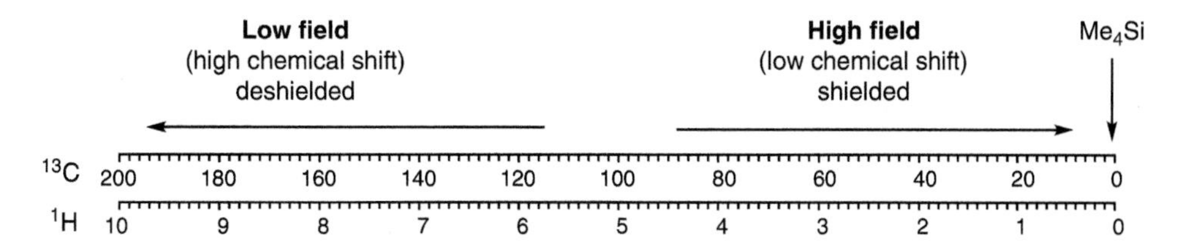

Fig. 4.11 The δ scales for ^{13}C and ^{1}H NMR spectra

quency. The high-frequency end of the spectrum, on the left with high δ values, is described as being *downfield*, and the right side, with low δ values, is said to be *upfield*. It is essential that you get used to this convention.

Most instruments only resolve signals when they are 0.5 Hz or more apart. In practice, ^{13}C signals, spread out over 20,000 Hz (in a 100 MHz instrument), rarely coincide, but ^{1}H signals, spread over a narrower range of 4000 Hz (in a 400 MHz instrument), do quite often give coincident, or near coincident signals from different protons. Resolution gets better with instruments working at higher field, but coincidence, or at any rate overlap, is still common.

Figure 4.12 shows the ^{13}C and ^{1}H NMR spectra of 3,5-dimethylbenzyl methoxyacetate **1**, a compound which gives rise to signals over most of the usual chemical shift range, free of the complication of coupling, which we shall come to later. In the ^{13}C NMR spectrum the nine different kinds of carbon atoms are fully resolved, each upward line corresponding to a carbon atom. Notice that C-2 and C-6 are in identical environments, because of free rotation about the single bond that joins the aromatic ring to the side-chain. Similarly, C-3 and C-5 are in identical environments, and so are the pair of methyl groups on C-3 and C-5. On the other hand, the ^{1}H NMR spectrum is not fully resolved, because the signal at δ 6.96 from the hydrogen atom on C-4 can only just be seen as a shoulder on the side of the signal from the two identical hydrogen atoms on C-2 and C-6 at δ 6.97. H-4 is in a

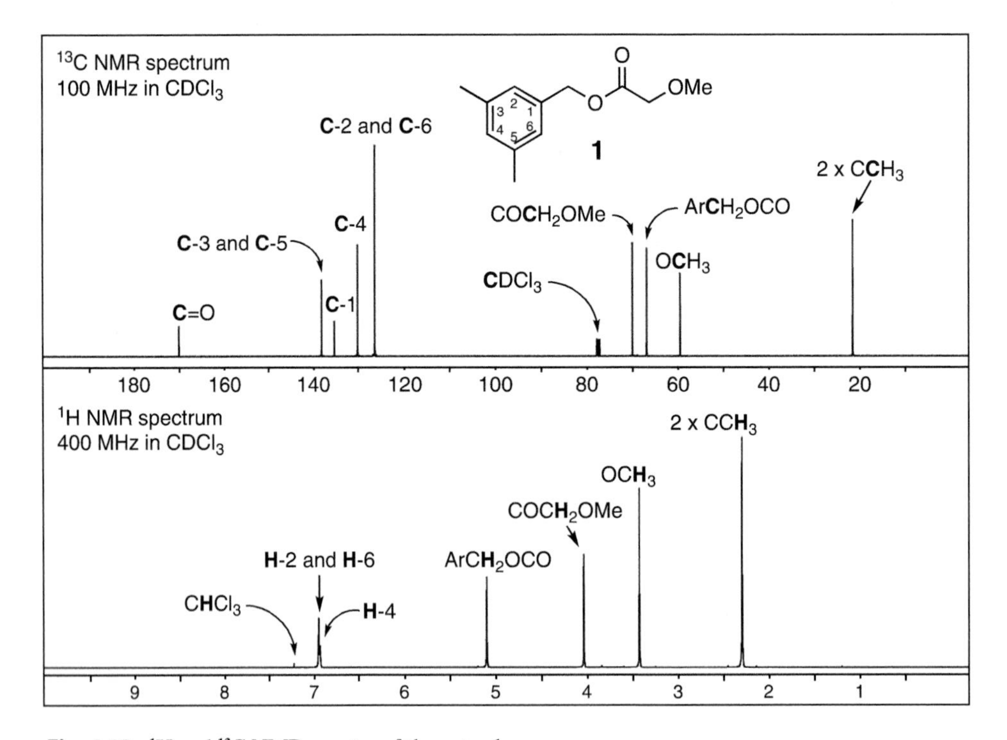

Fig. 4.12 ^{1}H and ^{13}C NMR spectra of the ester **1**

different chemical environment from H-2 and H-6, but the environments are so similar that their chemical shifts are barely resolved. Each of the remaining lines corresponds to one of the five significantly different kinds of hydrogen atoms.

The solvent in this case is deuterochloroform ($CDCl_3$). In the ^{1}H NMR spectrum, there is a small residual signal at δ 7.25 from the presence of incompletely deuterated chloroform. In the ^{13}C NMR spectrum, on the other hand, the deuterated chloroform is present in large molar excess, it has just as high a proportion of ^{13}C atoms as the ester **1**, and so the solvent signal is relatively intense, although a lot less intense than it is for solutions that are not as concentrated as this one was.

The absorption of a signal in a ^{1}H NMR spectrum is generally proportional to the number of protons coming into resonance at the frequency of that signal, because the FID has normally decayed completely before the next pulse is applied (Fig. 4.5). The area under the peaks is therefore proportional to the number of protons being detected. This is illustrated in Fig. 4.13, which shows two ways in which the area under the peak can be presented.

In the older method, the instrument has plotted an integration trace, starting as a horizontal line and rising from left to right as it passes each absorption. The extent of the rise is proportional to the area under the peak. Measured with a ruler on the trace, and normalised to add up to 16, the numbers, reading from left to right, are 2.91, 2.01, 2.01, 3.00 and 6.07, which shows that the number of hydrogens giving rise to each peak must be, since they can only be integers, 3, 2, 2, 3 and 6, respectively. More usually, the integration is presented as a number under the peak, with vertical lines indicating the limits of the integral. The actual numbers produced directly by the instrument, based on the assumption that there are three protons in the methoxy signal, are 2.98, 2.05, 2.03, 3.00 and 6.19, which gives some idea of how accurate a routine integration of a reasonably pure sample can be.

However, integrations are not always as good as this. The digitisation, for a start, inherently creates an imperfectly defined curve for each peak. The very act of apodization,

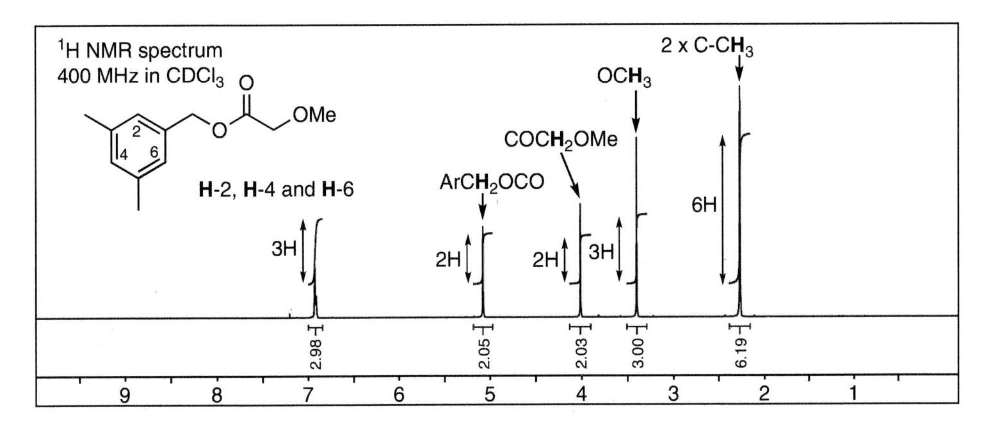

Fig. 4.13 Integration traces

described earlier, introduces errors in the integration, because a signal which decays fast, but suffers from apodization that limits the contribution from the early part of the FID, will be reported with a lower intensity than a more slowly decaying signal. The other parameters chosen during the taking and processing of the spectrum, the compromises made in an effort to get a linear baseline, or the best resolution or the best signal-to-noise ratio, also have differential effects on the integration of the signal from each proton. You cannot trust the integrations reported from spectra, but you can use them as a helpful guide, with the example above as good as it gets.

Since all the nuclei must have relaxed to their equilibrium distribution between successive pulses for integration to be reliable, integration is not normally useful in ^{13}C spectra, because ^{13}C nuclei do not completely relax within the time period used for collecting the FID. Figure 4.12 shows that the peaks in the ^{13}C spectrum are not proportional in intensity to the number of carbon atoms contributing to each signal. For relaxation to occur, the precessing magnetisation of Fig. 4.4 must interact with local fluctuations in the magnetic fields in the molecule, especially those caused by molecular tumbling near the other nuclear magnets. For this reason, the relaxation rate of carbon atoms directly bonded to hydrogen atoms is higher than for carbon atoms not so bonded. A second contribution to the intensity of the carbon resonances is from the nuclear Overhauser effect (NOE), which is discussed in Sect. 4.13. Those carbons carrying hydrogen atoms are not only enhanced in intensity by the fluctuating magnetic field helping them to relax, but also substantially by a relatively large nuclear Overhauser effect from the closeness in space of the directly bonded hydrogen atoms. The sum of these two effects can be seen in Fig. 4.12, where the two lowest-intensity peaks, at δ 170.1 and 135.3, correspond to the carbonyl carbon and C-1, neither of which has a hydrogen atom bonded to it to speed up the relaxation. The other fully substituted carbon atoms, the pair C-3 and C-5, give rise to a signal at δ 138.1 that is approximately twice as strong as the signal from C-1. It is not unusual to see that a pair of identical carbons, C-3 and C-5, gives rise to a signal twice as intense as an otherwise similar carbon—two carbons in a similar situation will have similar relaxation rates, and hence will give signals more or less in proportion to their abundance. But the signal from C-3 and C-5 at δ 138.1 is not twice as strong as the signal from C-4 at δ 130.1. On the other hand, the signal from C-2 and C-6 at δ 126.3 is about twice as intense as the signal from C-4 at δ 130.1. Because of this variability, integration is not used in routine ^{13}C NMR spectra, except in this limited kind of way, and it is not uncommon to find that some peak intensities are so low that they are barely discernible in the spectrum.

The ^{13}C signal from deuterochloroform in Fig. 4.12 is not nearly as intense as one might expect, given that the solvent is present in large excess. The reasons for this are: (1) that deuterium, with a low magnetogyric ratio, is much less effective than hydrogen at relaxing the ^{13}C signal; (2) there is no NOE from attached hydrogen atoms; and (3) the molecules of chloroform, being small, tumble with a frequency above that of the frequency of the NMR spectrometer. It is possible to increase the intensity of weak or absent signals by adding a paramagnetic salt such as $Cr(acac)_3$, which brings with it a powerful magnetic field to speed up the relaxation, but unfortunately it also broadens all the signals.

4.4 Factors Affecting the Chemical Shift

4.4.1 The Inductive Effect

In a uniform magnetic field, the electrons, which have a high magnetogyric ratio, surrounding a nucleus circulate, setting up a secondary magnetic field opposed to the applied field at the nucleus (Fig. 4.14, where the solid curves indicate the lines of force associated with the induced field and the dashed circle the circulating electrons). As a result, nuclei in a region of high electron population experience a field proportionately weaker than those in a region of low electron population. As a result, the levels N_α and N_β are not split as far apart, and the frequency of the transition between them is reduced. Equally one can say that a higher field has to be applied to bring them into resonance. However it is explained, whether by referring to the frequency or the field, such nuclei are said to be *shielded* by the electrons.

Thus, a high electron population shields a nucleus and causes resonance to occur at relatively high field (i.e. at low frequency, with low values of δ). Likewise, a low electron population causes resonance to occur at relatively low field (i.e. at high frequency, with high values of δ), and the nucleus is said to be *deshielded*. The extent of the effect can be seen in the positions of resonance of the ^1H and ^{13}C nuclei of the methyl group attached to the various atoms listed in Table 4.2. The electropositive elements (Li, Si) shift the signals upfield, and the electronegative elements (N, O, F) shift the signals downfield, because they supply and withdraw electrons, respectively.

Hydrogen is more electropositive than carbon, with the result that every replacement of hydrogen by an alkyl group causes a downfield shift in the resonance of that carbon atom and any remaining hydrogens attached to it. Thus, methyl, methylene, methine, and fully substituted carbons (and their attached protons) come into resonance at successively lower fields (**2**–**6**).

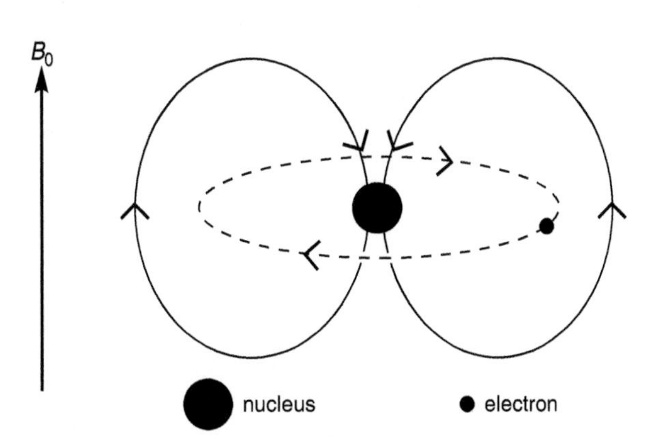

Fig. 4.14 Shielding of a nucleus by the current induced in an electron cloud

Table 4.2 Chemical shifts for methyl groups attached to various atoms in CH_3X

CH_3X	δ_C	δ_H	CH_3X	δ_C	δ_H
CH_3Li	−14.0	−1.94	CH_3OH	50.2	3.39
$(CH_3)_4Si$	0.0	0.0	$(CH_3)_2S$	19.3	2.09
CH_4	−2.3	0.23	$(CH_3)_2Se$	6.0	2.00
CH_3Me	8.4	0.86	CH_3F	75.2	4.27
CH_3Et	15.4	0.91	CH_3Cl	24.9	3.06
CH_3NH_2	26.9	2.47	CH_3Br	10.0	2.69
$(CH_3)_3P$	16.2	1.43	CH_3I	−20.5	2.13

δ_C −2.3	δ_C 8.4	δ_C 15.9	δ_C 25.0	δ_C 27.7
CH_4	$MeCH_3$	Me_2CH_2	Me_3CH	Me_4C
δ_H 0.23	δ_H 0.86	δ_H 1.33	δ_H 1.68	
2	**3**	**4**	**5**	**6**

More dramatically, every replacement of hydrogen by an electronegative element causes a larger downfield shift, more or less additive, in the resonance of the carbon atom and any remaining hydrogens attached to it. Thus, the carbon and the attached protons of methyl chloride, dichloromethane, chloroform, and carbon tetrachloride come into resonance at successively lower fields (**7**–**10**).

δ_C −2.3	δ_C 24.9	δ_C 54.0	δ_C 77.2	δ_C 96.1
CH_4	CH_3Cl	CH_2Cl_2	$CHCl_3$	CCl_4
δ_H 0.23	δ_H 3.06	δ_H 5.33	δ_H 7.24	
2	**7**	**8**	**9**	**10**

4.4.2 Anisotropy of Chemical Bonds

Chemical bonds are regions of high electron population that can set up magnetic fields. These fields are stronger in one direction than another (they are anisotropic), and the effect of the field on the chemical shift of nearby nuclei is dependent upon the orientation of the nucleus in question with respect to the bond. The anisotropy in π-bonds is especially effective in influencing the chemical shift of nearby protons, as illustrated in Fig. 4.15.

When a double bond (Fig. 4.15a) is oriented at right angles to the applied field B_0, the electrons in it are induced to circulate in the plane of the double bond (dashed ellipse in Fig. 4.15a), creating a magnetic field opposed to the applied field at the centre and augmenting it at the periphery. The effect of the induced field is to shift signals from any hydrogens in the augmented region downfield, identified with minus signs (−) in Fig. 4.15 to symbolise deshielding.

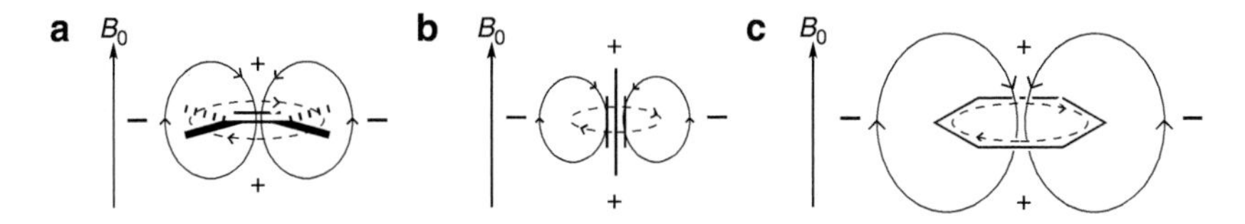

Fig. 4.15 Anisotropic fields created in π conjugated systems

Table 4.3 Chemical shifts for ^{13}C and ^{1}H in, on and near multiple bonds

Compound	δ_C	δ_H	Compound	δ_C	δ_H
$\mathbf{CH_3}$H	−2.3	0.23	$\mathbf{CH_3}$CHO	31.2	2.20
$\mathbf{CH_3}$CH=CH$_2$	22.4	1.71	$\mathbf{CH_3}$COMe	28.1	2.09
$\mathbf{CH_3}$C≡CH	3.7	1.80	$\mathbf{CH_3}$CN	1.30	1.98
$\mathbf{CH_2}$=CH$_2$	123.3	5.25	Me**CHO**	199.7	9.80
HC≡CMe	66.9	1.80	Me$_2$**CO**	206.0	–
Me**C**≡CMe	79.2	1.75	Me**C**N	117.7	–

Thus, olefinic hydrogen atoms are shifted downfield relative to their saturated counterparts. But the observed downfield shift is large (Table 4.3, column 3), and only partly because of the anisotropy effect; it is also caused by the fact that trigonal carbons are more electronegative (because of their higher s-character) than are tetrahedral carbon atoms. The anisotropic effect also shifts the signals from allylic carbons and hydrogens downfield, but to a lesser extent. A carbonyl group has a similar large effect on the hydrogen bonded to it in aldehydes, where the anisotropy is augmented by the inductive effect of the electronegative oxygen atom. The carbonyl group has a smaller effect upon the carbon atoms attached to it and on the protons attached to those carbon atoms (Table 4.3, columns 5 and 6). Olefinic and carbonyl carbons suffer large downfield shifts relative to their saturated equivalents, although simple models that rationalise these shifts are less easily conveyed. Nevertheless, it is clear that carbonyl carbons and their directly attached protons—which suffer not only the influences experienced by olefinic carbons and their directly attached protons, but also experience the electronegativity of the oxygen atom—exhibit remarkably large chemical shifts (Table 4.3, entry 10). Triple bonds are noticeably different: they experience an anisotropy set up by electrons circulating around the triple bond with cylindrical symmetry (Fig. 4.15b), which has the opposite effect from that produced by the π-bond of an alkene or a carbonyl group. The net effect on the chemical shift of acetylenic protons and carbons is for them to come into resonance in between the protons and carbons of alkenes and alkanes (Table 4.3, entries 3, 5 and 6).

An even larger anisotropic effect is produced by the π-system of aromatic rings. The circulating electrons are now called a *ring current*, and they create a relatively strong magnetic field (Fig. 4.15c). The effect of the induced field is to deshield substantially the hydrogens attached to the aromatic ring **12**, which generally come into resonance 1.5–2 p.p.m. downfield from the corresponding olefinic signals in ethylene **11**. The ^{13}C signal for benzene is similarly shifted downfield (by 5.2 p.p.m.), but the effect is much less noticeable because of the relatively large chemical shifts in ^{13}C spectra.

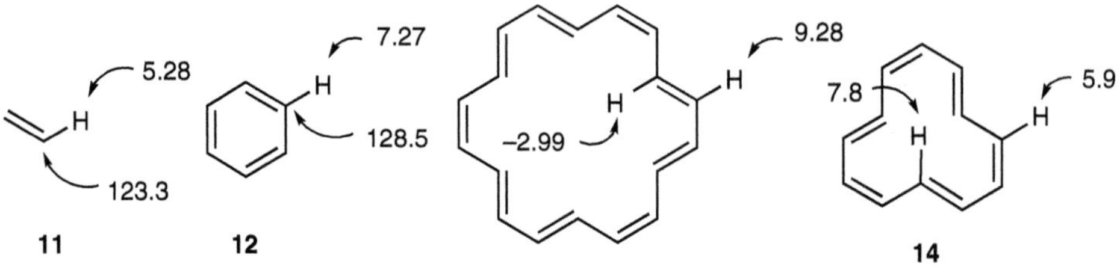

5.28 H 123.3 **11**

7.27 H 128.5 **12**

H −2.99 H 9.28 **13**

7.8 H H 5.9 **14**

In the special case of macrocyclic aromatic rings like [18]-annulene **13**, there are 'inside' as well as 'outside' hydrogens. The 'inside' hydrogens experience a weaker field than the applied field (the + signs in Fig. 4.15c signifying shielding) and come into resonance at a conspicuously high field, while the 'outside' hydrogens experience a stronger field and come into resonance at the low-field end of the aromatic region. Cyclic conjugated systems with 4n electrons are much less common, because they are usually unstable. They are called antiaromatic, and when they can be isolated, as [12]-annulene **14** can (at −170°), their ring current circulates in the opposite direction, leading to upfield shifts in the outside protons and downfield shifts in inside protons. The magnetic field induced in double bonds, and especially in aromatic rings, can have a profound effect on the chemical shift of nuclei held anywhere in their neighbourhood, not just those in and attached directly to them. The precise orientation can make a large difference, so that the NMR spectra of complex structures with aromatic rings in them can be hard to predict.

Cyclopropanes present a special case, because both the carbon and proton resonances appear at exceptionally high field, above the usual range for methylene groups, and even above the range for most methyl groups. One explanation is that it is a consequence of a ring current: in cyclopropane itself **15** the three *cis* vicinal C–H bonds are able to conjugate with each other, just as p-orbitals conjugate with each other. The cyclic six-electron conjugated system gives rise to a ring current, and both the carbons and the protons sit in the shielding region of the magnetic field induced by that ring current. Substitution by alkyl groups and by electronegative elements moves the resonances downfield in the usual way. The antiaromatic conjugation of four C–H bonds in cyclobutane **16** is much less effective, because of puckering in the ring, but the effect is discernibly in the opposite direction—cyclobutane protons come into resonance at slightly lower field than comparable methylene protons, as in cyclopentane **17**, which has chemical shifts similar to those of open-chain compounds. Cyclopentane is not flat but is constantly changing its conformation. All the protons and all the carbons during the time of the FID see an average of all the environments that they pass through, and all the protons (and all the carbons) come into resonance at the same place.

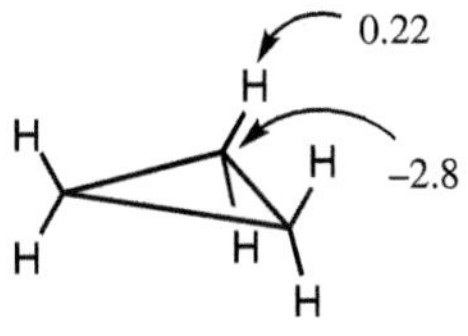

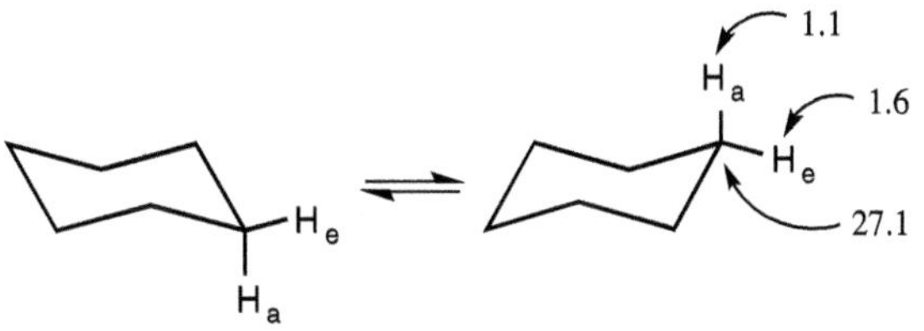

Similarly, cyclohexane **18** is not flat, but is constantly flipping between chair conformations that exchange the positions of the axial and equatorial protons. At room temperature the protons effectively experience the average field of the two environments, and come into resonance at δ 1.44 as though it were a flat molecule. At −100°, however, the flipping is slowed down so that the difference between the axial and the equatorial environments is revealed, with the axial protons coming into resonance upfield (δ 1.1) of the equatorial (δ 1.6). The upfield shift of the axial protons may reflect the presence of a weak ring current from the three axial C–H bonds, effectively placing the axial protons 'inside' and the equatorial protons 'outside'. Rigid cyclohexanes, where the axial and equatorial protons retain their distinction, show the same trend, with axial protons apt to be upfield of equatorial protons, but other influences can come into play.

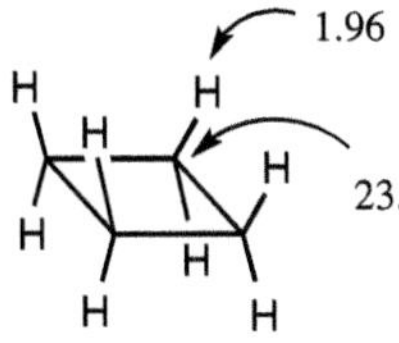

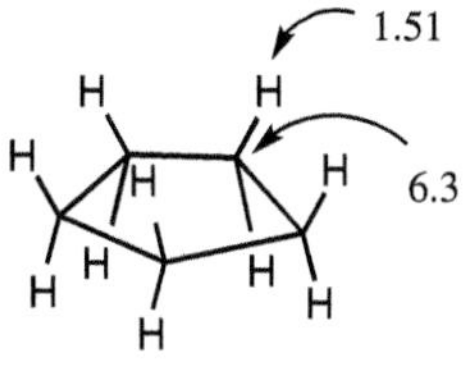

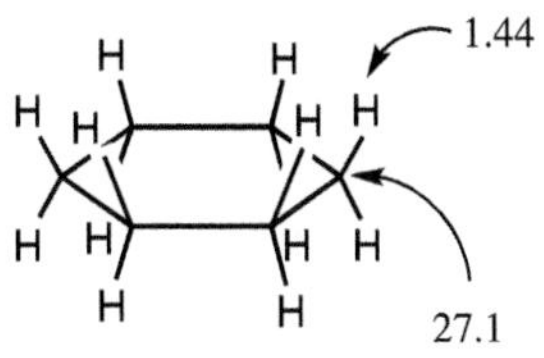

4.4.3 Polar Effects of Conjugation

When a double bond carries a polar group, the electron distribution is displaced. The displacement is usually understood as a combination of inductive effects, which operate in the σ-framework (and simply fall off with distance) and conjugative effects, which operate in the π-system (and alternate along a conjugated chain). The effects in the π-system can be illustrated simplistically with curly arrows on the canonical structures for methyl vinyl ether **19** and methyl vinyl ketone **20**. The curly arrows illustrate the effect on a C=C double bond with a π-donor **19** and a π-acceptor group **20**, and molecular orbital calculations of the electron distribution in π-systems support this simple picture.

Table 4.4 Conjugative effects on the chemical shifts of substituted alkenes

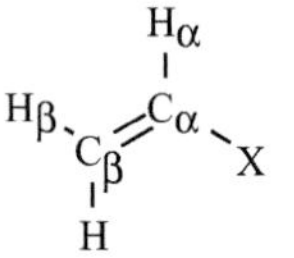

X	Electronic nature	$\delta_{C\beta}$	$\delta_{C\alpha}$	$\delta_{H\beta}$	$\delta_{H\alpha}$
H	Reference compound	123.3	123.3	5.28	5.28
Me	Weak π- and σ-donor	115.4	133.9	4.88	5.73
OMe	π-Donor, σ-acceptor	84.4	152.7	3.85	6.38
Cl	σ-Acceptor, weak π-donor	117.2	125.9	5.02	5.94
Li	π-Acceptor, σ-donor	132.5	183.4		
$SiMe_3$	π-Acceptor, σ-donor	129.6	138.7	5.87	6.12
$CH{=}CH_2$	Simple conjugation	130.3	136.9	5.06	6.27
COMe	π-Acceptor, weak σ-acceptor	129.1	138.3	6.40	5.85

19 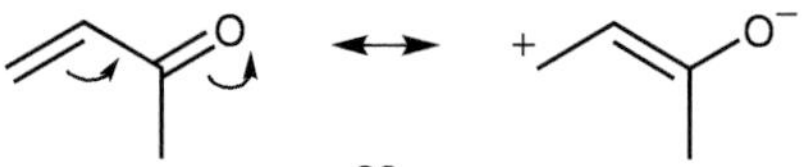 **20**

These displacements of electron population naturally affect the position of resonance of nearby nuclei, as shown in Table 4.4. In general, π-donor groups (Me < MeO < Me_2N) on π-systems shield the β-nuclei, as implied by the canonical structure on the right for **19**, causing an upfield shift relative to their position in ethylene. Largely because of the inductive effect, electronegative elements simultaneously induce a downfield shift of the α nuclei. The effects of π-acceptor groups (Li, $SiMe_3$ and COMe) on π-systems are not so easily explained using inductive effects and resonance structures like **20**. Because of the way the substituent interacts with the π-system, affecting the anisotropic field surrounding the π-bond, electropositive substituents give rise to downfield shifts on trigonal carbons, both α and β, in contrast to the upfield shifts they induce on tetrahedral carbons.

When the β carbon is unsubstituted, there are two β-hydrogen atoms, and they can experience different fields (the data in Table 4.4 are for the hydrogen *trans* to the substituent). A donor substituent, as in ethyl vinyl ether **22**, causes the β-hydrogen atom *trans* to the substituent to move further upfield, relative to ethylene **21**, than the hydrogen atom *cis* to it. A carbonyl group, as in methyl vinyl ketone **23**, causes the β-hydrogen *cis* to the carbonyl group to be shifted further downfield than the hydrogen *trans* to it, probably because of a direct contribution through space from the anisotropic field induced in the carbonyl group.

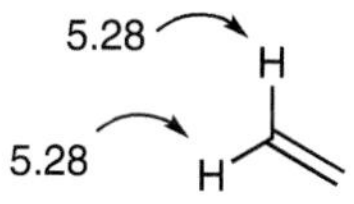

Polar groups attached directly to a benzene ring cause upfield and downfield shifts more or less in the same way as they do on a simple double bond. The effects of a σ-withdrawing and π-donor are seen in anisole **25**, where the signals of the *ortho* and *para* carbons and hydrogens are shifted upfield by the methoxy group, relative to the signals of benzene **24**, and the *ipso* carbon is shifted far downfield. The effect of the π-acceptor group in nitrobenzene **26** is less straightforward, just as the electron-withdrawing groups are on an alkene: the *ortho* hydrogen and the *para* carbon and hydrogen are shifted downfield, as one might expect, but the *ortho* carbon is shifted upfield. The downfield shift of the *ortho* hydrogen is probably greater than that on the *para* hydrogen because of a through-space effect, analogous to that of the acetyl group on the *cis* hydrogen in methyl vinyl ketone **23**.

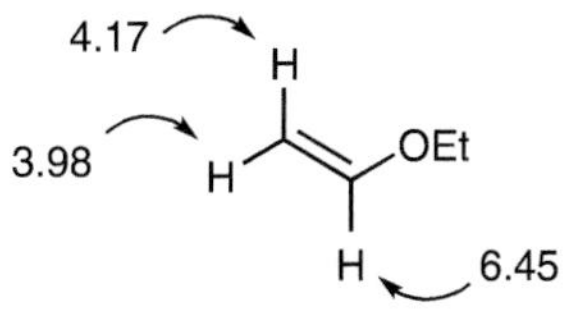

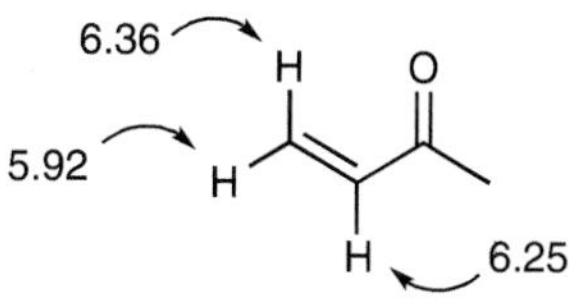

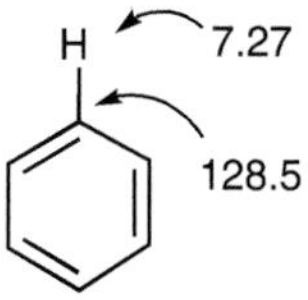

25 (OMe)

	δ_C	δ_H
i	158.7	
o	113.8	6.81
m	129.4	7.17
p	120.4	6.86

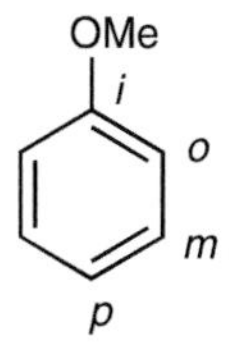

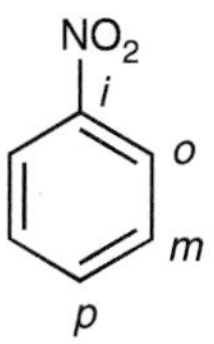

26 (NO_2)

	δ_C	δ_H
i	148.1	
o	123.2	8.21
m	129.3	7.45
p	134.5	7.66

Very approximately, a change in substitution pattern in any organic structure has a similar effect on both the carbon and the proton spectra: the δ value of the carbon signal is about 20 times the δ value of the proton signal. However, there are many large deviations from this general picture, including the opposite effect of the nitro group on the *ortho* carbons and hydrogens of nitrobenzene **26**, and the big effect of the aromatic ring current on protons compared with the appearance of a small effect on the carbon atoms.

The overall pattern that has emerged from the discussion so far is summarised in Fig. 4.16, which gives the approximate ^{1}H and ^{13}C chemical shifts in and adjacent to many of the common functional groups.

4.4.4 Van der Waals Forces

When a substituent, especially an electronegative element, is pressed close to a hydrogen atom (closer than the sum of their Van der Waals radii), the electron population around the proton is pushed away (it is a form of weak hydrogen bonding). The proton is overall

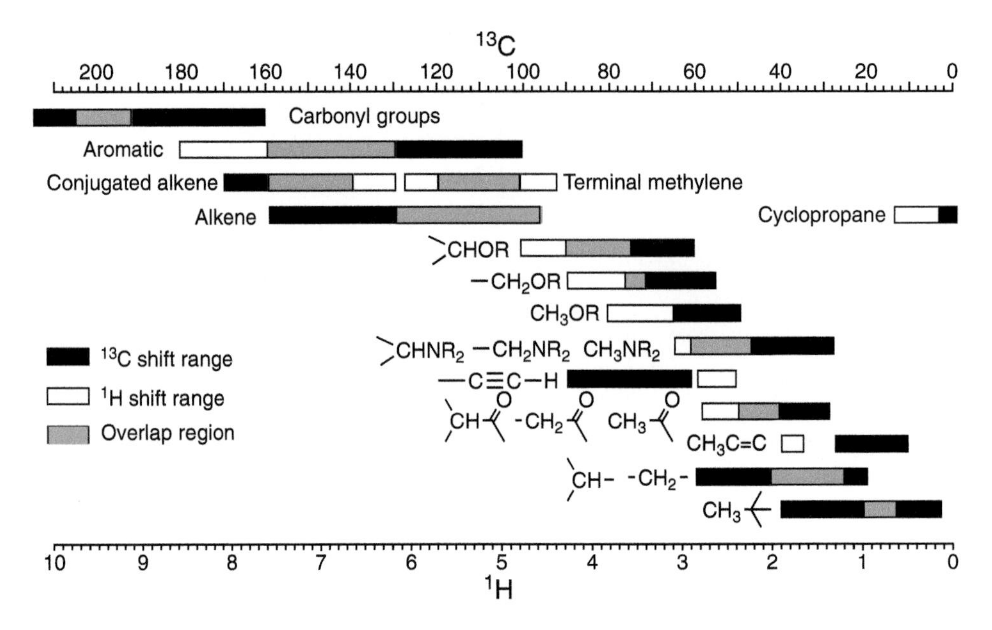

Fig. 4.16 Approximate chemical shift ranges

deshielded, and comes into resonance at unusually low field, as in the ^{1}H NMR spectrum of 2-adamantanol **27**, where the signal from the two protons H_b appears downfield from the broad, largely unresolved signal (δ 1.89–1.69) from all the other CHs (the 'methylene envelope'), except, of course, for H_c, which is bonded to a carbon carrying an electronegative element. The proximity-induced displacement of the electrons has the opposite effect on a second proton if there is one bonded to the same carbon atom. This proton is shielded, and as a result the signal from the two protons H_a in adamantanol appears upfield of all the other CHs.

δ2.06 H_b H_b OH H_c δ3.75 H_a δ1.53 H_a

27

4.4.5 Isotope Effects

Replacing a lighter isotope by a heavier one causes an upfield shift in the signals from nearby atoms. The effect is most easily seen when a proton is replaced by a deuterium

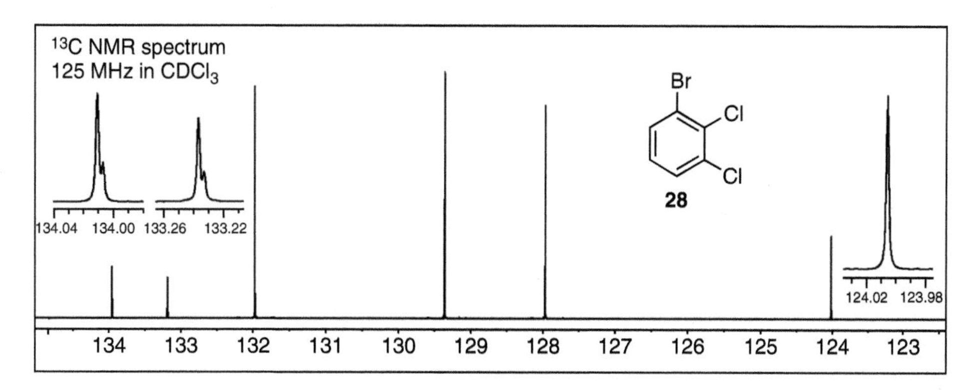

Fig. 4.17 ^{13}C NMR spectrum of 2,3-dichlorobromobenzene

atom: the ^{13}C chemical shift to which the deuterium is attached is shifted upfield typically by 10–30 Hz, and the next carbon in the chain is shifted upfield by 0–5 Hz. These shifts are especially useful in mechanistic and biosynthetic studies, where the carbon atoms attached to deuteriums, and one atom away from a deuterium, are easily picked out by the differences in their chemical shifts.

Recently it has become possible to detect much more subtle isotope effects. Figure 4.17 shows the ^{13}C NMR spectrum of 2,3-dichlorobromobenzene **28**. Both carbon atoms carrying the chlorine atoms show, in the expansions, two signals, separated by 0.4 Hz, in a 3:1 ratio, corresponding to the natural abundance of the ^{35}Cl and ^{37}Cl isotopes. In this case the signal at highest field from the carbons attached to ^{79}Br and to ^{81}Br is not resolved, but it has been (just) in some published cases.

4.4.6 Estimating a Chemical Shift

The inductive, conjugative and anisotropic effects in polyfunctional molecules are more or less additive, so that we can, for example, see that the proton signals in the ^{1}H spectrum of Fig. 4.12 are in appropriate places. The signal from the C-3 and C-5 methyl groups (at δ 2.31) appears downfield from that of the methyl group in propene (δ 1.71, Table 4.3), because the methyl groups are adjacent to a benzene ring and suffer some of the effect of the ring current, which is typically responsible for a downfield shift of about 0.6 p.p.m. The methylene group attached to C-1 of the benzene ring simultaneously suffers the effects of being adjacent to the benzene ring and of being adjacent to an electronegative element. We can expect it to be downfield by about 1.4 p.p.m. for the effect of being benzylic and about 2.8 p.p.m. for being next to an ester oxygen (see the tables at the end of this chapter for a tabulation of the effects of having various functional groups attached to methyl, methylene and methine carbons), making a total of 4.2 p.p.m. downfield from the position of a simple methylene group (δ 1.33, **4**). Thus, we can expect it to give a signal at δ 5.5, and it actually comes into resonance at δ 5.11. Similarly, using the tables at the end of this chapter, the methylene group between the carbonyl group and the methoxy group might be expected to be shifted downfield by the carbonyl group by 0.9 p.p.m. and by the methoxy group by 2.1 p.p.m., giving an estimate of δ 4.3, close to the observed value of δ 4.05. The methoxy group (at δ 3.45) is a little downfield of the position for methanol (at δ 3.39), principally because of the anisotropic effect of the nearby carbonyl group. Finally, the aromatic protons at δ 6.97 are a little upfield of the signal from benzene (δ 7.27), because they are *ortho* to two alkyl groups and *para* to another. Alkyl groups, which are mildly electron-donating, shift an *ortho* or *para* proton upfield by about 0.2 p.p.m.

But these are very crude estimates, and there are several sets of better empirical rules with which to estimate the chemical shifts commonly encountered for ^{13}C and ^{1}H nuclei. One such set of rules, and the tables of data needed to apply them, are grouped together at the end of this chapter, where they can easily be found when you use this book as a handbook in the laboratory. Using them, you can estimate the chemical shift of the different kinds of carbon atoms in simple aliphatic compounds (Eq. (4.19) and its accompanying tables), in simple alkenes (Eq. (4.20)), and in substituted benzene rings (Eq. (4.21)), and of the carbon atoms in the various kinds of carbonyl groups. There are similar rules and tables for estimating the proton chemical shifts of substituted alkanes (Eq. (4.23)), substituted alkenes (Eq. (4.24)), and substituted benzene rings (Eq. (4.25)), although estimates for protons are rarely as good as those for ^{13}C nuclei.

It is better still, and easier if you have access to the NMR processing programs like Topspin or Mestrenova, to draw the chemical structure, and have the program tell you what the probable chemical shifts will be. These programs use similar equations and empirical reference data to those given at the end of this chapter, and save you from doing the sums. The ^{1}H and ^{13}C chemical shifts for the ester **1**, measured by the spectrometer and estimated by the program ChemNMR incorporated into ChemDraw®, are shown in Fig. 4.18. This relatively simple compound is unlikely to have unexpected long-range effects, and as a result the estimates are as good as these programs get: the Mean Absolute Deviation (MAD) is 0.16 p.p.m. for protons and 0.66 p.p.m. for carbon, and the largest discrepancy is 0.34 p.p.m. for protons and 2.20 p.p.m. for carbon.

In general, ChemNMR gives ^{13}C chemical shifts with a standard deviation of 2.8 p.p.m. for 95% of compounds, and ^{1}H chemical shifts, rather less reliably, with a standard devia-

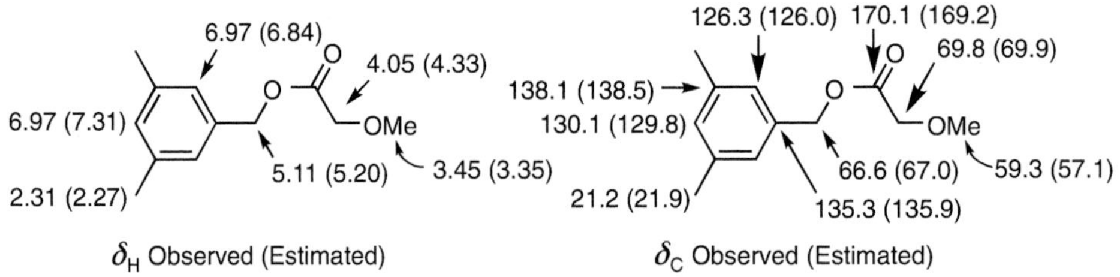

Fig. 4.18 Observed and estimated chemical shifts

tion of 0.3 p.p.m. for 90% of compounds. The conformation of one molecule may not be the same as the conformation of the several models on which the rules are based; the anisotropy of the field then causes the local field in the compound under investigation to differ from that of the model. The effects of distant groups are not included in the rules, usually because they are relatively unimportant, but, in some molecules, a distant group may fold back into the region of the nucleus under investigation and shift its resonance dramatically. This is especially the case when aromatic rings, with their powerful ring currents, are present.

Here is an example where estimating the chemical shift made it possible to correct a structure which had been wrongly assigned to a natural product. Aquatolide, formed by irradiation of asteriscunolide C **29**, had been assigned the structure **30**. An estimate of the ^{13}C chemical shifts gave a MAD of 7.23 p.p.m. with the largest deviation (bold in **30**) of 24.33 p.p.m. These numbers were well outside the normal range, and indicated that the structure was most probably wrong.

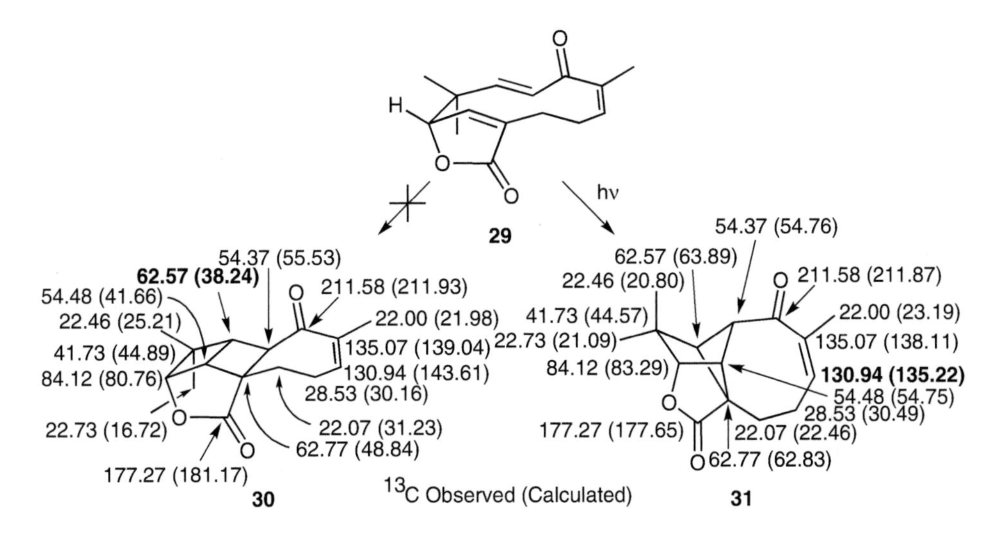

The revised structure **31**, in which the [2 + 2] cycloaddition has the opposite orientation, gave much better estimates: MAD 1.37 p.p.m. with the largest deviation (bold in **31**) of 4.28 p.p.m. The estimates in this work were derived from a high level quantum computation, and not from a program based on empirical data—none of them would have been good enough in this case.

4.4.7 Hydrogen Bonds

A hydrogen atom involved in hydrogen bonding is sharing its electrons with two electronegative elements. As a result, it is itself deshielded, and comes into resonance at low field. In water, in a very dilute solution in $CDCl_3$, with hydrogen bonding essentially absent, the protons come into resonance at δ ~1.5. The ^{1}H spectrum in Fig. 4.12 actually shows this signal at δ 1.66 when it is expanded 10×. This signal can often be seen in FT spectra taken in more dilute solution. In droplets of water, on the other hand, suspended in $CDCl_3$, the molecules are intermolecularly hydrogen bonded with each other, and they come into resonance at δ ~4.8.

The position of resonance of the OH and NH protons of alcohols and amines is unpredictable, because the extent to which the hydrogen atoms are involved in hydrogen bonding is both unpredictable and concentration dependent. The usual range is δ 0.5–4.5 for alcohols, δ 1.0–4.0 for thiols, and δ 1.0–5.0 for amines. The much stronger intermolecular hydrogen bonding in carboxylic acid dimers **32** leads to very low-field absorption in the δ 9–15 range, and the corresponding intramolecular hydrogen bonding of enolised β-diketones **33** and *o*-hydroxycarbonyl compounds **34** is similar (δ 15.4 for acetylacetone **33** itself). These are out of the range of the usual ^{1}H NMR spectral width, and may have to be looked for specially.

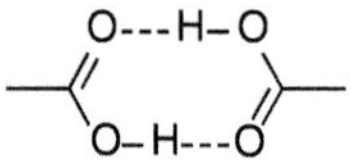

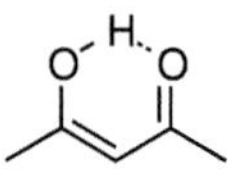

Fortunately, it is easy to identify the signal from a hydrogen atom bonded to an electronegative element, in spite of the uncertainty about where it will appear in a ^{1}H NMR spectrum: if the sample in $CDCl_3$ solution is shaken with a drop of D_2O the OH, NH, and SH hydrogens exchange rapidly with the deuterons, the HDO floats to the surface, out of the region examined by the spectrometer, and the signal of the OH, NH, or SH disappears from the spectrum (or, quite commonly, is replaced by a weak signal close to δ 4.8 from suspended droplets of HDO). This technique is known as a D_2O shake. It does not always work for amide NHs, which need some acid catalysis to exchange rapidly.

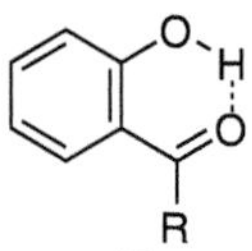

It is possible to determine the number of hydroxyl protons (and their multiplicity) in a molecule by taking the spectrum in d_6-DMSO under conditions that are mildly acidic ('pH' in the range 2–4). This solvent usually contains traces of water which resonate near δ 3.4. A normal spectrum is stored in the spectrometer's computer memory, and a second spectrum is similarly stored, but this time obtained with simultaneous irradiation of the water signal. In the latter spectrum, the OH proton signals have been removed by transfer of saturation from the water signal. The one spectrum can be subtracted from the other to give a difference spectrum, which contains only the signals from the water and the various OH groups.

4.4.8 Solvent Effects and Temperature

Chemical shifts are little affected by changing solvent from CCl_4 to $CDCl_3$ (±0.1 p.p.m.), but change to more polar solvents—such as acetone, methanol, or DMSO—does have a noticeable effect, ±0.3 p.p.m. for ^{13}C and for protons. Benzene can have an even larger effect, ±1 p.p.m. for ^{13}C and for protons, because it weakly solvates areas of low electron population; since the benzene has a powerful anisotropic magnetic field (Fig. 4.15), solute atoms lying to the side of or underneath the solvating benzene ring can experience significant shielding or deshielding relative to their position in an inert solvent like $CDCl_3$. Pyridine can be even more effective. This solvent-induced shift can be used to separate two signals which overlap, as we shall see in Sect. 4.10.2. More substantial shifts can be induced by complexation with paramagnetic salts, as discussed later in Sect. 4.10.3.

The resonance position of most signals is little affected by temperature. Hydrogen-bonded protons from OH, NH, and SH groups come into resonance at a higher field at higher temperatures, because the degree of hydrogen bonding is reduced. The major effect that temperature has is on signals from protons undergoing exchange from one environment to another during the time in which the FID is collected. If the exchange is rapid, the protons experience an average environment, and temperature has no effect. But if the exchange rate is slow, the proton signals are broadened and the degree of broadening is temperature dependent.

On the other hand, pH, or its equivalent in hydroxylic media, has a substantial effect when substituents like amino and carboxyl groups are present. The polar effects of an amino group with a lone pair of electrons at high pH and a protonated amino group at low pH are very different. Similarly, the polar effects of carboxyl groups (at low pH) and carboxylate ions (at high pH) are different. An α,β-unsaturated carboxylic acid will usually have the β-carbon downfield of the α-carbon in the ^{13}C spectrum (like the ketone in Table 4.4), but they are likely to be the other way round in the spectrum of the corresponding carboxylate ion.

The common solvents in NMR spectroscopy are used in deuterated form, in order not to introduce extra signals, but most of them have residual signals from incomplete deuteration. It is important to recognise these signals, listed in a table at the end of this chapter, in order to discount them from spectra that you are interpreting. The weak signal of $CHCl_3$ (close to δ 7.25 in ^{1}H spectra) is evident in many of the spectra taken in $CDCl_3$ used to illustrate this chapter. Fortunately, most of the common solvents introduce only one or two sharp signals, and they are easily recognised. The carbon of $CDCl_3$ is equally recognisable: it can also be seen in many of the spectra used to illustrate this chapter as the group of three lines centred at δ 77.3. The reason that this signal has three lines, and not one, takes us into the next section.

4.5 Spin-Spin Coupling to ^{13}C

We have left out of the discussion so far an important effect that neighbouring magnetic nuclei have on the signal we detect. If a nearby nucleus has a spin, that spin affects the magnetic environment of the nucleus we are observing, and the signal we detect is not a single peak, but a group of peaks, the complexity of which depends upon the nature and number of the nearby magnetically active nuclei.

A magnetically active nucleus experiences slightly different magnetic fields depending upon the spin state of neighbouring nuclei. If the neighbouring nucleus has a magnetic moment aligned with the applied field, the nucleus under investigation will experience a slightly enhanced magnetic field. If the former is opposed, the latter will experience a slightly reduced magnetic field. Since the difference in energy between the spin states is very small, there is an essentially equal probability that the magnetically active nucleus will be influenced by each of the spin states of its neighbour. The result is that the magnetically active nucleus comes into resonance at two slightly different frequencies with equal probabilities. The nucleus is said to be coupled to its neighbour, and the separation of the lines in Hz is called the coupling constant, J.

There are two pathways, both distance-dependent, by which the magnetic field from one nucleus affects the magnetic field of another close to it. One pathway is through space, and is called *dipolar coupling* when the neighbouring nucleus has a spin $I = 1/2$ (and quadrupolar coupling if the spin is greater than $I = 1/2$). This pathway has no effect on standard NMR spectra taken in solution. As the molecule tumbles in solution, the neighbouring nucleus takes up many orientations in space relative to the nucleus under investigation, and the effect of all the different magnetic fields it engenders is averaged out. This pathway comes into play only when the molecules are not tumbling isotropically, as in the solid state or in an oriented environment like a liquid crystal. The other pathway is relayed by the electrons in the bonds connecting the two nuclei. This pathway is called *scalar coupling* (or J coupling) and is not affected by the direct spatial arrangement of the nuclei. It is, however, affected by the orientation of the intervening orbitals, by the electron population in the intervening orbitals, and by the number of bonds.

The ester **1** was used specifically to postpone discussion of coupling until now, because it was designed so as not to exhibit it. Somewhat unconventionally, we shall begin by looking at coupling from ^{13}C to ^{1}H and ^{2}H. These couplings are not much used directly in structure determination, they are normally taken out of the picture, but when they are examined they tell a relatively simple story with which to introduce the subject.

4.5.1 ^{13}C–^{2}H Coupling

The carbon atom of $CDCl_3$ is attached to a deuterium. Deuterium has a spin $I = 1$, which means that there are three possible energy levels for a deuterium nucleus placed in a magnetic field. The carbon atom therefore experiences three slightly different magnetic fields depending upon the spin state of the deuterium nucleus to which it is attached. The result is that the carbon nucleus comes into resonance at three frequencies with equal probability, as we can almost see in the ^{13}C spectrum in Fig. 4.12, where the $CDCl_3$ signal is actually three equally spaced lines centred at δ 77.25. The carbon is said to be coupled to the deuterium, and the separation of the three lines in Hz is called the coupling constant, J. Because there is only one bond between the carbon and the deuterium, J is further qualified as $^{1}J_{CD}$. Carbon-deuterium coupling is much less important than carbon-hydrogen coupling, but we begin with it because it is visible in all ^{13}C spectra taken in $CDCl_3$. It is also visible in biosynthetic and other studies taking advantage of the isotope shift obtained by attaching a deuterium atom to a ^{13}C nucleus (Sect. 4.4.5). The signal from the ^{13}C atom attached to a deuterium atom is not only shifted upfield by a small but detectable amount but is also distinctively a 1:1:1 triplet.

4.5.2 ^{13}C–^{1}H Coupling

Why is there no comparable carbon-hydrogen coupling in the ^{13}C spectrum in Fig. 4.12? The answer is that, throughout the time in which that spectrum was being taken, the sample was irradiated with a strong signal (at 400 MHz) encompassing the whole range of frequencies within which the protons in the molecule came into resonance. This caused the N_α and N_β protons to be exchanging places rapidly several times during the measurement of the carbon signal. Each carbon atom, therefore, 'saw' only an average state for the protons near to it, just as it 'saw' an average field from the influences transmitted through space. Instead of being coupled, each ^{13}C atom simply gave rise to a single sharp line (and the spectrum, incidentally, was made more intense by the Overhauser effect, Sec 4.13). ^{13}C NMR spectra are usually taken in this way, and are described as *proton decoupled*. For this reason, we rarely encounter ^{13}C-^{1}H coupling, but we can get a massive amount of information from it if we care to look.

If we take the same spectrum without decoupling the protons, we see the spectrum in Fig. 4.19. Such spectra are actually weaker, because we no longer have the benefit of the Overhauser effect, and it is also much more complicated to analyse. Nevertheless, in this particular case, we can identify the signals from each carbon, and we can identify the multiplicity. The three carbon atoms to which no hydrogen is attached are still singlets, the two different kinds of carbon atom to which one hydrogen atom is attached are doublets, the two carbon atoms to which two hydrogens are attached are triplets and the carbon atoms to which three hydrogen atoms are attached are quartets.

The easiest signals to interpret are the carbon atoms to which no hydrogen is attached, one from the carbonyl carbon at δ 170.1, one from the pair of identical carbons, C-3 and C-5, at δ 138.1, and one from C-1 at δ 135.3: they are singlets because they have no hydrogens on them to couple with. The signals from the other carbon atoms in the aromatic ring, C-4 and the identical pair C-2 and C-6, are at δ 130.1 and 126.3. They have only one hydrogen atom attached to them, and in each case the hydrogen atom ($I = 1/2$) can take up two orientations with respect to the applied field, with essentially equal probability. The carbon atom therefore experiences two slightly different magnetic fields, and comes into resonance as two lines, which is referred to as a doublet (Fig. 4.20b). The two lines are separated by the difference in the resonance frequency, which is measured in Hz and is

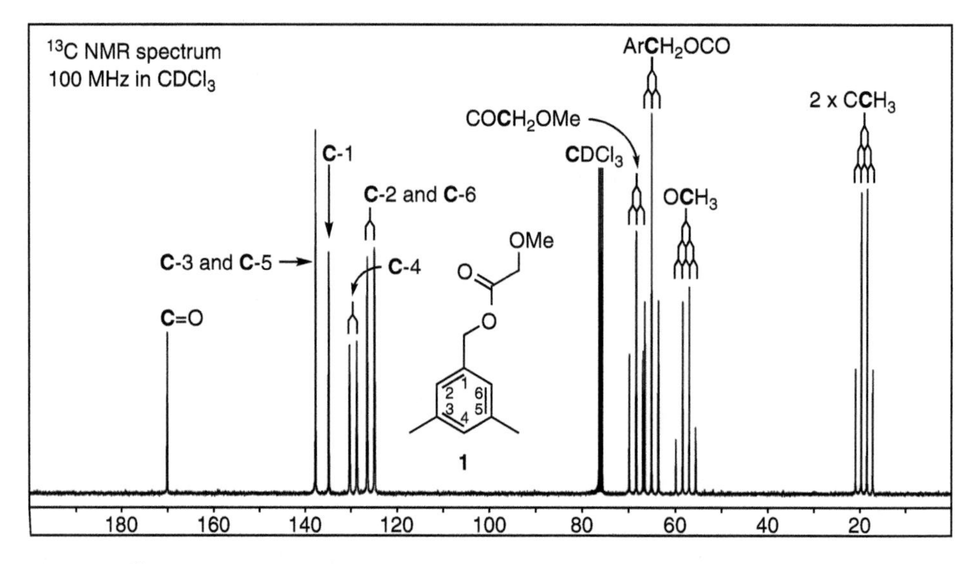

Fig. 4.19 ^{13}C NMR spectrum without proton decoupling

called the *coupling constant, J*. The two lines of each doublet are equally intense, as you can see in both doublets, but the doublet at δ 126.3 is approximately twice as intense as the doublet at δ 130.1.

The methylene groups in the ester **1** have two protons attached to each carbon. The easiest way to understand the consequence of having two neighbouring protons is to look at the effect of each in turn. The first proton would split the signal into two, and the second would then split it again by the same amount, as illustrated in Fig. 4.20c.

You can see, from simple geometry, that the consequence of having two equal coupling constants is a coincident line in the centre. The central line is therefore twice as intense as the two outer lines, and the resulting signal is called a 1:2:1 triplet. In Fig. 4.19 we can see the two triplets centred, like the corresponding lines in the decoupled spectrum in Fig. 4.12, at δ 69.8 and 66.6. Methyl carbons are each attached to three hydrogens. We can simply extend the argument of Fig. 4.20c to Fig. 4.20d: the first hydrogen splits the carbon into a doublet, the second splits each line of the doublet into a doublet, making a triplet as before,

Fig. 4.20 Singlets, doublets, triplets and quartets

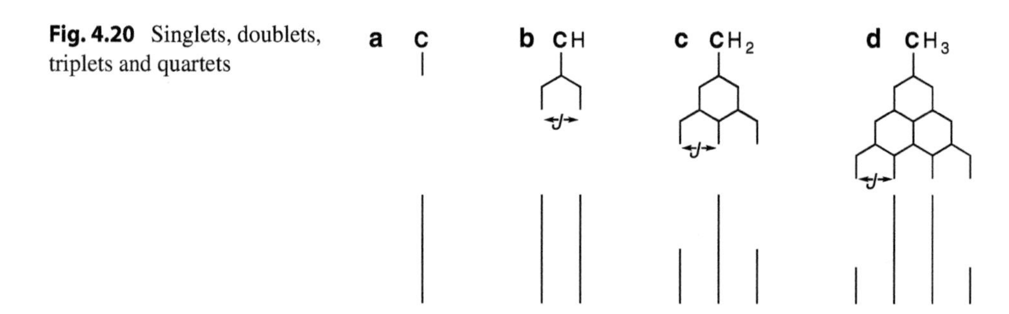

and the third splits each of the lines of the triplet into doublets, creating a quartet. Because the three hydrogens are identical, the coupling constants are identical, and the two central lines are therefore made up of perfect coincidences. The distribution of intensity within the signal creates a 1:3:3:1 quartet, which can be seen in the OCH_3 and the pair of identical CCH_3 signals in Fig. 4.19. In each of the signals shown in Fig. 4.19, the true chemical shift for that carbon atom is the centre of the multiplet.

In summary, the signals of fully substituted, methine, methylene, and methyl carbons are a singlet, a doublet, a triplet and a quartet, respectively, Fig. 4.20a–d. These patterns are quite general, and we shall meet them again in proton NMR spectra. The rule, for nuclei of $I = 1/2$, is that a nucleus, equally coupled to n others, gives rise to a signal with $(n + 1)$ lines, and the intensities are given by the coefficients of the terms in the expansion of $(x + 1)^n$ known as Pascal's triangle.

Clearly, the information contained in these multiplets is valuable, but interpretation can be made difficult by overlapping signals in complicated molecules, where it becomes nearly impossible to disentangle the multiplets. Even in the fully resolved spectrum of Fig. 4.19, the two triplets almost overlap. The situation can be saved by another technique: while the ^{13}C spectrum is being measured, the sample is irradiated at a frequency close to but not coinciding with the resonance frequency of the protons. This is called *off-resonance decoupling*, and has the effect of narrowing the multiplets, without removing them altogether, as in fully decoupled spectra. Thus, it is easily possible to find out how many of each kind of carbon is present in a molecule of unknown structure. This technique, however, may still produce overlapping multiplets if the molecule has several similar carbon atoms giving closely spaced signals. It has now been superseded by other techniques (see Sect. 4.15 later in this chapter) which completely overcome this problem, and you are not likely to come across off-resonance decoupling except in spectra taken in the past, where you will see the signals reported as singlets, doublets, triplets and quartets.

The magnitude of the coupling constant, J, is also informative. Illustrative examples can be found in several tables at the end of this chapter. J-Coupling is not mediated through space, but by the electrons in the bonds connecting the coupling partners. What is more it is only the s electrons that can report magnetic information to and from the nucleus, because only the s electrons have any probability of being found at the nucleus; p electrons have a node at the nucleus and zero probability of being found there. For this reason, tetrahedral (sp^3) carbons exhibit the lowest coupling constants (typically 120–150 Hz), trigonal (sp^2) carbons have higher coupling constants (155–205 Hz), and digonal (sp) carbons the highest (~250 Hz). The other major influence on J is the presence of electronegative atoms, which raise coupling constants sometimes outside these limits, so that, in an extreme case, chloroform has $^1J_{CH} = 209$ Hz, even though it has a tetrahedral carbon. The $^1J_{CH}$ values in the ester **1** are: 126 Hz for the CCH_3 groups and somewhat larger at 142 Hz for the OCH_3 group; 142 and 147 Hz for the two OCH_2 groups; and 155 and 156 Hz for the two trigonal carbons in the benzene ring. $^1J_{CH}$ coupling constants of tetrahedral carbon can be estimated using Eq. (4.22), given at the end of this chapter.

The $^1J_{CH}$ coupling constants can be measured from the ^{13}C spectrum, but they are more usually measured in the proton NMR spectrum, because fully coupled spectra like that in Fig. 4.19 are rarely taken. Most of the proton NMR spectrum is unaffected by the presence of ^{13}C: 99% of the signal from a proton is from those protons attached to ^{12}C, and these are not coupled because ^{12}C is magnetically inert; 1% of the signal, however, comes from protons attached to ^{13}C nuclei, and these are coupled, showing up—when the uncoupled proton signal is a singlet—as a weak doublet, with the two lines placed symmetrically about the strong signal from the protons attached to ^{12}C, and separated from each other by an amount in Hz equal to the coupling constant $^1J_{CH}$. Since each line of this doublet is only 0.5% of the intensity of the main signal, it often goes unnoticed, but it can be searched for, if that region of the spectrum is not crowded with other signals and if the S/N ratio is good enough. All of the ^{13}C satellites, as they are called, can just be seen in Fig. 4.12. When the proton signal is itself a multiplet, the ^{13}C satellites are even weaker, because they are also multiplets, and they are then harder to pick out from the noise. Strictly speaking the signals from the protons attached to ^{13}C are not symmetrically placed about the strong signal from those protons attached to ^{12}C—there is a small, and usually unnoticed, isotope effect (Sect. 4.4.5) that displaces the two sidebands minutely (typically 0.8 Hz) upfield.

The coupling of a ^{13}C nucleus to a proton through more than one bond has much smaller coupling constants, measurable in a full ^{13}C spectrum, when the signals are not too confused. Counter-intuitively, $^2J_{CH}$ and $^3J_{CH}$ coupling constants are rather similar, usually between 0 and 10 Hz and typically about 5 Hz. The factors that influence their magnitude are discussed in Sect. 4.7.5.

These longer-range couplings can be seen in expansions of the individual signals in the coupled ^{13}C spectrum of the ester **1**. Beginning at the high-field end with the two quartets from the methyl groups, each line of the 1:3:3:1 quartet from the *O*-methyl group (at δ 59.3 in Fig. 4.19) is actually a fine 1:2:1 triplet, which can be seen in Fig. 4.21a. The carbon atom of the OMe group is three bonds away from the methylene protons on the

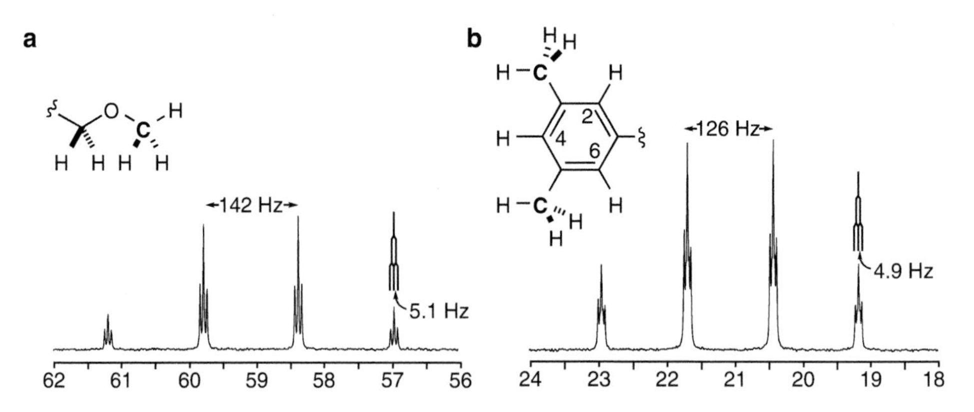

Fig. 4.21 $^3J_{CH}$ fine structure in the methyl quartets

other side of the oxygen atom, and we can detect this relationship by the $^3J_{CH}$ coupling. The coupling constant for the $^3J_{CH}$ triplet is 5.1 Hz, a typical value. Similarly, the *C*-methyl groups on C-3 and C-5, which give rise to the 1:3:3:1 quartet at δ 21.2 in Fig. 4.19, actually has fine $^3J_{CH}$ coupling to the two nearly equivalent *ortho* protons (on C-2 and C-4 for the C-3 methyl group and C-6 and C-4 for the C-5 methyl group), making each line of the quartet into a 1:2:1 triplet, which can be seen in Fig. 4.21b. These four triplets are not baseline resolved; although the coupling constants to the two *ortho* protons must be close to 4.9 Hz, they are not exactly the same. Both signals in Fig. 4.21 are either called quartets of triplets or triple quartets.

In Fig. 4.19, the two methylene carbons are triplets close to each other. When we expand them in Fig. 4.22, we can see that each line of the downfield triplet is a fine 1:3:3:1 quartet, whereas each line of the upfield triplet is a 1:2:1 triplet. The former would be described as a triple quartet and the latter as a triple triplet. The $^3J_{CH}$ coupling constants are 5.2 Hz and 4.6 Hz, respectively.

This pattern shows that we have assigned these two carbons correctly, because the downfield carbon is coupled ($^3J_{CH}$) to the three protons on the *O*-methyl group, and the upfield carbon is coupled ($^3J_{CH}$) to the two *ortho*-protons in the aromatic ring. Note that the predicted chemical shifts for these two carbons in Fig. 4.18 were only 0.9 p.p.m. apart, making it quite plausible that we could have assigned them the wrong way round, but the 3-bond couplings show that we have not

The carbon atoms having only one hydrogen attached, C-4 and the pair C-2 and C-6, are doublets in Fig. 4.19, but the expansion in Fig. 4.23 shows that they are more complicated, each line of the doublet in the signal for C-4 is made up of nine lines, and each line of the doublet in the combined signal for C-2 and C-6 is made up of eight lines. The carbon C-4 is three bonds away from the six methyl protons and three bonds away from the protons on C-2 and C-6. The total number of hydrogens is eight, and the multiplet is therefore made up of nine lines. Similarly, C-2 (and C-6) is three bonds away from the two methy-

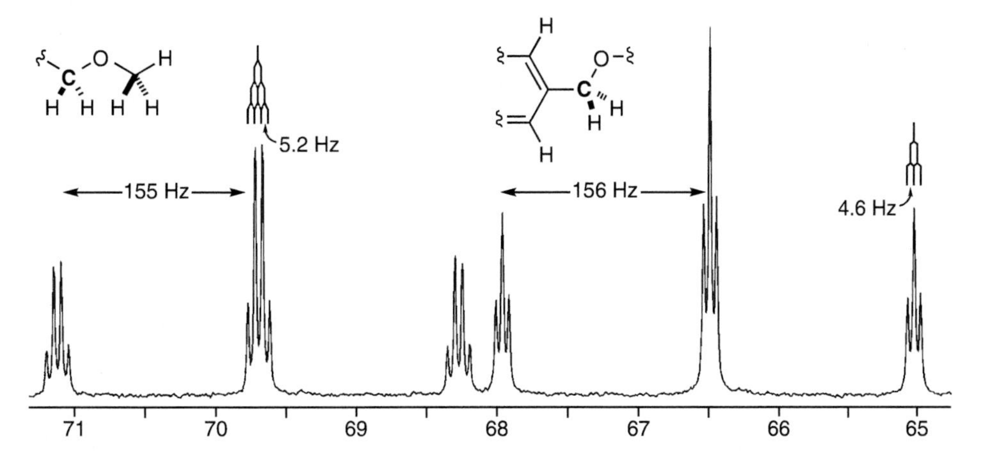

Fig. 4.22 $^3J_{CH}$ fine structure in the methylene triplets

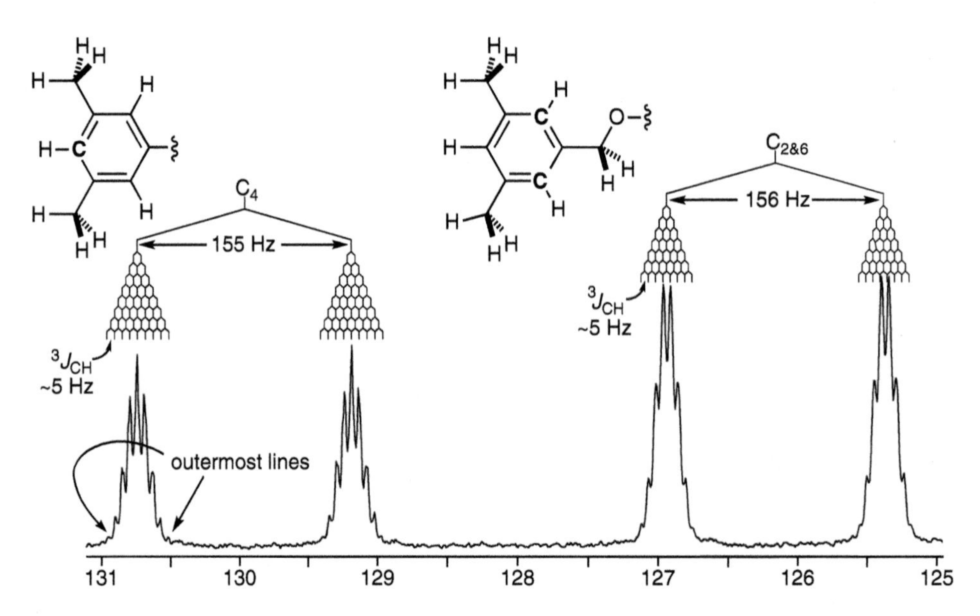

Fig. 4.23 $^3J_{CH}$ fine structure in the methine doublets

lene protons, three bonds away from the protons on C-4 and C-6 (C-2), and three bonds away from the methyl group on C-3 (C-5). The total is seven protons, and the multiplet is made up of eight lines.

The protons coupling to C-4 are not all identical. While the coupling constant to all six of the methyl protons will be identical, it will not be exactly the same as the coupling to the protons on C-2 and C-6. As a result, the internal lines of the nine-line pattern do not perfectly coincide, as the internal lines do for the quartets and triplets in Figs. 4.21a and 4.22. Similarly, the four different kinds of protons coupled to the carbon atoms C-2 and C-6 lead to even more broadened internal lines in the two octets. As a result, the multiplets in Fig. 4.23 are not as well resolved as they are in the two earlier figures, and the coupling constant deduced from the separation of any two lines (a little over 5 Hz in each case) is not an accurate measure of any of the component coupling constants. The only reliable number comes from the separation of the outer lines of the whole signal, which measures the sum of all the coupling constants. In the case of C-4, for example, the separation of the outer lines is 197 Hz, which must be ($^1J_{CH} + 6 \times J_{Me} + 2 \times J_{2,6}$). Also, note how small the outer lines are for the eight- and nine-line multiplets. The outer lines from the nine-line pattern within the doublet from C-4, for example, are only just discernible in Fig. 4.23, where they are picked out with arrows. A nine-line multiplet with exactly equal coupling constants would have intensity ratios in Pascal's triangle of 1:8:28:56:70:56:28:8:1, and the outermost lines might easily be overlooked, just as they can be here. The rest of the signal (8:28:56:70:56:28:8 = 1:3.5:7:8.75:7:3.5:1) could be mistaken for a septet, unless one remembered that a true septet would have a much steeper pattern of intensities (1:6:15:20:15:6:1).

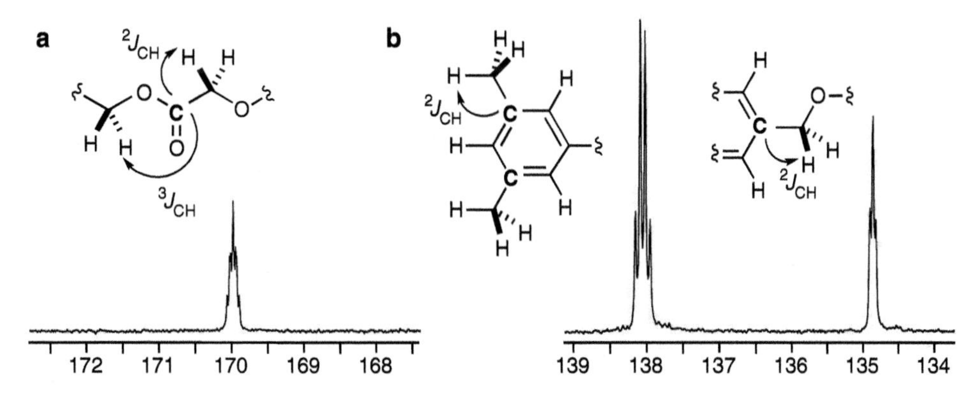

Fig. 4.24 $^3J_{CH}$ and $^2J_{CH}$ fine structure in the singlets from the fully substituted (quaternary) carbons

Finally, expansion of the singlets in Fig. 4.19 shows that they are also narrow multiplets, but this time with some $^2J_{CH}$ coupling as well as $^3J_{CH}$ coupling. The expanded singlet in Fig. 4.24a from the carbonyl carbon looks similar to a quintet, which indicates that it has $^2J_{CH}$ coupling to one pair of methylene protons and $^3J_{CH}$ coupling to the other pair of methylene protons with closely similar but not quite equal coupling constants of about 4 Hz. The signal from C-3 and C-5 looks like a 1:3:3:1 quartet, indicating that it has $^2J_{CH}$ coupling of about 6 Hz to the protons on the methyl group, but undetectable $^2J_{CH}$ coupling to the protons *ortho* to it. Similarly, the signal from C-1 looks like a 1:2:1 triplet, with $^2J_{CH}$ coupling of about 4 Hz to the protons on the benzylic methylene group, but essentially zero $^2J_{CH}$ coupling to the *ortho* protons. $^2J_{CH}$ coupling to *ortho* protons is typically 1 Hz, which was not resolved here.

^{13}C–^1H Coupling is rarely examined in the detail we have seen above, because it is rarely needed, but on this occasion it has allowed us to see the appearance of first-order doublets, triplets, quartets, quintets and even eight- and nine-line patterns, many of which we shall see repeatedly when we come to look at the much larger and more important subject of ^1H–^1H coupling.

4.5.3 ^{13}C–^{13}C Coupling

Because of the low natural abundance of ^{13}C, only one ^{13}C atom in every 100 is bonded to another ^{13}C. Enrichment with ^{13}C is possible for mechanistic and biosynthetic studies, whereupon ^{13}C–^{13}C coupling can be seen relatively easily. The amount of s character at each end of the C–C bond is the main factor affecting the coupling constant $^1J_{CC}$. The $^1J_{CC}$ between two carbon nuclei C^x and C^y is estimated using Eq. (4.5):

$$^1J_{C^xC^y} = 0.073(\%s_x)(\%s_y) - 17 \tag{4.5}$$

where $\%s_x$ and $\%s_y$ are the percentages of s character (using the sp^n notation) in C^x and C^y. Thus, the (tetrahedral) methyl group in toluene is estimated to be coupled to the (trigonal) ipso carbon with a coupling constant of 43 Hz; the observed value is 44 Hz. As with $^1J_{CH}$, neighbouring electronegative elements raise the coupling constants; for example, the methyl group of acetates is coupled to the carbonyl carbon with ^1J of 59 Hz, at the upper end of the range of C–C coupling constants for a tetrahedral carbon bonded to a trigonal carbon. Bonding between two trigonal atoms can be expected to lead to coupling constants near 64 Hz, and higher than this if electronegative elements are attached to one or more of the carbon atoms.

Any signals coming from the rare combination of a ^{13}C attached to a ^{13}C at natural abundance are usually too weak to use, but in a spectrometer fitted with a cryoprobe, an optional piece of equipment which cools the electronics, and significantly improves the sensitivity by reducing the noise level, it is becoming possible to see such coupling, and to use it to assign signals and connectivity in favourable cases. We can see one such case in the ^{13}C NMR spectrum of DDQ **35** in Fig. 4.25. The expansions show the greatly amplified region at the base of each singlet, in which the sidebands are from those atoms in which one ^{13}C atom is bonded to another.

This is a remarkable spectrum—all four carbon atoms inherently give weak signals, since none of them has a proton attached to speed up the relaxation, nor is there an Overhauser effect to enhance the signal intensity. Nevertheless, we can see at the base of the only signal we can safely assign by chemical shift, the carbonyl group at δ 170.0, a pair of just-resolved sidebands separated by 62.2 and 57.8 Hz. These match the sidebands at the base of the signals at δ 142.5 and δ 127.7, respectively, which must be the signals from the carbon atoms carrying the chlorine and the cyano group. That the latter carries the cyano group is proved by the sidebands it also shows with a separation of 90.8 Hz, which it shares with the signal at δ 109.5. The large coupling constant is consistent with

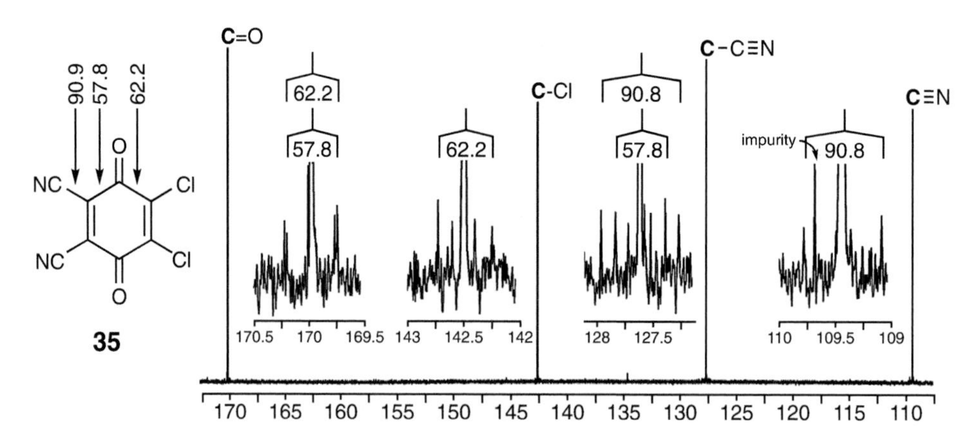

Fig. 4.25 ^{13}C NMR spectrum of DDQ showing sidebands from ^{13}C–^{13}C coupling

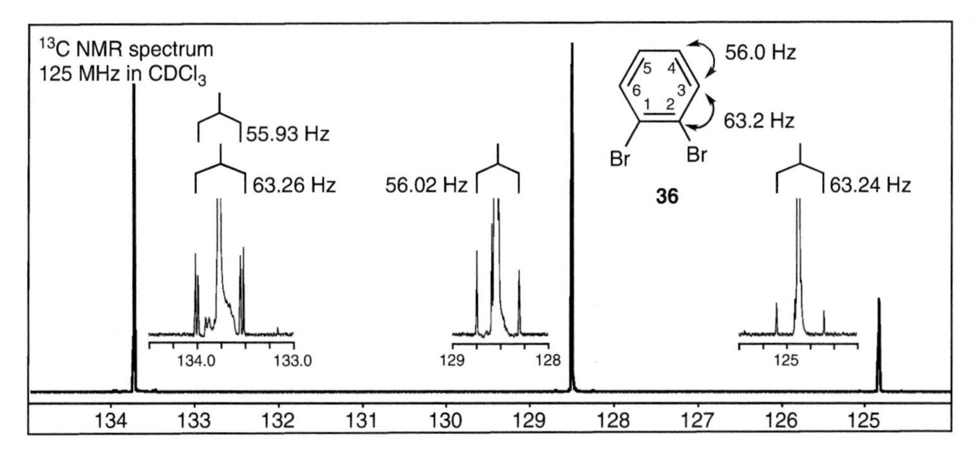

Fig. 4.26 ^{13}C NMR spectrum of *o*-dibromobenzene showing sidebands from ^{13}C–^{13}C coupling

expectation for a bond to the digonal cyano carbon. The full assignment is therefore complete, and has been achieved without any help from a ^{1}H NMR spectrum, since there is none.

Here is another example where the $^1J_{CC}$ coupling constants can be used to assign signals. *o*-Dibromobenzene **36** has three ^{13}C signals, one at δ 133.8, one at δ 128.5 and the third, weaker and clearly, because it is so weak, the one from the carbon atoms carrying the bromine atoms, at δ 124.9. The estimates from Mestrenova and ChemNMR are δ 133.5 and 133.8 for C-3/6 and δ 128.2 and 129.9 for C-4/5 (with C-1/2 at 124.7 and 124.5), reassuringly close to the observed values, but the uncertain pair are only 4 Hz apart. Assigning which of these two carbon signals is which with certainty is possible by looking at the sidebands on the carbon signals. Figure 4.26 shows that the downfield signal has *two* $^1J_{CC}$ couplings, which define it as that from carbons C-3 and C-6. The signal next further upfield is therefore that from the carbons C-4 and C-5, and the signal most upfield that from carbons C-1 and C-2. As expected, the larger $^1J_{CC}$ coupling constant, 63.2 Hz, is from the bond towards the carbon carrying the electronegative heteroatom.

4.6 ^{1}H–^{1}H Coupling: Multiplicity and Coupling Patterns

We have already seen with ^{13}C–^{1}H coupling what happens when two nuclei with $I = 1/2$ are coupled. Much the same is true for proton-proton coupling, except that we are no longer concerned with one-bond couplings. Two-bond coupling **37** is called geminal coupling, three-bond coupling **38** is called vicinal coupling, and coupling through more than three bonds **39** is called long-range coupling. In this section we shall look at the multiplicity, and the patterns that coupling creates in the proton signals, and leave until later (Sect. 4.7) the factors that affect the magnitude of the coupling constants.

geminal coupling $^{2}J_{HH}$ H_a H_b **37**

vicinal coupling $^{3}J_{HH}$ H_a H_b **38**

long-range coupling $^{n+3}J_{HH}$ H_a H_b n **39**

4.6.1 ^{1}H–^{1}H Vicinal Coupling ($^{3}J_{HH}$)

Let us begin with vicinal coupling, because coupling from H_a to H_b in **38** tells us something about connectivity when we are trying to put together a structure. In proton NMR spectra, as with ^{13}C spectra, we see doublets, triplets, and quartets whenever a proton is coupled equally to one, two or three protons, respectively. To take a simple case, the expanded signals in Fig. 4.27 show the coupling of the characteristically low-field aldehyde proton in diphenylacetaldehyde **40** with the proton on the α-carbon. Ignoring for the moment the confused-looking signals in the δ 7.5–7.0 region from the aromatic protons, we can see that the signal from each of the two protons is split equally into a doublet by the other, and the pattern is described as being that of a coupled AX system. The convention used is to label protons close in chemical shift with the letters A, B and C, those far away in chemical shift with the letters X, Y and Z, and those intermediate with the letters M, N and O.

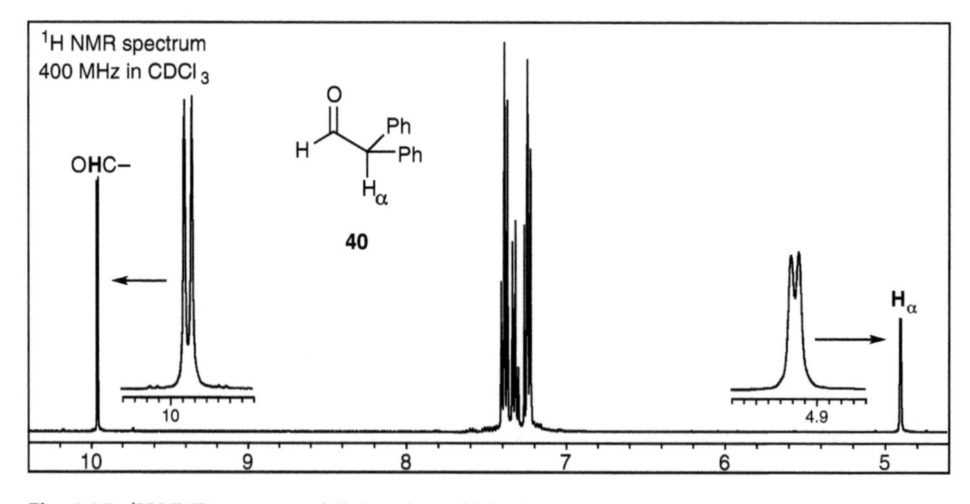

Fig. 4.27 ^{1}H NMR spectrum of diphenylacetaldehyde

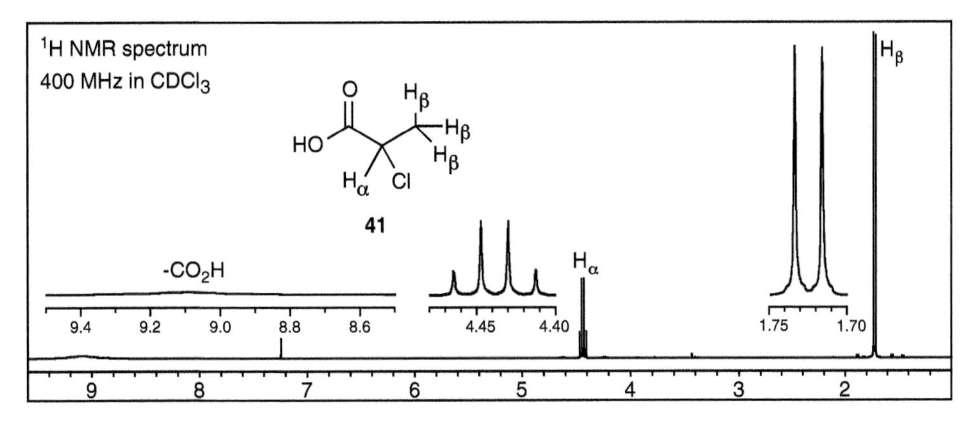

Fig. 4.28 ^{1}H NMR spectrum of 2-chloropropionic acid

Moving on to a slightly more complicated system, Fig. 4.28 shows the AX_3 system in the ^{1}H NMR spectrum of 2-chloropropionic acid **41**. The mid-field signal centred at δ 4.44 is the signal from the methine hydrogen H_α, downfield because it has a carbonyl group and an electronegative element attached to the methine carbon. It resonates as a 1:3:3:1 quartet because the methine hydrogen is coupled to the three identical hydrogens H_β of the methyl group. Likewise, the upfield signal centred at δ 1.725 is the signal from the methyl hydrogens H_β with a chemical shift slightly downfield from the position of a normal C-Me group, because it has an electronegative element on the next carbon. It appears as a doublet because the three hydrogens of the methyl group are coupled to the single methine hydrogen, and are split into two by it.

The three methyl hydrogens, because of the free rotation about the C–C bond, experience identical magnetic environments and they come into resonance at exactly the same place. The methyl protons are, in fact, coupled to each other, but *coupling between protons with identical chemical shifts does not show up in NMR spectra*. The upfield doublet is three times as intense as the downfield quartet and three times as intense as the broad signal from the carboxylic acid proton. In summary, the mid-field signal is a one-proton *quartet* because the proton that gives rise to it is equally coupled to *three* protons, and the upfield signal is a three-proton *doublet* because the protons that give rise to it are coupled to *one* proton. The rule is the same as that given in the section on ^{13}C–^{1}H coupling: a nucleus, equally coupled to n others, will give rise to a signal with $(n + 1)$ lines, and the intensities are given by the coefficients of the terms in the expansion of $(x + 1)^n$ (Pascal's triangle).

The spectrum in Fig. 4.29 of ethyl propionate **42** twice over illustrates the characteristic appearance of A_2X_3 signals from ethyl groups. The three protons of the methyl groups couple equally with the two protons of the neighbouring methylene groups. Likewise, the two protons of the methylene groups couple equally with the three protons of the neighbouring methyl groups. This pattern of an upfield three-proton 1:2:1 triplet and a downfield two-proton 1:3:3:1 quartet is characteristic of an ethyl group in which the methylene

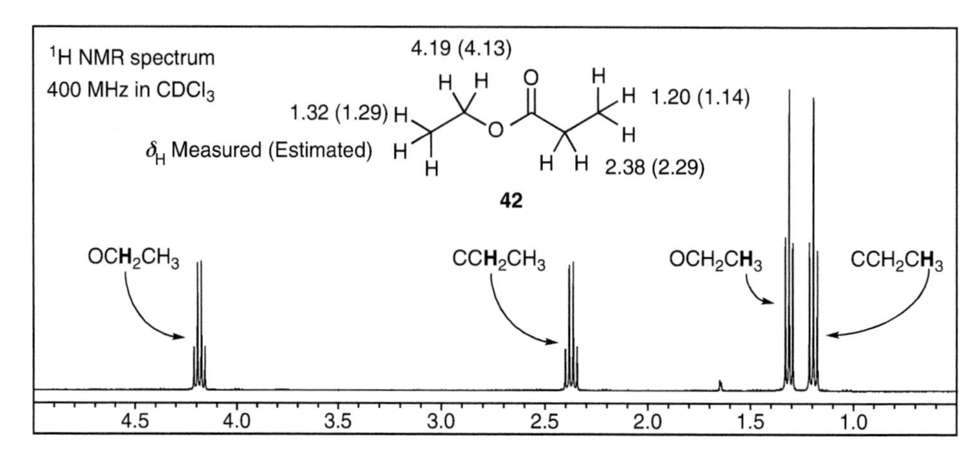

Fig. 4.29 ¹H NMR spectrum of ethyl propionate

protons are not coupled to anything else. The chemical shifts of the methylene groups δ 4.19 and 2.38 are strongly indicative of the nature of the atom to which they are bonded—oxygen for the former and carbon for the latter.

Although it is reasonably certain that the OCH_2 signal is downfield from the CCH_2 signal, the assignment of which signal comes from which methyl group is less secure. We have assumed, correctly as it happens, that the methyl group of the OEt group will give rise to the triplet at lower field δ 1.32 than the triplet from the methyl group of the CEt group δ 1.20. The ChemNMR estimates and the measured values are shown next to the structure in Fig. 4.29, where we can see that they support this expectation, but the difference in chemical shift values is not large enough for us to be completely confident. There are a number of ways in which we can confirm the assignment, as we shall see by matching coupling constants (Fig. 4.33), by selective decoupling (Sect. 4.11) and, best of all, by using COSY spectra (Sect. 5.2).

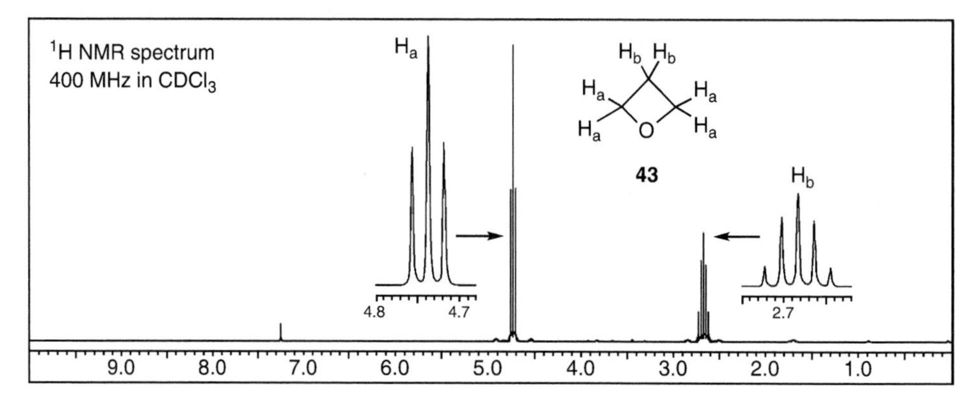

Fig. 4.30 ¹H NMR spectrum of oxetane

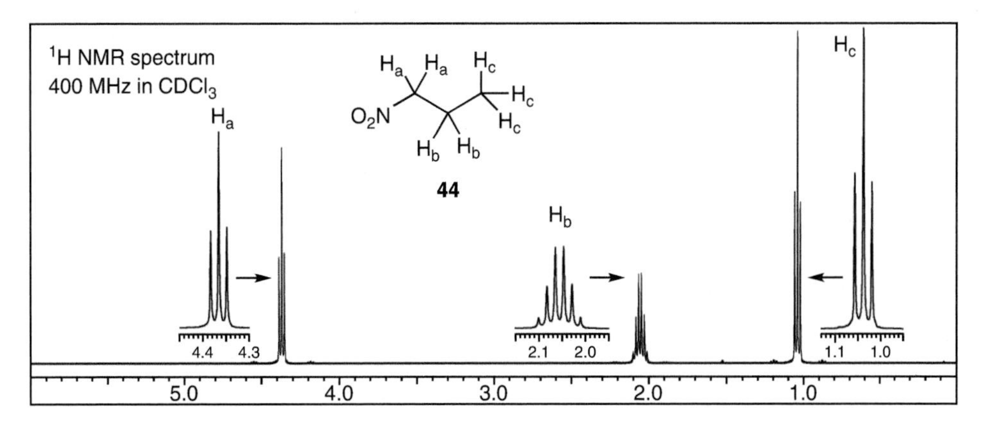

Fig. 4.31 ^{1}H NMR spectrum of 1-nitropropane

Moving on to larger multiplets, the $A_2X_2A_2$ spectrum of oxetane **43** in Fig. 4.30 shows the downfield four-proton triplet from the pair of identical methylene groups flanking the central methylene group, which gives rise to the clean 1:4:6:4:1 quintet. The triplet is downfield, because the methylene groups giving rise to this signal are adjacent to the oxygen atom. Note that the quintet here has base-line resolution and lines in the proper proportions, unlike the quintet-like signal in Fig. 4.24a. All the couplings in oxetane **43** are coincidentally equal, whereas the two couplings, $^{2}J_{CH}$ and $^{3}J_{CH}$, to the carbonyl carbon in ester **1**, although similar in magnitude, were not exactly equal.

Figure 4.31 shows a slightly more complicated example in the spectrum of 1-nitropropane **44**. The protons H_c on the methyl group give rise to a three-proton signal at high field (δ 1.04) as a triplet, because they are adjacent to a methylene group. The protons on the methylene group H_a give rise to a two-proton triplet at low field (δ 4.37). The chemical shift is appropriate for a methylene group next to an electronegative element, and the multiplicity is appropriate for protons coupling to another methylene group. The protons of the methylene group in the middle H_b give rise to a two-proton 1:5:10:10:5:1 sextet at δ 2.06. The chemical shift is appropriate for a methylene group between two alkyl groups, but not far from an electronegative group. The multiplicity is appropriate for protons coupling equally to a total of five protons. Actually the coupling constant J_{ab} (7.0 Hz) is slightly smaller than J_{bc} (7.5 Hz), but the difference is not resolved in the sextet, showing up only as a slight broadening of the lines and resolution that does not reach the base line.

In the spectra in Figs. 4.28, 4.29, 4.30, and 4.31 the coupling constants have all been very much the same, either inherently, as in the spectra of 2-chloropropionic acid **41** and ethyl propionate **42**, or accidentally, as in the spectra in oxetane **43**, where the *cis* and *trans* couplings were equal, and nitropropane **44**, where the coupling J_{ab} is almost the same as J_{bc}. The coupling constant $^{3}J_{HH}$ = 6–8 Hz in these four spectra is typical of coupling

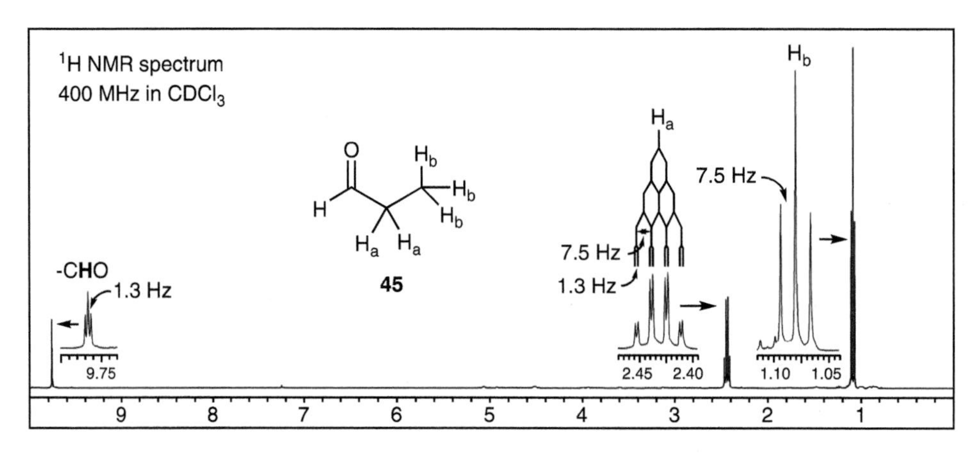

Fig. 4.32 ^{1}H NMR spectrum of propionaldehyde

constants in freely rotating alkyl chains. However, the coupling constant for the mutually coupled doublets in diphenylacetaldehyde **40** shown in Fig. 4.27 is noticeably smaller, $J = 2.6$ Hz. When a multiplet is split again by coupling to other protons with a different coupling constant, more complicated patterns emerge than the doublets, triplets, quartets, quintets and sextets that we have seen in Figs. 4.27, 4.28, 4.29, 4.30, and 4.31.

For example, in the spectrum of propionaldehyde **45** in Fig. 4.32 the methylene protons are not the quintet that would be produced by coupling equally to the four neighbouring protons. Instead, the coupling between the methylene protons and the aldehyde proton has a coupling constant of 1.3 Hz, whereas the coupling constant to the methyl protons is 7.5 Hz. The methylene signal is therefore a double quartet, made up in the pattern shown above the expanded signal for H_α. Note how the two coupling constants can be measured in each of the participating signals, the smaller coupling constant both in the aldehyde triplet and in the double quartet, and the larger coupling constant both in the methyl triplet and in the double quartet.

Proton-derived signals often have to be reported in the experimental sections of research papers and in text-based compilations of data. The form in which they are reported varies with the requirements of the journal or company policy, but a typical way of reporting the spectrum of the aldehyde **45** is:

δ_H($CDCl_3$, 400 MHz) 9.77 (1H, t, J 1.3), 2.43 (2H, qd, J 7.5 and 1.3) & 1.06 (3H, t, J 7.5)

The order in which the chemical shift, intensity, multiplicity and coupling constant(s) are printed might vary, but a system like this is concise and easily understood. The convention here is to start at the low-field end of the spectrum (with the larger chemical shifts) and read the spectrum from left to right; the coupling constants are reported in order of decreasing magnitude, with the designations s, d, t, etc. in the same order as the coupling constants (so that the quartet above is identified as having the J value of 7.5 Hz).

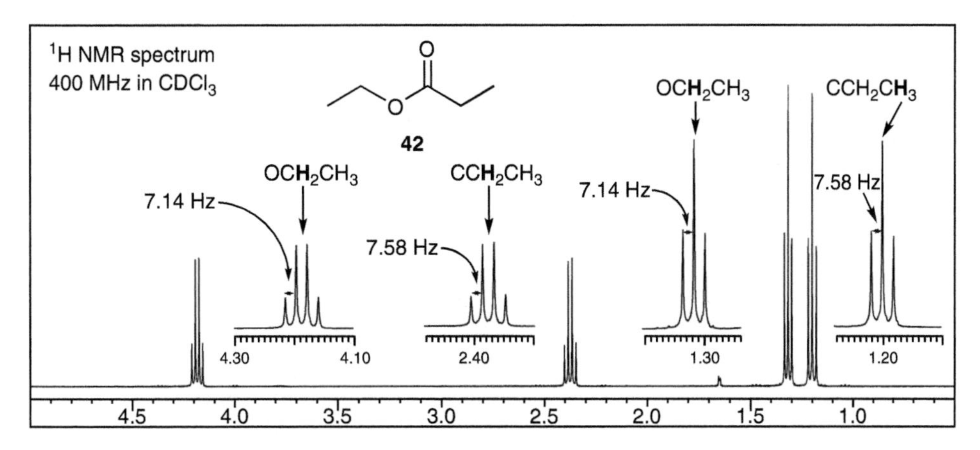

Fig. 4.33 ^{1}H NMR spectrum of ethyl propionate in more detail

Occasionally, coupling constants measured on the spectrum are not completely consistent. Instruments are not immaculate in this respect, especially in the second place of decimals, since they report what the computer produces from its algorithms identifying the midpoint of a series of data points. The problem is common when all the coupling is not perfectly resolved. It is wise to make it clear when reporting coupling constants whether you have rationalised them (i.e. made them match up in what seems to be the obvious way) or whether you are reporting exactly what the instrument gives you. Carefully matching the coupling constants can help in the assignment of signals in complicated spectra. As a simple example, we can go back to the spectrum of ethyl propionate **42**. In Fig. 4.29, both of the coupling constants look to be about the same, but enlarging them as in Fig. 4.33 (or reading the data from the NMR spectrometer) reveals that the downfield quartet has a coupling constant of 7.14 Hz, while the upfield quartet has a coupling constant of 7.58 Hz. If we look at the two triplets, we can see that the downfield triplet has the smaller coupling constant and the upfield triplet the larger coupling constant. Being able to pair up the signals like this shows that the assignment in Fig. 4.29 was correct.

In the double quartet in Fig. 4.32, the quartet has a large coupling constant and the doublet a much smaller one. In contrast, in the triple quartet in Fig. 4.22, the quartet has the smaller coupling constant and the triplet a much larger one. In both cases, the pattern is easy to discern—the coupling constants are so different from each other that the individual doublet and triplet components of the signal are well separated. But in many cases individual protons give rise to patterns of lines that are much less obvious. Thus, the proton H_c on the double bond of allyl bromide **46**, centred at δ 6.03, is doubled by coupling to the *trans* proton H_a, doubled again with a different coupling constant by coupling to the

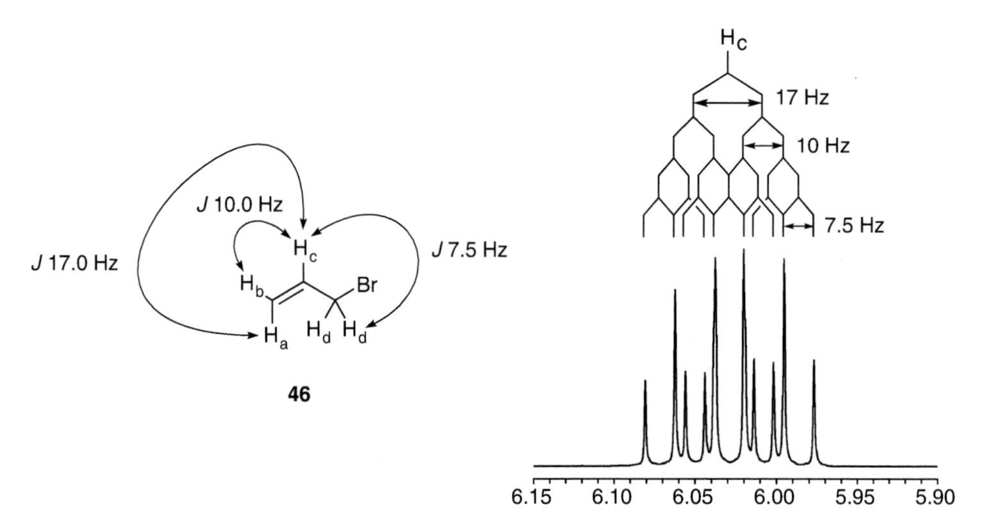

Fig. 4.34 ^{1}H NMR double double triplet from allyl bromide

cis proton H_b, and it is further coupled to the two protons H_d on the methylene group (Fig. 4.34). Since the three coupling constants are all different, it is a double, double, triplet, and could give rise to as many as 12 lines. The actual appearance of this signal is shown in Fig. 4.34, together with the analysis in the descending tree-like drawing above the spectrum. Because the couplings J_{cd} (7.5 Hz) and J_{ca} (10 Hz) add up to a number very close to J_{ca} (17 Hz), several lines almost perfectly coincide. Only 10 lines are resolved, instead of the full complement of 12 expected for a double-double triplet with three different coupling constants.

In general, if a proton has as neighbours sets n_a, n_b, n_c... of chemically equivalent protons, the multiplicity of its resonance will be $(n_a + 1)$ $(n_b + 1)$ $(n_c + 1)$..., but it is not uncommon for lines to coincide, as in the example above; nor is it uncommon for protons that are not chemically equivalent to have coincidental coupling constants, as in nitropropane (Fig. 4.31). In both cases, the observed number of lines is fewer than this formula suggests, and in the general case a large variety of patterns can emerge. Recognising them is a skill of great value in interpreting ^{1}H NMR spectra. It is lazy to report a signal like the one shown in Fig. 4.34 simply as a multiplet, when its components can be analysed with a little thought. Recognising the patterns in this signal is helped if you can pick out the 1:2:1 triplet feature repeated along the multiplet. It is most easily picked out as the first, second and fourth lines, reading from the left or from the right. Once a feature like the triplet is recognised, its doubling by the 10 Hz separation of the first and third lines, and redoubling by the 17 Hz separation of the first and fifth lines, is easier to build up into the full tree.

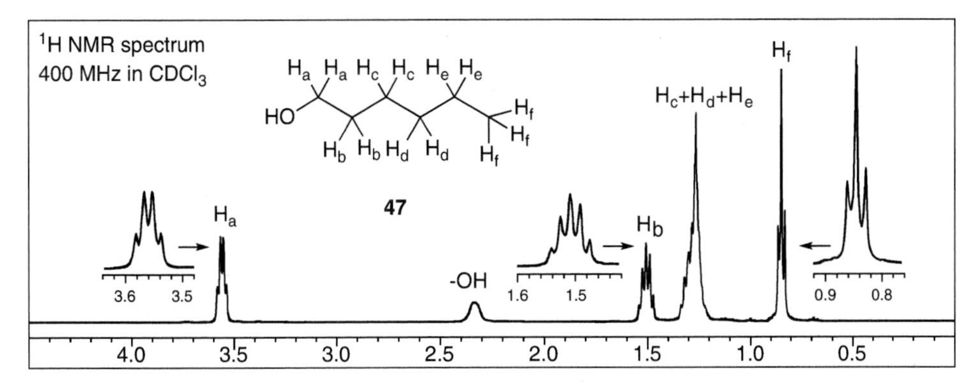

Fig. 4.35 ¹H NMR spectrum of n-hexanol

In contrast, when signals seriously overlap, so that they cannot be disentangled, reporting them as multiplets is perfectly acceptable. For example, n-hexanol **47** has the spectrum in Fig. 4.35. The downfield quartet at δ 3.56 is produced by the methylene protons H_a adjacent to the oxygen atom, and the next methylene protons H_b give rise to the quintet at δ 1.51. But the remaining three methylene groups, H_c, H_d and H_e, are so similar in environment that they are not resolved. Even though the first-order analysis used here (see below for further details) predicts that they will give rise to a quintet, a quintet and a sextet, none of these patterns can be discerned, and the signal must be reported as a multiplet. The broad unresolved signal that they give rise to in the range δ 1.35–1.18 is often called a *methylene envelope*. The protons H_f of the methyl group, giving rise to the triplet at δ 0.84, resonate outside the methylene envelope, because methyl protons are usually at higher field than methylene protons.

4.6.2 AB Systems

In all the spectra considered so far, the separation of the signals (in Hz) has been much greater than the coupling constants (in Hz)—they have all been A_nX_m systems, which have allowed us to interpret the spectra using what is called *the first-order approximation*. Coupling in which the coupling constant is small relative to the chemical shift difference between the coupling partners is called *weak coupling*, and gives rise to the first-order spectra we have seen so far. Coupling in which the coupling constant is large relative to the chemical shift difference between the coupling partners is called *strong coupling*, and leads to spectra that are not first-order. (Large coupling constants should not be identified as strong coupling, although this is a common misusage.) A_nB_m systems, where the chemical shifts are close together, are the most simple examples showing deviations from first-order spectra, and the most simple of all is an AB system, consisting of two mutually coupled protons

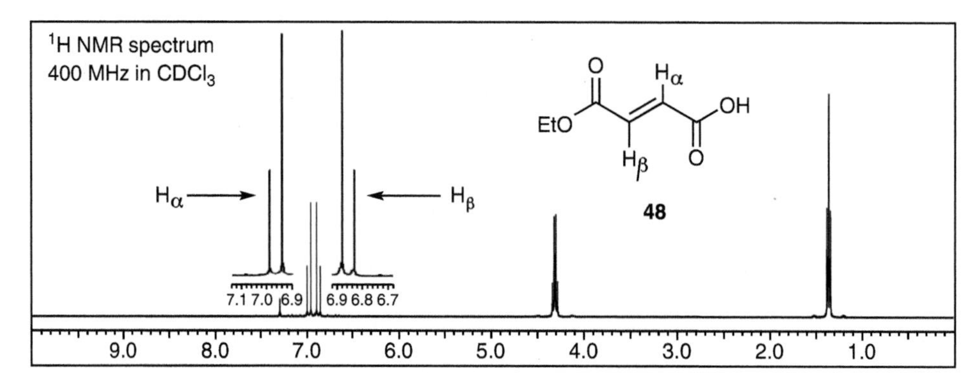

Fig. 4.36 ^{1}H NMR spectrum of ethyl fumarate

A and B, which are not coupled to any other protons. When the difference in chemical shift between the A and the B signal in frequency units ($\nu_A - \nu_B$) is comparable in magnitude to the coupling constant J_{AB}, the two lines of the doublets are not equal in intensity. As an example we can look at Fig. 4.36 showing the alkene signals from the mono-ethyl ester **48** of fumaric acid. The two protons, H_α and H_β, are in slightly different chemical environments, and as a result they have slightly different chemical shifts (δ 6.97 and 6.87). The difference in chemical shift in this 400 MHz spectrum is 40 Hz and the coupling constant of 17 Hz is not very different. In consequence, the 'inside' lines of the AB system are more intense, and the 'outside' lines less intense. The comparison is with a simple, first-order AX system, where the lines in the A and the X signals are essentially equal in intensity (Fig. 4.27).

The formulae governing the appearance of AB patterns are given in Eqs. (4.6), (4.7), and (4.8), with the symbols explained in Fig. 4.37. The smaller the chemical shift difference, the more the inside peaks, 3 and 2, grow in size, and the outside peaks, 4 and 1, get smaller. The relative intensity of the peaks, I, is given by:

$$\frac{I_3}{I_4} = \frac{I_2}{I_1} = \frac{(\nu_4 - \nu_1)}{(\nu_3 - \nu_2)} \tag{4.6}$$

The coupling constants are given in the same way as in an AX system:

$$J_{AB} = \nu_4 - \nu_3 = \nu_2 - \nu_1 \tag{4.7}$$

In an AX system, the chemical shifts of the A and the X signals are given by the frequency of the midpoints of each of the doublets. This is no longer the case with an AB system, where the chemical shifts δ_A and δ_B may be calculated from the expression:

$$\nu_A - \nu_B = \sqrt{(\nu_4 - \nu_1)(\nu_3 - \nu_2)} \tag{4.8}$$

As illustrated in Fig. 4.37c, this places the true chemical shifts, which can be converted to δ units by dividing the frequencies ν by the operating frequency, closer to the inside peaks than to the outside peaks.

Thus, the closer the signals are in chemical shift, the greater the perturbation, as we can see by comparing the AB systems in Fig. 4.38 given by the protons on the double bonds in three α,β-unsaturated carbonyl compounds. The protons in methyl 3-methoxyacrylate **49** have very different chemical shifts, because the β-proton is adjacent to a σ-withdrawing substituent, conjugated, and *cis* to a π-withdrawing substituent; in contrast, the α-proton is conjugated to a π-donor substituent (see Table 4.4 and the associated text for a revision of

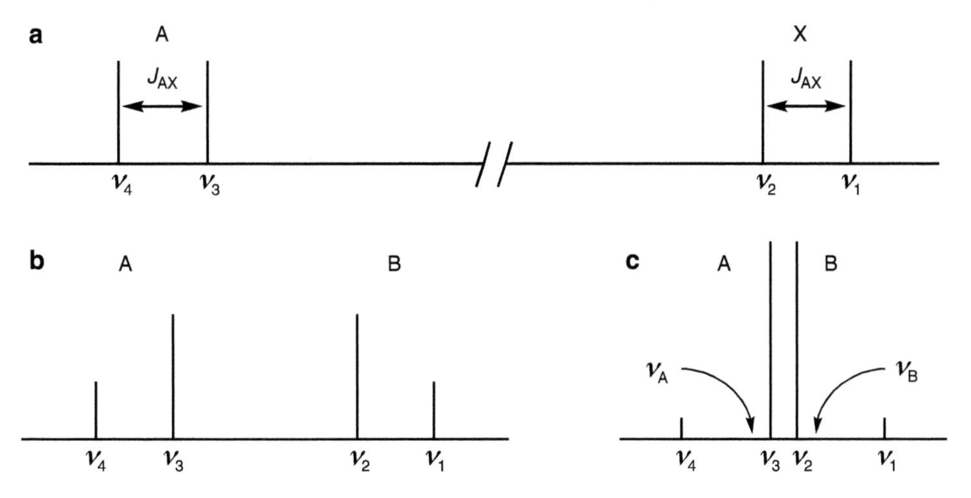

Fig. 4.37 Geometry of AB systems

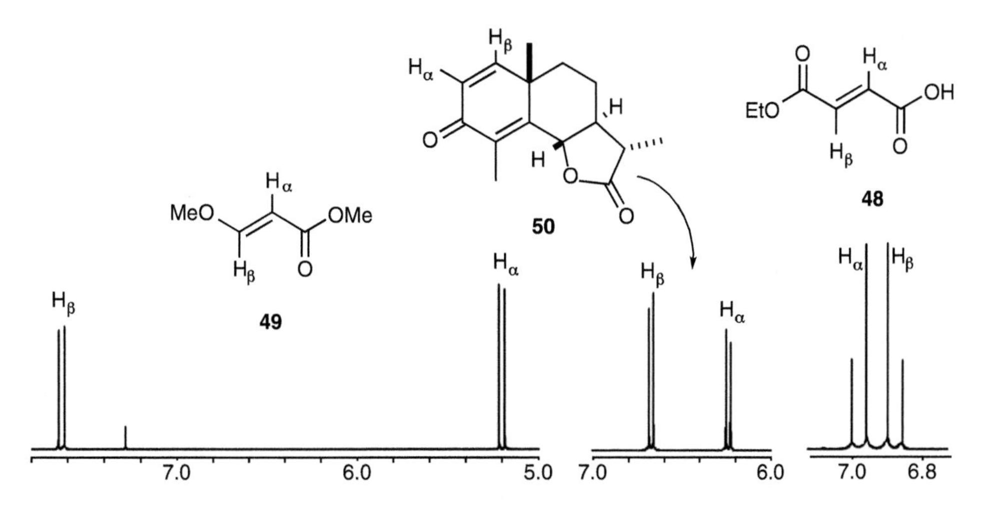

Fig. 4.38 Three AB systems with different separations of the A and B signals

these points). The values are δ 7.65 and 5.21, respectively, more or less at the extreme range for olefinic protons. At 400 MHz, this is a difference of 976 Hz, and the coupling constant, 13 Hz, is much smaller. As a result, the signals are only slightly perturbed from those of an AX system. Santonin **50** has the two signals separated by a much smaller amount, 172 Hz at 400 MHz, with a somewhat smaller coupling constant, 10 Hz, and the perturbation is more obvious. The ester **48** that we have already seen has a separation in chemical shift of only 40 Hz in the 400 MHz spectrum with a large coupling constant of 17 Hz, and the perturbation is considerable.

These patterns are helpful in identifying AB systems: a strongly perturbed doublet must be coupling to a proton close in chemical shift and a less perturbed doublet to one further away. Furthermore, the perturbation tells us in which direction to look for the coupling partners in the AB system. Nevertheless, it is always wise when you are assigning signals to measure up the doublets, in order to make sure that both halves of what you think are an AB system have matching coupling constants.

The doublet of each partner in an AB system is described as 'roofing' or 'pointing', where the metaphor comes from the picture in Fig. 4.39 with the 'roof' drawn from the top of the outside peaks through the top of the inside peaks and meeting, at the 'roofline' at the centre of the system.

At the extreme, when A and B have exactly the same chemical shift, the outside lines disappear, and the inside lines merge into a singlet. This is the situation with the methylene groups in the compounds **42–47** used in Figs. 4.29, 4.30, 4.31, 4.32, 4.33, 4.34, and 4.35. In these cases, the methylene hydrogens are *inherently* identical, but coupling also disappears when two chemically distinct protons *accidentally* come into resonance at the same frequency. The inside lines merge and the outside lines disappear. When the inside lines do not quite merge, the outside lines can be so small as to be overlooked, and the two central lines of the AB signal can easily be mistaken for a doublet.

The same type of perturbation occurs in all A_nB_m systems, and will be evident in many of the spectra in this chapter. Looking back, it is even possible, just, to see the effect in such spectra as that of ethyl propionate **42** in Fig. 4.29, where the roofing in the triplets of both A_2X_3 systems point to the quartets, with the triplet from the C-Et group, closer in chemical shift to its partner, pointing a little more strongly than the triplet from the O-Et group, further apart in chemical shift from its partner.

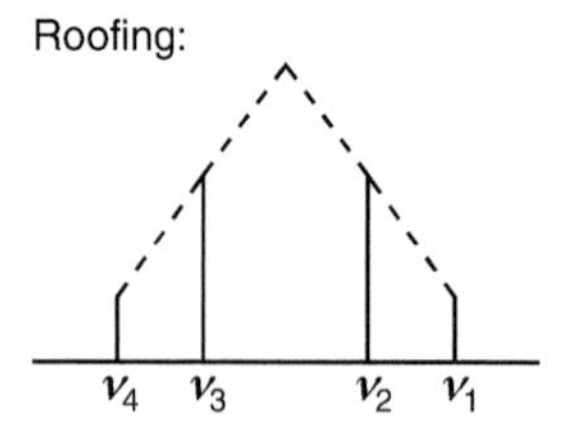

Fig. 4.39 Roofing

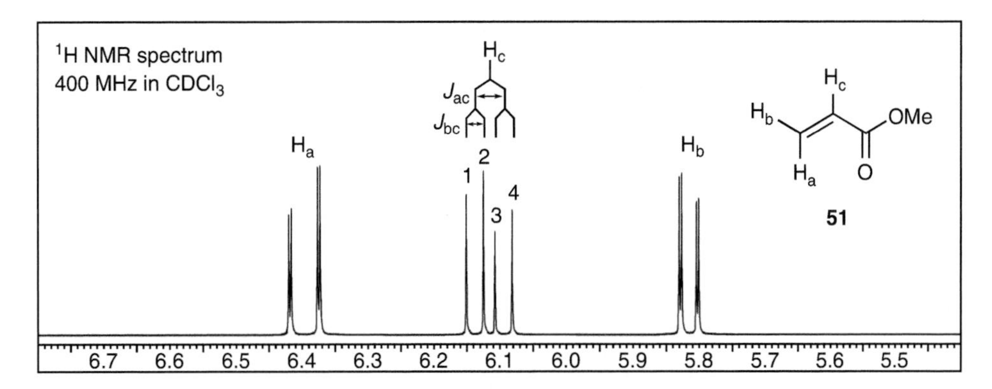

Fig. 4.40 Part of the ¹H NMR spectrum of methyl acrylate

Roofing can be helpful in making sense of a complex pattern, such as that from the three olefinic protons in methyl acrylate **51** shown in Fig. 4.40. We can expect all three protons to be downfield, and the *cis*-β-proton H_a to be the most downfield of all. Judging by methyl vinyl ketone **20** (Table 4.4), the α-proton H_c will be in between the two β-protons, as indeed it is. This assignment is reinforced by looking at the signal from the α-proton, which is a straightforward double doublet with coupling constants of 17.2 and 10.8 Hz. The larger separation of the lines 1 and 3, and the equally large separation of the lines 2 and 4, gives the coupling constant J_{ac}. This is confirmed by the roofing, since both these pairs point towards the downfield signal from H_a, which also has the larger coupling constant. Similarly, the smaller separation of the lines 1 and 2, and the equally small separation of the lines 3 and 4, gives the coupling constant J_{bc}, and this too is confirmed by the roofing, since both pairs point towards the upfield signal from H_b. In turn, the doublets given by H_a and H_b point back to the central signal given by H_c. Of course the same assignments can be made simply by looking at the coupling constants, but the roofing is a great help in quickly making sense of the appearance of the signal from H_c.

One should be cautious in assigning coupling constants using only the first-order analysis (with roofing) given here, because the separation of, say, lines 1 and 2 and lines 1 and 3 is not strictly an accurate measure of the two coupling constants. More often than not coupling constants are reported by measuring these separations, because J_{bc} is close to the separation of lines 1 and 2, and J_{ac} is close to the separation of lines 1 and 3. We can trust the numbers we measure for the coupling constant in this case, because the chemical shifts of the signals for H_c and H_a are 112 Hz apart in this 400 MHz spectrum, over six times the coupling constant between them. Likewise the separation of 121 Hz for the signals for H_c and H_b is over eleven times the coupling constant between them. In strong coupling, when the chemical shift separation in Hz is less than twice the coupling constant, serious errors would be made if we tried to measure the coupling constants by the frequencies of the lines displayed in the spectrum. Later (Sect. 4.6.5) we shall come across cases when the coupling constants cannot be measured in this way. Even in this case, it is true only that the separation of lines 1 and

4 is accurately the sum of J_{bc} and J_{ac}. This warning applies to all multiple spin systems, but we shall continue to use the first-order simplification, because the differences from the true coupling constants are only significant when the chemical shift separation in Hz approaches the value of the coupling constant, in other words when we have strong coupling.

Returning to the spectrum in Fig. 4.40 we might note that the signals from both H_a and H_b are actually double doublets with fine coupling of each to the other. This is the first example of $^2J_{HH}$ coupling that we have seen, and it takes us to the next section.

4.6.3 ^{1}H–^{1}H Geminal Coupling ($^2J_{HH}$)

Geminal coupling $^2J_{HH}$, also known as two-bond coupling, is found only in methylene groups **37** in which for some reason the two hydrogens H_A and H_B are not identical and do not therefore come into resonance at the same frequency. They give rise to multiplets, in the first-order approximation, with the same rules as for three-bond coupling, and the range of coupling constants is rather similar, 0–25 Hz. Not surprisingly, two hydrogens bonded to the same carbon atom are frequently close in chemical shift, and roofing is almost always visible.

The two hydrogens of a methylene group are different in terminal alkenes, as we have seen in methyl acrylate **51** in Fig. 4.40, where they split each other with a coupling constant of only 1.2 Hz. They are also different in cyclic compounds when one surface of the ring has different substituents from the other surface, as in the epoxide **52**, which has a *p*-nitrobenzoyloxymethyl group on one surface and a methyl group on the other. The methylene protons, H_c and H_d, are in different environments, and they give rise to an AB system with another small coupling constant, 2.6 Hz, and an appropriately small amount of roofing (Fig. 4.41).

It is not quite so obvious why, but the methylene group in the side chain, CH_aH_b, also gives rise to an AB system, and with the two signals actually further apart than the more obviously different pair H_c and H_d. At first sight, you might expect the two protons to be

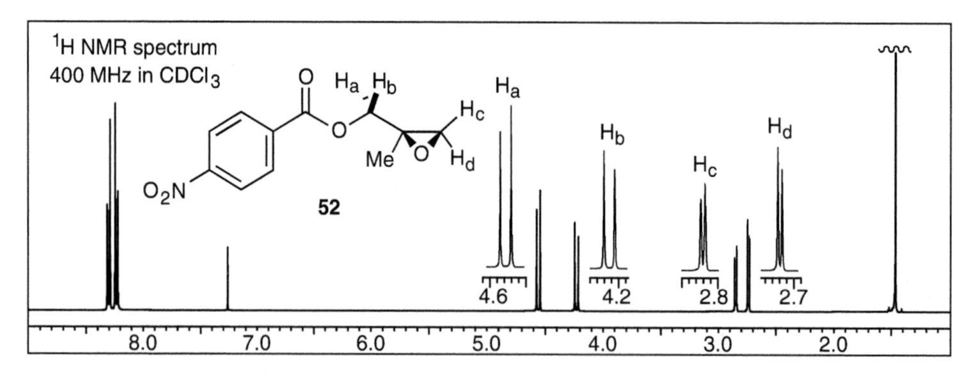

Fig. 4.41 ^{1}H NMR spectrum of an epoxide

chemically identical, especially when you allow for the free rotation of the side chain, but the presence of a stereogenic centre in the molecule has the effect of placing H_a and H_b in different environments. There are three conformations **53-55** in which all the groups are staggered about the bond connecting the methylene group to the ring.

53 **54** **55**

In the first place, the side chain will probably adopt one conformation in preference to any other, and in that conformation, say **53**, the two protons H_a and H_b are not in the same environment, and can come into resonance at different frequencies. Secondly, even if the rotation is completely free, and all three conformations are equally occupied, the average field experienced by H_a is not inherently the same as that experienced by H_b. At any one moment, say **53**, when H_a is in the top left segment, H_b will be placed between the epoxide methylene group and the methyl group, but when H_b comes to the top left **54**, H_a will be placed between the epoxide oxygen and the methyl group. Thus, H_b does not experience the same environment that H_a experienced when it was in the top left segment. The two conformations **53** and **54** are not identical, nor are they enantiomers. At no stage in any of the conformations **53–55** is either of the protons in the same environment as that which the other experiences as the side chain rotates. The two protons H_a and H_b are said to be *diastereotopic*. Only when the average field is the same by coincidence do diastereotopic protons, and diastereotopic methyl groups likewise, come into resonance at the same frequency.

The test for diastereotopic groups is to identify what kind of stereoisomers would be produced if first one of them were replaced by a completely different group, and then the other. There are three categories for a pair of identical groups on a tetrahedral carbon atom, illustrated in Fig. 4.42 using a methylene group. The three possibilities are homotopic, enantiotopic and diastereotopic. If the H atom in front in each of the methylene groups were to be replaced by an X group, the first would create an achiral molecule, the second would create one of a pair of enantiomers, and the third would create one of a pair of diastereoisomers. In each case changing the H at the back would give, reading from the top, the same compound, the enantiomer and the diastereoisomer, respectively, of the first set of compounds, thereby defining the topicity of each of the methylene groups. Thus in the *p*-nitrobenzoate **52**, if H_a were to be replaced by any group X, it would give the diastereoisomer of the compound produced by replacing H_b by an X group. H_a and H_b are diastereotopic.

Geminal coupling commonly occurs alongside vicinal coupling, and ABX patterns are often the result. They have a wide variety of appearances. The ABX system in diethyl

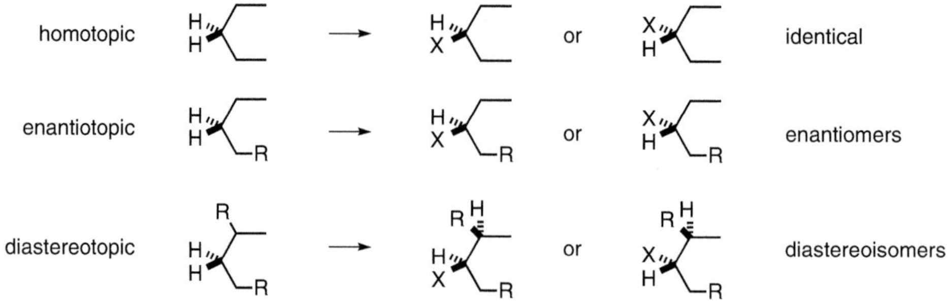

homotopic → or identical

enantiotopic → or enantiomers

diastereotopic → or diastereoisomers

Fig. 4.42 Definitions of homotopic, enantiotopic and diastereotopic

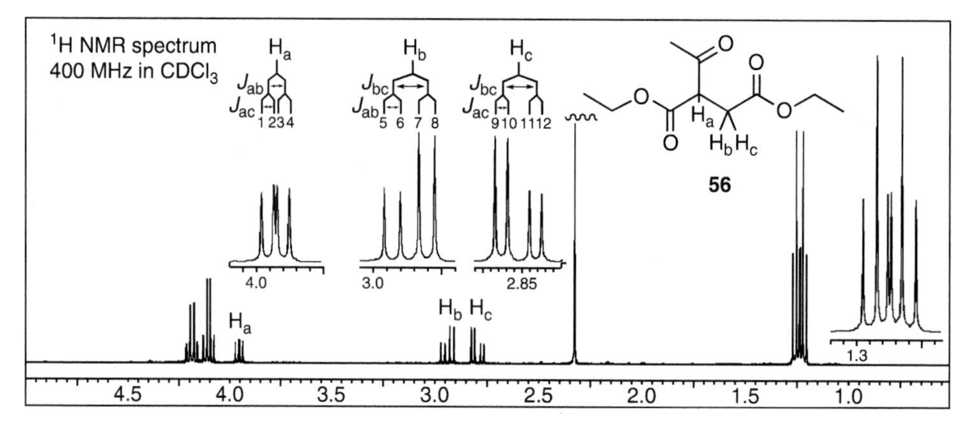

Fig. 4.43 ¹H NMR spectrum of diethyl acetylsuccinate

acetylsuccinate **56** in Fig. 4.43 is only one of many possible patterns. If we combine the treatment of an AB system from the previous section with a simple first-order prediction, we can expect that the four AB lines from H_b and H_c will each be split into doublets by coupling to the X proton H_a, as observed in Fig. 4.43. The proton H_a, being chemically well shifted from H_b and H_c, but coupled to both with different coupling constants, appears as a double doublet. The AB lines from H_b and H_c exhibit the differences in intensity stemming from their AB coupling, but the double doublet of the more distant X proton H_a has four lines nearly equal in intensity. This example shows the 12 lines expected from a first-order analysis (with roofing), since the signals are well spaced, and the coupling constants are all different: 8.3 (J_{ab}), 6.3 (J_{ac}) and 17.7 Hz (J_{bc}). There are many other possible patterns depending upon the individual chemical shifts and the magnitude of the coupling constants. Thus in a different compound, the X signal could be upfield of the AB system rather than downfield, and it could in its turn be coupled to other protons extending the spin system beyond that of an ABX.

If the separation of the A and B signals in Hz became closer to the coupling constants to the X signal, this first-order analysis would break down, and the two central lines in the X signal (lines 2 and 3 in the H_a signal in Fig. 4.43) would move closer together, and the coupling constants would no longer be measured accurately by the spacings of the 12 lines. Furthermore, more lines might appear than those in the first-order analysis, as we shall see in Sect. 4.6.5.

Separate signals from diastereotopic protons are common for a methylene group adjacent to a stereogenic centre, as in the compounds **52** and **56**, but it is even quite commonly observed when the stereogenic centre is further away. Equally, a pair of methyl groups can be diastereotopic just as hydrogen atoms are. Thus the t-butyldimethylsilyl ether **57** in Fig. 4.44 has diastereotopic protons H_a and H_b adjacent to the stereogenic centre, but it also has diastereotopic methyl groups Me_a and Me_b on the silicon atom, one further bond away from the stereogenic centre. However, the methylene protons in the *C*-ethyl group in the ketone **57**, although structurally diastereotopic, remain as a simple quartet at δ 2.45, because there is free rotation and several well populated conformers for the chain of four bonds between them and the stereogenic centre. Unsurprisingly they experience essentially identical magnetic environments.

But in a system where one conformation is heavily populated, the two protons in a methylene group even further away from the stereogenic centre may experience different magnetic fields. Thus the diastereotopic protons in the ethoxy group of the ester **58**, five bonds away from the stereogenic centre, appear as a pair of overlapping, heavily roofing, but resolved, double quartets (Fig. 4.45) with chemical shifts of δ 4.01 and 3.96.

Diastereotopic methyl groups frequently appear as separate signals in ^{13}C spectra, where the greater dispersion makes it easier to resolve the singlets from the two carbon atoms of an isopropyl group. Thus the diastereotopic methyl groups in

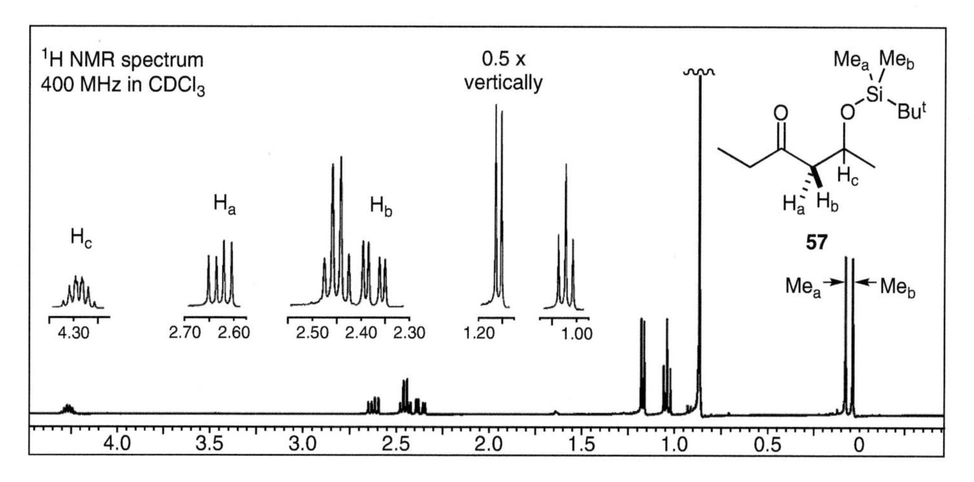

Fig. 4.44 ^{1}H NMR spectrum of a t-butyldimethylsilyl ether

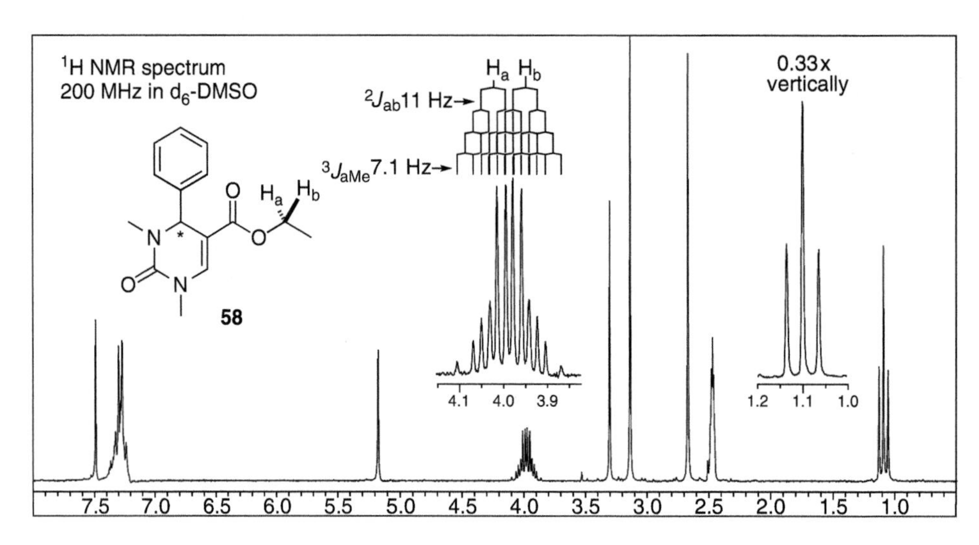

Fig. 4.45 Diastereotopic methylene signals five bonds from the stereogenic centre

4-methyl-2-pentanol **59** are resolved in the ^{13}C spectrum (Fig. 4.46) but not in the ^{1}H spectrum, which only resolves the diastereotopic protons adjacent to the stereogenic centre.

It is not even necessary for there to be a stereogenic centre. The two ethoxy groups in the diethylacetal **60** in Fig. 4.47 are *enantiotopic*—replacing one with another group would create the enantiomer that would be created by replacing the other ethoxy group. But the methylene hydrogens within the ethoxy groups are diastereotopic, even though the molecule is achiral. Taking one of the hydrogens, say H_a from the front ethoxy group, and replacing it with another group would create one diastereoisomer; replacing the other hydrogen H_b would create a different diastereoisomer. The same phenomenon is equally true of the enantiotopic ethoxy group at the rear, and the two diastereoisomers this time would be the enantiomers of the first pair. Thus, H_a and H_b are neither chemically nor magnetically equivalent, but the two H_as are chemically and magnetically equivalent, and the two H_bs likewise. In consequence, the ethoxy groups in the acetal **60** give rise to a complicated, highly symmetrical, but understandable set of signals in the range δ 3.85–3.68, expanded in Fig. 4.47. It consists of a pair of mutually coupled double quartets at δ 3.79 and δ 3.73, with strong roofing within the doublet component (J_{ab} of 9.6 Hz), and a smaller coupling constant ($J_{aMe} = J_{bMe} = 7.2$ Hz) for the coupling to the methyl group.

4.6.4 ^{1}H–^{1}H Long-Range Coupling ($^{4}J_{HH}$ and $^{5}J_{HH}$)

Coupling through four or more bonds is often called long-range coupling. The coupling constants are naturally quite small, rarely outside the range 0–3 Hz. They are at the higher

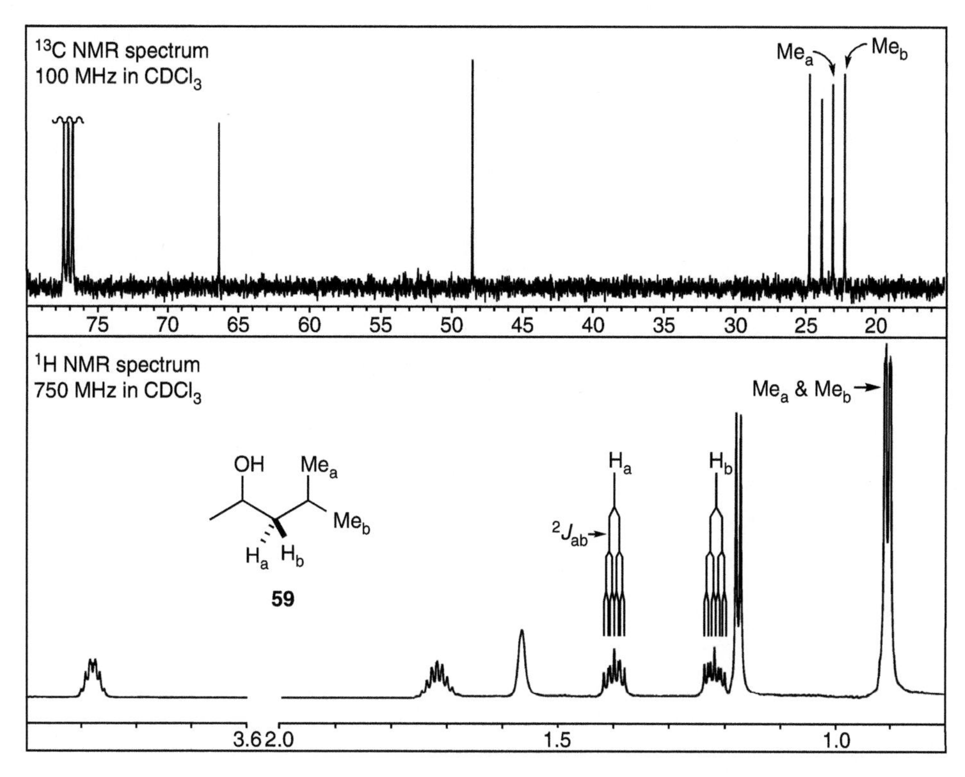

Fig. 4.46 ^{1}H and ^{13}C NMR spectra with diastereotopic methyl groups in the ^{13}C NMR spectrum

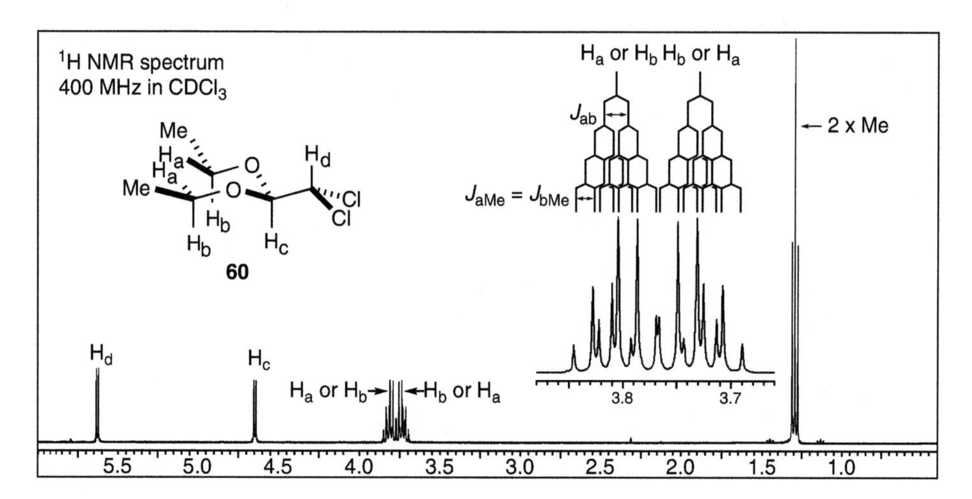

Fig. 4.47 ^{1}H NMR spectrum of the diethyl acetal of dichloroacetaldehyde

end of this range in two quite commonly encountered situations. The first is in unsaturated systems, when a double bond is oriented so that its π-system overlaps with a C–H σ-bond, as in the allyl, allene and propargyl systems **61–63**. The difference between the two allylic couplings shown in **61** is too small to permit a reliable assignment of geometry. Homoallylic coupling ($^5J_{HH}$ = 1–2 Hz) is sometimes resolved, but only when both allylic C–H bonds overlap with the double bond, as in the allyl and allene partial structures **64** and **65**, and is especially strong when the C–H bonds are rigidly held and doubly conjugated as in 1,4-cyclohexadienes **66**.

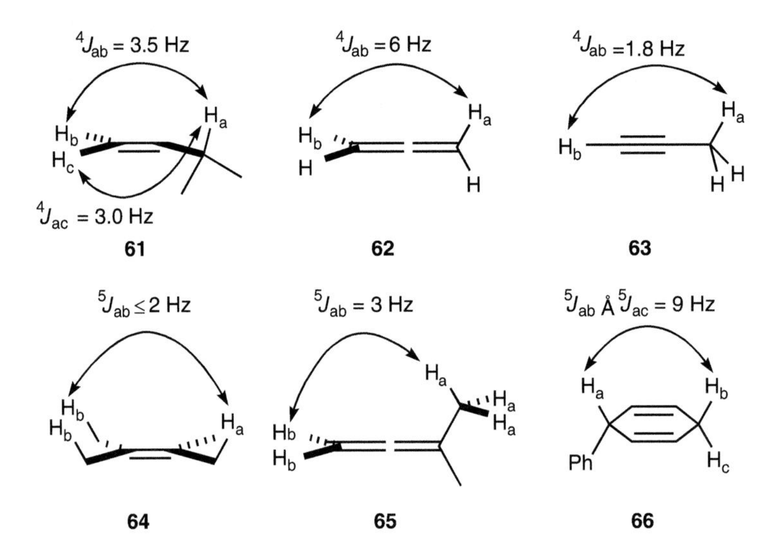

We can see an example of allylic coupling in the signal at *δ* 6.16 from the α-proton H_b of crotonaldehyde **67** in Fig. 4.48. This proton is vicinally coupled ($^3J_{bc}$) to the β-proton H_c with a coupling constant of 15.6 Hz; it is also coupled ($^3J_{ab}$) to the aldehyde proton H_a with a coupling constant of 7.9 Hz, nearly half as large. As a result, the α-proton H_b is a double doublet, consisting of four nearly equally spaced signals, but it is also allylically coupled ($^4J_{bMe}$) to each of the three protons of the methyl group, making each line of the double doublet a fine quartet with a coupling constant of 1.6 Hz. This signal, with resolved coupling from all of the other protons in the molecule, can be described as: *δ* 6.16 (1H, ddq, *J* 15.6, 7.9 and 1.6 Hz). The remaining signals match up: the aldehyde proton H_a at *δ* 9.52 is a doublet, with necessarily the same ^3J vicinal coupling of 7.9 Hz to the α-proton; the signal from the β-proton at *δ* 6.89, downfield from the α-proton, is a double quartet with coupling constants of 15.6 and 6.9 Hz; and the signal from the methyl group at *δ* 2.05 is a double doublet, with matching coupling of 6.9 Hz to the β-proton and allylic coupling of 1.6 Hz to the α-proton H_b.

The second commonly encountered case of long-range coupling is in rigid systems when the four bonds adopt a planar W arrangement, as emphasised for the 1,3-diequatorial protons in rigid cyclohexanes **68** and in bicyclo[2.2.1]heptanes **69.** Again, there are exceptionally high values when the overlap of the σ-bonds is especially favourable, as in

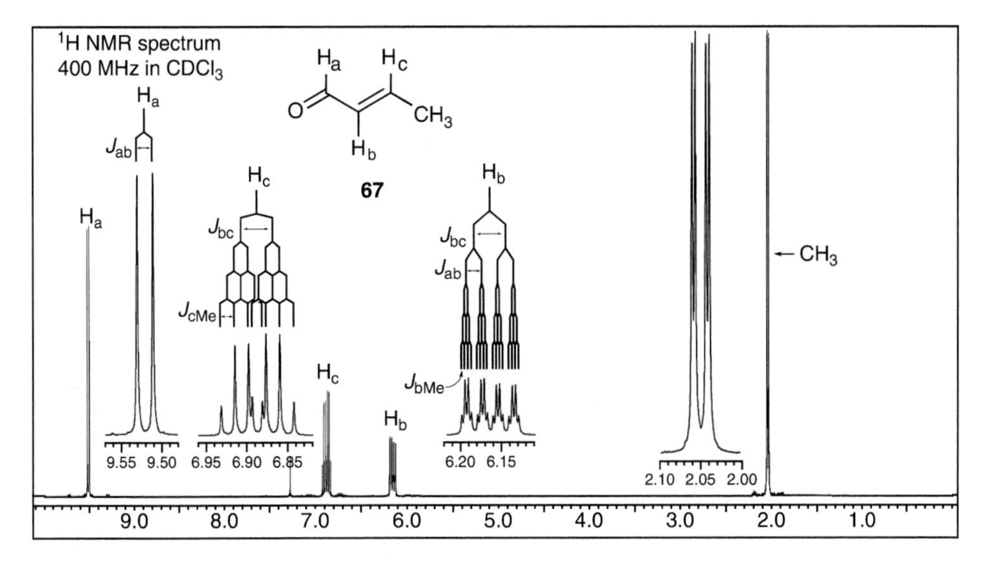

Fig. 4.48 ^{1}H NMR spectrum of crotonaldehyde

bicyclo[2.1.1]hexane **70**. W-coupling is also evident in the frequently resolved meta coupling in aromatic rings **71**, but five-bond para coupling is rarely resolved.

^{4}J = 1-2 Hz

^{4}J = 1-2 Hz

^{4}J = 7-8 Hz

H_o $^{3}J_o$ = ~8 Hz

H_m $^{4}J_m$ = 1-3 Hz

H_p $^{5}J_p$ = 0-1 Hz

68 **69** **70** **71**

We can see long-range W-coupling in an aromatic system in the spectrum of 3-chloropyridine **72** in Fig. 4.49. If we look only at the full spectrum and ignore the W-coupling revealed in the expansions, vicinal coupling leads H_a to be a singlet, H_b to be a doublet (^{3}J = 8.2 Hz), H_c to be a double doublet (^{3}J = 8.2 and 4.7 Hz) and H_d to be a doublet (^{3}J = 4.7 Hz). But each of these signals is split by meta coupling: H_a by H_b (^{4}J = 2.5 Hz), H_b by both H_a and H_d (^{4}J = 2.5 and 1.4 Hz), and H_d by H_b (^{4}J = 1.4 Hz). H_c even shows resolved para coupling (^{5}J 0.6 Hz), although it is not resolved in H_a, which is probably further broadened by unresolved meta coupling to H_d.

Long-range $^{5}J_{HH}$ coupling is most often resolved when the intervening chain of atoms is held in a planar zigzag conformation. If long-range coupling is present but not resolved, it leads simply to line broadening, as in the signal from H_a in 3-chloropyridine. Two or three of the earlier spectra used in this chapter show this phenomenon—look at the

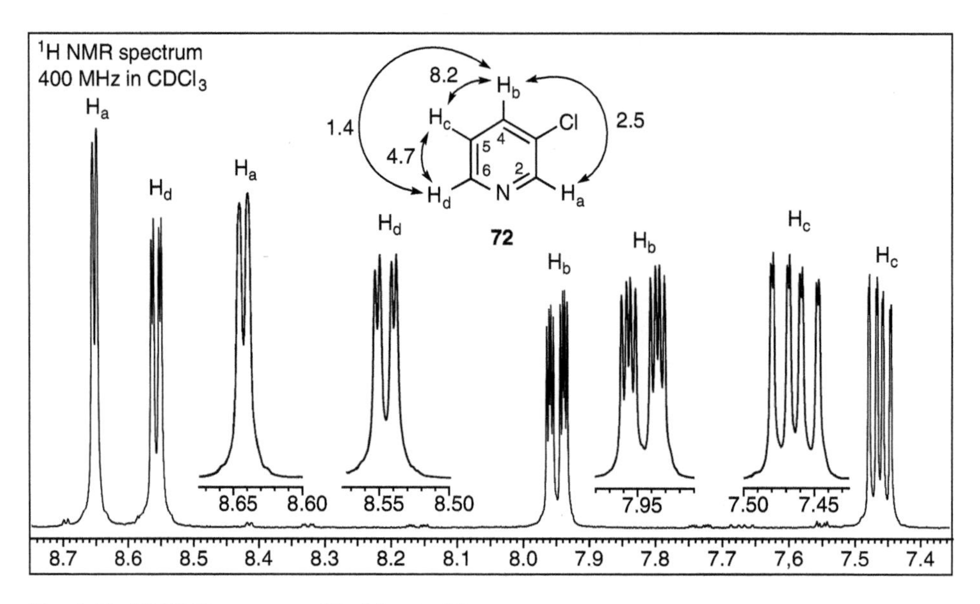

Fig. 4.49 ^{1}H NMR spectrum of 3-chloropyridine

spectrum of the ester **1** in Fig. 4.12: the 3-proton line from the methoxy group is actually taller than the 6-proton line from the two aromatic methyl groups. The integral, of course, reveals that the areas under the signals are in the proportion 3:6; but if the signals were expanded we would see that the width at half height of the *C*—Me signal was greater than that of the signal from the *O*—Me signal. The *C*—Me signal is broadened because the protons are coupled to the protons on C-2 and C-4 by ^4J coupling similar to allylic coupling. Similarly, we can now see why the doublet from H_α in diphenylacetaldehyde **40** in Fig. 4.27 is so broad, whereas the aldehyde proton gives rise to a sharp doublet: H_α is coupled with small and unresolved coupling to the four *ortho* protons in the aromatic rings.

An unusual example of long range coupling can be seen in the spectrum of pantolactone **73** in Fig. 4.50. The AB system of H_a and H_b with a coupling constant of 9.1 Hz is clear (δ 4.02 and 3.93). The doublet from H_b is notably shorter and broader than that of H_a. Similarly, the upfield singlet from one of the two methyl groups is shorter and broader than the downfield singlet; H_b is obviously coupled to Me_a. The expansion of the H_a and H_b signals in Fig. 4.50 is from a spectrum that has been optimised, both in the taking and the processing, in order to reveal the quartet structure within each line of the doublet, and a ^4J coupling constant of 0.6 Hz, matched in the expansion of the signal from Me_a. The most populated conformation of this molecule must have H_b held in a good W-arrangement with one of the hydrogen atoms of the methyl group *trans* to it Me_a. Since the methyl group is freely rotating, the observed coupling constant will have been reduced from the maximum value, because only one of the three hydrogen atoms can be in a W-arrangement at any one time, making the resolution of this signal all the more remarkable.

We can also see long-range coupling again in what we called the 'confused-looking signals in the δ 7.5–7.0 region' from the aromatic protons in diphenylacetaldehyde **40** in

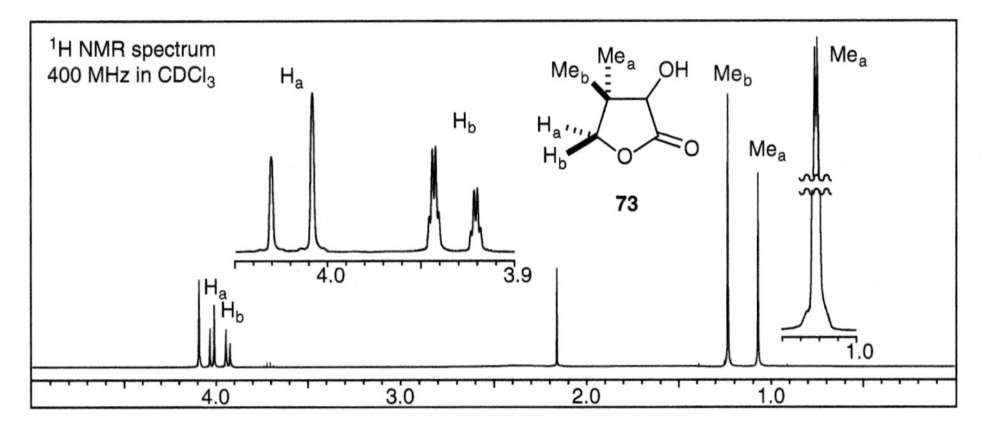

Fig. 4.50 ^{1}H NMR spectrum of pantolactone

Fig. 4.27. Monosubstituted aromatic rings like this often have overlapping signals, especially when the substituent is effectively an alkyl group. Indeed, the ChemDraw®-predicted chemical shifts for this compound are: *meta*: δ 7.33, *para*: δ 7.26, and *ortho*: δ 7.23, which would probably not be fully resolved. The effect through space from the anisotropy both of an aldehyde group and another phenyl ring has shifted the signals in ways that the program did not handle perfectly, and the three signals are in fact in the same order, but spread wide enough to allow us to analyse them: *meta*: δ 7.41, *para*: δ 7.34, and *ortho*: δ 7.26. They are expanded in Fig. 4.51, with a superimposed approximation to a first-order analysis.

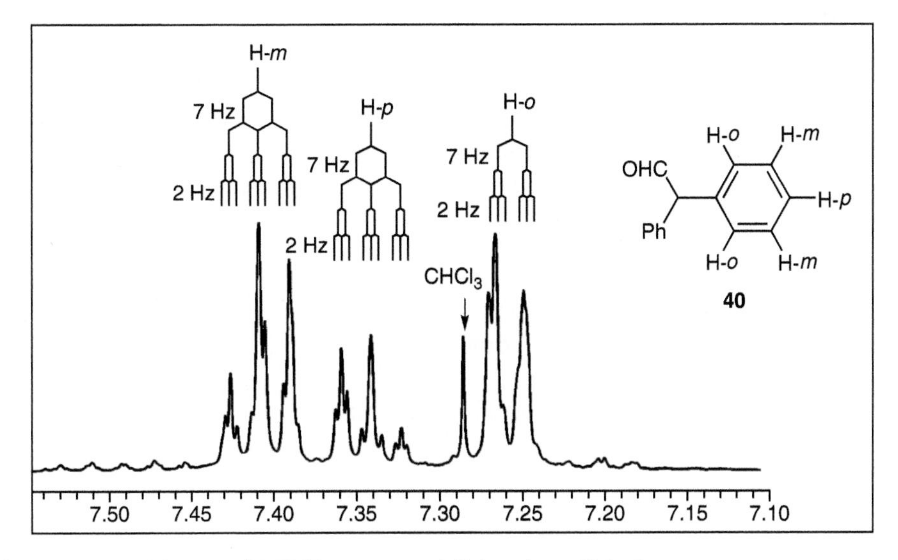

Fig. 4.51 Detail from the ^{1}H NMR spectrum of diphenylacetaldehyde

In a monosubstituted benzene ring like this, vicinal coupling will lead the *ortho* protons to give rise to doublets, the *meta* protons either to double doublets or to triplets (depending upon whether the coupling constants are equal or not), and the *para* proton, which will be half as intense, to a triplet. Whereas H_c in the pyridine **70** had two very different *ortho* coupling constants, the less polarised benzene ring in diphenylacetaldehyde has the two coupling constants equal, with both $^3J_{om}$ and $^3J_{mp}$ approximately 7 Hz. But in addition to the *ortho* coupling, these signals show fine structure that stems from *meta* coupling that is not fully resolved in all the signals. Thus, each of the lines in the strongly roofed 2-proton triplet in the middle from the *para* proton is split again into fine triplets, with $^4J_{op}$ of approximately 2 Hz, because of coupling to the *ortho* protons. The incomplete resolution may stem from para coupling, since the *meta* protons are coupled to the *ortho* in two ways, vicinal and long-range para. It may also stem from imperfect matching of the coupling constants to the *ortho* and *para* protons, and yet another explanation is covered in the next section.

Looking only at the couplings with the larger coupling constants, the two-proton doublet, two-proton triplet and one-proton triplet make a characteristic pattern for a monosubstituted benzene ring, with the smaller meta couplings not always resolved but supporting the assignment. Other patterns in the signals with the larger coupling constants identify the substitution patterns in the various di- and tri-substituted aromatic rings. Figure 4.52 illustrates examples of some of the most common patterns.

As with mono-substituted rings, the signals can appear in any order of chemical shift, depending upon the effects that the various substituents have in their different arrangements around the ring, but the pattern, ignoring the fine meta coupling, of two triplets and a doublet for a monosubstituted ring is definitive, as are two doublets and two triplets for an *ortho*-disubstituted ring, two doublets, a triplet and a singlet for a *meta*-disubstituted ring, two doublets for a *para*-disubstituted ring, two doublets and a triplet for a 1,2,3-trisubstituted ring, and two doublets and a singlet for a 1,2,4-trisubstituted ring, with, in this case, the meta coupling identifying which doublet is which. The patterns change of course if two of the substituents in any one ring are the same, and what is called a triplet in the list above may well be a double doublet if the two ortho coupling constants are not the same. Furthermore, signals can have such close and even identical chemical shifts that these patterns are obscured or rendered far from first-order.

4.6.5 Deviations From First-Order Coupling

We have not, so far, strayed far from the first-order approximation; we have only added the roofing that comes about with strong coupling, when the chemical shift difference and the coupling constant are similar, and added a cautionary word that the position of the lines may

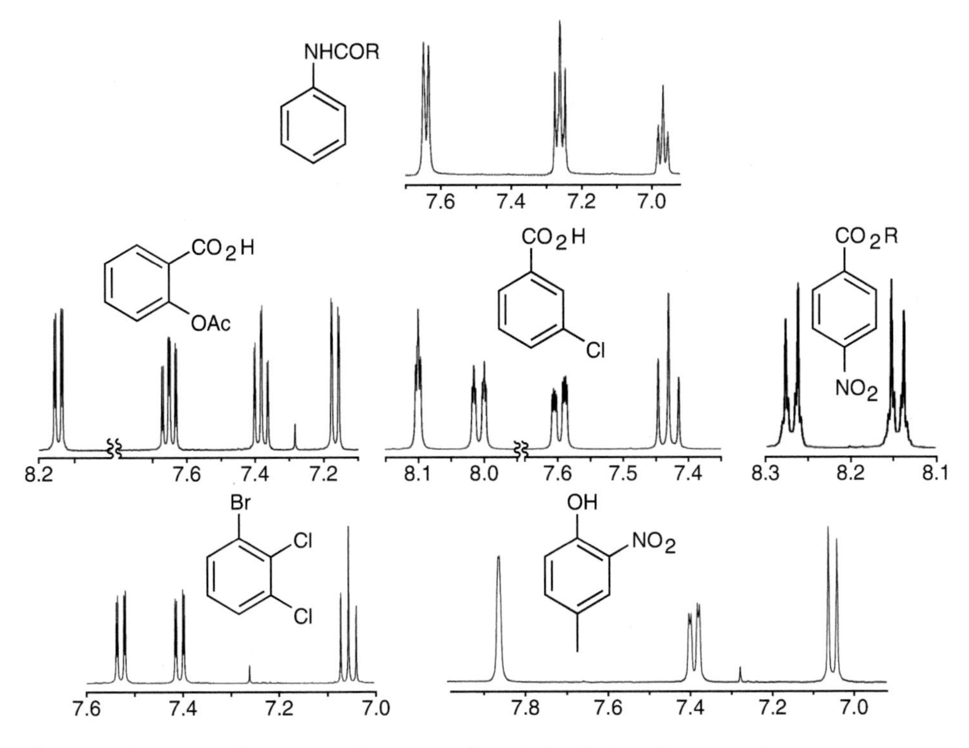

Fig. 4.52 Some coupling patterns for mono-, di- and tri-substituted benzene rings

not strictly measure the coupling constants. In a many proton spin system, however, the first-order analysis is not always adequate. It is usually possible to discern the essential pattern, but extra lines are not at all uncommon. They stem from the many energy levels populated in a multi-proton system, and the many transitions not taken into account in the first-order analysis.

In Fig. 4.53 upward-pointing arrows indicate nuclear magnets in their low-energy orientation with respect to the applied magnetic field, downward-pointing arrows indicate

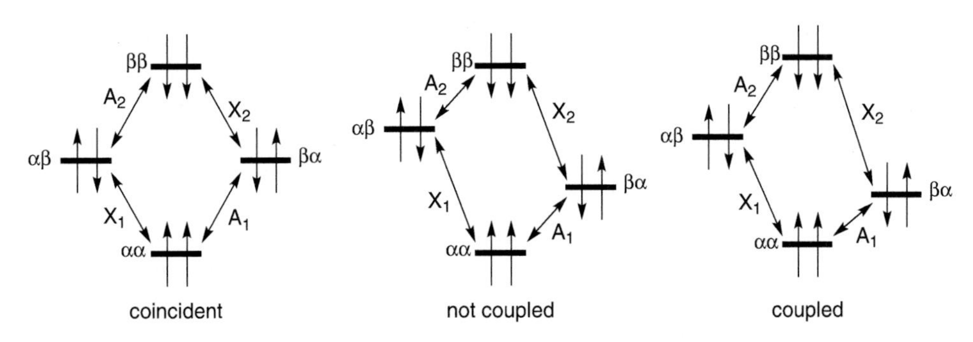

Fig. 4.53 Energy levels and one-quantum transitions for two spins

nuclear magnets in their high-energy orientation with respect to the magnetic field, and levels of higher energy are indicated by vertical upward displacement. The four energy levels in a two spin system can take one of three forms. If the chemical shifts of the A and X nuclei are identical, the four one-quantum transitions A_1, A_2, X_1 and X_2 are all equal, and the two nuclei will give rise to coincident singlets. If the chemical shifts of the A and X signals are different, and they are not coupled, there are two pairs of one-quantum transitions A and X, which will give rise to two singlets, since A_1 and A_2 are equal, and X_1 and X_2 are equal. If, on the other hand, they are coupled, the energy levels are all shifted so that the magnitudes of A_1 and A_2 (and of X_1 and X_2) are different from each other. The four different one-quantum transitions, A_1, A_2, X_1 and X_2, give rise to the four lines of a first-order AX system.

When we have more than two spins, the picture becomes more complicated, because there can be many more one-quantum transitions. Thus, as illustrated in Fig. 4.54, with three spins, there are 15 possible one-quantum transitions, and with four spins there are 56 possible one-quantum transitions.

In a first-order spectrum with three spins and no coincident lines, 12 of the 15 transitions are visible, as in the first-order (with roofing) spectrum of diethyl acetylsuccinate **56** in Fig. 4.43. Problems arise in an ABX system when the separation in chemical shift between the A and B signal measured in Hz is small compared to the coupling constant J_{AX} (or J_{BX}). If $|(\nu_A - \nu_B)|/J_{AX}$ is more than a factor of 2 or 3, an essentially first-order spectrum is observed, but if it is less, it may not be. A great variety of patterns emerge in these cases, and analysis by eye may not always be straightforward. A full theoretical treatment does account for the patterns observed, including for all those described below, but it is no longer a first-order analysis.

Take for example the ABX system in the ^{1}H NMR spectrum of the aromatic region of 2,3-dimethoxybenzoic acid **74** in Fig. 4.55. The hydrogen atom *ortho* to the carboxyl group gives rise to a clear downfield double doublet, seemingly first-order except that it appears to have a suspiciously large meta coupling constant, together with some extra, 'outside' lines from two of the fifteen transitions that do not show up in a first-order

Fig. 4.54 Energy levels for three- and four-spin systems

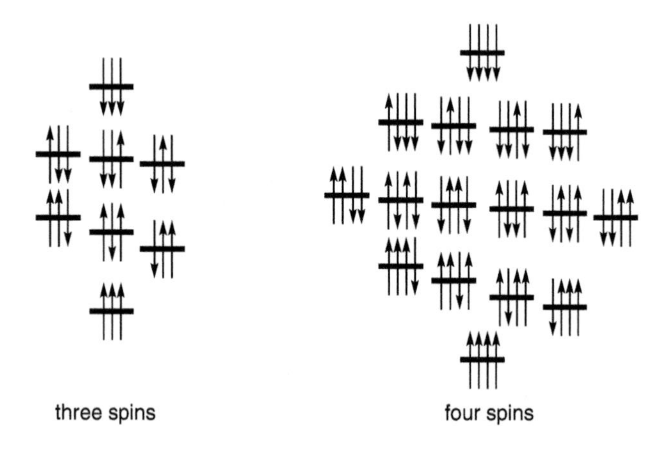

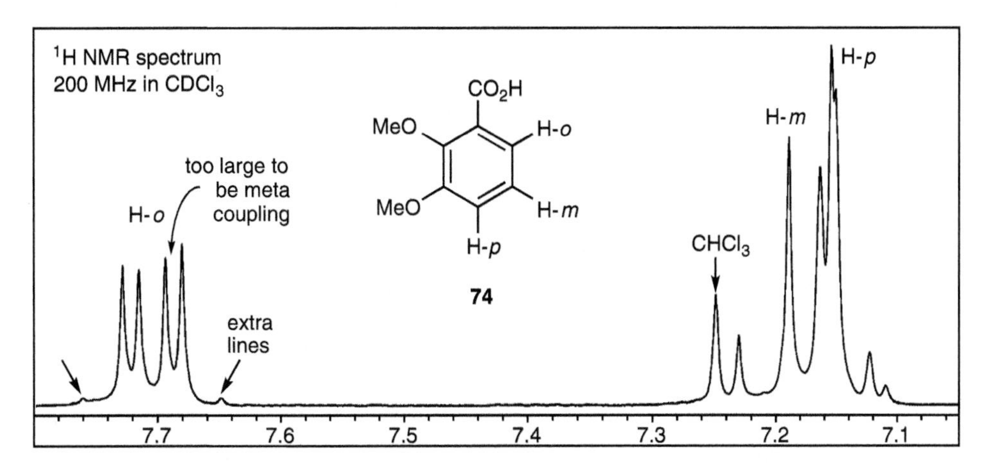

Fig. 4.55 ^{1}H NMR spectrum of 2,3-dimethoxybenzoic acid

spectrum. The signals from the other two aromatic protons are difficult to disentangle. We can expect them to be a double doublet and, more or less, a triplet. Although this pattern can be discerned in the δ 7.1–7.25 range, it is not easy to pick out.

The two overlapping signals are very close in chemical shift, and since they are coupled with steep roofing we should not expect to see a first-order spectrum from them. We cannot easily extract the coupling constants from this spectrum simply by measuring the separation of the peaks, although it is possible with a computation. However, we can use a computer program to simulate the spectrum, and extract the values used in the simulation. In this case, feeding in the values for $\delta_{\text{H-}o}$ 7.702, $\delta_{\text{H-}m}$ 7.181 and $\delta_{\text{H-}p}$ 7.146, and the coupling constants $^3J_{o\text{-}m}$ 7.75, $^4J_{o\text{-}p}$ −1.67 and $^3J_{m\text{-}p}$ 8.15 gave a good fit. The value of $|(\nu_A - \nu_B)|$ in this 200 MHz spectrum is only 7.0 Hz, from which we can confirm that $|(\nu_A - \nu_B)|/J_{o\text{-}m}$ at 0.9 is well below the value that can be expected to give rise to a first-order spectrum.

In the general case, as the chemical shifts of the A and the B protons approach each other, the two central lines of the X signal move towards each other, until, in the extreme, when the A and the B protons have identical chemical shifts, the central lines of the X signal would coincide—it would look exactly like a 1:2:1 triplet, even though the coupling constants $^3J_{AX}$ and $^3J_{BX}$ need not be identical.

Four spins, with their 56 possible transitions, can give rise to patterns of signals even more dramatically different from first-order. The most common examples are from AA′BB′ systems, where two identical A signals coupled to two identical B signals have more than one pathway between them, and more than one coupling constant. The three examples in the upper traces in Fig. 4.56 illustrate some patterns that turn up frequently in aromatic rings and dienes.

At first glance they look like doublets with roofing, but there are several more lines than the usual four for an AB system. The aromatic protons in *p*-bromophenetole **75** might be expected to show an AB pattern, and they more or less do, but there are extra lines, conspicuously inside each of the doublets, as a consequence of there being two identical protons *ortho* to each of the substituents and two identical protons *meta* to that substituent. This is a common type of AA′BB′ system characteristic of *para*-disubstituted benzenes. Another is that of 1,4-diphenylbutadiene **76**. The pattern for the 1,2-disubstituted benzene ring in 1,2-dibromobenzene **36** is more nearly that of a pair of double doublets, but the inside lines are too close together to be a measure of the meta coupling constant, and again there are extra lines. In order to measure the coupling constants it is helpful to simulate these spectra using one of the several available programs in NMR processing packages. The lower traces in Fig. 4.56 are simulations using Mestrenova, achieved by estimating the chemical shifts and trying plausible coupling constants, including all the long-range values and those for the coupling between identical spins, until a reasonably good fit is found. There are also more advanced packages that carry out the iterative process for you.

Another category of AA′BB′ systems is seen with a pair of adjacent methylene groups XCH_2CH_2Y having no further coupling. Quite often they appear in open-chain systems as a first-order pair of 1:2:1 triplets, as in the spectrum of dihydrocinnamic acid **77** in Fig. 4.57. Each of the protons H_a has two geometrical relationships with the protons H_b, and could have two different coupling constants. The simulation of the spectrum for this compound used two identical values, 8.0 Hz, presumably because the occupancy of the populated and rapidly interconverting conformations made the

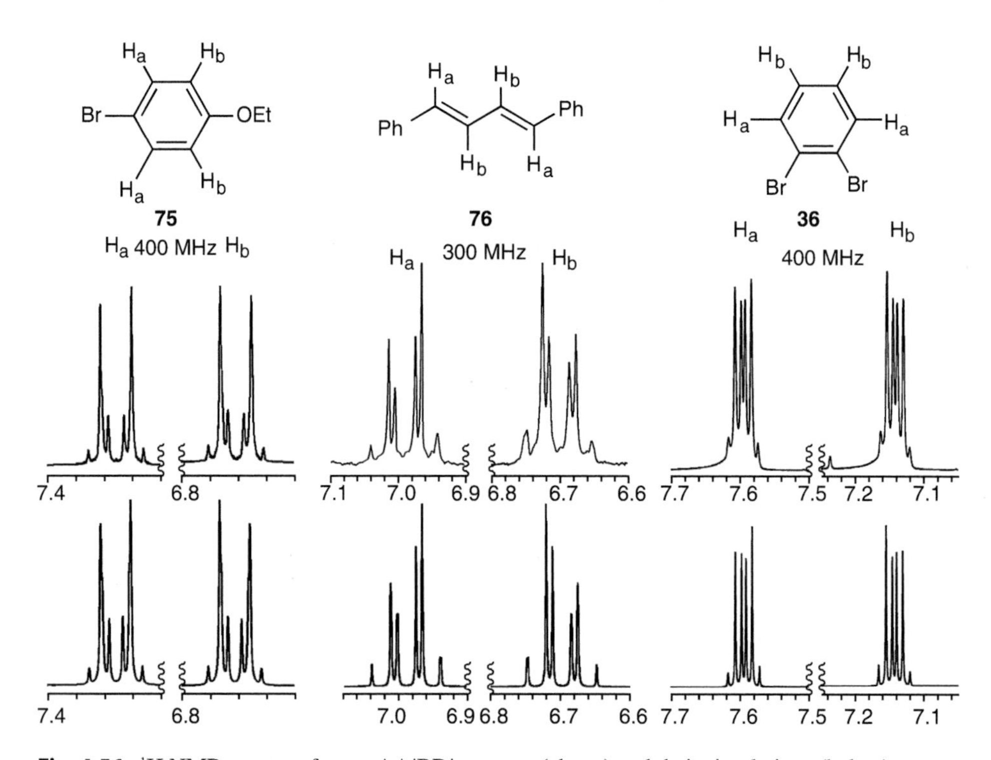

Fig. 4.56 ¹H NMR spectra of some AA′BB′ systems (above) and their simulations (below)

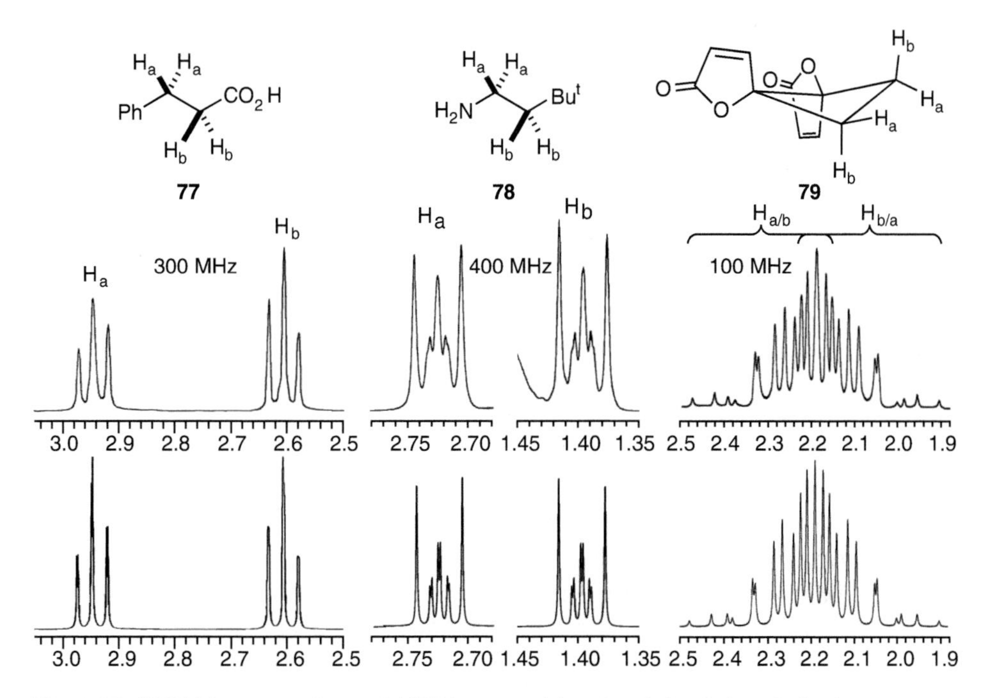

Fig. 4.57 ^{1}H NMR spectra of more AA′BB′ systems (above) and simulations (below)

two relationships average both coupling constants to the same value. On the other hand, even though they have rapid rotation about the C–C bond, open-chain compounds with adjacent methylene groups do not always give first order spectra. For example, the spectrum of 3,3-dimethylbutylamine **78** in Fig. 4.57 is clearly more complicated. It does bear some resemblance to a pair of triplets, but with the central line broken up. (The sloping line at the downfield end of the H_b signal is the shoulder of the broad signal of the NH_2 protons.) The simulation in this case needed two different values, 10.5 and 4.7 Hz, for the coupling constants between H_a and H_b to take account of the different relationships each atom H_a has to the two H_b atoms.

In AA′BB′ systems without free rotation, second-order spectra are much more common. Anemonin **79**, gave a 100 MHz spectrum that bears no resemblance to a pair of triplets. In this case the two H_a protons and the two H_b protons are so close in chemical shift that the two signals overlap. There are also two large coupling constants, one geminal $^2J_{ab}$ and the other vicinal $^3J_{ab}$, as well as the two values $^3J_{aa}$ and $^3J_{bb}$ combining to make the multiplets so spectacularly different from first-order.

For now, we need only accept that some splitting patterns are not readily analysed just by inspection and a first-order analysis, and to be aware that, should the need arise, it is possible to interpret these spectra with computer simulations or a full analysis of all the energy levels. In the case of anemonin **79**, for example, the full computer simulation found that the signals were separated by 0.176 δ units (17.6 Hz in this 100 MHz spectrum), and that the coupling constants were: $^2J_{AB}$ −12.15, $^3J_{AB}$ 10.19, $^3J_{AA}$ 2.24 and $^3J_{BB}$ 10.72 Hz. At 750 MHz, the signals would no longer overlap (they would be 129 Hz apart), but the spectrum would still be nothing like first-order.

4.7 ^{1}H–^{1}H Coupling: The Magnitude of Coupling Constants

4.7.1 The Sign of Coupling Constants

One of the ^{1}H–^{1}H coupling constants for anemonin was listed in the paragraph above as negative. In the course of the discussion so far, we have seen coupling constants as widely different as 0.6 Hz and 17.7 Hz, but some of them have been positive and some negative without this point being addressed. We have not needed to know about the sign of the coupling constant until now, because it does not affect the appearance of the spectrum, but we do need to know the sign, because it changes the way in which structural variations affect the magnitude of coupling constants, and we also need to know it in order to compute the appearance of spectra that are not first order.

To understand why coupling constants can be positive or negative, we need to look a bit more closely into the energetics of coupling. We saw on the left of Fig. 4.53 the four energy levels for two uncoupled nuclei with identical chemical shifts. The transitions which the instrument measures are those in which the alignment of one of the nuclei changes from the N_β to the N_α state of Fig. 4.1. There are four such transitions labelled A_1, A_2, X_1 and X_2 in Fig. 4.53, and all of them in the case on the left are equal in magnitude. The receiving coils detect only the one signal, and the output of the Fourier transform shows one line and no coupling.

If now we look at two different atoms A and X in the centre of Fig. 4.53, we have the same set-up, but this time the two energy levels in the middle are of different energy, the αβ level raised in energy and the βα level lowered in energy to the same extent. The α spin might be a ^{1}H, and the β spin a ^{13}C atom, but the picture is the same for all AX systems. If there is no coupling ($J = 0$), as when the nuclei are far apart, the αβ level will be as much above the mid-point as the energy level for the βα level is below it. There will again be four transitions, two equal for the A nucleus, labelled A_1 and A_2, and two equal for the X nucleus, labelled X_1 and X_2, giving rise to one line from each.

If, on the other hand, the two nuclei are bonded closely enough, their magnetic fields will affect each other and we have the picture on the right in Fig. 4.53 giving rise to four lines. Let us first look at how that picture arises in the case of two nuclei directly bonded, as they are in a ^{13}C–^{1}H bond. Figure 4.58 illustrates on the left the pairing of the nuclear and electron spins. In the lowest energy arrangement, the A spin will be aligned both with the applied field and with the magnetic field derived from the spin of the nearer s electron. This is not quite obvious, because normally spins are opposed in the lower energy arrangement. In this case, however, the lower energy arrangement is for them to be aligned, because the magnetogyric ratio of the nucleus is positive in sign whereas that

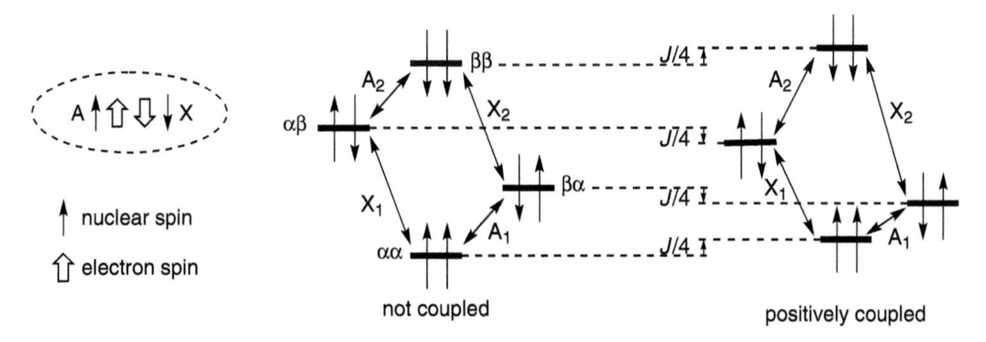

Fig. 4.58 The energy levels for coupling given a positive sign

of the electron is negative. Likewise, the X spin will be aligned with the magnetic field derived from the spin of its nearer s electron (only s orbitals have an electron population at the nucleus). The spin of the s electrons will necessarily be opposed to each other in the bonding orbital, and therefore the lower energy arrangement is that in which the magnetic field of the A nucleus is opposed to the magnetic field of the X nucleus. Conversely, whenever the nuclear spins are aligned, the system will be higher in energy. Thus, the two energy levels, αα and ββ, in which the A and X nuclei have parallel spins, will be raised in energy, and the two energy levels in which they are opposed, αβ and βα, will be lowered (Fig. 4.58, right).

There are now four new energy levels, four different transitions, A_1, A_2, X_1 and X_2, and four lines in the AX spectrum. The A signal is a doublet and the X signal is a doublet, with the same separation between the lines, because $(A_2 - A_1) = (X_2 - X_1) = J_{AX}$. Thus, the extent of the raising and lowering of each of the energy levels is $J_{AX}/4$. The expression $(A_2 - A_1) = (X_2 - X_1)$ arbitrarily defines the sign of the coupling constant. In the arrangement in Fig. 4.58, with the αα and the ββ levels raised in energy and the αβ and βα lowered in energy, J_{AX} has a *positive* sign. More complicated versions of this kind of diagram are needed to analyse spin interactions beyond the AX system, and even more complicated ones to make sense of those spectra that are not first order.

If instead of being directly bonded, the A and X nuclei are separated by two bonds, as they are in a diastereotopic pair of methylene hydrogens, the transmission of information through the s electrons leads the two nuclei to be parallel in the *lower*-energy arrangement on the left in Fig. 4.59, because Hundt's rule says that the two s electrons on the central atom will be lower in energy if they are aligned. The energy levels will have the αα and ββ energy levels lowered by the interaction of the two spins through the bonds, and the αβ and βα levels raised, in contrast to the arrangement in Fig. 4.58. If the coupling constant is the same as that in Fig. 4.58, the two transitions for the A nucleus, A_1 and A_2, are of the same magnitude as before but have changed places, and similarly for X_1 and X_2. The appearance of the spectrum will not have changed, but the coupling constant J, defined as before by $(A_2 - A_1) = (X_2 - X_1)$, is *negative* in sign.

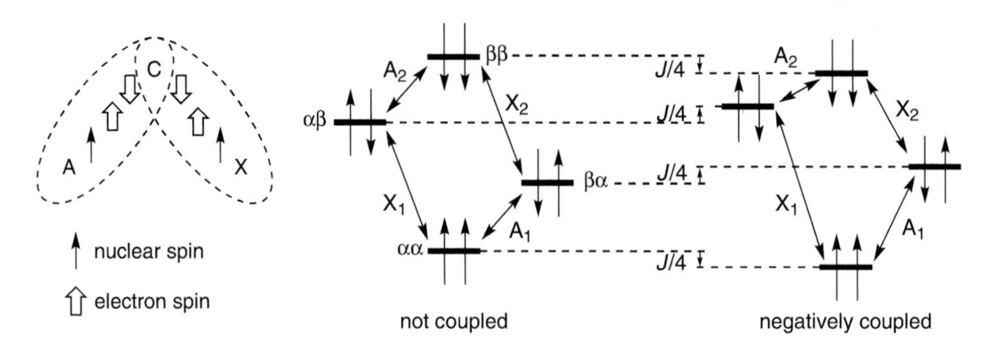

Fig. 4.59 The energy levels for coupling given a negative sign

In general, although not always, one-bond couplings ($^1J_{CC}$ and $^1J_{CH}$) and three-bond couplings ($^3J_{CC}$, $^3J_{CH}$ and $^3J_{HH}$) are positive in sign, and two- and four-bond couplings ($^2J_{CC}$, $^2J_{CH}$ and $^2J_{HH}$, $^4J_{CC}$, $^4J_{CH}$ and $^4J_{HH}$) are negative in sign. Given this understanding, we now discuss separately the factors affecting coupling constants for two-, three- and four-bond coupling.

4.7.2 Vicinal Coupling ($^3J_{HH}$)

The Dihedral Angle Coupling is mediated by the interaction of orbitals within the bonding framework. It is therefore dependent upon the overlap of the orbitals, and hence upon the dihedral angle between the bonds that are involved. Although the s orbitals are the only ones with a presence at the nucleus, information about the nuclear spin can be transmitted from s to p orbitals and hence back to the s orbital on the coupling partner. The relationship between the dihedral angle and the vicinal coupling constant ^3J is given by the theoretically derived Karplus equations:

$$^3J_{ab} = J_0 \cos^2 \phi - 0.28 \quad \left(0^\circ \le \phi \le 90^\circ\right) \tag{4.9}$$

$$^3J_{ab} = J_{180} \cos^2 \phi - 0.28 \quad \left(90^\circ \le \phi \le 180^\circ\right) \tag{4.10}$$

where J_0 and J_{180} are constants which depend upon the substituents on the carbon atoms and ϕ is the dihedral angle defined in the drawing **80** in Fig. 4.60.

The Karplus equations are plotted in Fig. 4.60 using $J_0 = 8.5$ and $J_{180} = 9.5$, the standard values when no better estimate is available. Coupling constants observed experimentally follow this relationship well, but it is not always easy to choose values of J_0 and J_{180}. The

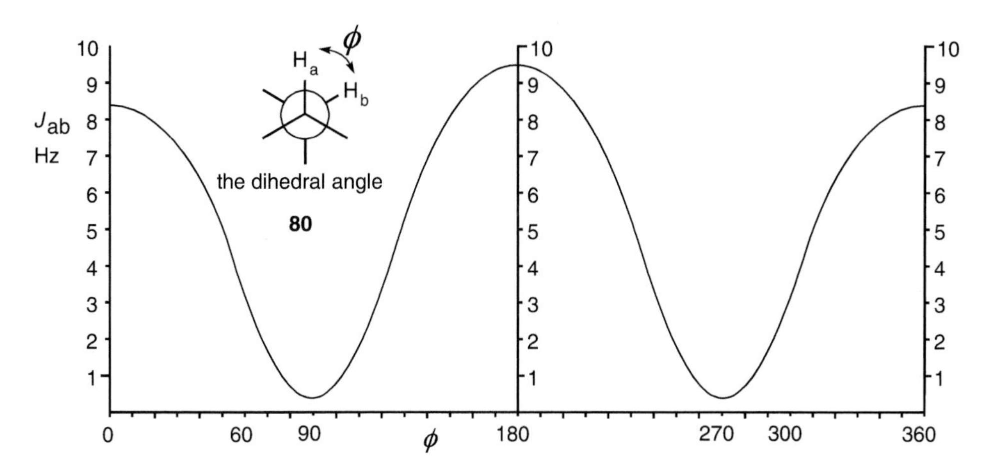

Fig. 4.60 Plot of the Karplus equation

main point to notice is that the coupling constant is at its largest when the dihedral angle is 180°, in other words, when the hydrogens are antiperiplanar and the p orbitals on the two carbon atoms are overlapping most efficiently; slightly smaller when it is 0°, when they are syncoplanar; and at its lowest when the dihedral angle is 90° and the orbitals are orthogonal. In an ethyl group, the free rotation allows the vicinal hydrogens to pass through all these angles, but they will spend most of their time in the usual staggered conformation, with dihedral angles of 60°, 180° and 300°. The coupling constants for an ethyl group that we have seen in Figs. 4.29, 4.31, 4.32, 4.35, 4.42, 4.43 and 4.44 are all close to 7 Hz, which is near the average of the coupling constants given by the Karplus equation for these three angles.

In rigid systems, where averaging is not possible, we frequently see both larger and smaller values. In rigid cyclohexanes **81**, for example, the axial-axial coupling constant, J_{aa}, is usually large, in the range 9–13 Hz, because the dihedral angle is close to 180°. The axial-equatorial and equatorial-equatorial coupling constants, J_{ae}, and J_{ee}, are much smaller, usually in the range 2–5 Hz, because the dihedral angles are close to 60°. The dihedral angles are clearer on the Newman projections **82** and **83**, but it should be remembered that the bond angles are not always so perfectly bisected in real systems. Nevertheless, these differences in a well-behaved and relatively rigid system are large enough to make this a powerful tool in assigning stereochemistry.

J_{aa} = 9-13 Hz J_{ae} = 2-5 Hz J_{ee} = 2-5 Hz

81   **82**   **83**

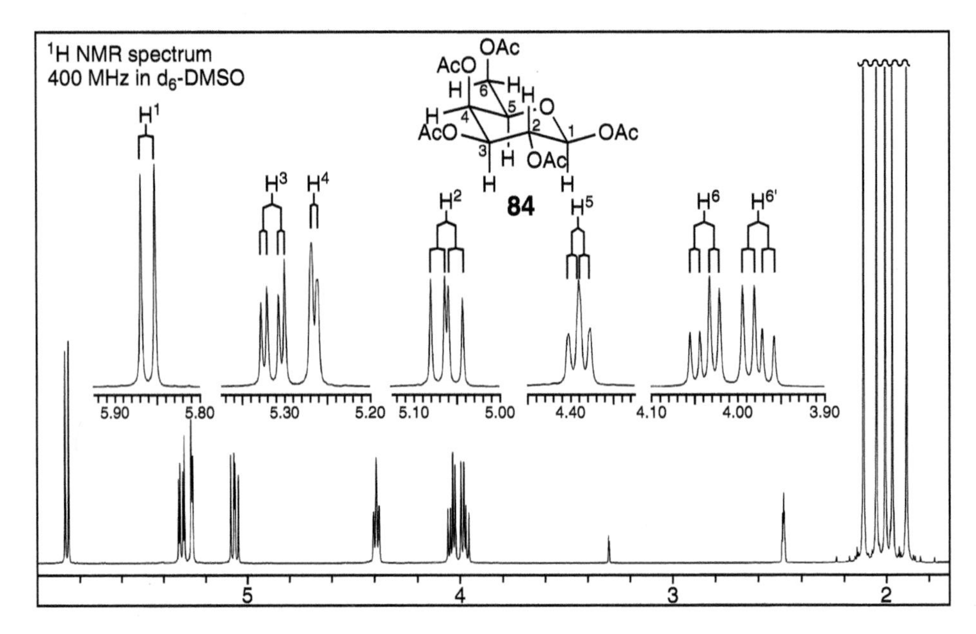

Fig. 4.61 ¹H NMR spectrum of β-D-galactose pentaacetate

For example, each of the hexoses has a characteristic pattern of coupling that makes assignment of structure relatively easy. Figure 4.61 shows the ¹H NMR spectrum of β-D-galactose pentaacetate **84**. The signal furthest downfield, a doublet, is from the proton on the anomeric carbon C-1, downfield because it has two electronegative atoms attached to the carbon atom. The coupling constant is 8.3 Hz, which matches one of the coupling constants in the double-doublet at δ 5.06. The other coupling constant in that signal is 10.3 Hz, which matches one of the coupling constants in the double-doublet at δ 5.33. Thus these are both axial-axial couplings and the acetoxy groups are therefore equatorial on C-1, C-2 and C-3.

The second coupling constant in the signal at δ 5.33 is 3.5 Hz, matching the doublet at δ 5.26. Clearly this is an axial-equatorial coupling, and the acetoxy group on C-4 must be axial. The coupling from the proton on C-4 to the proton on C-5, which gives the triplet at δ 4.38, has such a small coupling constant that it isn't resolved. We see it only as a broadening of both the H-4 and H-5 signals. The H-5 signal couples to the diastereotopic signals of the protons on C-6 with coupling constants of 5.7 and 6.9 Hz, and they are coupled to each other with a geminal coupling of 11.4 Hz. These numbers are at the low end of the ranges in the Karplus equation because of the presence of the electronegative acetoxy substituents (see below).

The conformations of 5-membered rings are much less predictable than 6-membered rings. The angles between *trans* and *cis* protons depend upon where the fold in the ring places them, and several conformations may be significantly populated. Arrangements close to antiperiplanar **85** are rare and eclipsing **88** is more common than in six-membered rings, with the result that *cis* coupling constants **87** and **88** are sometimes higher but sometimes lower than *trans* coupling constants **85** and **86**—making stereochemical assignments based on coupling unreliable in 5-membered rings.

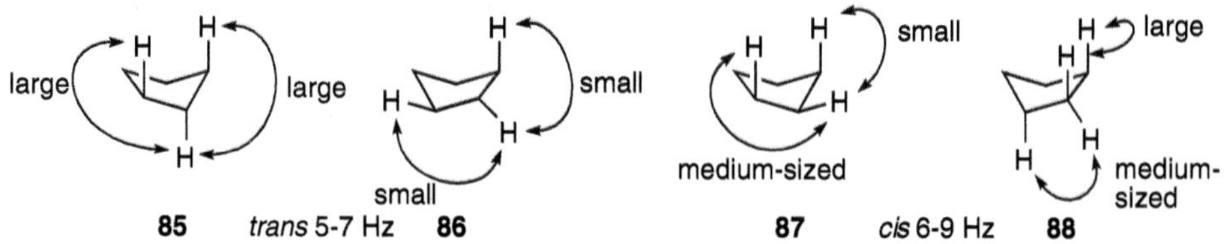

large large small small small medium-sized large medium-sized

85 *trans* 5-7 Hz **86** **87** *cis* 6-9 Hz **88**

Here is a specific example in a group of 5-membered rings—pyrazolines **89** have *cis* coupling constants larger than the *trans*, whereas pyrazolines **90** have *trans* coupling constants larger than the *cis*. The electron-withdrawing group Z in **89** makes the nitrogen atom to which it is attached trigonal, and the whole ring is close to planar. As a result the *cis* protons have dihedral angles close to 0°. The electron-donating group X in **90** makes the nitrogen atom to which it is attached tetrahedral and the ring is buckled. As a result the *trans* protons can have larger dihedral angles approaching 180°.

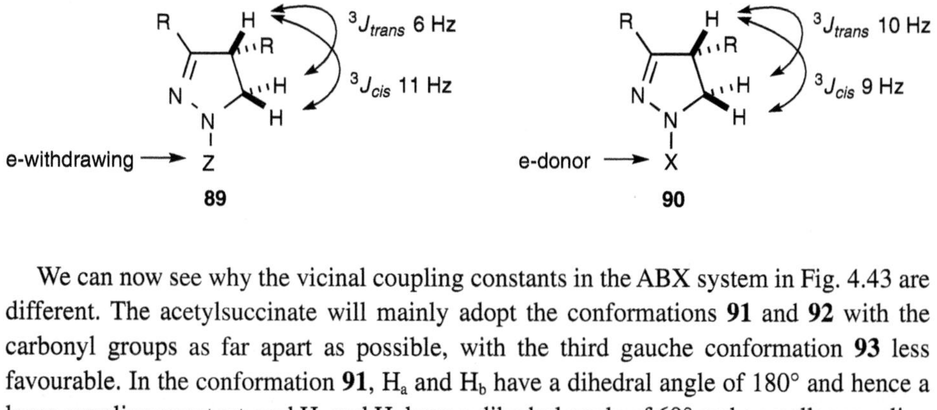

$^3J_{trans}$ 6 Hz, $^3J_{cis}$ 11 Hz; e-withdrawing → Z; **89**

$^3J_{trans}$ 10 Hz, $^3J_{cis}$ 9 Hz; e-donor → X; **90**

We can now see why the vicinal coupling constants in the ABX system in Fig. 4.43 are different. The acetylsuccinate will mainly adopt the conformations **91** and **92** with the carbonyl groups as far apart as possible, with the third gauche conformation **93** less favourable. In the conformation **91**, H_a and H_b have a dihedral angle of 180° and hence a large coupling constant, and H_a and H_c have a dihedral angle of 60° and a smaller coupling constant. In the alternative conformation **92**, these relationships are inverted. As long as one of these two conformations is more populated than the other, the coupling constants will be different. That they have similar values, 8.3 and 6.3 Hz, indicates that both conformations are populated, but not quite equally.

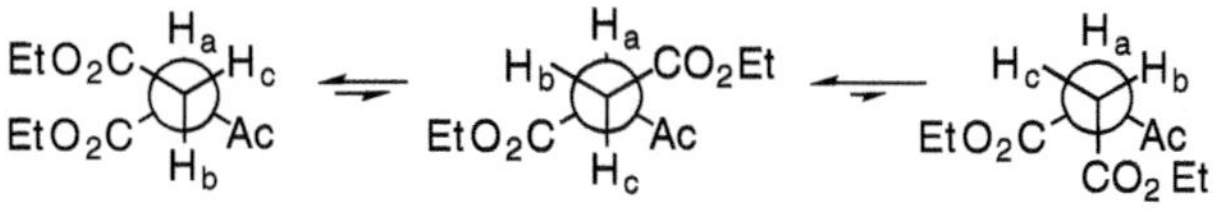

91 ⇌ **92** ⇌ **93**

A modified Karplus equation can be applied to vicinal coupling in alkenes; the numbers are slightly different, but the conclusion is the same. A dihedral angle of 180° is found in *trans* double bonds **94**, where the coupling constants are large, and a dihedral angle of 0° is found in *cis* double bonds **95**, where the coupling constants are smaller. This is exemplified in the coupling constants of 17.2 and 10.8 Hz found in methyl acrylate **51** in Fig. 4.40.

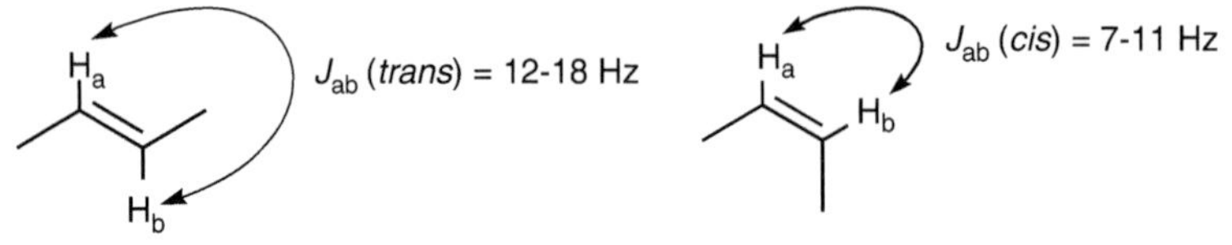

J_{ab} (*trans*) = 12-18 Hz

94

J_{ab} (*cis*) = 7-11 Hz

95

The Presence of Electronegative or Electropositive Elements An electronegative element directly attached to the same carbon atom as one of the vicinally coupled protons reduces the coupling constant, because it reduces the electron population responsible for transmitting the coupling information. Electropositive elements raise the coupling constant. For freely rotating chains, the effect is small **96–98**. The effect of electronegative elements is cumulative, as we can see in the spectrum of the acetal **60** in Fig. 4.47, where the coupling constant between H_c and H_d, with two electronegative substituents each, has dropped to 5.2 Hz in spite of the likelihood that the two protons are antiperiplanar much of the time, whereas those in the structures **96–98** are not.

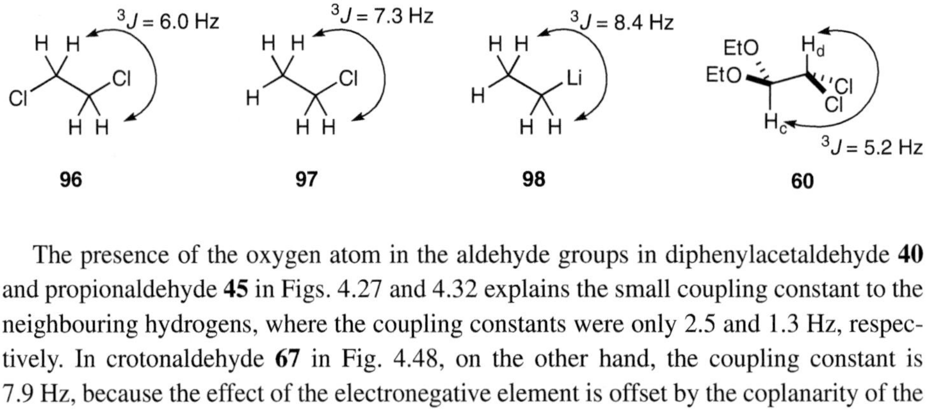

^3J = 6.0 Hz

96

^3J = 7.3 Hz

97

^3J = 8.4 Hz

98

^3J = 5.2 Hz

60

The presence of the oxygen atom in the aldehyde groups in diphenylacetaldehyde **40** and propionaldehyde **45** in Figs. 4.27 and 4.32 explains the small coupling constant to the neighbouring hydrogens, where the coupling constants were only 2.5 and 1.3 Hz, respectively. In crotonaldehyde **67** in Fig. 4.48, on the other hand, the coupling constant is 7.9 Hz, because the effect of the electronegative element is offset by the coplanarity of the conjugated system that keeps the aldehyde proton H_a and the α-proton H_b antiperiplanar most of the time. We also saw the effect of an electronegative element in the spectrum of ethyl propionate **42** in Fig. 4.33, where the coupling constant in the *O*-Et group was smaller (7.14 Hz) than in the *C*-Et group (7.58 Hz). The former has an electronegative element attached and the latter does not.

When the electronegative element is held rigidly antiperiplanar with respect to one of the protons (heavy outline in **99**), then the effect is larger. Thus, J_{ae} is only 2.5 ± 1 Hz when X (OH, OAc or Br) is axial, but it is 5.5 ± 1 when X is equatorial **100**, even though the dihedral angles are close to 60° in both cases. This angular dependence in the effect of an electronegative element explains the exceptionally low coupling constant between the protons on C-4 and C-5 in the spectrum of galactose pentaacetate **84** in Fig. 4.61. The proton on C-5 is axial and antiperiplanar to the acetoxy group on C-4, and the proton on C-4 is antiperiplanar to the ring oxygen, with the result that the coupling between the protons on C-4 and C-5 is so small as to be unresolved.

^3J = 2.5±1 Hz
X = OH, OAc or Br
99

^3J = 5.5±1 Hz
X = OH, OAc or Br
100

We can see a further example of the lowering of the coupling constant by an electronegative element in the spectrum in Fig. 4.62. 3,4-Epoxytetrahydrofuran **101** has a plane of symmetry, and therefore shows only three resonances: an AB system from the methylene protons H_a and H_b at δ 4.04 and 3.68 (not necessarily respectively), with a coupling constant of 10.4 Hz; and a sharp singlet in between at δ 3.81 from the methine proton H_c.

The dihedral angle between H_a and H_c is close to 90°, and it is not surprising that there is no coupling between them. However, the dihedral angle between H_b and H_c is somewhere between 0° and 30°, and a coupling constant of 6–8 Hz might be expected on the basis of the Karplus equation, whereas in fact the only sign of coupling is the slight broadening of the upfield signal. A major influence is the anti arrangement of the epoxide oxygen to proton H_b. Furthermore, the tetrahydrofuran ring oxygen is aligned to some extent anti to the proton H_c, with angle strain contributing to the extraordinary disappearance of vicinal coupling to both protons.

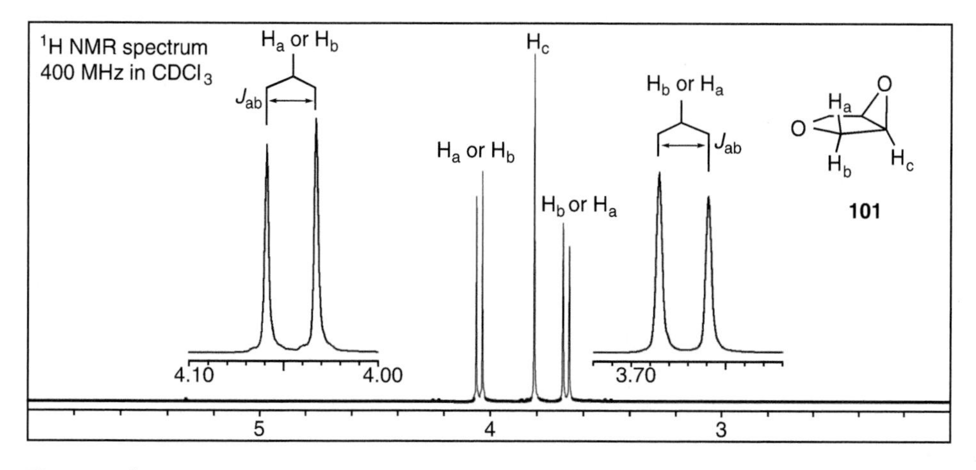

Fig. 4.62 ^{1}H NMR spectrum of 2,3-epoxytetrahydrofuran

On double bonds, both the antiperiplanar and the syncoplanar protons are affected. An electronegative element substantially lowers both the *cis* and the *trans* coupling constants in vinyl fluoride **102** relative to propene **103**, but the *cis* coupling constant, from the proton antiperiplanar to the fluorine, is the more affected. An electropositive element has the opposite effect: the *cis* coupling constant for vinyl-lithium **104** is raised so that it is higher even than the normal value for a *trans* coupling, and the *trans* coupling is raised too, but not by quite so much. Vinylsilanes, where the silicon atom, like the lithium atom, is electropositive relative to carbon, has *cis* coupling constants (typically 15 Hz) similar to the usual *trans* coupling constants in alkenes, and the *trans* coupling (typically 20 Hz) is larger still.

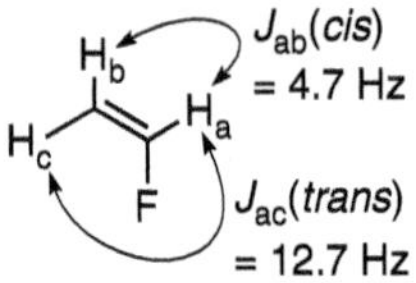

$J_{ab}(cis)$ = 4.7 Hz, $J_{ac}(trans)$ = 12.7 Hz (**102**, F); $J_{ab}(cis)$ = 10.0 Hz, $J_{ac}(trans)$ = 16.8 Hz (**103**, Me); $J_{ab}(cis)$ = 19.3 Hz, $J_{ac}(trans)$ = 23.9 Hz (**104**, Li)

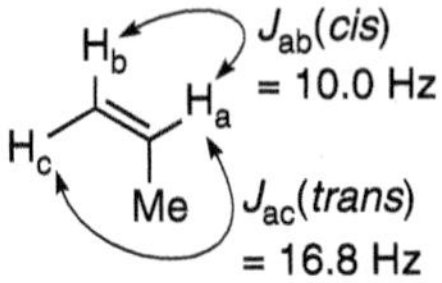

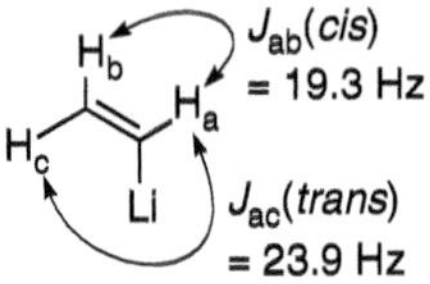

Angle Strain In the fragment **105**, orbital overlap, and hence ^3J, decreases as θ and θ' increase. This effect is most noticeable in the *cis* coupling constants between olefinic protons in cycloalkenes **106–109**: as the ring size increases, the coupling constant increases. It is therefore possible to tell in many cases into what size ring, from three- to six-membered, a double bond is incorporated.

H_a θ θ' H_b

105

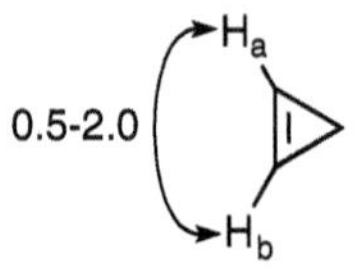

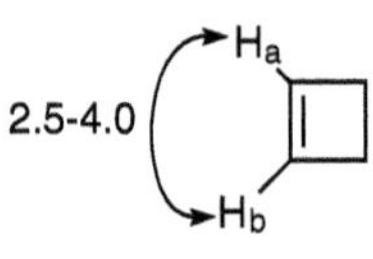

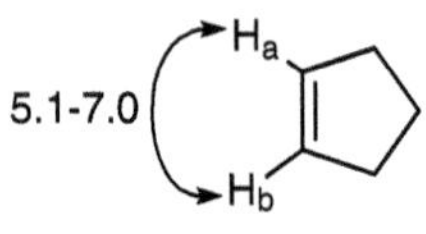

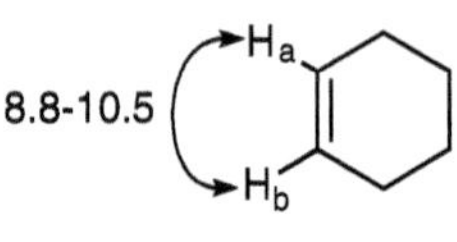

0.5-2.0 **106** 2.5-4.0 **107** 5.1-7.0 **108** 8.8-10.5 **109**

Bond-Length Dependence Double bonds are shorter than single bonds, vicinal overlap is better, and the coupling constants are larger, other things being equal. Thus, cyclohexadiene **110** has similar dihedral angles for all the adjacent olefinic C–H bonds; but the coupling constant is greater across the double bonds than across the intervening single bond. Open-chain dienes like butadiene **111** exist mainly in the s-*trans* conformation, and the intermediate coupling constant is larger than usual, but not as large as the *trans* coupling constant for the double bond. Aromatic carbon-carbon bonds have bond lengths intermediate between the single and double bonds of alkanes and alkenes. In consequence, *ortho* coupling constants are typically rather lower than *cis* olefinic coupling constants: about 7–8 Hz in benzene rings compared to 8.8–10.5 Hz in cyclohexenes.

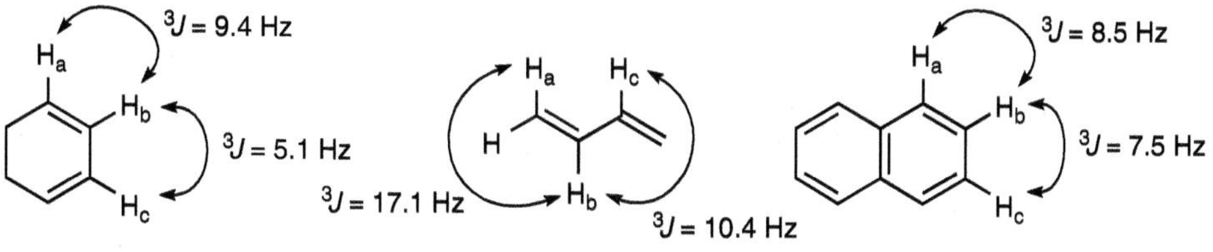

We have seen some representative numbers for aromatic compounds in the spectrum of diphenylacetaldehyde **40** in Fig. 4.51 where the *ortho* coupling constant is close to 7 Hz, and in the spectrum of 3-chloropyridine **72** in Fig. 4.49, where the vicinal coupling constants are 8.3 and 4.6 Hz, with the latter reduced in magnitude because of the influence of an electronegative element, the nitrogen atom, antiperiplanar to H_c. In contrast, alkenes like allyl bromide **46** in Fig. 4.34 and methyl acrylate **51** in Fig. 4.40 have *cis* coupling constants of 10.0 and 10.8 Hz, respectively. Polycyclic aromatic rings have unequal bond lengths, and unequal coupling constants, as in naphthalene **112**.

4.7.3 Geminal Coupling ($^2J_{HH}$)

Geminal coupling can only be seen in a spectrum when the two protons attached to the same carbon atom resonate at different frequencies. However, the coupling constants can be measured, even in molecules such as methane, by introducing a deuterium atom, and measuring the geminal coupling from H to D. The value obtained is related to the proton-proton coupling constant by Eq. (4.11), which applies to all H-D coupling, whether geminal or not.

$$J_{\mathrm{HH}} = 6.5 J_{\mathrm{HD}} \tag{4.11}$$

Adjacent π-Bonds The ^2J coupling constant for a simple hydrocarbon, such as methane (measured from a partially deuterated methane), is −12 Hz **113**. When the C–H bonds are able to overlap with neighbouring π-bonds, as in toluene **114** or acetone **115**, orbital overlap, known as hyperconjugation, is promoted by the adjacent π-system. The effect of hyperconjugation is to withdraw electrons from the C–H bonds and to lengthen them. In vicinal coupling longer bonds led to smaller (less positive) coupling constants. Just so here, they are more negative and now, in absolute magnitude, larger. The effect is even greater when the hyperconjugation is with the π-bond of a carbonyl group than when it is simply with a C=C double bond—in toluene **114** the coupling constant is −14.3 Hz and in acetone **115** it is −14.9 Hz. The methyl group in toluene and in acetone is freely rotating, and the measured coupling constants are weighted averages of the coupling between the geminal hydrogens for all the conformational relationships in which they find themselves. In rigid, and especially cyclic, systems in which the conformation is held favourably for overlap, with one C–H bond above the π-bond and one below, it commonly reaches −16 or −18 Hz, as it evidently does in the ester **56** with a geminal coupling constant of −17.7 Hz. If the hyperconjugative overlap is with two double bonds flanking a methylene group, the coupling constant can be close to −20 Hz.

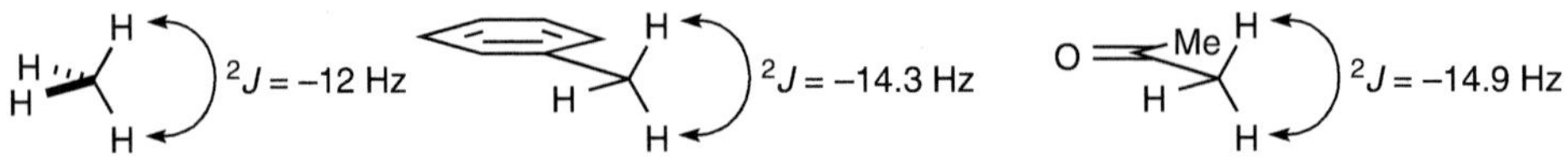

^2J = −12 Hz ^2J = −14.3 Hz ^2J = −14.9 Hz

113 **114** **115**

Adjacent Electronegative and Electropositive Elements In contrast to a π-bond, which is effectively electron-withdrawing, an electronegative element directly attached to a methylene group is effectively a π-donor with respect to the C–H bond, donating electrons into the antibonding C–H $\sigma*$ orbitals symbolised by the dashed line in **116** (**116** is the acetal **60**), increasing the electron population. The coupling constant is now more positive than the −12 Hz for methane **113**, in other words effectively smaller. It is −10.8 Hz in methanol, −9.6 in the acetal **60** (Fig. 4.47) and −9.1 Hz in pantolactone **73** (Fig. 4.50). In the epoxide **52**, the coupling constant in the side-chain is −12.2 Hz, combining the effect of the electronegative element and the effect of the adjacent π bond. The effect of an electropositive element is in the opposite direction. This can be seen in the silane **117**. The Si–Me bonds conjugated with the C–H bonds are polarised from Si towards C making them effectively electron-withdrawing, removing electron population from the C–H bonds, and making the coupling constant more negative.

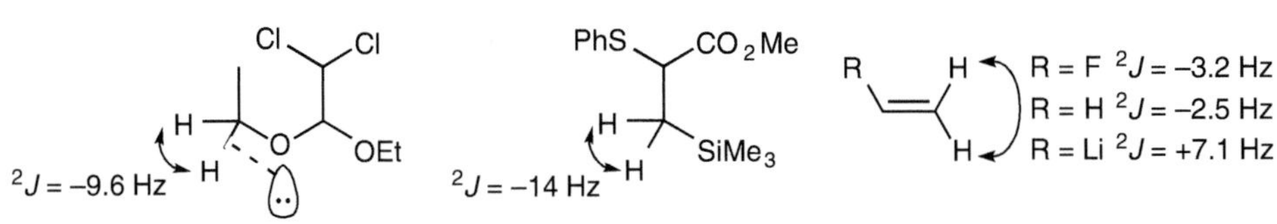

^2J = −9.6 Hz ^2J = −14 Hz R = F ^2J = −3.2 Hz R = H ^2J = −2.5 Hz R = Li ^2J = +7.1 Hz

116 **117** **118**

Note the contrast between the effect of electronegative and electropositive elements in vicinal and geminal coupling, and the apparent similarity of the outcome. In the former the antiperiplanar relationship with the electronegative element removes electrons and reduces the magnitude of the positive coupling constant; in the latter the conjugation with the lone pair on an electronegative element adds electrons and makes the coupling more positive. The apparent effect is the same—decreasing the observed coupling constant. With electropositive elements, the opposite obtains, but again the observed effect is the same—the apparent coupling is increased in both cases, more positive in vicinal coupling and more negative in geminal coupling. The general guide that overall electron withdrawal from the bonds connecting coupled protons makes coupling less positive, and electron donation makes coupling more positive still holds.

Angle Strain An increase in the H–C–H angle makes ^{2}J more positive, in other words smaller. This effect is most noticeable in the methylene groups of terminal alkenes **118**, where the angle is close to 120° and the coupling constant is close to zero, as we saw earlier in methyl acrylate **51** (Fig. 4.40), which had a geminal coupling constant of −1.2 Hz. This coupling is dependent upon the nature of substituents at the other end of the π-bond which are conjugated antiperiplanar to one of the protons: electronegative elements, like fluorine, make them more negative and electropositive elements, like lithium, actually make the coupling positive in sign and quite large. Angle strain in small rings is similar, as in the epoxide **52** (Fig. 4.41), which has a geminal coupling constant of −2.6 Hz. The effect of the H–C–H angle is also seen in the ranges of ^{2}J values for cycloalkanes.

4.7.4 Long-Range Coupling ($^{4}J_{HH}$ and $^{5}J_{HH}$)

The coupling constants for allylic, W and other long-range coupling were discussed earlier, where the main influence was the degree of overlap through the intervening bonds. The coupling constants are usually small, and the influence of substituents not all that noticeable. Most visible allylic coupling, ^{4}J, is negative in sign, but passes zero to low positive as the degree of overlap of the C–H bond with the π-bond changes from maximum (at 0°, as in the drawings **59** and **60**) to minimum (at 90°). Most five-bond coupling, ^{5}J, is positive in sign, and is at a maximum (2 Hz) when the protons are connected in a planar zigzag arrangement. It is rarely resolved, but its effect is still visible as line broadening.

4.7.5 C–H Coupling ($^{1}J_{CH}$, $^{2}J_{CH}$ and $^{3}J_{CH}$)

Finally, let us remember the pattern for $^{1}J_{CH}$ coupling, which is always positive in sign. We saw in Sect. 4.5.2 that electronegative elements raised the coupling constants, that is they made them more positive, and effectively larger. Electropositive elements make them less positive, effectively smaller. Figure 4.63 gives some examples that match this picture, with the degree of influence greater on trigonal carbons than on tetrahedral carbons.

This general rule based on the inductive effect is modified in detail by a stereoelectronic effect known as the Perlin effect. Axial C—H bonds in cyclohexanes have a slightly smaller $^{1}J_{CH}$ value than equatorial C—H bonds, typically 122 Hz and 126 Hz, respectively. This stereoelectronic effect is more conspicuous when a C—H bond is held antiperiplanar to the lone pair of electrons of an electronegative element. Thus the α-anomer of a hexose typically has a $^{1}J_{CH}$ value 10 Hz greater at C-1 than that of the β-anomer. Antiperiplanar

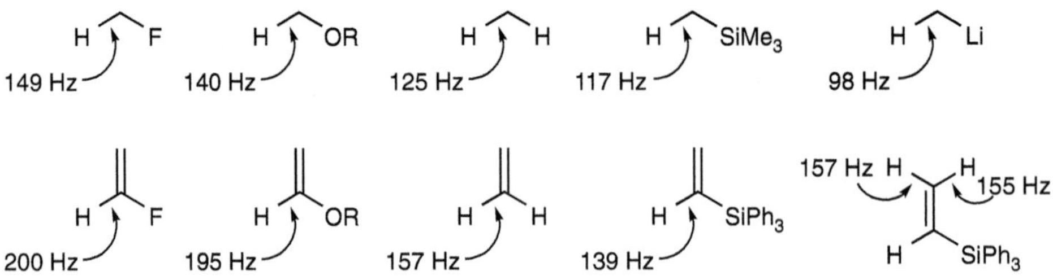

Fig. 4.63 Some $^1J_{CH}$ coupling constants

alignment of a C—H bond with the bond to an electropositive element has the opposite effect, increasing the $^1J_{CH}$ value, as illustrated by vinyltriphenylsilane in Fig. 4.63.

Two-bond coupling constants $^2J_{CH}$ are usually quite small, in the range 0–6 Hz, with exceptions from aldehyde protons (20–50 Hz) and when the carbon atoms are digonal (10–50 Hz). They are significantly lower (0–3 Hz) when an alkene or arene carbon is coupled to a proton on the adjacent carbon. These couplings are rarely used in structure determination, but are helpful when interpreting HMBC spectra (Sect. 5.6), where it can be useful to know that small coupling constants lead to a weak or invisible crosspeak. They are not always negative in sign, especially when the carbon atoms are trigonal. Small structural changes can make the base value more positive, sometimes reducing them close to zero, as we saw in the absence of visible coupling from C-1, C-3 and C-5 to the protons *ortho* to them in the ester **1** in Fig. 4.24. When an electronegative element is antiperiplanar to the hydrogen atom or attached to an intervening trigonal carbon atom, $^2J_{CH}$ coupling is even larger and becomes positive (7.5 Hz).

Three-bond coupling constants $^3J_{CH}$ are usually positive, usually a little larger than $^2J_{CH}$ values, and are governed mainly by a Karplus-like relationship, being large (9 Hz) when the angle is 180°, smaller (6 Hz) when the angle is 0°, and zero when the angle is 90°. It is even larger (10–16 Hz) in those alkenes and arenes holding the carbon and hydrogen rigidly antiperiplanar. To a rough approximation, they are close to 0.6× the more familiar $^3J_{HH}$ values that most nearly resemble them. Tables at the end of this chapter give representative examples of coupling constants for protons to carbon, illustrating the influence of substituents on these geometrically induced patterns.

4.8 Coupling From ^{1}H and ^{13}C to ^{19}F and ^{31}P

4.8.1 ^{13}C NMR Spectra of Compounds Containing ^{19}F and ^{31}P

While ^{13}C spectra are routinely taken with proton decoupling, they are not usually taken with decoupling to fluorine and phosphorus. The carbon signals in compounds containing one or more of these elements show the usual multiplicity—1:1 doublets if there is one nearby ^{19}F or ^{31}P nucleus, 1:2:1 triplets if there are two identical ^{19}F or ^{31}P nuclei and 1:3:3:1 quartets if there are three identical ^{19}F or ^{31}P nuclei. As a result, the total number of signals in a ^{13}C spectrum is no longer a measure of the number of different carbon atoms in the molecule.

In fluorobenzene, for example, each of the four different carbon atoms gives rise to a 1:1 doublet from coupling to the fluorine. The directly bonded fluorine has a coupling constant of 245 Hz, and the *ortho*, *meta* and *para* carbons have coupling constants of 21, 8 and 3 Hz, respectively, reflecting the number of bonds separating the two nuclei. Likewise, in triphenylphosphine, the *ortho*, *meta* and *para* carbons have coupling constants of 19.5, 6.9 and 0.3 Hz, respectively, reflecting the number of bonds separating the nuclei. In contrast, the one-bond coupling is anomalously small at 10.8 Hz, although it is in fact in the normal range for P(III) compounds. In trifluoroethyl trifluoroacetate **119** in Fig. 4.64, all four carbons give rise to 1:3:3:1 quartets, with the two carbons directly bonded to the fluorine atoms coming into resonance well downfield, and having large coupling constants of 280 and 285 Hz, and the other two carbons showing smaller, two-bond couplings of 45 and 35 Hz. C-1 is evidently too far from the fluorine atoms on C-2′, and C-2′ is too far from the fluorines on C-1 for either to show any coupling.

In contrast to the apparently anomalous low value for the one-bond coupling in triphenylphosphine, in triethyl phosphonoacetate **120** in Fig. 4.65 it is much larger at 134 Hz, showing that coupling constants from carbon to phosphorus are very dependent upon the oxidation level of the phosphorus.

The coupling through two and three bonds is weaker, but, small though it is, the two-bond coupling of 6.3 Hz to C-1″ makes it possible to identify which of the two signals at δ 62.7 and 61.6 is which, and the three-bond coupling of 6.2 Hz makes it possible to identify which of the two methyl signals at δ 16.3 and 14.3 is which. In other compounds, especially those giving rise to many other carbon signals, picking out the doublet, triplet or quartet of a carbon carrying one or more fluorine atoms can be difficult. It is important

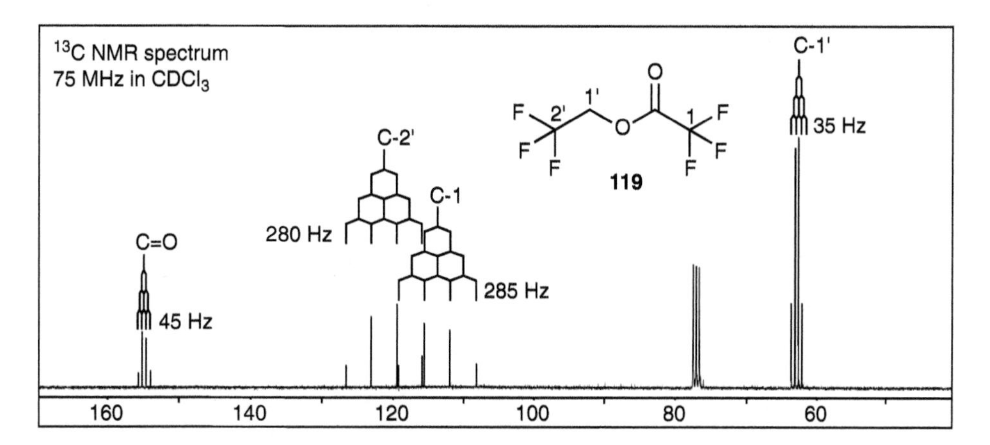

Fig. 4.64 ^{13}C NMR spectrum of 2,2,2-trifluoroethyl trifluoroacetate

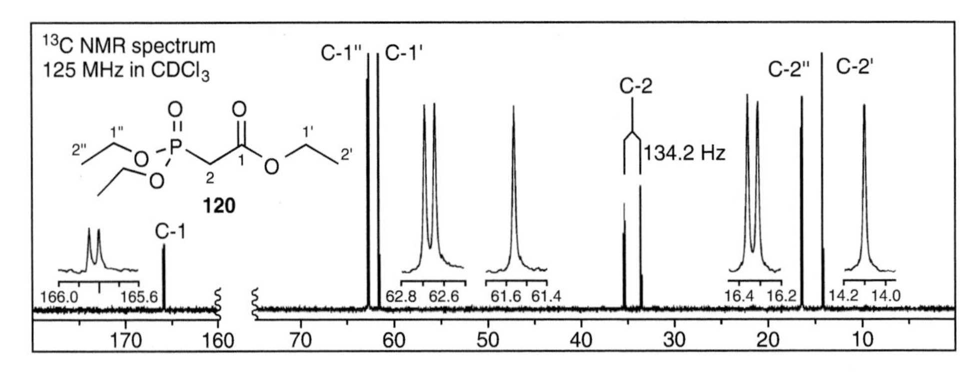

Fig. 4.65 ^{13}C NMR spectrum of triethyl phosphonoacetate

to know whether phosphorus or fluorine is present in a molecule before trying to interpret a ^{13}C NMR spectrum.

With a compound having both phosphorus and fluorine, the spectrum simply shows the appropriate coupling to both elements. Methyl bis(trifluoroethyl)phosphonoacetate **121** in Fig. 4.66 shows a simple doublet from C-2 at δ 33.8 with a large, one-bond coupling constant to the phosphorus of 145 Hz, a doublet at δ 165 from the carbonyl group with two-bond coupling to the phosphorus of 4 Hz, and a singlet from the OMe group. The other two signals have couplings to both fluorine and phosphorus: the signal from C-2′ is a quartet at δ 122 with a large, one-bond coupling constant of 277 Hz to the three equivalent fluorines, each line of which is split by 8.3 Hz from the three-bond coupling to the phosphorus; the signal from C-1′ at δ 62.7 is a quartet with a large, two-bond coupling constant of 120 Hz to fluorine, each line of which is doubled by 5.6 Hz from two-bond coupling to the phosphorus.

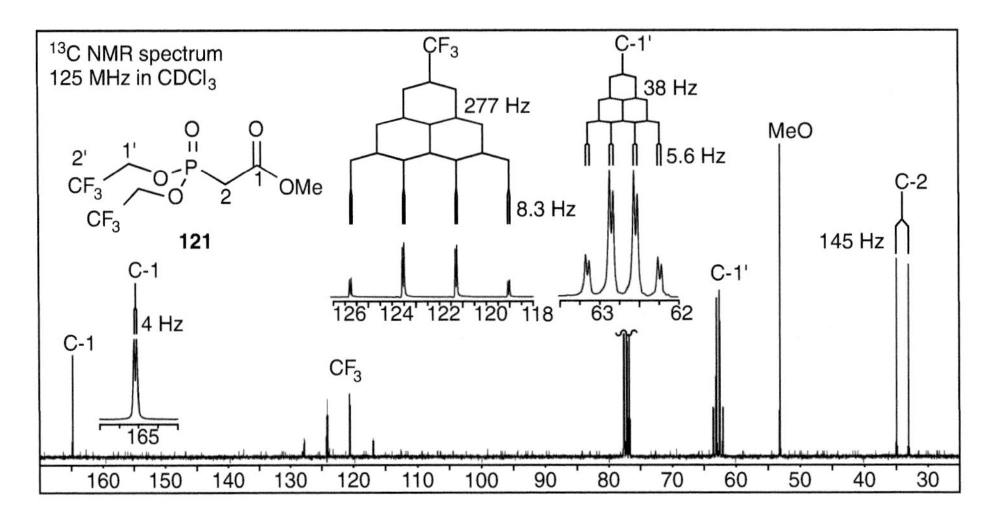

Fig. 4.66 ^{13}C NMR spectrum of methyl bis(2,2,2-trifluoroethyl)phosphonoacetate

4.8.2 ^{1}H NMR Spectra of Compounds Containing ^{19}F and ^{31}P

In ^{1}H NMR spectra, coupling from protons to fluorine and to phosphorus adds to the existing coupling of proton to proton. Again, to see the coupling partner it would be necessary to take ^{19}F or ^{31}P spectra, but this is often unnecessary. The coupling constants from protons to fluorine and to phosphorus are comparable to or rather larger in absolute magnitude than those in the corresponding proton to proton connections.

The ^{1}H spectrum of triethyl phosphonoacetate **120** in Fig. 4.67 shows the methylene protons on C-2 next to the phosphorus atom as a doublet with a two-bond coupling constant of 21.6 Hz. In contrast the terminal methyl groups give rise to the two high-field triplets, essentially unaffected by the relatively distant phosphorus. However, the signals from the methylene protons in the ethoxy groups overlap; one of them CH_2-1′ is a simple quartet from $^{3}J_{HH}$ coupling together with an only just discernible $^{5}J_{PH}$ coupling of 0.35 Hz, but the other is a more complicated pattern. First of all, it has proton-proton coupling of about 7 Hz to the methyl group leading to a quartet, but it also has a three-bond coupling to phosphorus close to 7 Hz. If this were all that were going on, the signal would look like the quintet drawn above the expansion of the 1″ signal. But this signal is not just a quintet, probably because the protons on the two 1″-CH_2 methylene groups are diastereotopic, like those of the ethoxy groups in the acetal **60** in Fig. 4.47.

Coupling from protons to fluorine is similar, but with somewhat larger coupling constants; they differ only in that most couplings from protons to fluorine are positive in sign, whether two-bond, three-bond or four-bond, because of the closeness of the lone pairs of electrons on the fluorine to the nucleus. The ^{1}H NMR spectrum of 4-(1-fluoroethyl)phenyl acetate **122** in Fig. 4.68 shows the three-bond coupling from the fluorine atom to the methyl protons of 23.9 Hz and the two-bond coupling from the fluorine atom to the methine proton of 47.5 Hz, while the methyl and methine $^{3}J_{HH}$ coupling is only 6.4 Hz, slightly lower than usual for vicinal coupling in an open-chain, because of the presence of the electronegative element.

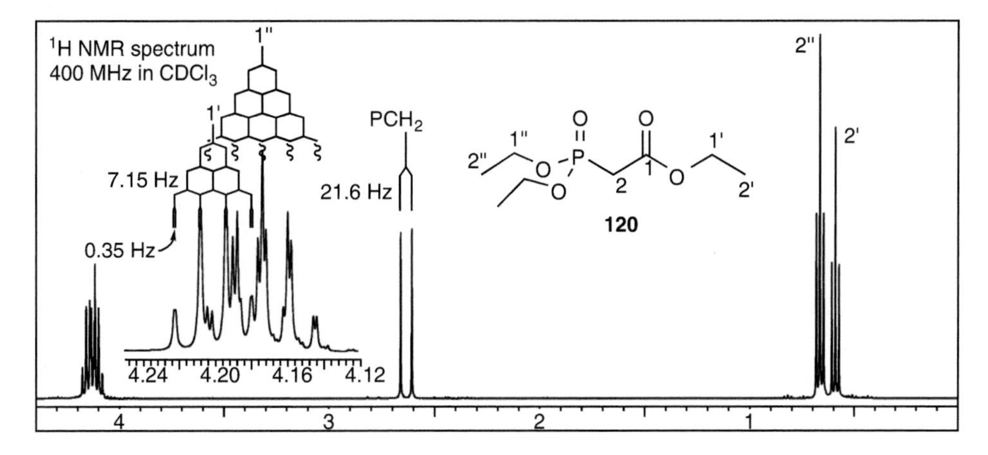

Fig. 4.67 ^{1}H NMR spectrum of triethyl phosphonoacetate

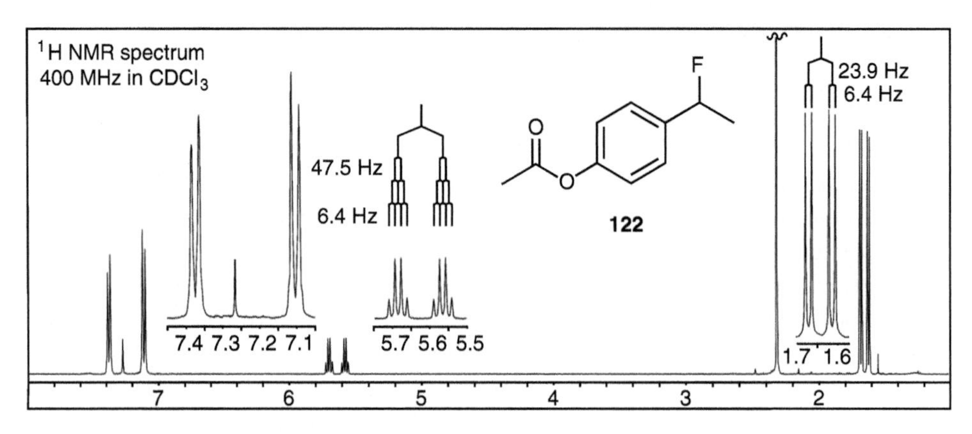

Fig. 4.68 ^{1}H NMR spectrum of 4-(1-fluoroethyl)phenyl acetate

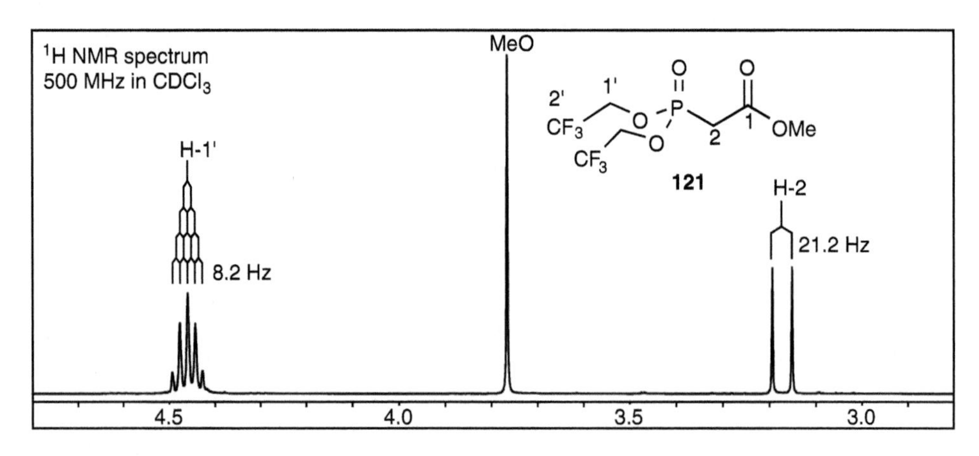

Fig. 4.69 ^{1}H NMR spectrum of methyl bis(2,2,2-trifluoroethyl)phosphonoacetate

With both fluorine and phosphorus in the molecule, the methyl bis(trifluoroethyl)phosphonoacetate **121** in Fig. 4.69 shows the signal from the methylene protons close to both fluorine and phosphorus as a quintet with three-bond coupling of 8.2 Hz to the three fluorines and, coincidentally the same, to the phosphorus atom. The methylene protons on C-2, two bonds away from the phosphorus but five bonds away from the fluorine, show coupling only to the phosphorus.

4.9 Relaxation and Its Consequences

Even in the absence of spin-spin coupling, the signals in NMR spectra are not sharp lines—they all have appreciable, although often very narrow, half-height widths. There are four common reasons for broadening: unresolved coupling, the more or less inevitable small inhomogeneities in the magnetic field, efficient relaxation and environmental exchange. In this section we shall deal with the mechanism of relaxation and its consequences in the appearance of the spectra.

4.9.1 Longitudinal Relaxation

When a nucleus loses energy on passing from the high-energy N_β state to the low-energy N_α state, it is said to *relax* (Fig. 4.1). The energy that is lost passes to the surrounding environment, which is called the lattice, and the equilibrium Boltzmann population, which existed before excitation, is restored through relaxation. Just as excitation requires the matching of the frequency of irradiation with the natural frequency of the nucleus, so relaxation requires the fluctuation of a local magnetic field matching the Larmor frequency of the nucleus. This can be achieved if the molecule containing the nucleus tumbles in the applied field at a rate that is close to the Larmor frequency (75–750 MHz, depending upon the nucleus and the field strength). Thus, molecular tumbling rates of ca. 10^8 s^{-1} promote efficient relaxation. If a C–H bond is oriented parallel to the applied magnetic field B_0, and the nuclear magnets of the ^{13}C and ^{1}H nuclei are each oriented in their arrangement of lowest energy, the lines of force associated with the magnetic field from the nuclear magnet of one nucleus reinforces the applied magnetic field at the other. If the molecule containing this C–H bond rotates it through 90° as it tumbles in solution, the notional 'bar magnets' associated with the nuclei are still oriented along the direction of B_0, but now the lines of force from the one nucleus oppose the applied field at the other. Thus, as the molecule tumbles in solution, the local field at each nucleus fluctuates at the frequency of the rotation. This magnetic field fluctuation provides an efficient mechanism for the relaxation of ^{13}C by ^{1}H when the frequency of tumbling has components close to the resonance frequency of ^{13}C. In the same way, protons that are close in space to each other in a molecule can relax each other.

This mechanism for relaxation works whether the nuclei are bonded or not, they just have to be close enough in space for the one to experience the field of the other. Quantitatively, there is an r^{-6} dependence upon the distance apart (r) of the dipoles—the effect falls off rapidly as the separation of the nuclei increases, as it does for a ^{13}C not directly bonded to a proton. It also depends upon the square of the magnetogyric ratios of each nucleus. For this reason protons relaxing each other or relaxing other nuclei are nearly always the major stimulus for relaxation. Deuterium, with a γ value 1/6.5 [$(1/6.5)^2 = 0.024$], is much less effective than hydrogen in relaxing the ^{13}C signal of $CDCl_3$, which in any case suffers because chloroform tumbles too fast for a good match with the frequency of the spectrometer.

The γ value for an electron is about 1000 times that of the proton, and so radicals suffer relaxation so fast that useful NMR spectra are rarely obtained. In rare cases, a radical or radical ion can be added in small quantities to a molecule (as a paramagnetic salt, for example, or adventitiously or deliberately created from a molecule by oxidation or reduction), and the effect of the single electron distributed in a conjugated system within the molecule can be mapped. This has been done for several large conjugated systems, and the sites where the single electron population is highest can be seen by the degree of broadening imparted to the nearby protons.

The form of relaxation we have discussed so far is called spin-lattice relaxation or more helpfully longitudinal relaxation, because it is concerned with the rate at which the magnetisation vectors relax from their tilt towards the xy plane back to the vertical z axis. Longitudinal relaxation has a first-order rate constant R_1 for each nucleus, which is associated with its reciprocal, the time constant T_1 (Eq. (4.12)). T_1 is typically a second or two for ^1H, but it is usually much longer for ^{13}C, even as much as 300 s. The decay is exponential, and so there is no such thing as a time of decay, only a time constant.

$$R_1 = 1/T_1 \tag{4.12}$$

In practice, it would be ideal if we could wait about 5 × T_1 between successive pulses for relaxation to have reached near enough to completion. This is especially impractical with the long relaxation times for ^{13}C, and so relaxation is always incomplete for that nucleus before the next pulse is applied, and that is why the intensity of ^{13}C signals is unreliable for integration. The absolute values for T_1 are too much influenced by the presence of paramagnetic materials and other impurities to be useful numbers, but we can compare T_1 values for the different protons within a molecule and extract useful information, such as the proximity of features helping the protons to relax. The method by which T_1 values are measured is fairly simple to understand, and can be our first pulse sequence more complicated than the single pulse-acquire we saw earlier in Fig. 4.5. The two-pulse sequence shown in Fig. 4.70 begins with a pulse continued for long enough to tip the magnetisation through 180°. This is followed by a series of incrementally lengthening delays τ before the usual $\pi/2$ pulse precedes acquisition.

The π pulse puts the magnetisation onto the $-z$ axis, and this cannot be detected in the coils along the x axis. The magnetisation relaxes back to the equilibrium position at the rate we are trying to measure, and as it does so the vector gets shorter. The second pulse, the $\pi/2$ pulse, brings whatever magnetisation is still left into the xy plane where it can be detected as an FID in the usual way. A series of values of τ will give a series of spectra after FT. The first few will be strong but upside down because the $\pi/2$ pulse will have

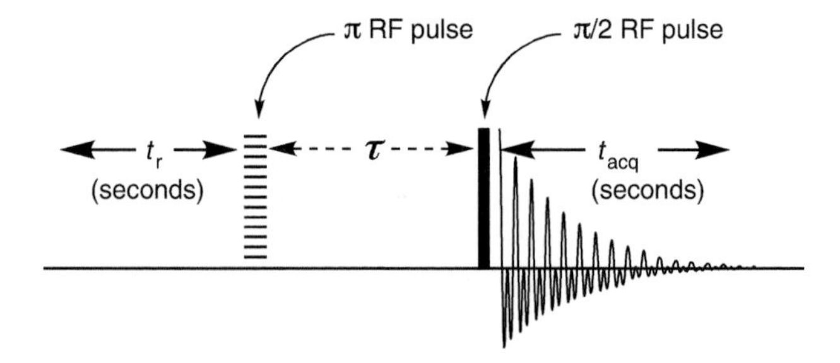

Fig. 4.70 Pulse sequence for measuring T_1

tipped the magnetisation clockwise from the $-z$ axis to point towards the $-y$ axis, and the next few will have signals of successively decreasing intensity. One by one the individual signals will pass through a null, and then increase until they give the normal positive spectrum when the $\pi/2$ pulse comes after the spins have completely relaxed. Each spin will have its own T_1 value, and will produce a plot of its intensity M_1 starting negative, passing through the null point, and back to the full value M_0. T_1 values can be extracted from these plots graphically by fitting them to the formula in Eq. (4.13), or by using the time τ_n to reach the null point for each signal given in Eq. (4.14), which is usually accurate enough.

$$M_1 = M_0\left(1 - 2e^{-\tau/T1}\right) \tag{4.13}$$

$$\tau_n = 0.693T_1 \tag{4.14}$$

Structural information can occasionally be gleaned from T_1 values, because T_1 values are affected by the number of nearby protons helping the signal to relax. It is not often needed for this purpose, but the example of *o*-dibromobenzene **36** shows how it can be used. The ^{1}H NMR spectrum of this compound has two multiplets, one centred at δ 7.60 and the other at δ 7.14, which can be seen in Fig. 4.56, one pair from the protons H_a on C-3 and C-6 and the other from the protons H_b on C-4 and C-5; but how can we assign which is which? The T_1 values are 6.19 ± 0.05 s for the downfield signal and 5.64 ± 0.04 s for the upfield, different enough to be reliable. The protons H_b on C-4 and C-5, with two neighbouring protons each, relax more rapidly than the protons H_a on C-3 and C-6, which have only one neighbour each.

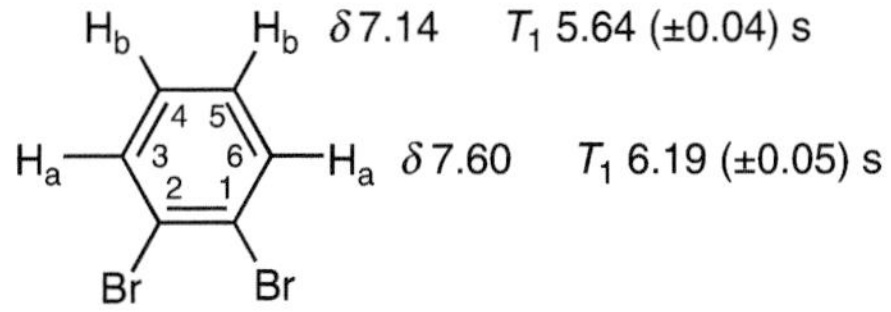

4.9.2 Transverse Relaxation and Exchange

Signal intensity falls as the spins in the N_β level return to the N_α level to restore the Boltzmann distribution, but signal intensity can also be lost independently by other means. The magnetisation vector from one nucleus can get out of phase for various reasons with the magnetisation vector from the same nucleus in other molecules. Aggregated over the whole sample, this nucleus would be represented by many vectors pointing in many different directions in the xy plane, and eventually the signal derived from them would vanish altogether as the vectors cancelled each other out. This form of relaxation takes place within the xy plane, and is therefore called transverse relaxation. The Boltzmann distribution has not changed, we still have spins in the xy plane, but they are no longer coherent, and we detect no signal. The overall rate constant for these pathways for relaxation is R_2^*, which is associated with its reciprocal, the time constant T_2^*. R_2^* can never be lower than R_1, since it cannot come into play once the Boltzmann distribution has been restored. R_1 and T_1 are associated with signal intensity, since signals are weak when longitudinal relaxation has not gone to completion during the acquisition. R_2^* and T_2^* are associated with the width of the signal, since loss of phase coherence in the magnetisation vectors leads successively to more and more broadening before it vanishes. The width of a singlet at half-height is given by Eq. (4.15).

$$\Delta v_{\frac{1}{2}} = \frac{1}{\pi T_2^*} \tag{4.15}$$

There are two main mechanisms for transverse relaxation, one associated with the spectrometer and the other with the compound under study, each with its own T_2 value. The former arises from inhomogeneity in the applied magnetic field. If a nuclear spin on one molecule experiences a different applied magnetic field from that experienced by the same nuclear spin on a different molecule, their magnetisation vectors will get out of phase as they precess at different rates. It is this problem that is minimised with a high quality magnet and by thorough shimming. The T_2 value for this mechanism is called $T_{2(B0)}$. The T_2 value derived from the intrinsic properties of the molecule is called simply T_2, and the two together make up the observed value T_2^*, related by Eq. (4.16).

$$\frac{1}{T_2^*} = \frac{1}{T_2} + \frac{1}{T_{2(B0)}} \tag{4.16}$$

The intrinsic mechanisms by which spins get out of phase are from interactions with magnetic anisotropy within the molecule and by exchange. Anisotropy is associated with the ellipsoidal electrical fields from nuclear quadrupoles, with the most common in organic chemistry being that from the ^{14}N nucleus, which has a spin I of 1. Protons directly attached to nitrogen atoms suffer T_2 relaxation, and are often broader than the signals from carbon-bound protons. The broadening is exacerbated, further reducing the T_2 value, when the protons exchange from one molecule to another during the time of an FID. For this reason amine NH signals are often broader than those from amide NH signals, because they exchange more rapidly.

In more detail, if a proton spends some of the time during the acquisition in one environment, and some of the time in a different chemical and hence magnetic environment, it will have two frequencies, and both will appear in the FID. The critical factor affecting how they appear in the spectrum is the rate of exchange from the one environment to the other. If it is slow relative to the difference in frequency between the two signals, each of the signals will appear in the spectrum. If, on the other hand, it is fast relative to the difference in frequency between the two signals, the output will be one signal with a chemical shift half way between the two. In between, the signals will be broadened. If k is the rate constant for the exchange and $\Delta\nu$ is the difference in frequency between the two signals, the signals will coalesce when the condition of Eq. (4.17) obtains.

$$k = \frac{\pi \Delta \nu}{\sqrt{2}} = 2.221 \Delta \nu \tag{4.17}$$

If we know the temperature at which this condition is met, known as the coalescence temperature T_c, we can calculate the free energy of activation for the process from Eq. (4.18).

$$\Delta G^{\ddagger} = RT_c \left[23 + \ln \frac{T_c}{\Delta \nu} \right] \tag{4.18}$$

Another class of exchange is the physical departure of an atom such as a proton from one molecule and its arrival at another, as occurs when protons leave one OH or NH group and become bonded to another OH or NH group. In consequence, OH, NH and SH signals are frequently broad, as they are in the spectra of hexanol **47** in Fig. 4.35, in pantolactone **71** in Fig. 4.50, and in the spectrum of hydroxymethylfurfural **123** on the left in Fig. 4.71. The chemical shift and the appearance of OH and NH signals in NMR spectra is affected by the rate of exchange, which is, in turn, affected by the concentration, the temperature, the nature of the solvent, and by the presence or absence of acid or base catalysis. Coupling to OH, NH and SH protons disappears if the rate of exchange is fast, but it can be present when the rate of exchange is low, as it is when the sample is in dilute

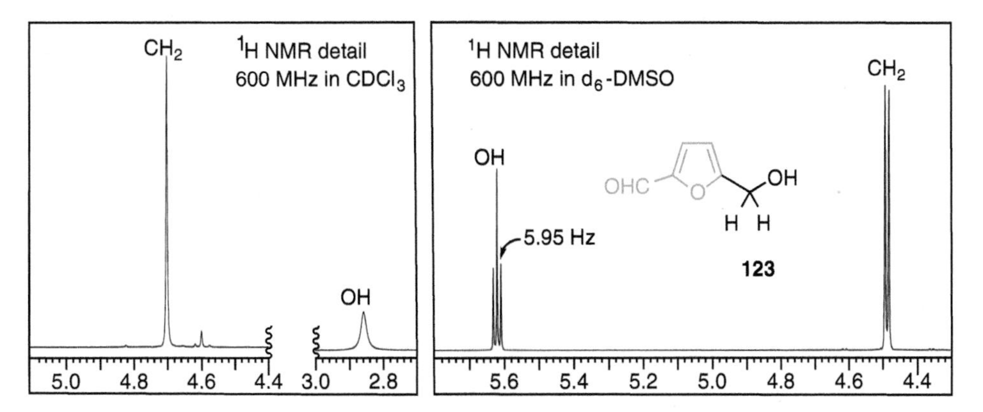

Fig. 4.71 Coupling to OH signals visible in spectra taken in DMSO

solution, is exceptionally pure, or when the solvent is d_6-DMSO. In very pure samples and in d_6-DMSO, the rate constant for the exchange falls below 6 s^{-1}, and coupling to neighbouring carbon-bound protons (for which J is ~6 Hz) is then visible. The methylene signal in the spectrum taken in DMSO on the right in Fig. 4.71 appears as a doublet, and the OH signal as a triplet, with characteristic changes in chemical shift compared to the spectrum in deuterochloroform.

The more and more frequent use in recent times of very dilute solutions is making coupling with OH protons increasingly more often observed in spectra taken in deuterochloroform, especially if the solvent has been dried using molecular sieves. The multiplicity in the OH signals, often well resolved from each other, gives useful information about their neighbours. Coupling involving NH protons in amines and SH protons in thiols is invisible, more often than not. However, CH groups adjacent to amide NH groups usually show coupling (J = 5–9 Hz), even when the amide NH signal is broad. In such cases, the broadness of the NH signal is not primarily because of exchange-induced broadening, but rather from fast relaxation of the NH proton caused by the quadrupole moment of the ^{14}N nucleus.

Other kinds of exchange affecting T_2 are when the molecule changes shape to move one nucleus into the position occupied by another. Molecules will also experience inhomogeneity if some of them are associated with a paramagnetic salt, which has its own relatively strong magnetic field, while others are not, and the salt is moving from one molecule to another. The rate constants for events like these can be measured by changing the temperature and looking for coalescence. The limitation is that the difference in chemical shift in Hz of the signals undergoing coalescence and the rate should lead to a coalescence temperature within the range under which NMR spectrometers can operate. Here are two examples, one from each end of the accessible range.

The ^{1}H NMR spectrum of dimethylformamide **124**, taken at room temperature, shows two singlets for the N-methyl groups at δ 3.0 and 2.84. This is because π-overlap between the nitrogen lone pair and the carbonyl π-bond (**124**, arrows) slows the rotation about this bond. On warming, however, the lines broaden and coalesce. The coalescence temperature T_c of 337 K was measured long ago on a 60 MHz instrument. The difference in frequency $\Delta\nu$ is 0.16 × 60 = 9.6 Hz, from which we can calculate $\Delta G^{\ddagger}$ for the rotation at this temperature as 74 kJ mol^{-1}. It is ironic that this particular experiment would be out of easy range today, because the coalescence temperature would be closer to 360 K on a 400 MHz instrument, on which the $\Delta\nu$ value becomes 64 Hz.

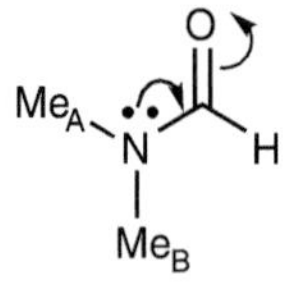

124 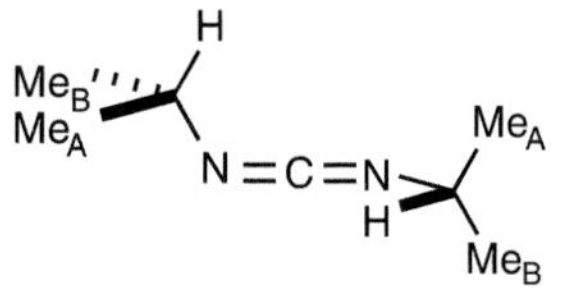 **125**

At the other end of the temperature range, the carbodiimide **125** racemises relatively easily at temperatures above 143 K, exchanging the environments of the diastereotopic methyl groups, which give rise simply to a doublet in the ^{1}H spectrum. At 128 K the methyl groups are no longer exchanging rapidly, and they give rise to a broad pair of doublets approximately 6 Hz apart. The coalescence temperature in the 240 MHz spectrometer that was used for this experiment was in between, at approximately 133 K, giving a $\Delta G^{\ddagger}$ value of 28 kJ mol^{-1}. This experiment was only possible using an unusual solvent in order to avoid freezing: a 1:1 mixture of vinyl chloride and dichlorofluoromethane.

T_2 values can be measured, but they are rarely useful in structure determination. Of the two components of T_2^* only the intrinsic T_2 is of interest. Measurement of T_2 values must therefore remove the component $T_{2(B0)}$ from inhomogeneities in the magnetic field using a pulse sequence called a spin-echo that has many other uses. Its basic form is given in Fig. 4.72 along with a picture of what happens to the vectors at each stage. The vectors are shown in Fig. 4.72 in the rotating frame of reference, which is described more fully later in the chapter (Sect. 4.14). It is not necessary to understand it at this stage—simply allow for the change of perspective from the three-dimensional views we have used so far, repeated at the left in Fig. 4.72, to the view from above in the other drawings of the vectors.

The first $\pi/2$ pulse tips the magnetisation vectors onto the y axis, and they begin to precess. After time t_D the spins will have reached various stages in the xy plane depending upon their Larmor frequencies. In Fig. 4.72, the A spin has a higher frequency than the B spin. The effect of the π pulse, also on the x axis, is to tip the vectors across the xz plane, which in the picture places them on the negative side of the y axis, where they continue to precess at the same rate, the A spin faster than the B spin. After the *same* time t_D, they meet on the $-y$ axis, providing an echo of their original state. The FID acquired at this stage will match the FID that would have been obtained in the usual way immediately after the $\pi/2$ pulse, but it will be upside down. Any loss of phase coherence stemming from the

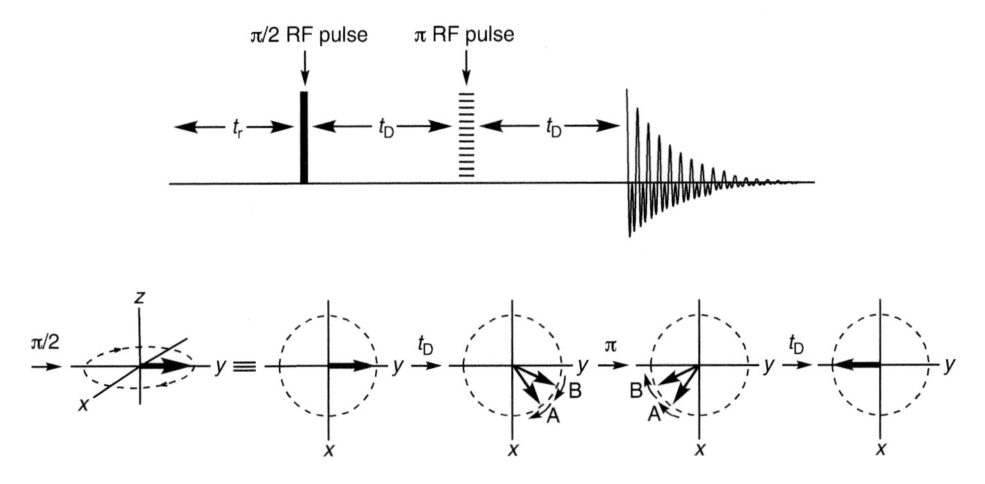

Fig. 4.72 Basic pulse sequence for a spin-echo

inhomogeneity of the applied field is refocused by this means, and the only loss of intensity comes from whatever depredations the intrinsic T_2 mechanisms have caused. By repeating the sequence with incrementally increasing values of t_D the rate of exponential decay in the signals can measure the intrinsic T_2 value for each signal. A series of spectra would be produced, initially upside down, but gradually returning to the normal spectrum, with each individual spin following its own curve determined by T_2.

The full sequence is more complicated, and one of the complications is inherent in multi-pulse sequences like this. In a single-pulse sequence, the tip angle does not need to be exactly 90°. It is too difficult to get it exactly right every time, and we do not need even to try, because all that the coils on the x and y axes detect is the component of the vector in the xy plane—any residual magnetisation on the z axis plays no further part. With multi-pulse sequences, we have to consider what happens to the residual z magnetisation that is present as a consequence of an imperfect $\pi/2$ pulse. In the spin-echo sequence above, any residual z magnetisation, will be tipped by the π pulse onto the $-z$ axis, but this will also be an imperfect pulse, just as the original $\pi/2$ pulse was imperfect. As a consequence, this vector, although largely on the $-z$ axis, will have a small projection onto the xy plane, which would add inappropriately to the signals we want to detect.

These unwanted signals can be removed by another manipulation of the spins—the sequence is repeated, but with alternate π pulses given on the $+x$ axis and $-x$ axis. Any xy magnetisation stemming from the imperfections in the pulses will alternate in phase, and *adding* the FIDs of alternate sequences will cause them to cancel in the xy plane, leaving the spin echoes alone (Fig. 4.73). This form of phase alternation, like the phase cycling to be discussed later (Sect. 4.14), is standard practice in many multi-pulse sequences. We shall return to this problem, with another solution to it, in Sect. 5.2.6.

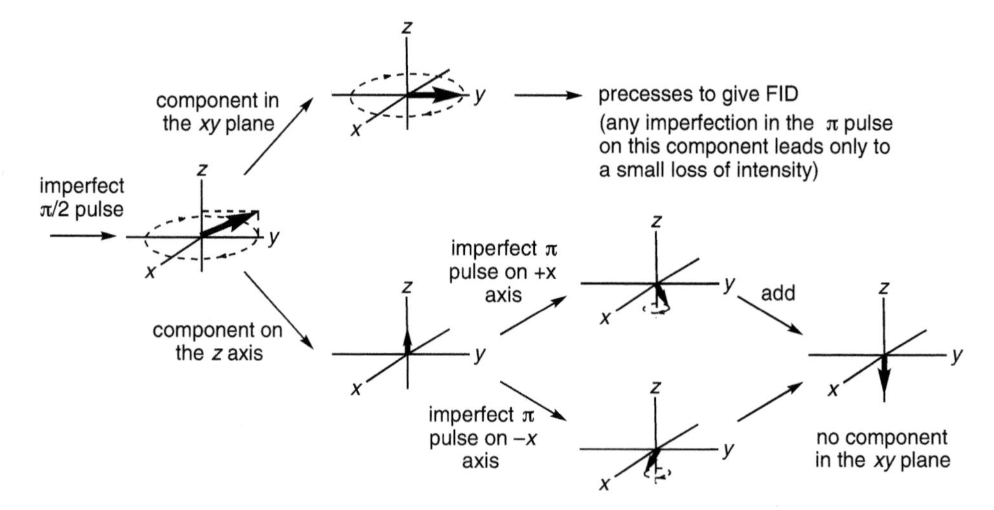

Fig. 4.73 Vectors showing how xy magnetisation cancels with phase alternation

Although there are mechanisms for T_2 relaxation that are independent of T_1, the value of T_2 is not. T_2 is also affected (like T_1) by fluctuations in the applied magnetic field experienced as the molecules tumble. If we imagine the local field as being aligned with the x axis, only the y components can relax and the x component does not; tumbling is only half as effective within the xy plane as it is for longitudinal relaxation. In small molecules, with molecular weights up to a few hundred, the tumbling frequency is close to 10^{11} s^{-1}, which is too fast to promote efficient relaxation. R_1 is small and T_1 is long, and so therefore is T_2, making signals sharp. However, larger molecules, with molecular weights of, say, 1000 or more, tumble at a frequency of approximately 10^9 s^{-1} or less (the precise value depending upon the viscosity of the solvent). R_1 is large and T_1 is now short, and so inevitably is T_2, since it can never be greater than T_1. As a result the NMR signals are broader, and the higher the molecular weight, the broader the signals are.

4.10 Improving the NMR Spectrum

4.10.1 The Effect of Changing the Magnetic Field

NMR spectrometers are available with a variety of magnetic fields; thus 300, 400, 500, 600 and 750 MHz instruments are in common use (the resonance frequency for protons is used to identify them), and instruments with ever higher fields are steadily being introduced. These high-field instruments have a number of substantial advantages, in spite of the cost of superconducting circuitry and the attendant support costs.

The frequency of a resonance changes as the field is changed, but the chemical shift value δ does not. This means that the separation between, for example, $\delta = 2$ and $\delta = 3$ is 200 Hz on a 200 MHz instrument, but 400 Hz on a 400 MHz instrument (remember that δ is expressed in p.p.m.). Coupling constants do not change as the magnetic field changes, but their appearance in the spectrum does. Thus, a doublet with a coupling constant of 18 Hz occupies 30% of the space between δ values differing by 1 p.p.m. in the spectrum from a 60 MHz instrument, but only 4.5% of the space on the spectrum from a 400 MHz instrument. This means that multiplets which overlap when the spectrum is taken at low field are less likely to do so at higher field—the multiplet is effectively narrower. The signals are more easily recognised, and otherwise unresolved signals can come out of methylene envelopes.

The effect can be quite dramatic, as seen in the spectra of aspirin **126** taken on two different instruments (Fig. 4.74). The lower trace shows an old spectrum of the aromatic region measured at 60 MHz; the four different protons are all coupled to each other, and give rise to a spectrum which, except for the double doublet from H_a, cannot easily be analysed. The upper trace shows the same part of the spectrum measured on a 400 MHz instrument, where the signals are now clearly separate and easily analysed.

Another advantage of high-field instruments is that second order effects, which can disguise the coupling patterns, are sometimes, but not always, reduced to the point where

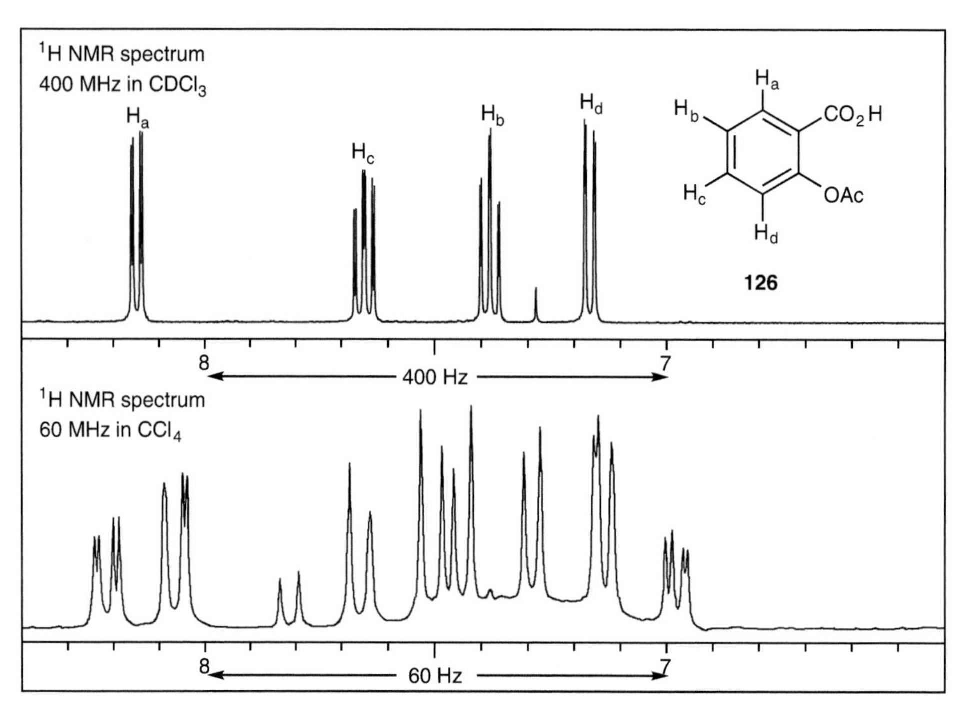

Fig. 4.74 Details of ^{1}H NMR spectra of aspirin at different field strengths

they no longer impede a first-order analysis. Thus the spectrum of the aromatic region of 2,3-dimethoxybenzoic acid **74** taken on a 200 MHz spectrometer, Fig. 4.55, and repeated at the bottom in Fig. 4.75, shows the downfield double-doublet from H-*o* with both coupling constants appearing to be large, even though one was from ortho coupling and the other from meta coupling. This problem arose because the coupling partners H-*m* and H-*p* were close in chemical shift, only 7.0 Hz apart in the 200 MHz spectrometer, and H-*m* was coupled to H-*o* with a coupling constant of 7.75 Hz, similar in magnitude to the separation in chemical shift. In a higher-field spectrometer, the separation of the H-*m* and H-*p* signals is effectively greater, 21.0 Hz at 600 MHz, while the coupling constant remains the same at 7.75 Hz, making the value of $|(\nu_A - \nu_B)|/J_{AX} = 2.7$, safely above the value that leads to second-order spectra. The signals for these three protons become essentially first-order in appearance (with roofing) at 600 MHz, shown in the upper trace in Fig. 4.75, and all the coupling constants, ortho and meta, could now be measured accurately enough for most purposes by measuring the separation of the peaks.

A third advantage of high-field instruments is their greater sensitivity. At higher field strengths, there is a bigger separation in energy between the two spin states, and a bigger difference in the numbers of the nuclei N_α and N_β. As engineers have been able to design and manufacture instruments operating at higher and higher fields, ever smaller samples can be used, a factor of especial importance in ^{13}C NMR, where the low sensitivity and low natural abundance of the nucleus had limited its use.

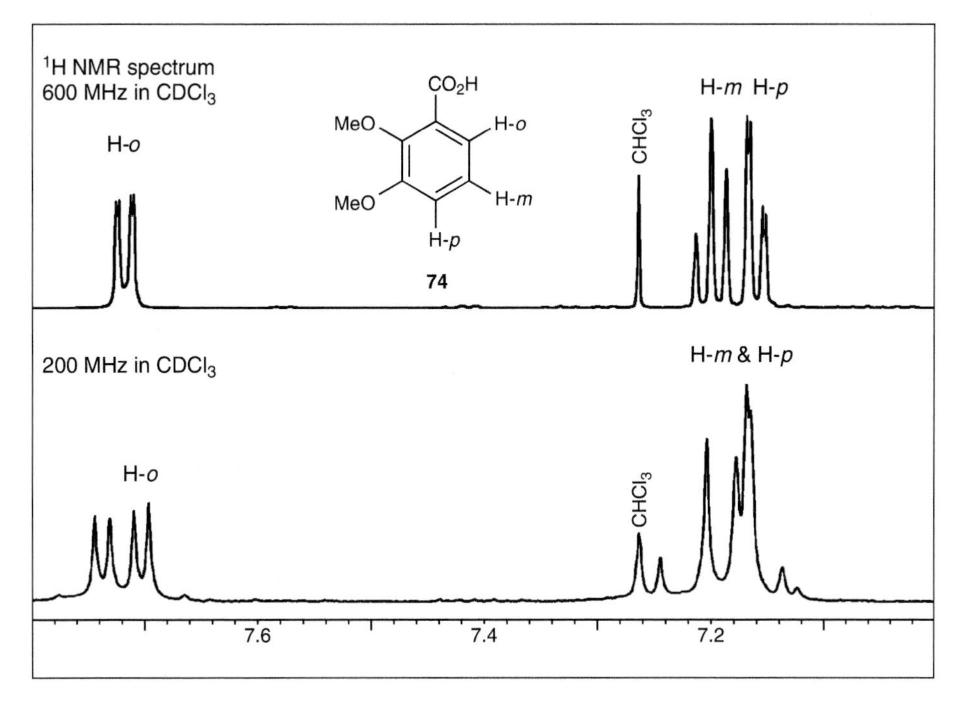

Fig. 4.75 Details of the ^{1}H NMR spectra of 2,3-dimethoxybenzoic acid at 200 and 750 MHz

The older spectrometers with permanent magnets, restricted to 100 MHz or less, are clearly at a disadvantage. For many years they went out of use. More recently, the fact that they are relatively inexpensive, both to manufacture and to service, has seen them return to many laboratories as robust bench-top instruments for use in the many ^{1}H-analyses where the best separation of overlapping signals and the highest sensitivity are not needed. They have other advantages too: signals with very small coupling constants, taking up more of the spectral width, are quite often better resolved at 60 MHz than they are in higher field instruments. Furfural **127** in Fig. 4.76 has a clean looking routine spectrum at 600 MHz, but a distinctly better resolved spectrum in a continuous wave spectrometer at 60 MHz, where H-4, for example, can be seen as a double-double-doublet with coupling constants of 3.7, 1.7 and 0.3 Hz. The last of these coupling constants, 0.3 Hz, would be difficult to measure on a high-field instrument without extensive optimisation. It is a 5-bond coupling to the aldehyde proton, where it is also resolved, along with another 5-bond coupling of 0.65 Hz from the aldehyde proton to H-5. The 3.7 and 1.7 Hz couplings are from H-4 to H-3 and H-5, respectively, at the very low end of the range for vicinal coupling constants, because both H-3 and H-4 are antiperiplanar to the oxygen atom.

4.10.2 Solvent Effects

By changing the solvent, the chemical shifts of individual signals can change. Figure 4.76 shows the small shift of all the signals to slightly higher field in going from carbon

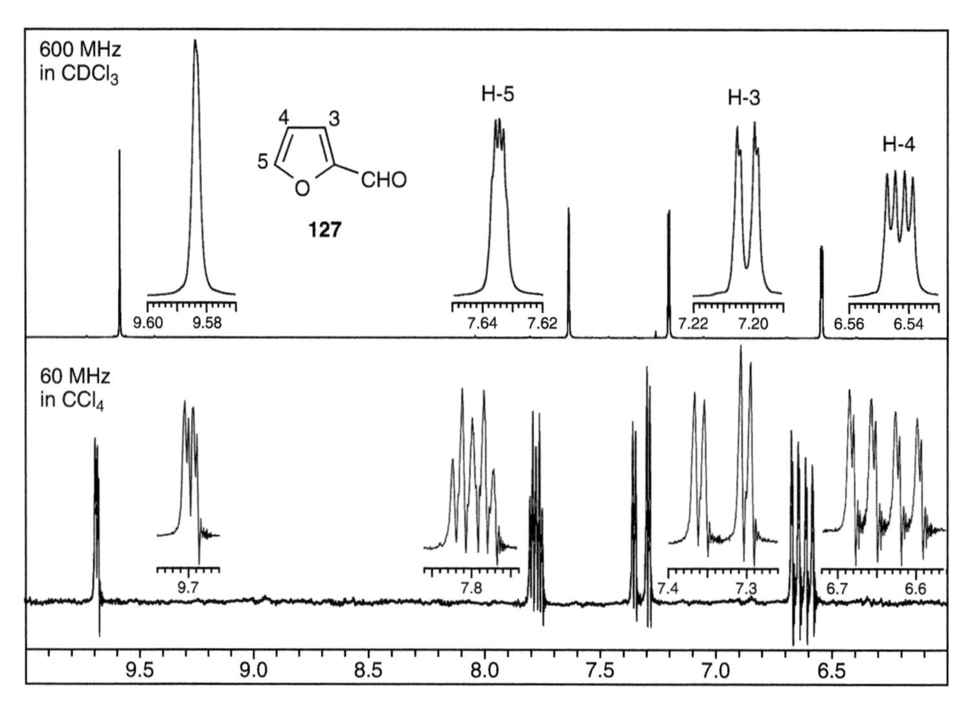

Fig. 4.76 600 MHz and 60 MHz ^{1}H-NMR spectra of furfural compared

tetrachloride to the slightly more polar deuterochloroform. If the change of solvent is from the standard deuterochloroform, to a polar solvent like d_6-acetone or d_6-DMSO, or to an aromatic solvent like d_6-benzene or d_5-pyridine, overlapping multiplets can often be resolved. These solvent-induced shifts, which rely on solute-solvent interactions are rather unpredictable, and have to be searched for by trial and error. Figure 4.77 shows the effect of different solvents on the spectrum of eugenol **128**, taken because the doublet-doublet-singlet pattern for the 1,2,4-trisubstituted benzene ring was not clear in deuterochloroform. In fact all that was needed to clarify this spin system was to add a few drops of d_6-benzene to the chloroform. The aromatic proton signals are also clear in d_6-acetone and pure d_6-benzene, and the terminal methylene signals are easier to see in d_6-acetone and d_5-pyridine.

The strong methoxy singlet has dramatically moved on top of the side-chain methylene signal in d_6-benzene, although it had not when only a few drops of benzene had been added to the chloroform. Unsurprisingly the hydroxyl signal is different in each solvent (in d_5-pyridine it is at δ 10.7).

Enantiomerically pure chiral solvents are also used to measure the proportions of enantiomers present in an incompletely resolved mixture. Mildly acidic solvents like the fluorinated alcohol **129** can be used to analyse basic compounds like amines, and chiral amines can be used for acidic substances. More commonly, the analysis of mixtures of enantiomers is carried out by attaching a chiral auxiliary covalently, and integrating the NMR spectrum

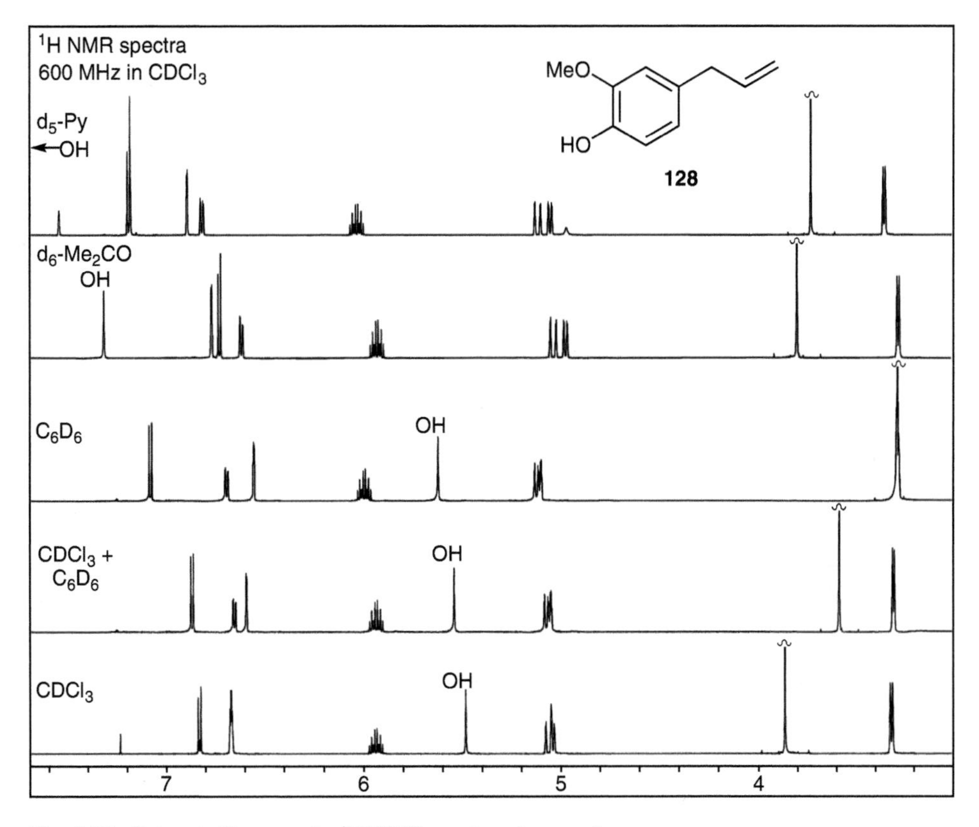

Fig. 4.77 Solvent effects on the ^{1}H NMR spectra of eugenol

of the mixtures of diastereoisomers. The most frequently used chiral auxiliary is Mosher's acid **130**, the esters and amides of which give sharp and frequently well-separated signals in the ^{1}H NMR spectrum from the methoxy group, or in the ^{19}F NMR spectrum from the trifluoromethyl group. It is even possible, because the diastereoisomers are so similar, to integrate corresponding pairs of signals in the ^{13}C NMR spectrum.

H OH CF$_3$ **129** F$_3$C OMe CO$_2$H **130**

Mosher's esters have a further use: if the esters from the *S*-acid and the *R*-acid are prepared separately, characteristic chemical shifts in the signals from the substituents adjacent to the alcohol function can be used to identify the *absolute* configuration of the alcohol [1].

4.10.3 Shift Reagents

In the spectrum of n-hexanol **47** in Fig. 4.35, signals from six of the protons in similar environments overlap in the methylene envelope between δ 1.35–1.18, and their multiplicity can no longer be seen, even in a 400 MHz spectrum. This is a much more serious problem in larger molecules when signals often overlap, causing useful information to be buried. The addition of a shift reagent alters this picture dramatically. Shift reagents are usually β-dicarbonyl complexes of a rare earth metal, the commonest being $Eu(dpm)_3$ **131**, $Eu(fod)_3$ **132** (M = Eu), and $Pr(fod)_3$ **132** (M = Pr). These complexes are mild Lewis acids, which attach themselves to basic sites such as hydroxyl and carbonyl groups. They are also paramagnetic, and have the effect of changing substantially the magnetic field in their immediate environment. The result is a shift of the signals coming from the protons near the basic site in the organic molecule. The shift, downfield with the two europium reagents but upfield with the praseodymium reagent, falls off, with angular variation, as the inverse cube of the distance from the metal. Thus, the spectrum of n-hexanol is spread out when $Eu(dpm)_3$ is added to the solution, and the resonances of each of the three methylene groups can be seen as two quintets and a sextet.

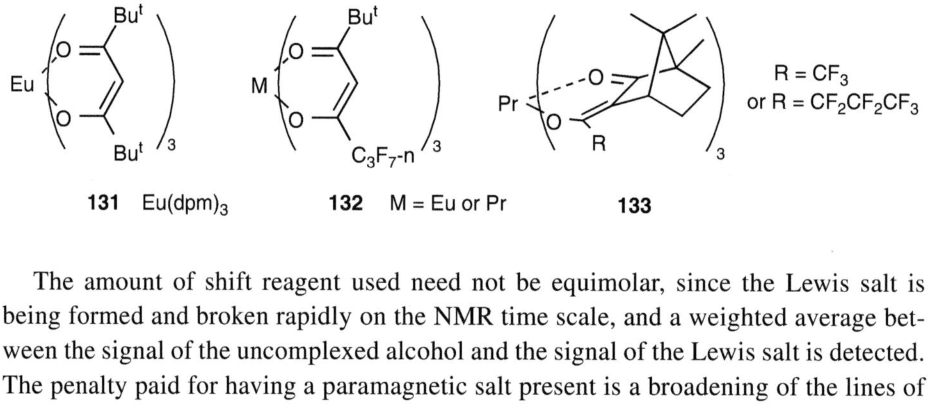

131 $Eu(dpm)_3$ **132** M = Eu or Pr **133** ($R = CF_3$ or $R = CF_2CF_2CF_3$)

The amount of shift reagent used need not be equimolar, since the Lewis salt is being formed and broken rapidly on the NMR time scale, and a weighted average between the signal of the uncomplexed alcohol and the signal of the Lewis salt is detected. The penalty paid for having a paramagnetic salt present is a broadening of the lines of the multiplets.

The multiplicity of the signals is usually clear enough in spectra taken at 100 MHz or less. Unfortunately, the broadening is a function of the square of the operating field strength, and so shift reagents are much less useful on modern instruments, where broadening obscures the multiplicity. Nevertheless, the camphor derivative **133** can still be used to measure the proportions of enantiomers present in an incompletely resolved mixture. As long as the two enantiomers have binding constants towards the chiral reagent that are different enough, the signals may separate. They may be integrated, and the proportions of the two enantiomers measured, even if their signals are individually ill resolved.

4.11 Spin Decoupling

4.11.1 Simple Spin Decoupling

In earlier sections we have seen how a proton with neighbouring protons can give rise to a multiplet, how the multiplet pattern can be recognised and how coupling constants can be measured. However, in more complicated molecules than the ones we have seen so far, we want to be able to pair up and extend spin systems that are not obvious. The multiplicity and the chemical shift of a signal from one or more protons may be visible, but it is not always obvious to which of the other signals it is coupled. We were able to pair up the signals from ethyl propionate **42** in Fig. 4.33 using the coupling constants, but we cannot use this technique to pair up the quartets and triplets in diethyl acetylsuccinate **56** in Fig. 4.43, because the coupling constants are the same for both ethoxy groups. Fortunately there is a powerful technique, *spin decoupling*, for making this type of connection unambiguously.

If, during the time that a signal is being collected, the proton (or any other magnetic nucleus) has a neighbour coupling with it that is exchanging its spin state rapidly, the proton we are observing will experience an average of all the states. We have seen this already in the loss of coupling to OH protons, where the exchange was a chemical exchange, the OH protons moving from molecule to molecule with a rate constant greater than the coupling constant. The same loss of coupling occurs when the exchange is between spin states stimulated by irradiating the coupling partner at its resonance frequency, as we have seen in all proton-decoupled ^{13}C spectra. In those spectra the decoupling was unselective, but it is also possible to decouple selectively.

Thus, we can look again at the two triplets and quartets from diethyl acetylsuccinate **56** (Fig. 4.43, with the undisturbed spectrum repeated in Fig. 4.78). If we measure the

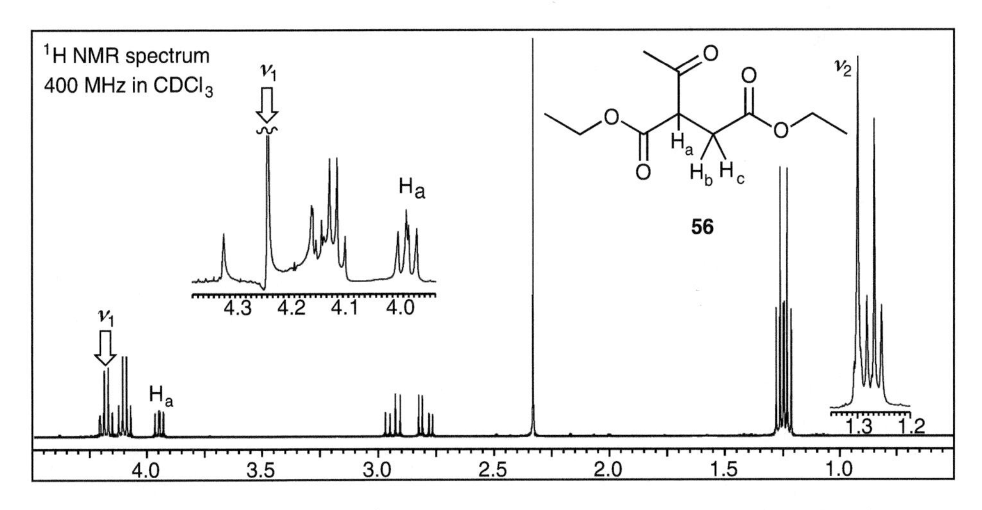

Fig. 4.78 Irradiation at one signal selectively decoupling its partner in diethyl acetylsuccinate

resonance frequency ν_1 of the downfield quartet, and then irradiate the sample at precisely that frequency with a soft pulse (low power for a longer time) at the same time as we collect the FID, the spin states of these methylene protons will be rapidly exchanging places with each other. The signal at ν_1 appears as a strong distorted singlet with side bands spaced equally above and below this singlet by 0.08 p.p.m., as seen in the expansion in Fig. 4.78. The methyl protons to which they are coupled will have their coupling to the methylene group 'turned off', and they come into resonance at their usual frequency ν_2, but as a singlet instead of a triplet (right-hand enlargement in Fig. 4.78). In this way we can be sure that the methylene group giving rise to the downfield quartet is connected to the methyl group giving rise to the downfield triplet. The irradiation at ν_1 leaves other signals more or less undisturbed (such as the signals illustrated in the enlargements in Fig. 4.78 from the other methylene group, from H_a and from the other methyl group), because they are not coupled to the protons being irradiated. There is one limitation: the relevant signals must be reasonably well separated in chemical shift. Thus, in the case of the ester **56**, ν_1 is only just well enough separated from the frequency of the other quartet, but the downfield triplet is perilously close in chemical shift to the upfield triplet, and it would probably not be possible to carry out the experiment the other way round—irradiating one of the triplets to see which quartet collapsed. When the nuclei are the same element, typically both ^1H, this technique is called *homonuclear decoupling*. When the experiment involves nuclei of different elements, it is called *heteronuclear decoupling*.

Selective decoupling is used when only a few of the coupling relationships are in doubt. It is now more usual simply to run a COSY spectrum, which can identify all the coupling relationships in one experiment (see the next chapter Sect. 5.2). However, selective decoupling of a multiplet will reveal the remaining coupling pattern, so that a difficult to interpret double-double-quartet, for example, will collapse to a recognisable double-doublet when the sample is irradiated at the resonance frequency of the protons on the methyl group to which it is coupled. Deconvoluting a puzzling multiplet like this gives useful information which a COSY spectrum does not provide.

4.11.2 Difference Decoupling

It is possible to use selective decoupling to reveal a buried signal. Figure 4.79c shows a narrow part of the methylene region, between δ 1.14 and 0.9, of the spectrum of the steroid **134**. This signal is a composite of the multiplets from four protons, one of which is $H_{7\alpha}$, all overlapping inextricably. When the signal of $H_{6\alpha}$, which is further downfield than the ones in Fig. 4.79, is irradiated, the signal from $H_{7\alpha}$ loses one of its couplings, and the signal changes to that in Fig. 4.79b. The multiplets are just as impossible to analyse as before, but now, because the spectrum is in the computer in digital form, it is possible to subtract the original spectrum (c) from the decoupled spectrum (b). The result is plotted in Fig. 4.79a. Fortunately, the other three protons in this signal were not coupled to $H_{6\alpha}$, and were unaffected by the decoupling; the subtraction therefore removed them from the signal and left

Fig. 4.79 Difference decoupling to extract a signal from overlapping multiplets [2]

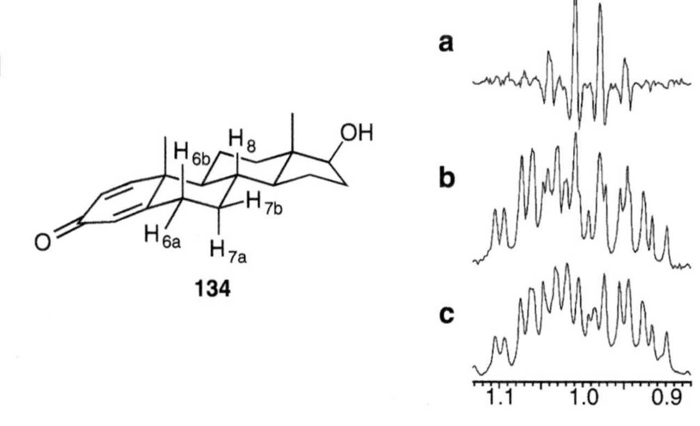

only a signal from $H_{7\alpha}$. The signal left has both the coupled and the decoupled signal in it, the original coupled signal is a double quartet pointing down and the partly decoupled signal is a quartet pointing up, since (c) was subtracted from (b). Evidently $H_{7\alpha}$ is coupled equally to each of the protons $H_{6\beta}$, H_8, and $H_{7\beta}$, with a coupling constant of 13 Hz, leading to the quartet, and to $H_{6\alpha}$ with a coupling constant of 4.3 Hz, doubling that quartet.

4.12 Identifying Spin Systems: 1D-TOCSY

Decoupling identifies those protons that are coupled to each other, but it is often helpful to be able to identify all the coupling connections within a *spin system*. In sucrose octaacetate **135** there are three spin systems: (a) the protons on the carbon atoms numbered 1–6 in the glucose ring; (b) the protons on the carbon atoms numbered 3´–6´ in the fructose ring; and (c) the methylene protons on C-1´ in the fructose. Within each of these groups all the protons are coupled to at least one other proton in the system, but the coupling from one system to the others is interrupted by the oxygen atom that joins the two monosaccharides together and by the fully substituted atom C-2´ in the furanose ring. Negligible coupling takes place across these barriers. The ^{1}H NMR spectrum of sucrose octaacetate has all these signals in the same chemical shift region δ 6.0–4.0, shown at the bottom in Fig. 4.80, but we cannot easily see which signals belong to which ring.

The solution to this type of problem is provided by a 1D-TOCSY spectrum (TOtal Correlation SpectroscopY), also called a HOHAHA spectrum (HOmonuclear HArtmann-HAhn). It requires that a signal from one of the protons in a spin system be separated far enough from the signals from protons in the other spin systems, for it to be irradiated selectively. The pulse sequence begins with a long irradiation of one such signal within a spin system, followed by a delay called the mixing phase, which causes any protons which are coupled to it to be raised to the N_β level. Depending upon how long the mixing phase is maintained, any protons coupled to this proton are the next to be raised, and so on down a chain of coupled protons, with the protons that have passed on their high-energy spins,

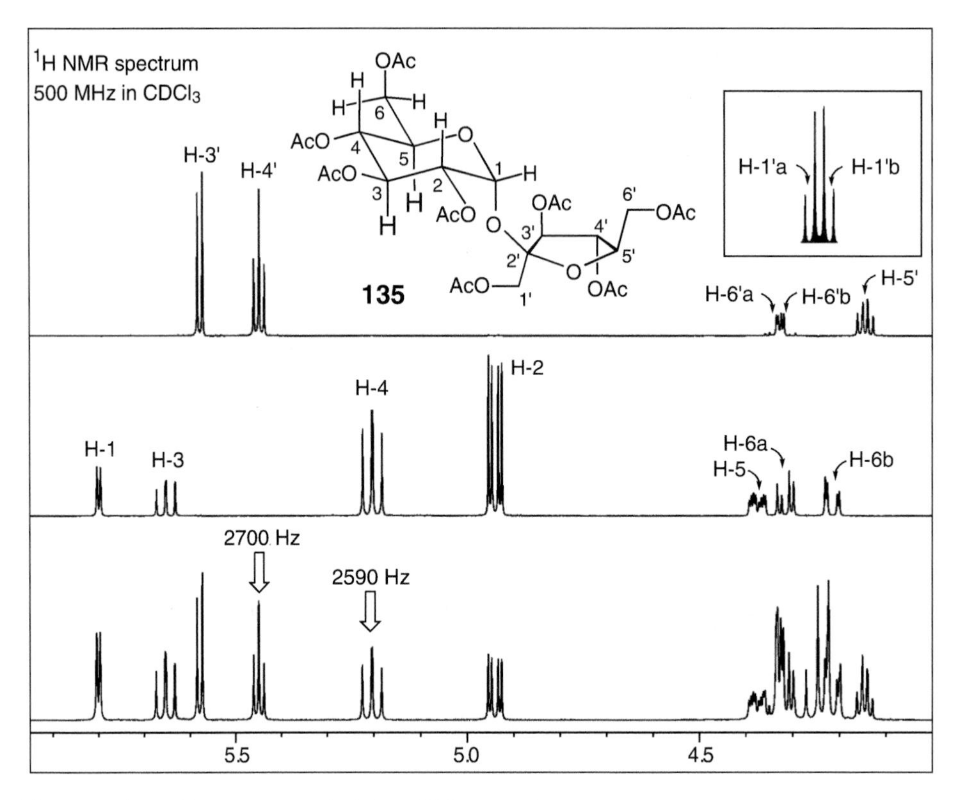

Fig. 4.80 ^{1}H NMR spectrum and 1D-TOCSY spectra from sucrose octaacetate

getting weaker and weaker. The FID is then collected in the acquisition phase, and processed to give a spectrum showing signals from all the protons within that spin system. It is usual to try several lengths of time in the mixing phase, and then pick the spectrum giving the best picture of the multiplets within the spin system. The mixing phase needs to be quite long (hundreds of milliseconds) if the spin system is extended over several atoms, but can be quite short (20–100 ms) if it is compact.

In sucrose octaacetate all the signals in the δ 5.9–4.9 region are well enough resolved to use in a 1D-TOCSY experiment. Irradiating the triplet at 2590 Hz gives the spectrum in the middle in Fig. 4.80 showing only the seven mutually coupled protons from the glucose ring. Irradiating instead the triplet at 2700 Hz gives the spectrum at the top in Fig. 4.80, showing only the five mutually coupled protons from the fructose ring. As well as separating the spin systems, these two spectra clarify the δ 4.4–4.1 region in which the signals overlap and are hard to analyse: the glucose protons clearly give rise to a double-double-doublet from H-5 and a mutually coupled pair of double-doublets from the diastereotopic methylene protons H-6; the fructose protons give rise to a 2-proton multiplet from the diastereotopic methylene protons H-6′ and a quartet from H-5′. (The multiplet is easily analysed, if this signal is enlarged—it is the 8-line AB portion of an ABX system, with very strong roofing from A to B.) The methylene protons on C-1′ are also in the δ

4.4–4.1 region, but all the signals in this narrow region are too close in chemical shift to allow us to produce a 1D-TOCSY spectrum by irradiating either of them. Nevertheless we can easily pick out the methylene protons: if we mentally subtract the two 1D-TOCSY spectra from the full spectrum, we can recognise the AB system from the protons on C-1′ shown filled in in the box at the top in Fig. 4.80.

Unfortunately we must do this subtraction by eye, and not digitally, because the intensities of the signals in a 1D-TOCSY experiment are not in proportion to each other. The intensity of any signal is dependent upon how well coupled the proton is to its nearest neighbour in the spin system, how close it is along that chain to the proton being irradiated, and how long the mixing phase is. This feature can be turned to our advantage: the protons giving the signal being irradiated are the first to appear at short mixing times, and the signals from those protons structurally closest to it are the next to appear. It is possible to use a short mixing time to bring up a signal out of an overlapping set, even when several of the signals are in the same spin system. Figure 4.81 illustrates an example in which this techniques allows us to expose an obscured signal.

A fragment with the structure **136** from a larger molecule gave rise to the overlapping pair of signals close to δ 1.6 on the right in the lower spectrum in Fig. 4.81. They came from the methyl group and the protons on C-2, both of which were in the same spin system. Irradiation at 1815 Hz, the frequency of the signal from the protons on C-1, followed by 100 ms of mixing, gave the spectrum shown in the middle of Fig. 4.81, with the signal from the protons on C-2 free of the signal from the methyl group, which was further away along the chain of atoms in the spin system. Repeating the experiment with a longer mixing time of 400 ms brought up both signals in the top spectrum in Fig. 4.81, more like

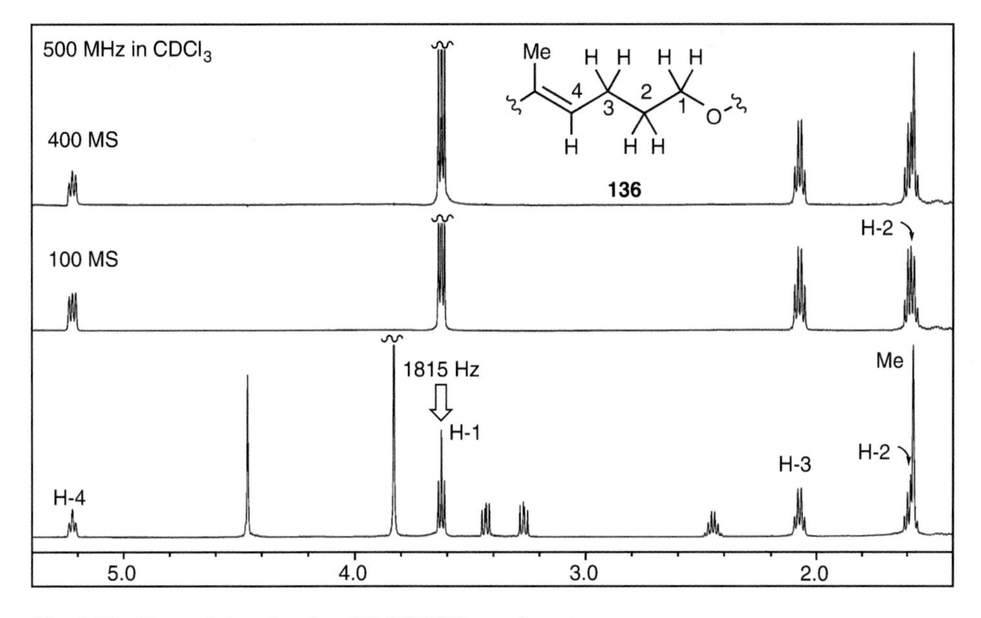

Fig. 4.81 Short mixing time in a 1D-TOCSY experiment

the original overlapping pair, demonstrating that both were in the same spin system. The quintet in the central spectrum is clearly the appropriate signal for the protons on C-2, which have four neighbours. It is even easier to expose a signal when it is not obscured by protons in the same spin system—the offending signal simply disappears in a 1D-TOCSY experiment, as the methyl singlet would have disappeared if the methyl group had not been connected by allylic coupling.

Not only does a 1D-TOCSY spectrum separate one spin system from another, but it also separates them from signals given by impurities. 1D-TOCSY spectra can therefore be used on mixtures to pull out at least part of the spectrum of one component. Here is an example using a sample taken from a bottle labelled 'limonene.' It gave the lower ^{1}H NMR spectrum in Fig. 4.82, but it was immediately clear that limonene **138** was a minor component, since it was known to give rise to the distinctive, but relatively weak, signals marked by the arrows. A 1D-TOCSY experiment irradiating the methyl singlet at 814 Hz (δ 1.628) brought up the upper spectrum, which must belong to one of the major components. This spectrum showed a pair of methylene signals (δ 2.1–1.6) with geminal coupling constants of 19 and 20 Hz, an olefinic signal at δ 5.275, and a pair of high-field signals at δ 0.7–0.5, characteristic of protons on a cyclopropane ring. These features were definitive for the structure of the main component, 3-carene **137**, with only the signals from the two methyl groups missing, because they are insulated from the spin system by the quaternary centre. Each of the other recognisable components in this mixture, α- and β-pinene and *p*-cymene, could equally easily have been selected.

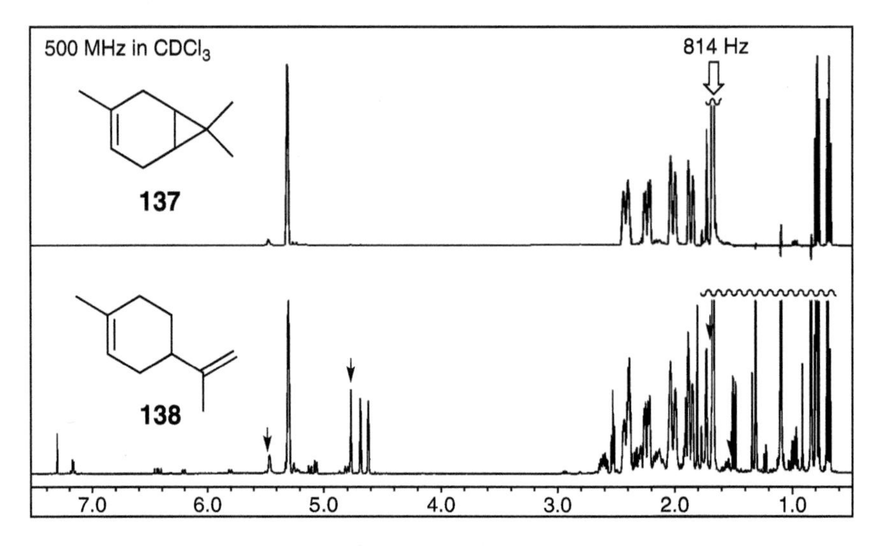

Fig. 4.82 1D-TOCSY spectrum of the contents of a bottle mislabelled limonene

4.13 The Nuclear Overhauser Effect

4.13.1 Origins

The interaction of one magnetic nucleus with another leading to spin-spin coupling takes place through the bonds of the molecule. The information is relayed by electronic interactions, as one can see from the dependence of the coupling constant on the geometrical arrangement of the intervening bonds. Magnetic nuclei can also interact through space, but the interaction does not lead to coupling. The interaction is revealed when one of the nuclei is irradiated at its resonance frequency and the other is detected as a more intense or weaker signal than usual. This is called the nuclear Overhauser effect (NOE or nOe). In an NOE experiment we do not see decoupling, because the irradiating signal is applied before the FID is collected, and is turned off during the FID. The radio frequency irradiation used in selective decoupling (Sect. 4.11.1) is turned on only during the time that the FID is collected. The decoupling used in the NOE experiment is called gated, and that used in selective decoupling is sometimes called inverse gated.

The NOE is only noticeable over short distances, generally 2–4 Å, falling off rapidly as the inverse sixth power of the distance apart of the nuclei. This is because the interaction is dependent upon the relaxation of the observed nucleus by the irradiated nucleus (Sect. 4.9.1). Two nuclei A and X relaxing each other, whether coupling or not, interact to set up four populated energy levels, as we saw in Fig. 4.53. The NOE is through space, and nuclei can show NOEs whether they are coupled or not. The energy levels in Fig. 4.83 are the

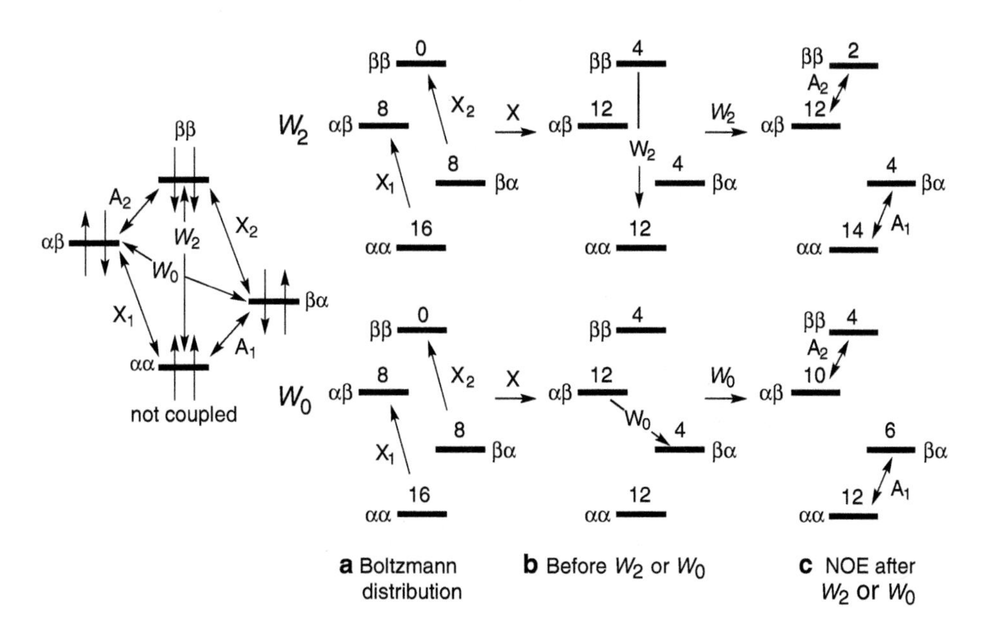

Fig. 4.83 The origin of the Nuclear Overhauser effect

same as those in Fig. 4.53 for the case with no coupling but with the addition of two more transitions, labelled W_2 and W_0.

The populations in Fig. 4.83a are arbitrary, but in proportion for the Boltzmann distribution of the excess spins to start with. They show how the populations change with the different transitions. Transitions A_1 and A_2 give rise to the line we associate with the A nucleus and the transitions X_1 and X_2 give rise to the line we associate with the X nucleus. When the sample is irradiated at the resonance frequency of the X nucleus, the transitions X_1 and X_2 take place rapidly in both directions. The population levels ββ and αβ grow at the expense of levels βα and αα, respectively, so that levels αβ and αα become equally populated, and so do levels ββ and βα Fig. 4.83b. There is still no obvious effect on the intensity of the A signal, because it is produced by transitions A_1 and A_2. The intensity of the A signal is dependent upon the difference between the sum of the populations of levels ββ and βα and of αβ and αα, and this has not been affected. However, there is another relaxation pathway, W_2, which does not lead to observable signals but does affect the populations of the four energy levels. W_2 is a two-quantum process (two nuclei change their spins) between well-separated energy levels, and relaxation by this pathway is stimulated by the more rapid (higher-frequency) tumbling with a frequency close to 10^{10}–10^{11} s^{-1} of molecules with a molecular weight in the region of roughly 100–400 in a solvent of low viscosity. The effect is to increase the population of energy level αα at the expense of energy level ββ. The sum of the populations of the energy levels ββ and βα has decreased, and the sum of the populations of the energy levels αβ and αα has increased; Fig. 4.83c. the difference between the populations of the energy levels is greater than it was before, and the signal from the A nucleus is, therefore, more intense. This effect is called a *positive NOE*.

In contrast, the other relaxation pathway W_0 is a zero-quantum process (no net change of spin) between energy levels close in energy. Relaxation by this pathway is stimulated by the slower (lower-frequency) tumbling of larger molecules with molecular weights ≥1000. The effect is to increase the population of energy level βα at the expense of energy level αβ. The sum of the populations in energy levels ββ and βα has now increased and the sum of the populations of levels αβ and αα has decreased; the difference between the populations of the energy levels is smaller, and the signal from the A nucleus is, therefore, less intense—*a negative NOE*. Molecules with intermediate molecular weight fall between two stools, and show weak or non-existent NOEs.

In ^{13}C spectra, the maximum possible NOE produced by irradiating at the proton frequency is nearly 200%. NOEs of this order are found in the signals from ^{13}C atoms directly bonded to protons in proton-decoupled spectra, for which the irradiation is no longer gated, but continued into the time of the FID. It helps to increase the intensity of the otherwise inherently weak signals. In ^{1}H spectra, the maximum enhancement can only be 50% of the usual intensity of the signal, but the typically observed range is only 1–20%. NOEs are weakened when the proton being observed is being relaxed by protons other than the irradiated proton. Thus, a methyl group, in which each proton already has two nearby protons to speed up relaxation, often shows very little NOE when a nearby proton is irradiated. NOEs are most easily detected, therefore, in methine groups when a nearby methyl signal is irradiated. It is possible to measure NOEs by integrating signals with the irradiation on and then with it off, and measuring the difference by integration. The accuracy of integration is such that this method can only be used reliably if the NOE is at least 10%, and it can only be detected in the signals from methine and, occasionally, methylene protons. Nevertheless, even in this form, the NOE has been a useful method for detecting which groups are close in space to each other, providing valuable information about stereochemistry, but it has been superseded by NOE-difference and NOESY spectra.

4.13.2 NOE-Difference Spectra

NOEs are much more easily detected by subtracting, in the computer, the normal spectrum from a spectrum taken after gated irradiation, and printing only the difference between the two spectra. All the unaffected signals simply disappear, and all that shows is the enhancement itself, together with an intense negative signal at the irradiating frequency. The upper half of Fig. 4.84 shows the complex ^{1}H spectrum of the oxindole **139**, and above it

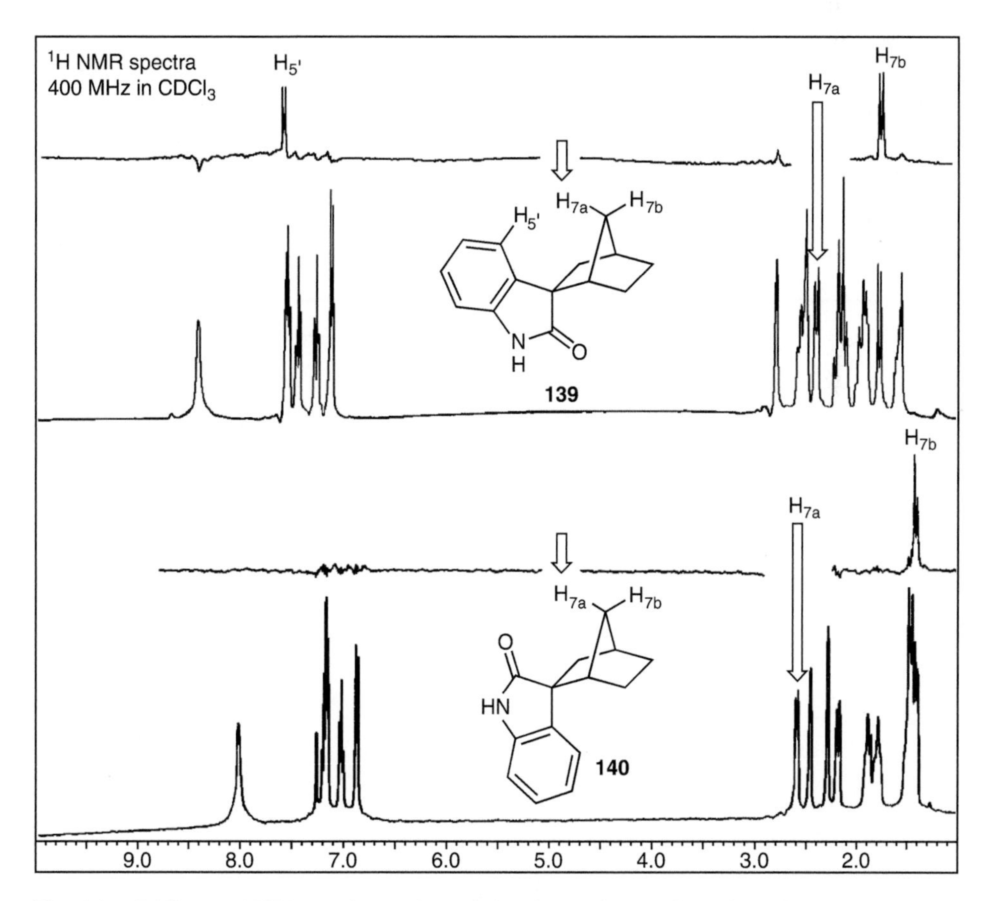

Fig. 4.84 Difference NOE experiment determining the relative configuration of two oxindoles

the difference spectrum created after irradiating the sample at the frequency of the heavy downward-pointing arrow. This frequency is that of proton H_{7a}, which is close in space both to its neighbour H_{7b} and to $H_{5'}$ on the benzene ring. Only signals from these two protons appear in the difference spectrum, and demonstrate that the stereochemistry of the oxindole **139** is that shown, and not the isomer **140** with the spiro-oxindole ring the other way up. A similar experiment on that isomer, illustrated in the lower half of Fig. 4.84, showed only noise in the aromatic region of the difference spectrum. The signals from H_{7b} in both of the difference spectra still show the coupling to H_{7a}, because the signal used to create the NOE is applied before the acquisition pulse, but is turned off during acquisition. Coupling is therefore unaffected.

Using difference spectra, it is easy to detect 1% enhancements, or even less, with the result that NOEs in methyl groups are now measurable. In consequence, it is usually possible to detect the NOE in both directions—not only, for example, from a methyl group to a nearby methine, but also back from the methine to the methyl group, a procedure that greatly increases one's confidence that the groups are indeed close to each other. Furthermore, the distance over which the NOE can now be detected is much greater using difference NOE.

Figure 4.85 shows a pair of NOE difference spectra from camphor **141**. Irradiating at the frequency of the most upfield of the three methyl singlets, δ 0.847, brings up the signals from the exo proton on C-3 and the bridgehead proton on C-4. Irradiating at δ 0.971, the most downfield of the three methyl singlets, brings up the signals from the exo protons on C-5 and C-6, as well as the same bridgehead proton on C-4. This establishes the signal at δ 0.847 as that from the methyl group C-8 and the signal at δ 0.971 as that from the methyl group C-9.

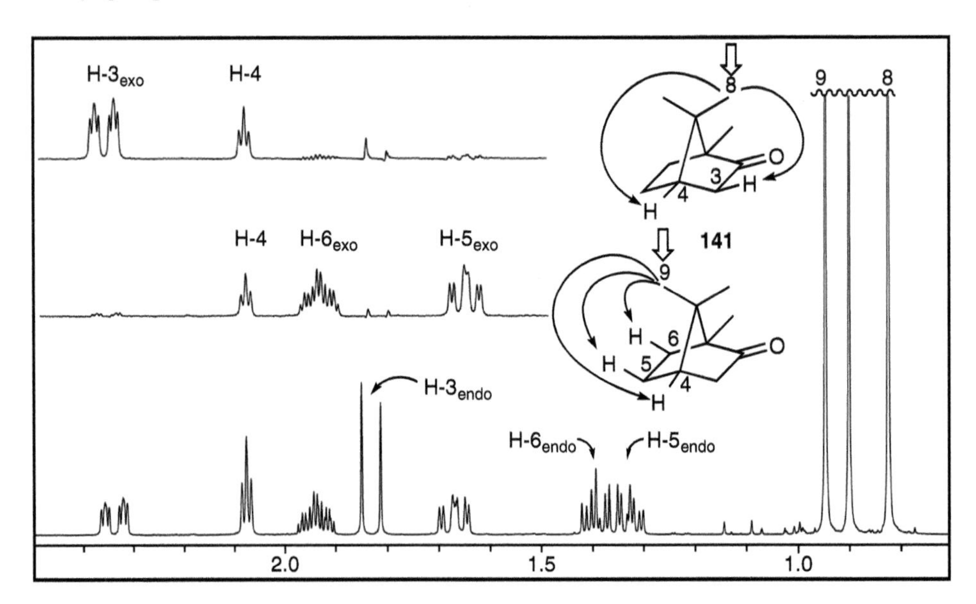

Fig. 4.85 Difference NOE spectra from camphor

There are several more complex pulse sequences that produce NOE difference spectra, and one is designed so that the protons giving signals of enhanced intensity, are themselves the source of further NOEs. Thus with camphor, irradiation at δ 0.847 using one of these pulse sequences enhances the signal, not only for H-3_{exo} as in Fig. 4.85, but also for H-3_{endo}, producing a sharp doublet. Similarly, irradiation at δ 0.971 produces the signals for H-5_{exo} and H-6_{exo}, but also produces signals for H-5_{endo} and H-6_{endo}. These new signals are called relayed NOEs, and they are frequently negative, so that the signal from H-3_{endo}, for example, would be a downward-pointing doublet, instead of the weak upward pointing one that is visible in Fig. 4.85. The signals from H-5_{endo} and H-6_{endo} would similarly be downward-pointing multiplets. It is even possible by adjusting the mixing time to move further out through space to the next set of nearby protons, mapping the spatial distribution of a large set of protons. Although there is no firm convention about which way up the signals are displayed, it is usual to show positive NOEs with the enhanced signals pointing upwards, while the irradiated signals, which are usually too strong to be shown, and the weaker relayed signals would point down.

NOEs are not restricted to proton-proton interactions, as we have seen with the NOEs from protons to the carbons they are attached to, enhancing the signals in ^{13}C spectra. Proton-to-carbon NOEs can be used in structure determination too: irradiating at the frequency of a specific proton signal and recording the difference in the carbon spectrum can identify a connection between protons and carbons that are close in space but not directly bonded.

A difference NOE spectrum also allows one to deconvolute overlapping signals—in other words, to extract a signal from under several others as we saw earlier for TOCSY and for decoupling–difference spectra. The ^{1}H-NMR spectrum of naloxone **142** has several well resolved signals, a selection of which are shown in Fig. 4.86. But two pairs of protons within this range, one pair labeled J and K, and the other L and M, overlap in a way that makes it difficult to analyse the two multiplets. Fortunately, the proton labeled I is close in space to J but not to K. Irradiation of I enhances the intensity of the J doublet, which can be seen cleanly free of the K signal. The K proton, on the other hand, is close in space to the F proton through their 1,3-diaxial relationship, but the J signal is not. The K double triplet is revealed by irradiating the F signal. These are both relatively weak NOEs, but the full structure of both components making up the multiplet is clear. The other pair are both strong NOEs. The intensity of the L double doublet is enhanced by irradiating at the frequency of its geminal partner P, and the M double doublet is enhanced by irradiating at the frequency of its geminal partner I. The appearance of roofing is not the usual roofing, and has to be ignored—it comes from the proximity of the irradiating signal rather than from the proximity of any of the coupling partners, and it 'points' in the opposite direction to that which we are used to seeing in AB roofing.

Which technique to use to deconvolute signals is dependent upon how the overlapping spins are connected. If they are in separate spin systems, TOCSY is most likely to be the best. If they are in the same spin system, but some signals arise from protons closer in

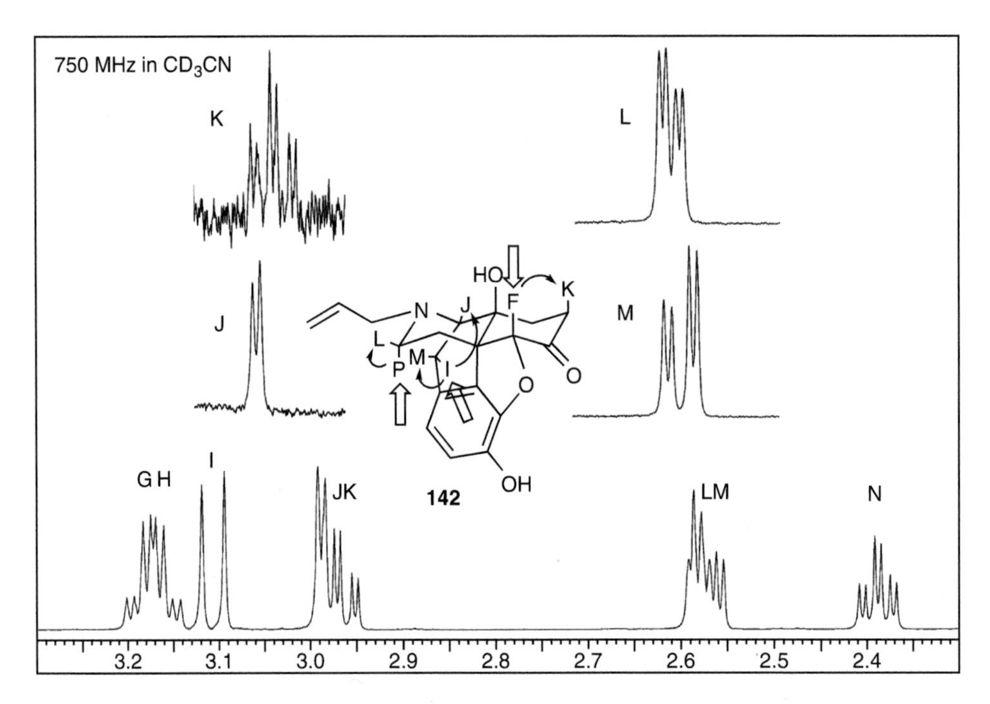

Fig. 4.86 NOE-difference spectra of naloxone

space to another signal than other protons, then NOE-difference might work better. Whichever technique works to give a clean picture of a multiplet makes it possible to recognise the pattern, count the number of different connected spins, and measure all the coupling constants.

4.14 The Rotating Frame of Reference

The pulse and acquisition technique used in FT spectroscopy makes it possible to carry out much subtler experiments than the simple one described in Sect. 4.2 using the basic $\pi/2$ pulse followed by acquisition illustrated in Figs. 4.4 and 4.5. As we have already seen in Sect. 4.9, it is possible to give a π pulse as well as a $\pi/2$ pulse, and to give them in either order with various lengths of time between pulses. It is not possible to describe comprehensively in this and the next chapter all the available pulse sequences, nor can we explain fully here what their effect is on the precessing vectors, but we do need to look at some of the more simple pulse sequences to see how they work in outline. Several books listed in the section on further reading at the end of this chapter go beyond the level we shall use here. However deeply we delve into their workings, the most important lesson that we can take from the output of these experiments is to recognise them for what they are designed to do, and to appreciate what kind of information they give us.

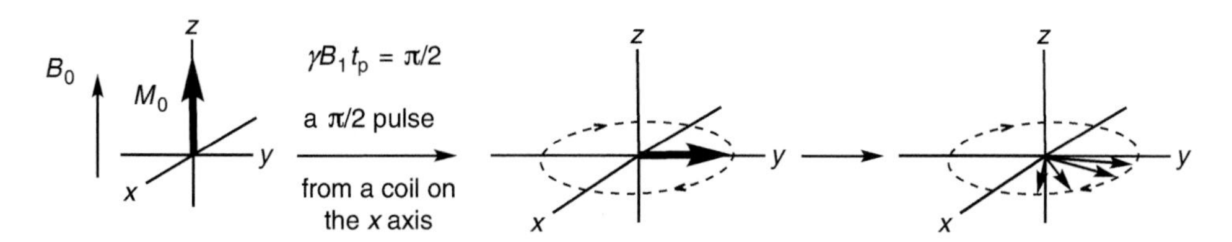

Fig. 4.87 The laboratory frame of reference

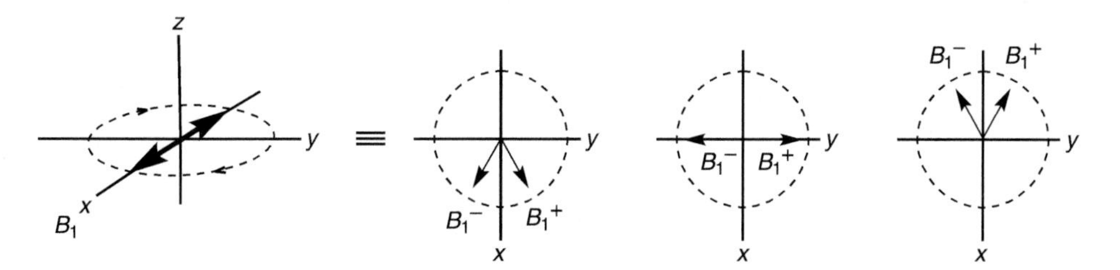

Fig. 4.88 Vectors replicating the oscillations on the *x* axis

In order to begin it is helpful to look more closely at the spin physics of the standard sequence. To repeat the picture in Sect. 4.2, the initial π/2 pulse tips the magnetisation towards the *y* axis. The vectors, one for each frequency in the NMR spectrum, precess clockwise with a component in the *xy* plane, cutting the *x* and *y* axes at the Larmor frequency for each vector. This is illustrated in Fig. 4.87 for a perfect π/2 pulse with only four vectors moving with four different frequencies. We describe this picture as being in the *laboratory frame of reference*.

The spectrum is derived from the oscillating magnetic field detected on both the *x* and *y* axes, where coils pick up the frequencies of all the vectors as they cut through the axes. In order to appreciate what is happening to all the vectors, it is helpful to separate ourselves from our view of the vectors whirling around the *z* axis at very high frequencies, in the hundreds of megahertz range, and to analyse only the oscillations on the *x* and *y* axes of each vector relative to the others. Any vector precessing in the *xy* plane cuts through the *x* axis to give an oscillating vector shown on the left in Fig. 4.88. This one-dimensional vector can be duplicated as a pair of two-dimensional precessing vectors B_1^+ and B_1^- on the right in Fig. 4.88, moving with the same frequency as each other in the *xy* plane viewed from above but in opposite directions. The projection of these vectors onto the *x* axis gives the same oscillating vector pictured on the left.

We can ignore the anti-clockwise vectors B_1^+, because only the clockwise vectors B_1^- will be in resonance with the Larmor frequencies. To recognise how all the different frequencies in the spectrum relate to each other, we choose a fixed reference frequency labelled $-\nu_r$ in Fig. 4.89, and compare them all with it. In order to picture what is happening, we mentally place ourselves on this vector, and whirl round the *z* axis at the frequency $-\nu_r$, dragging the *x* and *y* coordinates with us. We should give them new symbols x' and y', as

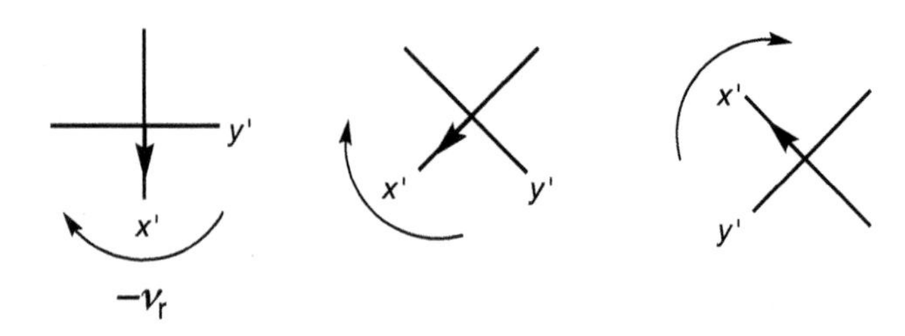

Fig. 4.89 The rotating frame of reference

shown in Fig. 4.89, but in practice we simply keep the former designation from the laboratory frame, and call them x and y. As we sit on the vector with the reference frequency $-\nu_r$, we can forget about the frequency of this rapid rotation, and look only at how much faster or slower the other vectors around us are moving. This gives us a *rotating frame of reference*. The effect is analogous to the effect of subtracting the high frequency of hundreds of megahertz before the digitisation described in Sect. 4.2. That was carried out in the electronics in order to make digitisation practical, whereas we are now doing something similar in order to make sense of what is happening—in both cases the numbers are brought down to workable values.

A helpful analogy to what we 'see' in the rotating frame of reference comes from our experience of the earth's rotation, which we are effectively unaware of (except by observation of the fixed celestial bodies). But if we look up at satellites orbiting the earth above us, we can easily tell which ones are rotating faster about the polar axis than we are and which ones are rotating slower. The faster satellite moves east and the slower west, and we could easily measure the relative rates of rotation. Our experience on earth provides us with a frame of reference.

Let us now use the rotating frame to see what happens when we have spins with different frequencies. Imagine first a spectrum with a single spin having a Larmor frequency that coincides precisely with that of the reference frequency $-\nu_r$; let us choose the resonance frequency of the protons on tetramethylsilane (TMS) as our signal and as the reference frequency. In the laboratory frame, the vector representing the bulk magnetisation tips onto the y axis, and immediately begins to precess at the reference frequency. In contrast, in the rotating frame it arrives on the y axis (strictly the y' axis, but we shall drop the prime symbol) and stays there unmoved, illustrated at the top left in Fig. 4.90. It remains fixed because it is at the reference frequency that defines the rotating frame. Fourier transform of the signal would give us a single frequency of 0 Hz, at the top right in Fig. 4.90, that of our reference TMS.

Now imagine a pair of singlets, one of which has a frequency ν_X 2800 Hz higher than that of TMS and another ν_A that has a frequency 400 Hz higher than TMS. They would first be tipped in the laboratory frame onto the y axis, and they would precess at their Larmor frequencies. In the rotating frame they would be perceived as vectors moving away from the y axis in a clockwise direction as shown at the bottom left of Fig. 4.90. They move clockwise because both frequencies are higher than the reference frequency that we have chosen to define the rotating frame. FT would convert these two signals into singlets at 2800 Hz and 400 Hz downfield of the TMS reference. In a 400 MHz instrument this would place the lines at δ 7.0 and 1.0 respectively, as shown at the bottom right in Fig. 4.90.

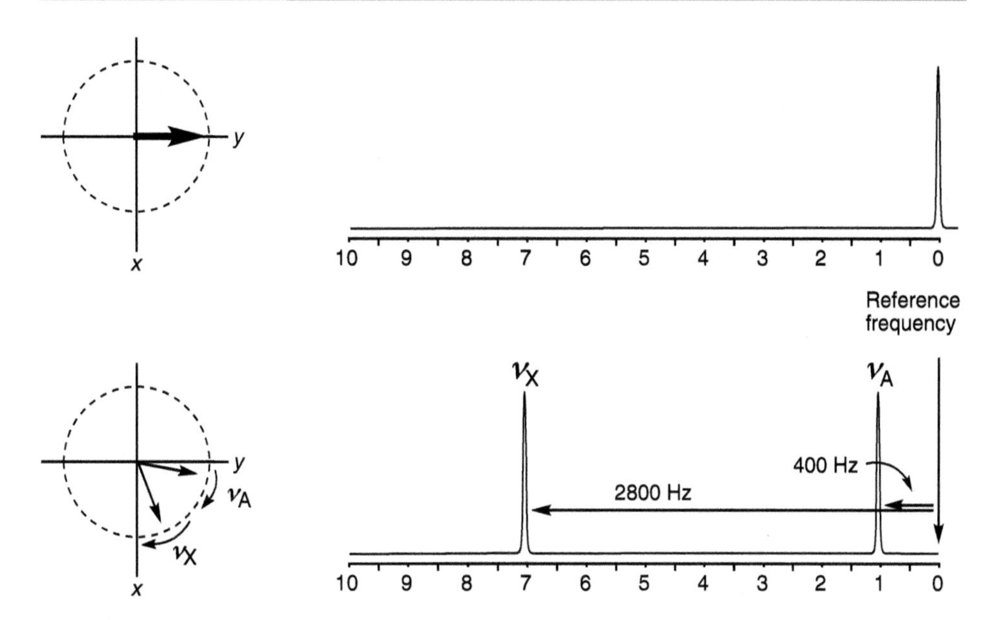

Fig. 4.90 Two singlets in the rotating frame with TMS as the reference frequency ν_r in a 400 MHz instrument

In practice the reference frequency against which the frequencies of the signals in the spectrum are measured is not likely to be that of TMS. Indeed it is a poor choice, because frequency information is unavoidably collected from the same range above and below the reference frequency. There is rarely anything to observe at δ values <0, and only noise would be collected there. It is usual instead to choose a reference frequency ν_r, called the offset frequency, somewhere near the middle of the range of frequencies expected in the spectrum, and to choose a spectral width (SW) that extends only over the range on either side of that frequency within which signals are likely to be found. As long as we know its frequency, the spectrum can be placed onto the δ scale.

Let us repeat the experiment in Fig. 4.90, but use as the reference frequency, arbitrarily, one which is 1200 Hz higher than that of TMS. The result is illustrated at the top of Fig. 4.91. With a reference frequency 1200 Hz above that of TMS, the frequency of the signal ν_A is now 800 Hz less than that of the reference, and the signal ν_X is 1600 Hz greater than that of the reference. The vector ν_A will move anti-clockwise at 800 Hz in the rotating frame, and the vector ν_X will move clockwise at 1200 Hz, while the reference frequency ν_r remains on the y axis. The FID and FT will plot exactly the same spectrum as before, with lines at δ 7.0 and δ 1.0 in a 400 MHz instrument, since we know what the reference frequency is relative to that of TMS. The picture doesn't change fundamentally when we consider two spins that are coupled as shown in the lower half of Fig. 4.91. In place of one vector for each spin we have two, one $J/2$ slower and one $J/2$ faster than the frequencies of the vectors for the singlets. The FID and FT will plot two doublets, each of which will be centred at the same frequencies ν_X and ν_A as before, with each vector corresponding to a line in the spectrum.

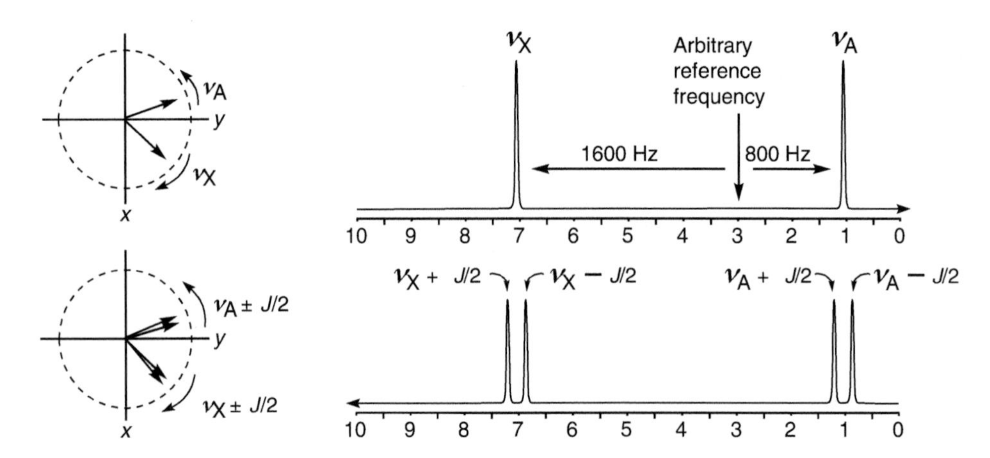

Fig. 4.91 Two singlets (top) and two doublets (bottom) in the rotating frame with ν_r 1200 Hz above that of TMS in a 400 MHz instrument

There is one complication from choosing, as we have in Fig. 4.91 and as we do in practice, a reference frequency from within the range of interest. We need to know whether the frequency of any signal is higher or lower than the reference frequency. As mentioned in Sect. 4.2, quadrature detection solves this problem by measuring the oscillating magnetic field on both the x and y axes. They both give the 800 and 1200 Hz frequency differences from that of the reference frequency, and having data from both axes, 90° out of phase with each other, allows the program that processes the FID to extract the sign of the frequency differences, just as we can define an angle if we know both its sine and cosine.

It is even more complicated than this, because a more elaborate sequence is used called *phase cycling* to remove unwanted signals such as rogue frequencies in the circuits, and other unwanted artefacts such as those stemming from small inequalities in the coils on the x and y axes and the imperfect alignment of their phases. It uses alternating pulses on the x and y axes, together with alternating this pair of pulses on the x and y axes with pulses on the $-x$ and $-y$ axes (in other words 180° out of phase with the first pair)—a four-step procedure (Fig. 4.92). The outputs of these four sets of detected signals are then combined, with and without sign changes, and routed to the appropriate memory channel for the sine

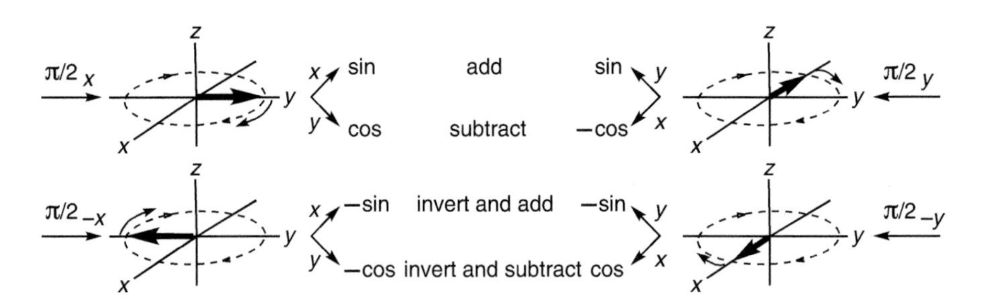

Fig. 4.92 Phase cycling

or cosine wave to make the two sine curves add up (and the two cosine curves add up) rather than subtract. The result is a signal derived only from the precessing magnetisation, evened up in the two channels, identified as to whether it is precessing faster or slower than the reference frequency, and free of the rogue frequencies.

We actually have four reference frequencies in play in the discussion, and we need to be clear about which is which. The TMS frequency defines the scale onto which we place the frequencies detected in the FT, so that we can compare one spectrum with another—all other frequencies are relative to it. A second reference frequency is a lock, which the spectrometer records from the sample, to allow it to locate the spectrum relative to TMS; it is often the deuterium signal in the $CDCl_3$, when that is the solvent, but it used to be that of TMS itself, which was added to the sample to provide the lock. The third reference frequency, the frequency for the rotating frame ν_r, comes in two forms. In the spectrometer itself ν_r is the actual frequency used to compare the precession frequency of each vector with the others without having to compare the differences between huge numbers. As we have just seen with the two versions of the same experiment in Figs. 4.90 and 4.91, ν_r can be placed anywhere, as long as the spectrometer has been told where it is, and the spectrum will be the same. When setting up each NMR run, the operator chooses a frequency somewhere near the middle of the expected spectral width, and the computer does the rest. In the second version of the reference frequency ν_r, regardless of where it has actually been placed when the spectrum is being taken, we can, for purposes of discussion, place it anywhere we like to make the discussion simple, knowing that it will not affect the appearance of the spectrum.

The pictures in Fig. 4.70 for the spin-echo experiment may now be reinterpreted: they are a view of the rotating frame, although it was not necessary to explain what that meant at that stage in the development of the story. However, we do now need to return to the spin-echo, in order to understand how it is used in a substantially useful experiment.

4.15 Assignment of CH_3, CH_2, CH and Fully Substituted Carbons in ^{13}C NMR

4.15.1 The Attached Proton Test (APT)

In structure determination, it is extraordinarily helpful to identify the number of methyl (CH_3), methylene (CH_2), methine (CH) and fully substituted carbon atoms (C, with no attached hydrogen atoms; they are often called quaternary carbons, although this term ought to be restricted to C atoms with four carbon substituents). We have already seen that the distinction can be achieved by off-resonance decoupling in a ^{13}C spectrum (Sect. 4.5.2), but a multi-pulse sequence known as the attached proton test (APT) is more effective.

Returning to the spin-echo pulse sequence of Fig. 4.72, and adopting the rotating frame perspective, let us look again at two uncoupled signals, A and B—two independent singlets with different chemical shifts. For purposes of discussion, we can place our reference

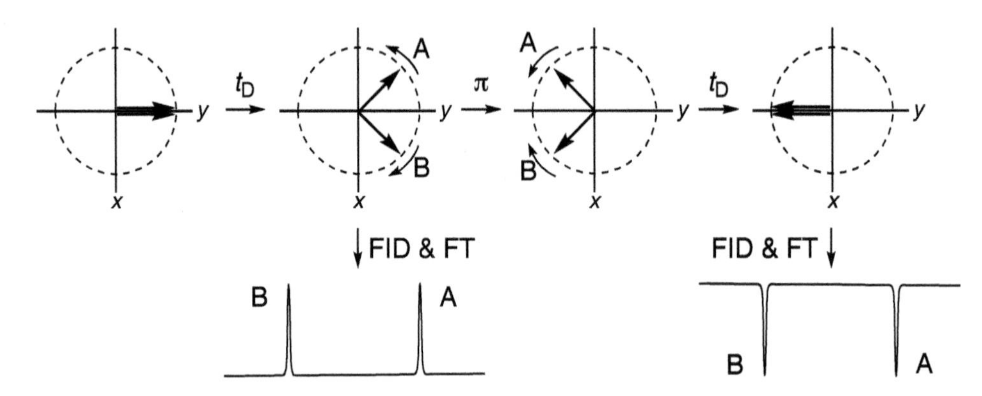

Fig. 4.93 Two singlets in the rotating frame of reference with a spin-echo applied

frequency wherever we like within the spectral range. In Fig. 4.72 we effectively placed it at a point with a lower frequency than either the A or B vector. It is easier to understand if instead we choose in Fig. 4.93 a reference frequency half way between the frequencies of the A and the B nuclei. The upfield A nucleus has a lower frequency and the downfield B nucleus has a higher frequency. As a result the A vector precesses anticlockwise with a frequency 1/2 of the difference in chemical shift and the B proton precesses clockwise with 1/2 of the frequency of the difference in chemical shift. Remember that the reference is precessing in the laboratory frame at a high frequency; in the rotating frame it is static, but the A nucleus has a lower frequency and the B nucleus a higher. After the time t_D, the vectors will be splayed out on either side of the reference in Fig. 4.93. If we were to take the FID at this stage, and carry out the Fourier transformation, we would get the normal spectrum.

When we apply the π pulse from Fig. 4.72, both vectors would flip across to the other side of the *xz* plane, but they would continue to precess in the same direction, and if we wait exactly the same length of time t_D as before they will both be pointing along the $-y$ axis, just as if they had started life there, and just as they did in Fig. 4.72. They are said to be an echo, and if the FID were to be collected at this stage they would give the same signals as usual, two single lines, but upside down. In the picture, the value of t_D was evidently just enough to move the vectors $\pi/2$ radians apart, $\pi/4$ in each direction. But it would not matter how far they had precessed, they would be refocused if the second delay t_D was the same as the first. Remember that it makes no difference to the outcome where we place our reference frequency, as we saw in the discussion for Figs. 4.90 and 4.91.

Now let us see what happens with the two spins of a doublet from a ^{13}C nucleus with one attached proton. If we use as our reference frequency the ^{13}C chemical shift of the doublet, the picture in Fig. 4.94 will look similar to that in Fig. 4.93. The vector for the signal affected by the proton in the β state (arising from the transition X_2 in Fig. 4.58) will precess ahead of the reference frequency at a frequency of $+J/2$, and the vector for the signal affected by the proton in the α state (arising from the transition X_1 in Fig. 4.58) will lag behind with a frequency of $-J/2$. If we were to take an FID at this stage and carry out the Fourier transformation, we would see the doublet in the usual way. The π pulse at the

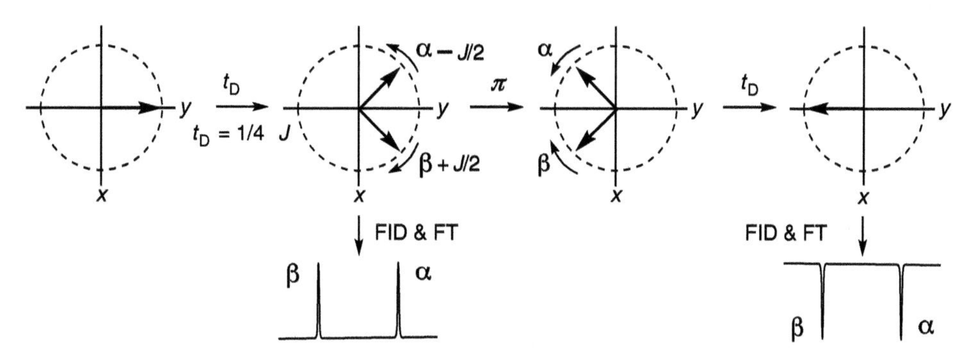

Fig. 4.94 Heteronuclear spin-echo for a doublet

carbon frequency will tip the carbon vectors across the xz plane, but the proton to which this nucleus is coupled will be unaffected, since protons have a very different Larmor frequency. As a result, the carbon vector labelled α will continue under the influence of the proton in the α state, precessing in the same direction. Likewise the carbon vector labelled β will continue to precess in the same direction as before, and after the second delay t_D the two vectors will be realigned along the $-y$ axis. FT of the FID will then create an upside down doublet.

To take an APT spectrum, we introduce a small change to the spin-echo pulse sequence—we decouple the protons at the end of the first delay t_D, as illustrated in Fig. 4.95. Although the spins will have evolved at the usual rate ($\pm J/2$) during the first t_D, turning on the decoupler, will stop all further development. The chemical shift will

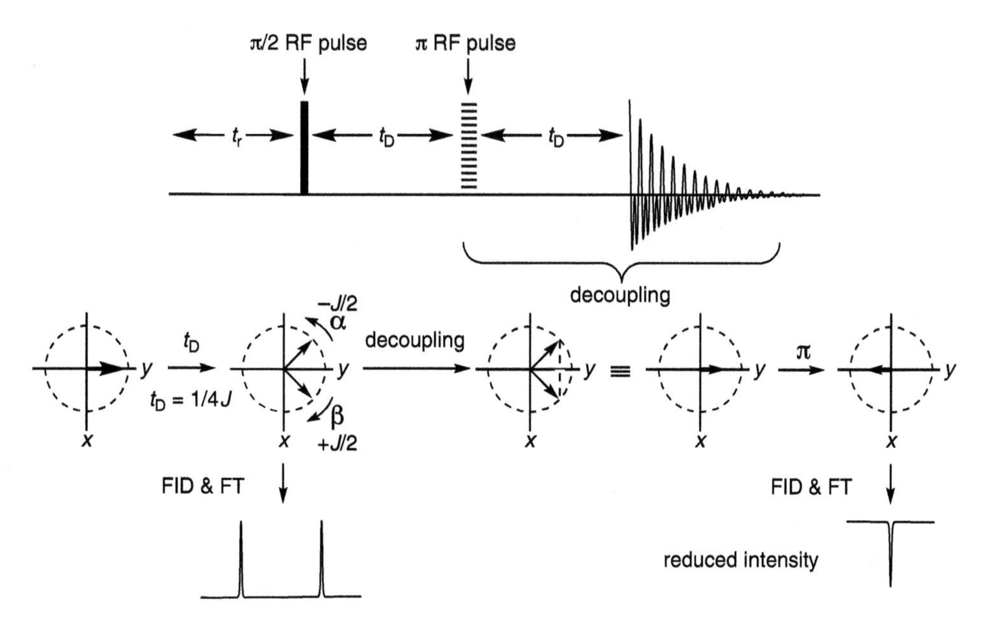

Fig. 4.95 Pulse sequence for an APT spectrum

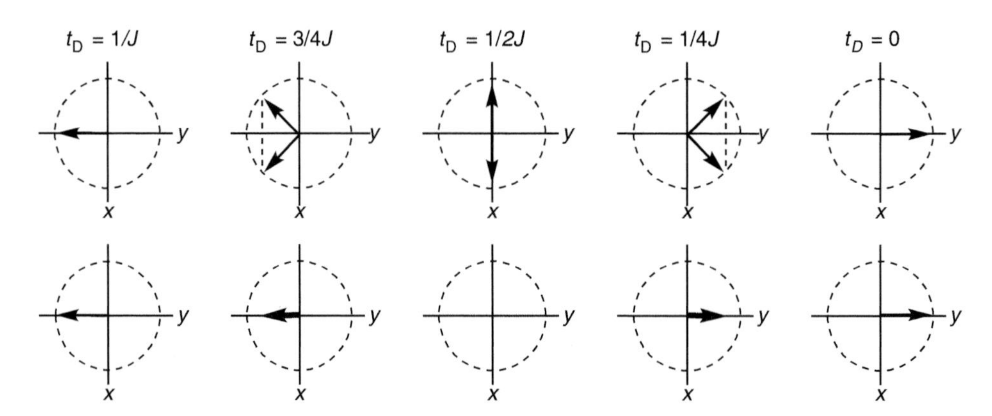

Fig. 4.96 Intensity change as a function of the delay t_D

continue to precess as usual, but the spins collapse to a single vector with an intensity corresponding to the projection onto the y axis of the angle they have reached after the time t_D. When this is flipped onto the $-y$ axis by the π pulse, it will not have the same intensity as usual, and Fourier transformation of the FID will give a singlet (all coupling removed) with reduced intensity. The intensity will depend upon how far the precession had taken place in the time t_D.

If we call the angle through which the precession has taken place ϕ, we can get some sense of the intensity of the signal as a function of t_D, and hence of ϕ, by looking in Fig. 4.96 at where the vectors have reached at different times. Reading from right to left, the top row shows the two vectors at each of the values of t_D, and the lower row shows the resultant vector after the decoupling. If the time t_D were essentially zero, the vector would not have moved, and would be at full strength. In a time $1/4J$, the vector will have moved by $\pi/4$ radians, and the net strength would be proportional to the cosine of 45° subtended onto the y axis. In a time $1/2J$, the vectors will have moved $\pi/2$ radians; the one vector will cancel the other, and the net vector will be a null. When t_D is $1/J$, the angle will be π, and the vector will be back at full strength pointing along the $-y$ axis. If we plot the strength of the singlet against ϕ following the π pulse which flips the vectors to the other side of the x axis, the FID and FT would give us the negative cosine curve in Fig. 4.97, in which the intensity of the signal from the carbon atom is proportional to $-\cos(\pi t_D J)$.

This has shown what happens to a ^{13}C atom carrying one hydrogen. If it carried two, it would give rise to a triplet, and there would be three vectors as seen in Fig. 4.98. Using the chemical shift as the reference frequency, one vector representing the central line of the triplet, that experiencing the field of the αβ and the βα protons, will carry half the total intensity, and will precess at the chemical shift, which means that it stays on the y axis in the rotating frame. One of the outer lines of the triplet, that experiencing the field of the αα protons, will lag behind the chemical shift by $-J$ and the other, that experiencing the field of the ββ protons, will move ahead by $+J$. Between them they carry the other half of the total intensity.

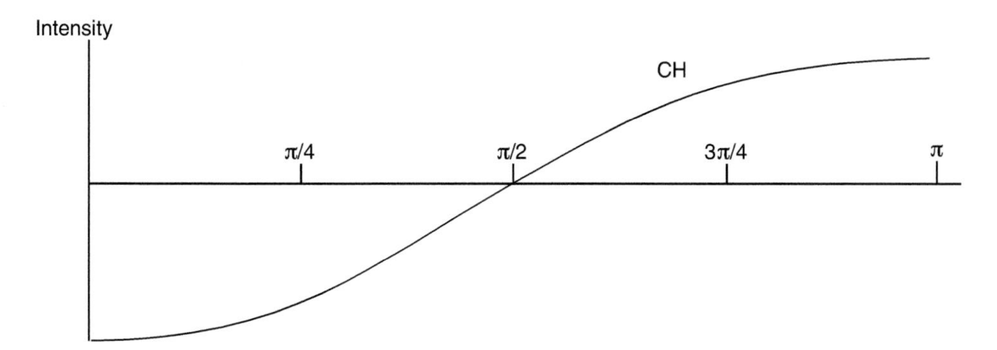

Fig. 4.97 The intensity of the signal from a methine carbon as a function of t_D

The rate of precession has doubled, and in consequence the intensities after the same intervals t_D are different from those for a methine carbon. Figure 4.99, reading from right to left, shows the critical times 0, $1/4J$, $1/2J$, $3/4J$ and $1/J$. After $1/4J$ the rotating vectors will have reached the point where they cancel each other, the resultant vector would be the remaining central line carrying half the strength of the original signal. After $1/2J$ the rotating vectors will have reached the $-y$ axis, where they cancel the central line giving a null. After $1/J$ the rotating vectors will have reached the starting point, where they will add to the central line and be back at full strength.

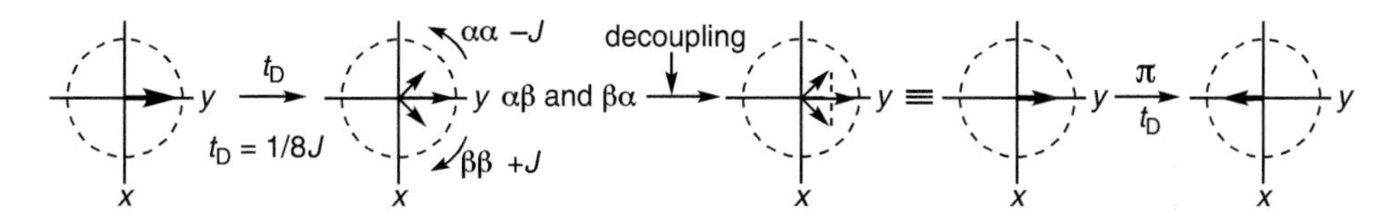

Fig. 4.98 A triplet in the spin-echo with decoupling

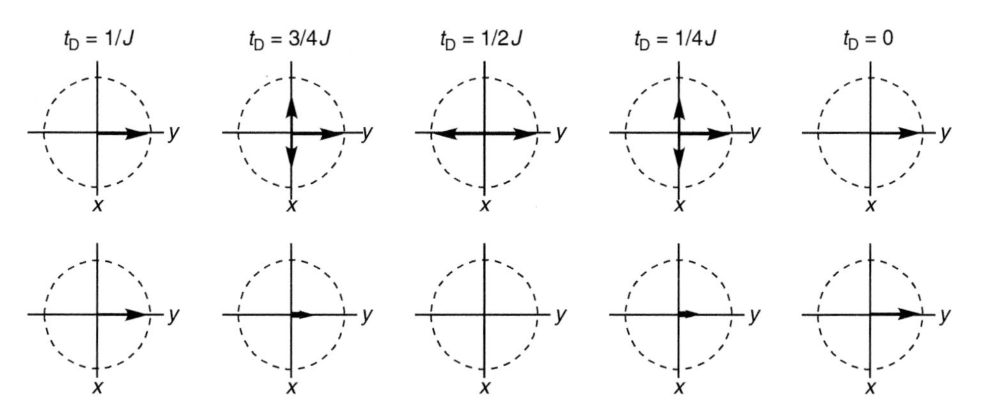

Fig. 4.99 The intensity of the signal from a methylene carbon as a function of t_D

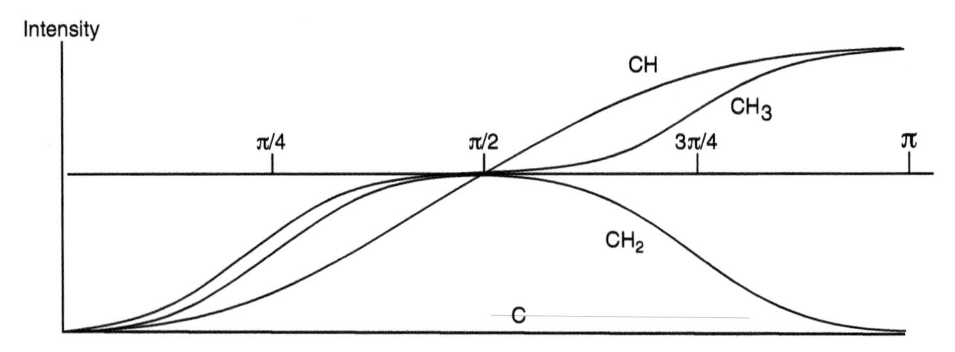

Fig. 4.100 The intensity of the carbon signals from C, CH, CH_2 and CH_3 groups as a function of t_D

After the π pulse putting all the resultant vectors onto the $-y$ axis, the FID and FT will show a negative intensity rising to zero and back down to full negative strength, a different result from that for a carbon carrying one hydrogen. For a methyl carbon, another story rather more tricky to draw, adds to the graph in Fig. 4.100, which is completed by including fully substituted carbons as a flat line unaffected by coupling or decoupling.

This gives us a method of counting the number of hydrogen atoms attached to a carbon atom. If we collect our signal at a time t_D close to $1/J$, on the right of Fig. 4.100, the signals from those carbons carrying zero and two hydrogens will be negative, and those from the carbons carrying one and three hydrogens will be positive. A good choice for t_D might be ~7 ms (1/140 since most one-bond ^{13}C–^1H J values are between 120 and 170 Hz). This is the basis of the Attached Proton Test (APT), although the details of the pulse sequence are rather more sophisticated. Figure 4.101 shows the APT spectrum of our ester, 3,5-dimethylbenzyl methoxyacetate **1**.

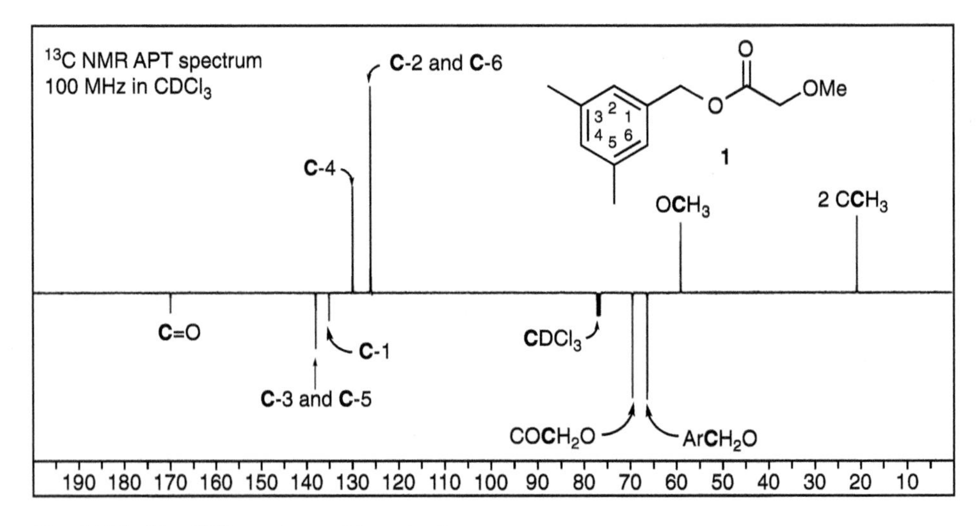

Fig. 4.101 The APT spectrum of the ester **1**

Although APT spectra do not distinguish between methyl and methine, nor between methylene and fully substituted carbons, it is rare for there to be much ambiguity. The upward-pointing methyl and methine signals can often be distinguished by their chemical shift, since most methyl signals are well upfield; similarly, the downward pointing quaternaries are often distinguishable from methylenes by their low intensities, and by their accumulation downfield in many spectra. The digonal carbon of a terminal acetylene does not always appear on the methine (CH) side of the APT spectrum—the usual t_D value of ~7 ms is set to optimise the reliability of the data for tetrahedral and trigonal carbons, which have $^1J_{CH}$ coupling constants of ~140 Hz, whereas the $^1J_{CH}$ value for an acetylenic CH is typically close to 250 Hz.

There is one source of confusion: it is just as easy to plot APT spectra the other way up, with C and CH_2 signals up and the CH and CH_3 signals down (instead of the way they are consistently presented in this book). It is not usually difficult to detect which way an APT spectrum has been plotted, since methyl signals, for example, many of which are at the high-field end of the spectrum, often reveal which convention has been used. The best guide, when in doubt, is from the $CDCl_3$ signal present whenever it has been used as the solvent—$CDCl_3$ has no protons on the carbon atom, and so the signal from it will have been plotted in the same direction as the methylenes and fully substituted carbons.

4.15.2 DEPT

Nevertheless, critical ambiguities do occur from time to time in APT spectra. A multi-pulse sequence which leaves us with no ambiguity is called DEPT (Distortionless Enhancement through Polarisation Transfer). This experiment uses multiple quantum coherence, and is not easily explained with vector diagrams. In outline, it uses the pulse sequence in Fig. 4.102, which incorporates an RF pulse with an angle θ on the y axis.

The effect of choosing three different values for the angle θ is to create three DEPT spectra called DEPT-45, DEPT-90 and DEPT-135. The three spectra have different degrees

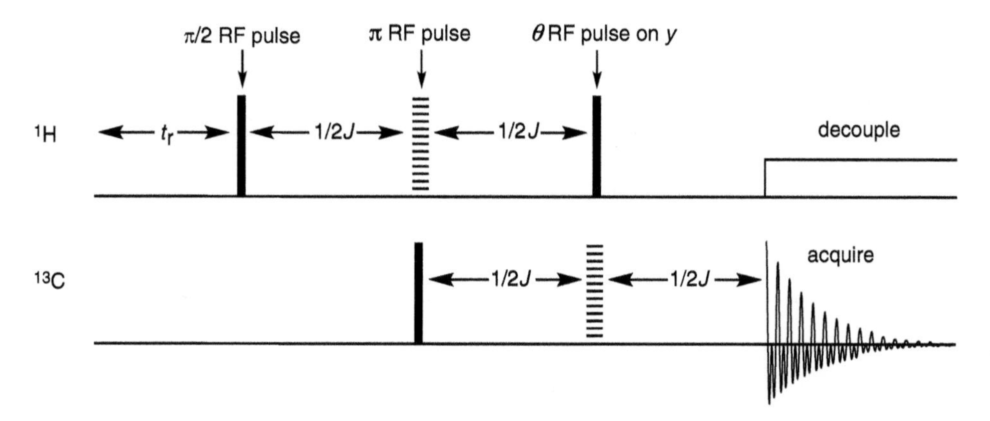

Fig. 4.102 Pulse sequence for DEPT

Table 4.5 Consequences from the values of θ in DEPT

	$\theta = 45°$	$\theta = 90°$	$\theta = 135°$
	DEPT-45	DEPT-90	DEPT-135
CH_3	+	0	+
CH_2	+	0	–
CH	+	+	+
C	0	0	0

of response from the methyl, methylene, methine and fully substituted carbons, giving rise to three spectra with the features shown in Table 4.5.

The three spectra between them identify which carbons in the full ^{13}C-spectrum belong to which class as we can see in Fig. 4.103, which shows the three DEPT spectra for the ester **1**. The DEPT-45 spectrum shows positive CH_3, CH_2, and CH signals; in other words everything except the fully substituted carbons: the C=O carbon, C-1, C-3 and C-5. The DEPT-90 spectrum shows only the CH signals, or rather shows only those signals strongly, since there are small signals from other carbons stemming from a not uncommon but imperfect choice of parameters in the pulse sequence. And the DEPT-135 spectrum shows the CH and CH_3 signals positive, but the CH_2 signals negative. Both the DEPT-90 and DEPT-135 spectra reveal that those signals still left unaccounted for in the DEPT-45 spectrum are derived from the CH_3 groups, and those signals still left unaccounted for in the full spectrum, not appearing in any of the three spectra in Fig. 4.103, are derived from the

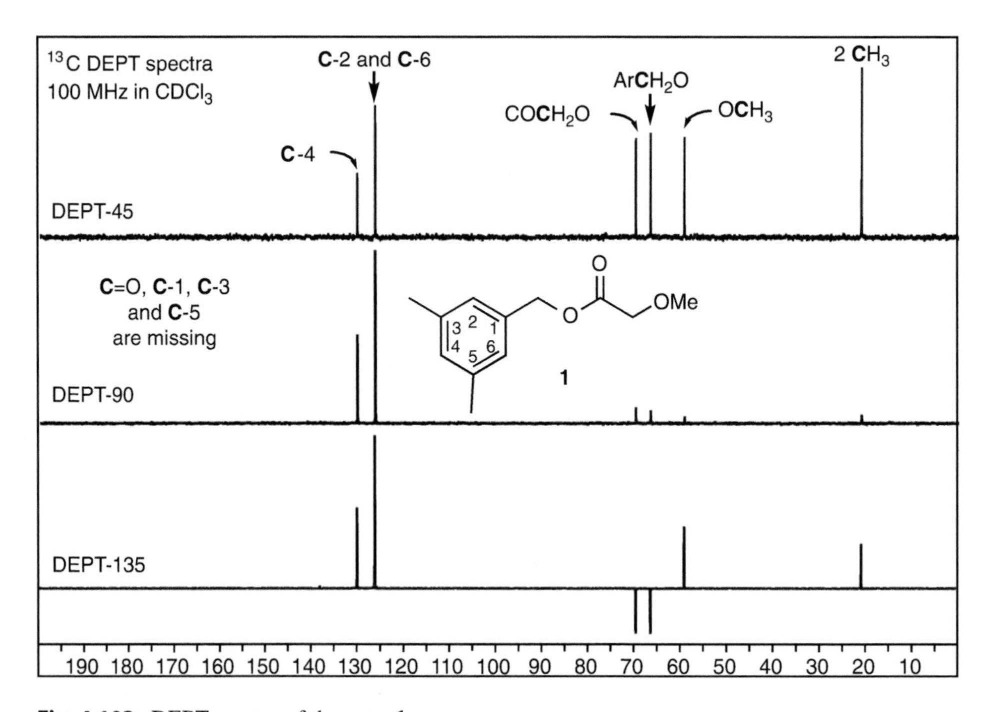

Fig. 4.103 DEPT spectra of the ester **1**

fully substituted carbons. It is also possible, but rarely worth the effort, to take weighted sums and differences to create spectra showing each kind of carbon alone.

In practice it is the DEPT-135 spectrum that gives us quickly most of what we need to know, and this is often all that is used unless a more substantial problem demands a full analysis. It has all the information of an APT spectrum, except for the absence of the fully substituted carbons, which can be identified in the full spectrum, making it marginally less ambiguous than the APT spectrum itself. Furthermore deuterated solvents give no signal, so that CH_3, CH_2 and CH signals that have coincidentally the same frequency as a solvent signal appear in the DEPT but may disappear under the solvent signal in the APT spectrum.

4.16 Hints for Structure Determination Using 1D-NMR

The simple one-dimensional ^1H and ^{13}C NMR spectra, with a wealth of information in the chemical shift, the coupling constants and the connectivity they identify, will take you a long way towards determining the structure of an unknown compound, or confirming the structure of the product of a reaction, especially if they are augmented by one or more of the pulsed techniques described in this chapter: spin decoupling, 1D-TOCSY, NOE and APT or DEPT. If these are insufficient, one or more of the two-dimensional techniques (COSY, NOESY, 2D-TOCSY, HSQC and HMBC) described in the next chapter, are the tools that most often complete the picture. Hints for the deployment of these techniques are given in the next chapter, but it is usual to go as far as you can with the 1D spectra. There is no fixed protocol—it all depends upon what you know about the compound and its provenance before you start. Some compounds yield their structure by accumulating evidence in a sequence like that suggested below; another problem might be unravelled more expeditiously with a quite different sequence.

Chapter 6 contains some worked examples demonstrating how spectra can be interpreted to deduce structures. Chapter 7 has a graded series of problems, from relatively easy to quite challenging, for you to try your skills at structure determination using combinations of all the techniques described in this book. The following sections indicate the kind of reasoning that you can use. They are presented here so that you can review what you have learned from this and the earlier chapters in preparation for applying them to structure determination yourself.

4.16.1 Carbon Spectra

1. Identify and discount the peaks caused by the solvent.
2. Check whether the number of ^{13}C signals is consistent with the molecular formula, if one is available from HRMS (Chap. 1). If it is larger than that required by the molecular formula, be aware that there may be weak signals from traces of impurities, and remember that fluorine and phosphorus couplings, when present, are not removed by

decoupling. If, on the other hand, the number of ^{13}C peaks is smaller than that required by the molecular formula, look for weak signals from carbon atoms having no attached protons, and remember that symmetry within the molecule leads to duplication of some of the signals.

3. Using an APT or DEPT spectrum, determine the relative numbers of CH_3, CH_2, CH and fully substituted carbons.
4. Attempt to estimate the relative numbers of carbonyl and other trigonal and tetrahedral carbons, noting whether the latter have chemical shifts suggesting that they are near double bonds or electronegative elements.
5. Using the chemical shift values, together with information from the IR spectrum, identify which kind of carbonyl groups, if any, are present.

4.16.2 Proton Spectra

1. Identify and discount the peaks caused by the solvent.
2. Search for a proton signal that is plausibly caused by one proton (or, less desirably, two or three protons). From this, using the integrals, make a best possible estimate of the number of protons in each resolved singlet or multiplet and in the spectrum as a whole, and check if this estimate is consistent with the molecular formula determined by HRMS.
3. Put labels (A, B_2, C_2, D, E…X_3, for example) on each proton signal as far as you can.
4. Look for protons that can be exchanged for deuterium (NH or OH identified as being present from the IR spectrum). These may be broad, or temperature or pH dependent. If necessary, take the spectrum again after a D_2O shake.
5. Estimate the numbers of protons in aliphatic, electronegative-influenced, and trigonal (sp^2) environments. Note the signals given by primary, secondary, or tertiary methyl groups, and be aware that signals given by vinyl methyl resonances are, because of allylic coupling, frequently somewhat broader than those associated with tertiary methyl groups.
6. Analyse any resolved coupling patterns, using the magnitudes of coupling constants to pair up signals. A TOCSY experiment can be helpful in separating one spin system from another, especially when the signals overlap. Keep in mind that CH_2 groups, the number of which may be known from the APT or DEPT experiment, can have diastereotopic protons, and that some of the couplings in the proton spectrum may be geminal. They are usually the larger ones. If necessary, run specific decoupling experiments to resolve ambiguities about which protons are coupled with which.
7. Use the labels (A, B, C…X) to catalogue your conclusions. Identifying coupling connections, and building structural fragments from them, is often the most enjoyable stage of the quest for a chemical structure. Practice, and a familiarity with the appearance of coupling patterns and roofing, is a great help in juggling all the pieces of evidence you have to hold in your head before enlightenment dawns.

8. Use the Karplus equation and NOEs to add information about the spatial relationships between protons—the conformation and the relative configuration of stereogenic centres.
9. If you have been able to deduce a plausible structure, use ChemNMR, Mestrenova or any other program you have available, to predict the ^{13}C and ^{1}H chemical shifts, and look for serious discrepancies from the experimental data. Prediction programs are not perfect, but ignoring a serious discrepancy, especially in the ^{13}C spectrum, can lead you astray.

4.17 Further Information

4.17.1 The Internet

The Internet is an evolving system, with links and protocols changing frequently. The following information is inevitably incomplete and much of it may no longer apply by the time you read this. Some websites require particular operating systems and may only work with a limited range of browsers, some require payment, and some require you to register and to download programs before you can use them.

Many websites, especially those reached by way of Google, will carry advertisements to commercial organisations measuring and interpreting NMR spectra.

The Chemical Abstracts Service SciFinder: <http://www.cas.org/-products/scifinder> includes spectroscopic data in the Registry Database.

The Japanese Spectral Database for Organic Compounds (SDBS): <http://sdbs.db.aist.go.jp/sdbs/cgi-bin/cre_list.cgi> with a choice of text in English or Japanese, has free access to a vast store of IR, Raman, ^{1}H- and ^{13}C-NMR and MS data.

A German site listing many databases for IR, NMR and MS is: http://www.lohninger.com/spectroscopy/dball.html

A German site called NMRShiftDB is an open-source database for chemical structures and their NMR data: <http://sourceforge.net/projects/nmrshiftdb/

The paper by S. Kuhn and N. E. Schlörer, *Magn. Reson. Chem.*, 2015, **53**, 582–589 (wileyonlinelibrary.com DOI 10.1002/mrc.4263) describes its capacities.

Bio-Rad Laboratories:<http://www.bio-rad.com/en-uk/product/nmr-spectral-databases> has, for a price, all types of spectra of organic compounds; follow the leads to their Spectroscopy Software KnowItAll®.

The Wiley-VCH website gives access to the SpecInfo data: <http://onlinelibrary.wiley.com/book/10.1002/9780471692294>

Sigma-Aldrich: <http://www.sigmaaldrich.com/> have an online chemical products catalogue with access to many NMR, IR & Raman spectra; search by product name, catalogue number, CAS number, MDL number, or molecular (empirical) formula.

ChemDraw, containing ChemNMR for chemical shift predictions, is produced by Cambridge Soft: <www.cambridgesoft.com/>

Mestrelab Research: <http://mestrelab.com> offer an NMR processing program called Mestrenova. It has been extensively used, along with PhotoShop and ChemDraw, in the preparation of this book.

4.17.2 Further Reading

Books and articles on NMR are rarely nowadays restricted to 1D-NMR, but a list of useful books from the last 25 years is appropriate here, even though much of the material in them applies both to 1D- and 2D-NMR.

How NMR Works

- Claridge TDW (2016) High-resolution NMR techniques in organic chemistry, 3rd edn. Elsevier
- Hore PJ, Jones JR, Wimperis S (2015) NMR: the toolkit: how pulse sequences work, 2nd edn. OUP, Oxford
- Benesi AJ (2015) A primer of NMR theory, with calculations in Mathematica. Wiley
- Keeler J (2010) Understanding NMR spectroscopy, 2nd edn. Wiley, Chichester
- Levitt MH (2008) Spin dynamics: basics of nuclear magnetic resonance. Wiley, Chichester
- Izydore RA (2007) Fundamentals of nuclear magnetic resonance spectroscopy. Durham Eagle Press, Durham, NC
- Freeman R (1997) A handbook of nuclear magnetic resonance, 2nd edn. Longman
- Grant DM, Harris RK (eds) (1996) Encyclopedia of nuclear magnetic resonance, 8 Vols. Wiley-VCH, Weinheim
- Sanders JKM, Hunter BK (1993) Modern NMR spectroscopy: a guide for chemists, 2nd edn. OUP, Oxford
- Ernst RR, Bodenhausen G, Wokaun A (1987) Principles of nuclear magnetic resonance in one and two dimensions. OUP, Oxford

Structure Determination

- Lambert JB, Mazzola EP (2018) Nuclear magnetic resonance spectroscopy: an introduction to principles, applications, and experimental methods, 2nd edn. Wiley
- Simpson J (2017) NMR case studies: data analysis of complicated molecules. Elsevier, Cambridge, MA
- Jacobsen NE (2017) NMR data interpretation explained: understanding 1D and 2D NMR spectra of organic compounds and natural products. Wiley
- Field LD, Li HL, Magill AM (2015) Organic structures from 2D NMR spectra. Wiley, Chichester
- Günther H (2013) NMR spectroscopy: basic principles, concepts and applications in chemistry, 3rd edn. Wiley-VCH, Weinheim

- Simpson JH (2012) Organic structure determination using 2-D NMR spectroscopy: a problem-based approach, 2nd edn. Elsevier, Amsterdam
- Friebolin H (2011) Basic one- and two-dimensional NMR spectroscopy, 5th edn. Wiley-VCH, Weinheim
- Crews P, Rodrigues J, Jaspars M (2010) Organic structure analysis, 2nd edn. OUP, Oxford
- Berger S, Sicker D (2009) Classics in spectroscopy: isolation and structure elucidation of natural products. Wiley-VCH
- Mitchell TN, Costisella B (2007) NMR—From spectra to structures, 2nd edn. Springer, Heidelberg
- Berger S, Braun S (2004) 200 and more NMR experiments: a practical course. Wiley-VCH, Weinheim
- Breitmaier E (2002) Structure elucidation by NMR in organic chemistry, 3rd edn. Wiley, Chichester
- Akitt JW, Mann BE (2000) NMR and chemistry, 4th edn. CRC Press, Boca Raton
- Sanders JKM, Constable EC, Hunter BK (1993) Modern NMR spectroscopy: a workbook of chemical problems, 2nd edn. OUP, Oxford

Data

- Pretsch E, Bühlmann P, Badertscher M (2009) Structure determination of organic compounds, tables of spectral data, 4th English edn. Springer, Berlin, is an exceptionally useful compilation of UV, IR, NMR and MS data
- Bruno TJ, Svoronos PDN (2006) CRC handbook of fundamental spectroscopic correlation charts. CRC Press, Boca Raton
- Pouchert CJ (1993) The Aldrich library of ^{13}C and ^{1}H FT NMR spectra, 3 vols. Aldrich Chemical Company Inc.; also available on CD
- Breitmaier E, Haas G, Voelter W (1979) Atlas of ^{13}C NMR data. Heyden

Specialised Aspects of NMR

- Kaupp M, Bühl M, Malkin VG (eds) (2004) Calculation of NMR and EPR parameters. Wiley, New York
- Neuhaus D, Williamson MP (2000) The nuclear Overhauser effect, 2nd edn. Wiley-VCH, New York
- Pregosin PS (2012) NMR in organometallic chemistry. Wiley-VCH, Weinheim
- Nöth H, Wrackmeyer B (2011) Nuclear magnetic resonance spectroscopy of boron compounds. Springer-Verlag

- Dolbier WR (2009) Guide to fluorine NMR for organic chemists. Wiley, Hoboken
- Iggo JA (2000) NMR spectroscopy in inorganic chemistry. OUP, Oxford
- Evans EA (1985) Handbook of tritium nuclear magnetic resonance spectroscopy. Wiley
- Apperley DC, Harris RK, Hodgkinson P (2012) Solid state NMR: basic principles and practice. Springer-Verlag, Berlin
- Axelson DE (2012) Solid state nuclear magnetic resonance: a practical introduction. CreateSpace Independent Publishing Platform
- Duer MJ (2005) Introduction to solid-state NMR spectroscopy. Blackwell Publishing
- Dong RY (ed) (2010) Nuclear magnetic resonance spectroscopy of liquid crystals. World Scientific, Singapore
- Berliner L (ed) Protein NMR: modern techniques and biomedical applications. Springer
- Hoch JC, Poulsen FM, Redfield C (eds) (2013) Computational aspects of the study of biological macromolecules by nuclear magnetic resonance spectroscopy, NATO ASI series. Springer Science+Business Media LLC
- Bertini I, McGreevy KS, Parigi G (2012) NMR of biomolecules. Wiley-Blackwell
- Roberts GCK, Lian L-Y (eds) (2011) Protein NMR spectroscopy: practical techniques and applications. Wiley, Chichester
- Cavanagh J, Fairbrother WJ, Palmer AG III, Rance M, Skelton NJ (2007) Protein NMR spectroscopy, 2nd edn. Academic Press, Burlington, USA
- Pham QT, Pétiaud R, Waton H, Llauro-Darricades, M-F (2002) Proton and carbon NMR spectra of polymers, 5th edn. Wiley
- Jimenez-Barbero J, Peters T (eds) (2002) NMR spectroscopy of glycoconjugates. Wiley, New York
- Markley JR, Opella SJ (eds) (1997) Biological N.M.R. spectroscopy. OUP
- Oschkinat H, Müller T, Dieckmann T (1994) Protein structure determination with three- and four-dimensional NMR spectroscopy. Angew Chem Int Ed Engl 33:277–293
- Roberts GCK (ed) (1993) NMR of macromolecules. OUP, Oxford
- Wüthrich K (1986) NMR of proteins and nucleic acids. Wiley, New York
- Everett JR, Harris RK, Lindon JC, Wilson ID (eds) (2015) NMR in pharmaceutical science. Wiley
- Bushong SC, Clarke G (2014) Magnetic resonance imaging: physical and biological principles, 4th edn. Elsevier
- Constantinides C (2014) Magnetic resonance imaging: the basics. CRC Press, Boca Raton
- Blumich B, Haber-Pohlmeier S, Zia W (2014) Compact NMR. De Gruyter Textbook. (This book deals with aspects of NMR on portable NMR spectrometers)

4.18 Tables of Data

Table 4.6 Some parameters of magnetic nuclei

Isotope	NMR frequency (MHz) at 9.396T	Natural abundance (%)	Relative sensitivity	Spin (I)[a]
^{1}H	400.00	99.98	1.00	$^{1}/_{2}$
^{2}H	61.40	0.015	0.00965	1
^{3}H	426.80	0	1.21	$^{1}/_{2}$
^{7}Li	155.44	92.58	0.293	$^{3}/_{2}$
^{11}B	128.32	80.42	0.165	$^{3}/_{2}$
^{13}C	100.56	1.11	0.0159	$^{1}/_{2}$
^{14}N	28.92	99.63	0.00101	1
^{15}N	40.52	0.37	0.00104	$-^{1}/_{2}$
^{17}O	54.24	0.037	0.0291	$-^{5}/_{2}$
^{19}F	376.32	100	0.833	$^{1}/_{2}$
^{23}Na	105.80	100	0.0925	$^{3}/_{2}$
^{27}Al	104.24	100	0.206	$^{5}/_{2}$
^{29}Si	79.44	4.70	0.00784	$-^{1}/_{2}$
^{31}P	161.92	100	0.0663	$^{1}/_{2}$
^{35}Cl	39.20	75.53	0.0047	$^{3}/_{2}$
^{39}K	18.68	93.10	0.000508	$^{3}/_{2}$
^{41}K	10.24	6.88	0.000084	$^{3}/_{2}$
^{51}V	105.12	99.76	0.382	$^{7}/_{2}$
^{53}Cr	22.60	9.55	0.000903	$^{3}/_{2}$
^{55}Mn	98.64	100	0.175	$^{5}/_{2}$
^{57}Fe	12.92	2.19	0.000034	$^{1}/_{2}$
^{59}Co	94.44	100	0.277	$^{7}/_{2}$
^{65}Cu	113.60	30.91	0.114	$^{3}/_{2}$
^{79}Br	100.20	50.54	0.0786	$^{3}/_{2}$
^{81}Br	108.00	49.46	0.0985	$^{3}/_{2}$
^{85}Rb	38.60	72.15	0.0105	$^{5}/_{2}$
^{113}Cd	88.72	12.26	0.0109	$^{1}/_{2}$
^{119}Sn	149.08	8.58	0.0518	$-^{1}/_{2}$
^{133}Cs	52.48	100	0.0474	$^{7}/_{2}$
^{195}Pt	86.00	33.80	0.00994	$^{1}/_{2}$
^{207}Pb	83.68	22.60	0.00916	$^{1}/_{2}$

[a]A minus sign in the Spin column signifies that the magnetogyric ratio γ, and hence the magnetic moment, is negative

Table 4.7 ^{13}C Chemical shifts in some alkanes

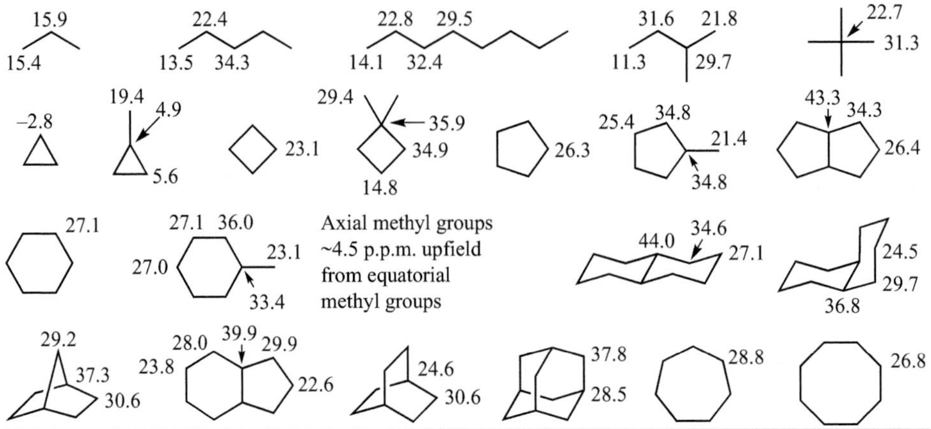

Axial methyl groups ~4.5 p.p.m. upfield from equatorial methyl groups

Estimation of ^{13}C Chemical Shifts in Aliphatic Chains

$$\delta_C = -2.3 + \Sigma z + \Sigma S + \Sigma K \qquad (4.19)$$

where −2.13 is the ^{13}C chemical shift for methane, z is the substituent constant (Table 4.8), S is a 'steric' correction (Table 4.9), and K is a conformational increment for γ-substituents (Table 4.10).

Example of Application of Eq. (4.19)

Diethyl butylmalonate has ^{13}C signals at δ 13.81, 14.10, 22.4, 28.5, 29.5, 52.03, 61.12 and 169.32.

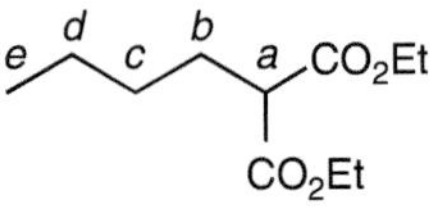

Take the methine carbon *a*:

Base value	−2.3	(methane)
1 α-alkyl group	9.1	(carbon *b*)
1 β-alkyl group	9.4	(carbon *c*)
3 γ-alkyl groups	−7.5	(carbon *d* and two Et groups of the OEt groups)
3 δ-alkyl groups	0.9	(Me and two Me groups of the OEt groups)
2 α-CO_2R groups	45.2	
S	−3.7	(*a* is a tertiary carbon bonded to two CO_2Et groups, which count as primary)
K	0	(open-chain compound with free rotation)
Calculated shift	51.5	Observed value 52.03

Table 4.8 Substituent constants z for Eq. (4.19)

	Substituent	z			
		α	β	γ	δ
H	H–	0	0	0	0
C	alkyl–	9.1	9.4	−2.5	0.3
	–C=C–	19.5	6.9	−2.1	0.4
	–C≡C–	4.4	5.6	−3.4	−0.6
	Ph–	22.1	9.3	−2.6	0.3
	OHC–	29.9	−0.6	−2.7	0.0
	–CO–	22.5	3.0	−3.0	0.0
	$-O_2C-$	22.6	2.0	−2.8	0.0
	N≡C–	3.1	2.4	−3.3	−0.5
N	R_2N-	28.3	11.3	−5.1	0.0
	O_2N-	61.6	3.1	−4.6	−1.0
O	–O–	49.0	10.1	−6.2	0.0
	–COO–	56.5	6.5	−6.0	0.0
Hal	F–	70.1	7.8	−6.8	0.0
	Cl–	31.0	10.0	−5.1	−0.5
	Br–	18.9	11.0	−3.8	−0.7
	I–	−7.2	10.9	−1.5	−0.9
S	–S–	10.6	11.4	−3.6	−0.4
	–SO–	31.1	9.0	−3.5	0.0

Table 4.9 'Steric' constants S for Eq. (4.19)

	Number of substituents other than H on the atoms directly bonded to the observed ^{13}C[a]			
Observed ^{13}C atom	1	2	3	4
Primary	0.0	0.0	−1.1	−3.4
Secondary	0.0	0.0	−2.5	−7.5
Tertiary	0.0	−3.7	−9.5	−15.0
Quaternary	−1.5	−8.4	−15.0	−25.0

[a]Except that CO_2H, CO_2R and NO_2 groups are counted as primary (column 1), Ph, CHO, $CONH_2$, CH_2OH and CH_2NH_2 groups as secondary (column 2), and COR groups as tertiary (column 3)

Table 4.10 Conformational correction K for γ substituents in Eq. (4.19)

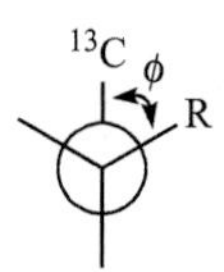

ϕ	0°	60°	120°	180°	Freely rotating
K	–4	−1	0	+2	0

There is no difficulty in assigning this signal, because it is the only methine and is easily identified as such. However, the three methylenes at 22.4, 28.5 and 29.5 are less securely identifiable.

The corresponding calculation for carbon *b* is:

Base value	−2.3	(methane)
2 α-alkyl group	18.2	[carbon *c* and $(EtO_2C)_2CH$]
1 β-alkyl group	9.4	(carbon *d*)
2 β-CO_2R groups	4.0	
1 γ-alkyl groups	−2.5	(Me)
2 δ-alkyl groups	0.6	(CH_2 groups of the OEt groups)
2 β-groups	4.0	
S	−2.5	(*b* is secondary and bonded to a carbon, namely *a*, with three groups on it other than hydrogen)
Calculated shift	28.9	

Similar calculations for carbons *c* and *d* give calculated values of 29.1 and 22.8. It is therefore possible, although not certain, that the signals at 22.4, 28.5 and 29.5 can be assigned to *d*, *b* and *c*, respectively. ChemNMR in ChemDraw®, using very similar protocols to Eq. (4.19) and the data in Tables 4.8, 4.9, and 4.10, gives the values on the left below, and Mestrenova gives the values on the right:

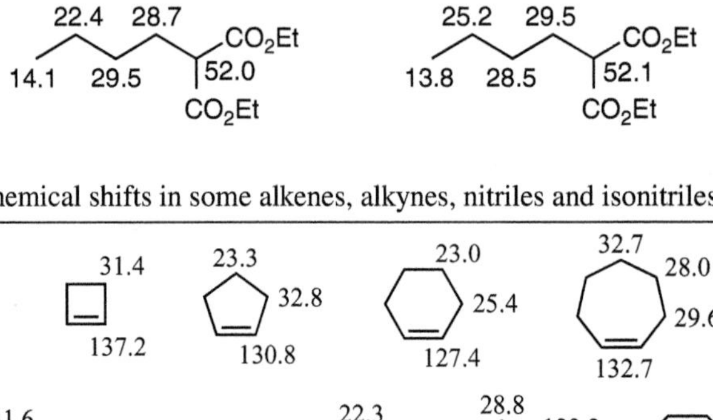

22.4 28.7 CO_2Et 14.1 29.5 52.0 CO_2Et

25.2 29.5 CO_2Et 13.8 28.5 52.1 CO_2Et

Table 4.11 ^{13}C Chemical shifts in some alkenes, alkynes, nitriles and isonitriles

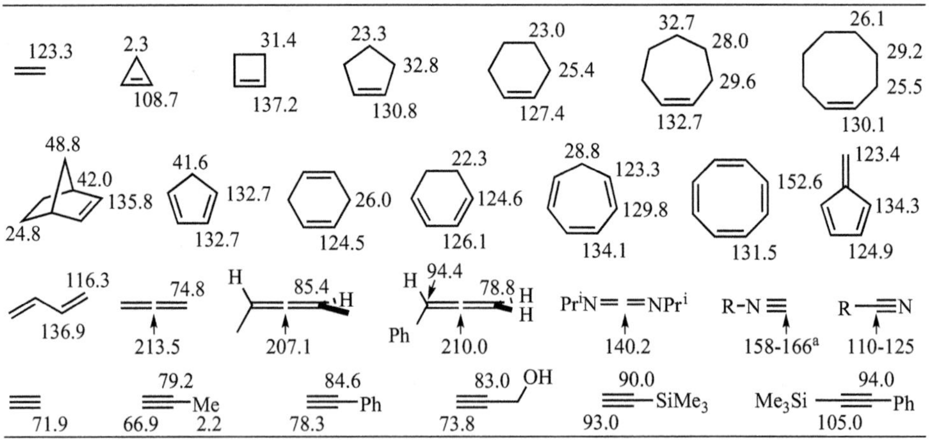

123.3; 2.3, 108.7; 31.4, 137.2; 23.3, 32.8, 130.8; 23.0, 25.4, 127.4; 32.7, 28.0, 29.6, 132.7; 26.1, 29.2, 25.5, 130.1

48.8, 42.0, 135.8, 24.8; 41.6, 132.7, 132.7; 26.0, 124.5; 22.3, 124.6, 126.1; 28.8, 123.3, 129.8, 134.1; 131.5; 152.6, 123.4, 134.3, 124.9

116.3, 136.9; 74.8, 213.5; H 85.4 H, 207.1; H 94.4 78.8 H H Ph, 210.0; Pr^iN=C=NPr^i 140.2; R-N≡C 158-166[a]; R-C≡N 110-125

71.9; 79.2 Me 66.9 2.2; 84.6 Ph 78.3; 83.0 OH 73.8; 90.0 $SiMe_3$ 93.0; Me_3Si 94.0 Ph 105.0

[a]The electron distribution around the N allows $^1J_{NC}$ coupling to show up as 1:1:1 triplet

Estimation of ^{13}C Chemical Shifts in Substituted Alkenes

R–C(1)=C(2)

$$\delta_C = 123.3 + \Sigma z_1 + \Sigma z_2 + \Sigma S \quad (4.20)$$

where 123.3 is the ^{13}C chemical shift for ethene, z_1 and z_2 are the substituent constants (Table 4.12) and S is a 'steric' correction for alkyl substituents:

For each pair of *cis* substituents	$S = -1.1$
For a pair of geminal substituents on C-1	$S = -4.8$
For a pair of geminal substituents on C-2	$S = 2.5$

Table 4.12 Substituent constants z for Eq. (4.20)

	Substituent R	z_1	z_2		Substituent R	z_1	z_2
H	H–	0	0	N	RAcN–	6.5	−29.2
C	Me–	10.6	−7.9		Me_2N–	28.0	−32.0
	Et–	15.5	−9.7		O_2N–	22.3	−0.9
	Pr^n–	14.0	−8.2	O	RO–	29.0	−39.0
	Pr^i–	20.4	−11.5		AcO–	18.4	−26.7
	Bu^t–	25.3	−13.3	Hal	F–	24.9	−34.3
	$ClCH_2$–	10.2	−6.0		Cl–	2.6	−6.1
	$HOCH_2$–	14.2	−8.4		Br–	−7.9	−1.4
	Me_3SiCH_2–	12.5	−12.5		I–	−38.1	7.0
	CH_2=CH–	13.6	−7.0	Si	Me_3Si–	16.9	6.7
	Ph–	12.5	−11.0	P	$Ph_2P(=O)$–	8.0	11.0
	OHC–	13.1	12.7	S	RS–	18.0	−16.0
	RCO–	15.0	5.8				
	RO_2C–	6.3	7.0				
	N≡C–	−15.1	14.2				

Example of Application of Eq. (4.20)

a, *b*

a	Base value	123.3	*b*	Base value	123.3
	1-Me	10.6		2 × 1-Me	21.2
	2 × 2-Me	−15.8		2-Me	−7.9
	1 *cis* pair	−1.1		1 *cis* pair	−1.1
	1 gem pair on C-2	2.5		1 gem pair on C-1	−4.8
	Calculated	119.5		Calculated	130.7
	Observed	118.5		Observed	131.8

Table 4.13 ^{13}C Chemical shifts in some arenes and heteroarenes

Estimation of ^{13}C Chemical Shifts in Substituted Benzenes

$$\delta_C = 128.5 + \sum z_i \tag{4.21}$$

Table 4.14 Substituent constants z_i for Eq. (4.21)

	Substituent R	z_1	z_2	z_3	z_4
H	H–	0	0	0	0
C	Me–	9.3	0.6	0.0	–3.1
	Et–	15.7	–0.6	–0.1	–2.8
	Pr^n–	14.2	–0.2	–0.2	–2.8
	Pr^i–	20.1	–2.0	0.0	–2.5
	Bu^t–	22.1	–3.4	–0.4	–3.1
	$ClCH_2$–	9.1	0.0	0.2	–0.2
	$HOCH_2$–	13.0	–1.4	0.0	–1.2
	CH_2=CH–	7.6	–1.8	–1.8	–3.5
	Ph–	13.0	–1.1	0.5	–1.0
	HC≡C–	–6.1	3.8	0.4	–0.2
	OHC–	9.0	1.2	1.2	6.0
	MeCO–	9.3	0.2	0.2	4.2
	RO_2C–	2.1	1.2	0.0	4.4
	N≡C–	–16.0	3.5	0.7	4.3

(continued)

Table 4.14 (continued)

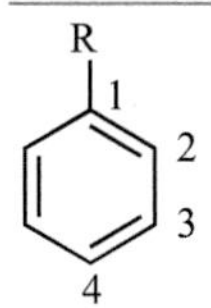

	Substituent R	z_1	z_2	z_3	z_4
N	H_2N–	19.2	−12.4	1.3	−9.5
	Me_2N–	22.4	−15.7	0.8	−11.8
	AcNH–	11.1	−16.5	0.5	−9.6
	O_2N–	19.6	−5.3	0.8	6.0
O	HO–	26.9	−12.7	1.4	−7.3
	MeO–	30.2	−14.7	0.9	−8.1
	AcO–	23.0	−6.4	1.3	−2.3
Hal	F–	35.1	−14.3	0.9	−4.4
	Cl–	6.4	0.2	1.0	−2.0
	Br–	−5.4	3.3	2.2	−1.0
	I–	−32.3	9.9	2.6	−0.4
Li	Li–	−43.2	−12.7	−2.2	−3.1
Si	Me_3Si–	13.4	4.4	−1.1	−1.1
P	Ph_2P–	8.7	5.1	−0.1	0.0
As	Ph_2As–	10.1	5.2	0.1	−0.1
S	MeS–	9.9	−2.0	0.1	−3.7

Table 4.15 ^{13}C chemical shifts of carbonyl carbons

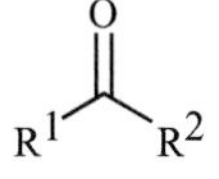

R^1–	–R^2	δ_C	R^1–	–R^2	δ_C
Me–	–H	199.7	Me–	–OH	178.1
Et–	–H	206.0	Et–	–OH	180.4
Pr^i–	–H	204.0	Pr^i–	–OH	184.1
			Bu^t–	–OH	185.9
CH_2=CH–	–H	192.4	CH_2=CH–	–OH	171.7
Ph–	–H	192.0	Ph–	–OH	172.6
Me–	–Me	206.0	Me–	–OMe	170.7
Et–	–Me	207.6	Et–	–OMe	173.3
Pr^i–	–Me	211.8	Pr^i–	–OMe	175.7
Bu^t–	–Me	213.5	Bu^i–	–OMe	178.9
$ClCH_2$–	–Me	200.7	CH_2=CH–	–OMe	165.5
Cl_2CH–	–Me	193.6	Ph–	–OMe	166.8
Cl_3C–	–Me	186.3			

Table 4.15 (continued)

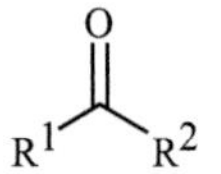

$R^1C(=O)R^2$

R¹–	–R²	δ_C	R¹–	–R²	δ_C
CH_2=CH–	–Me	197.2	$-(CH_2)_3O-$		177.9
Ph–	–Me	197.6	$-(CH_2)_4O-$		175.2
			Me–	$-NH_2$	172.7
$-(CH_2)_3-$		208.2	CH_2=CH–	$-NH_2$	168.3
$-(CH_2)_4-$		213.9	Ph–	$-NH_2$	169.7
$-(CH_2)_5-$		208.8	$-(CH_2)_3NH-$		179.4
$-(CH_2)_6-$		211.7	$-(CH_2)_4NH-$		173.0
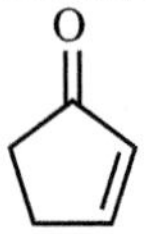 cyclopent-2-enone		209.0	Me—	—OAc	167.3
			Ph—	—OAc	162.8
			Me—	—Cl	168.6
			CH_2=CH—	—Cl	165.6
			Ph—	—Cl	168.0
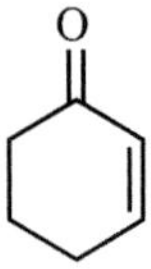 cyclohex-2-enone		198.0	Me—	$-SiMe_3$	247.6
			Ph—	$-SiMe_3$	237.9
			Me—	$-SiPh_3$	240.1

Table 4.16 ^1J $^{13}C-^{13}C$ Coupling constants in Hz

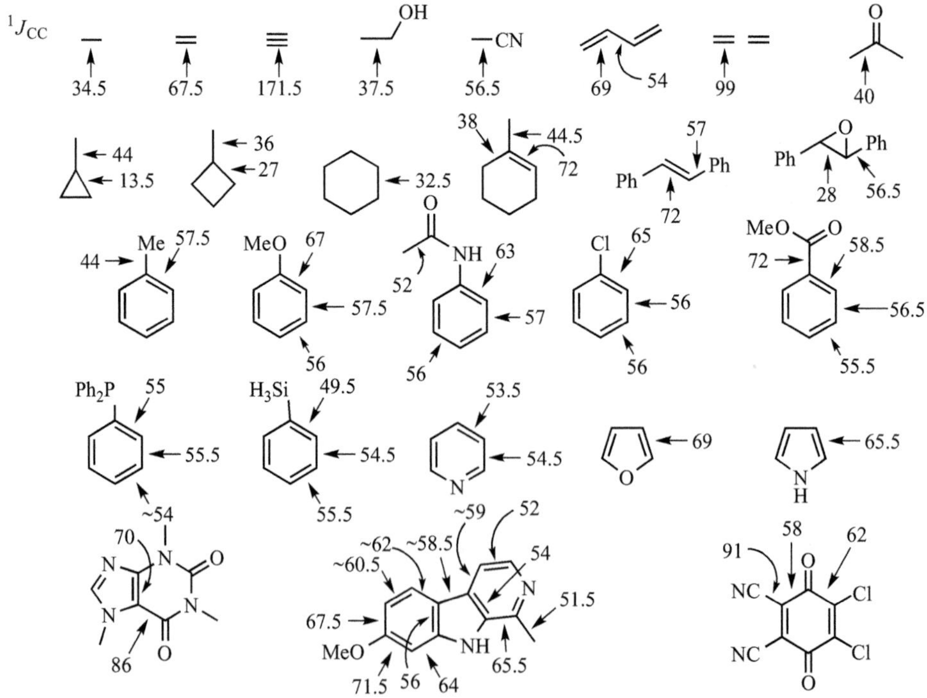

$^1J_{CC}$

Table 4.17 ^2J and ^3J ^{13}C–^{13}C Coupling constants in Hz

$^2J_{CC}$

<1

2.5

Cl Cl Br 3

Cl Cl Br 5.5

3

<1 OH 4.5

O 16

O 2.5

O H N 0.5 5

$^3J_{CC}$

9

OH 4.5[a]

4

O NC Cl NC Cl O 26

10

7.5 Cl Cl Br

6.5 Cl Cl Br

7 Cl Cl Br

8.5 Br Br

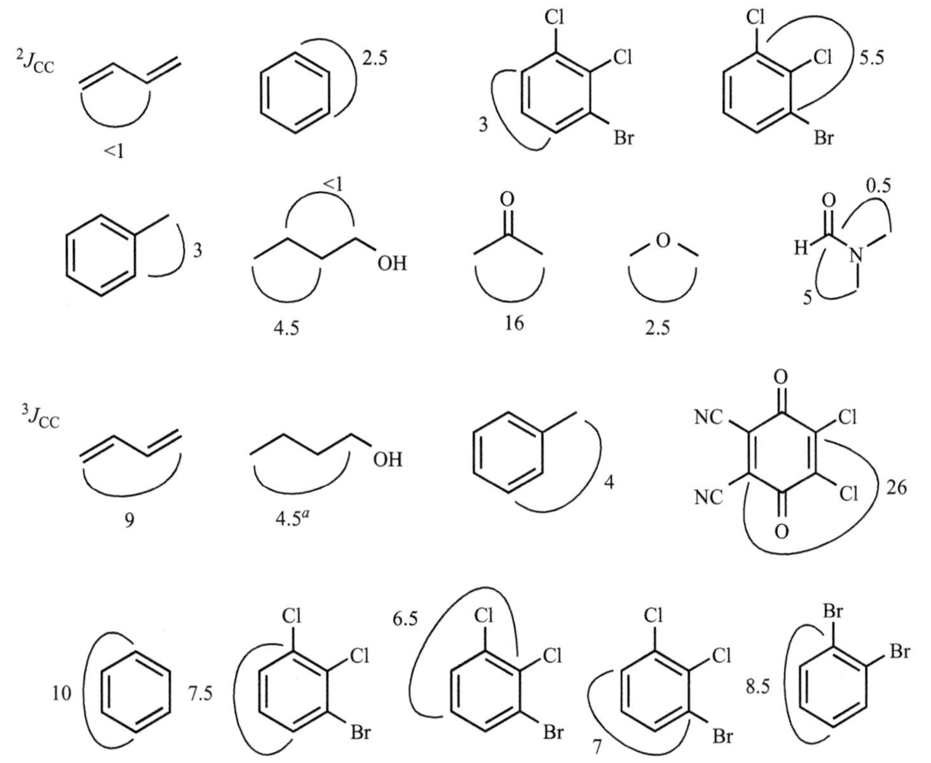

[a]Subject to a Karplus equation for the dihedral angle

Table 4.18 ^1J ^{13}C–^1H Coupling constants in Hz

$^1J_{CH}$ CH_4 125 CH_2=CH_2 156 (benzene)–H 159 HC≡CH 249 CH_2=C=CH_2 168

161 134 128 H 122 H 126 N 152 163 179

O=C(Ph)–H 174 N H 170 182 N N H 206 189 189 O 167 198 Cl N 170 167 181 186

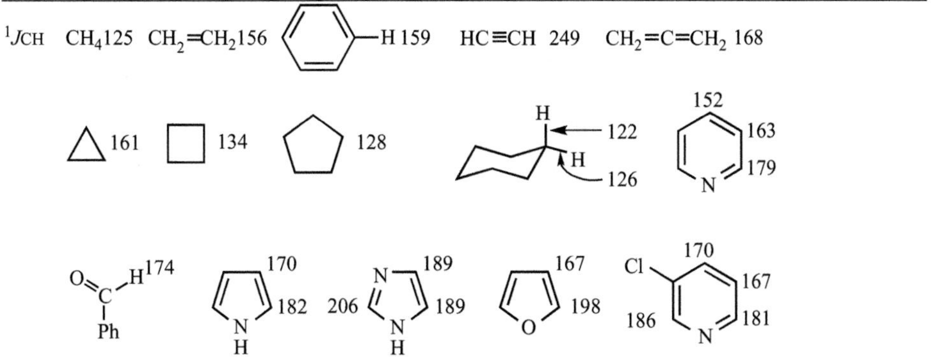

Estimation of $^1J_{CH}$ in Alkanes

$$\text{For } R^1R^2R^3C{-}H, \quad {}^1J_{CH} = 125 + \sum z^i \tag{4.22}$$

Table 4.19 Substituent constants z^i for Eq. (4.22)

	Substituent R^i	z		Substituent R^i	z
H	H–	0	N	H_2N–	8
C	Me–	1		Me_2N–	6
	Bu^t–	−3	O	HO–	18
	$ClCH_2$–	3	Hal	F–	24
	HC≡C–	7		Cl–	27
	Ph–	1		Br–	27
	OHC–	2		I–	26
	MeCO–	−1	S	MeSO–	13
	HO_2C–	6	Si	Me_3Si–	−8
	NC–	11			

Table 4.20 ^2J and ^3J ^{13}C–^1H Coupling constants in Hz[a]

$^2J_{CH}$: 1-6; 20-50; 5-8; 50 (H–≡C–); −10

Substituent effects in alkenes: −2.5; +7.5 (Cl); −8 (Cl); +7 (Cl); +16 (Cl, Cl)

−1 (Me); −2.5 (Me); +0.5 (Me); +5 (CH_3); 0 (Ph); −4.5 (Ph)

−1 (Ph); −7 ($Si(CH{=}CH_2)_3$); −0.5 (CHO); −3.5 (CHO); +0.25 (CHO); +2 (CHO)

+1 (benzene); +1 (pyridine); +3 (pyridine); +7.5 (pyrrole); +4.5 (pyrrole)

$^3J_{CH}$

Substituent and dihedral angle effects: 9 (antiperiplanar); 6 (eclipsed); 3 (gauche)

14 (CO_2H); 7.5 (CO_2H); 11 (CH_3); 7 (CH_3)

7.5 (benzene); 4 (–C≡C–); 8.5 (CH_3, Cl); 4 (CH_3, Cl)

[a]Signs are given only when they have been determined

Table 4.21 $^{13}C–^{19}F$ Coupling constants in Hz

Structure	$^1J_{CF}$	Structure	$^1J_{CF}$	$^2J_{CF}$	$^3J_{CF}$	$^4J_{CF}$
CH_3F	162	$H_2C{=}CF_2$	267	25		
CH_2F_2	235	HO_2CCH_2F	177	22		
CHF_3	274	HO_2CCHF_2	240	28		
CF_4	260	HO_2CCF_3	283	44		
$MeCF_3$	281	$RCH_2CH_2CH_2F$	165	18	6	<2
CCl_3F	337	$(RCH_2)_2CHF$	170	20		
CCl_2F_2	325	c-$C_6H_{11}F$	170	19	8	1.5
$CClF_3$	299	$PhCH_2F$	166	16.5	6	1
MeC(=O)F	353	Py-4-F	262	16	6	
$(CF_3)_2O$	265	Py-3-F	255	22.5 and 18	4	4
$H_2C{=}CF_2$	287	Py-2-F	236	37	7.5 and 15	4
		PhF	245	21	8	3
		$PhCF_3$	272	37	4	1

Table 4.22 $^{13}C–^{31}P$ Coupling constants in Hz[a]

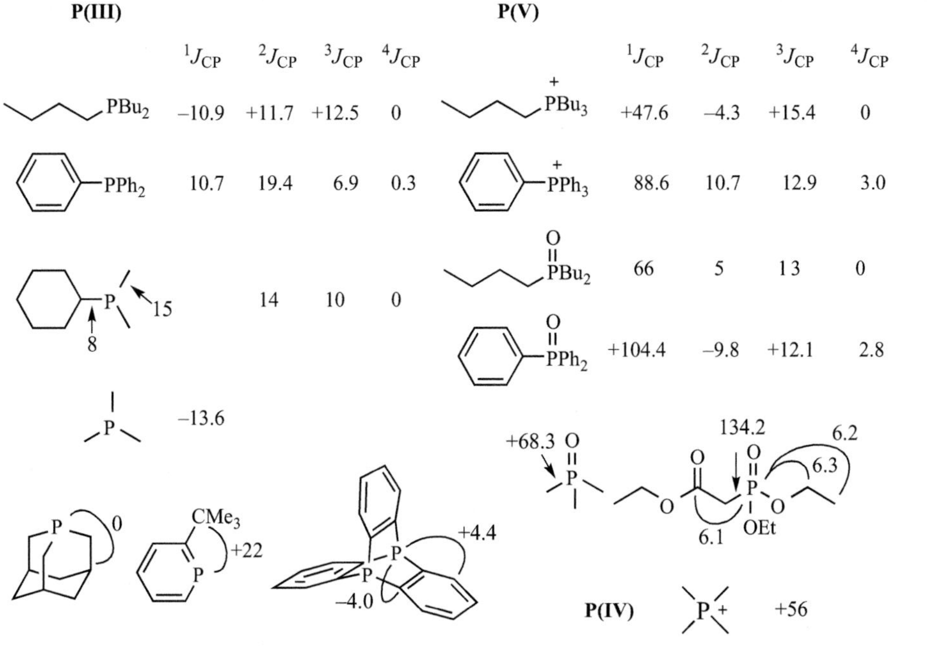

P(III)

Structure	$^1J_{CP}$	$^2J_{CP}$	$^3J_{CP}$	$^4J_{CP}$
PBu_2	−10.9	+11.7	+12.5	0
PPh_2	10.7	19.4	6.9	0.3
P (arrows: 15, 8)		14	10	0
P	−13.6			

P 0; CMe_3 P +22; P P +4.4 −4.0

P(V)

Structure	$^1J_{CP}$	$^2J_{CP}$	$^3J_{CP}$	$^4J_{CP}$
$\overset{+}{P}Bu_3$	+47.6	−4.3	+15.4	0
$\overset{+}{P}Ph_3$	88.6	10.7	12.9	3.0
O=PBu_2	66	5	13	0
O=PPh_2	+104.4	−9.8	+12.1	2.8

+68.3 O=P; O, O, O=P, O, OEt: 134.2, 6.2, 6.3, 6.1

P(IV) P^+ +56

[a]Signs are given only when they have been determined

Table 4.23 [1]H Chemical shifts in methyl, methylene and methine groups

	Methyl protons	δ_H	Methylene protons	δ_H	Methine protons	δ_H
C	R–**CH**$_3$	0.9	R–**CH**$_2$–R	1.4	R–**CH**R$_2$	1.5
	C=C–C–**CH**$_3$	1.1	C=C–C–**CH**$_2$–R	1.7		
	O–C–**CH**$_3$	1.3	O–C–**CH**$_2$–R	1.9	O–C–**CH**R$_2$	2.0
	N–C–**CH**$_3$	1.1	N–C–**CH**$_2$–R	1.4		
	O$_2$N–C–**CH**$_3$	1.6	O$_2$N–C–**CH**$_2$–R	2.1		
	C=C–**CH**$_3$	1.6	C=C–**CH**$_2$–R	2.3		
	Ar–**CH**$_3$	2.3	Ar–**CH**$_2$–R	2.7	Ar–**CH**R$_2$	3.0
	O=CC=C–**CH**$_3$	2.0	O=CC=C–**CH**$_2$–R	2.4		
	O=CC(**CH**$_3$)=C	1.8	O=CC(**CH**$_2$–R)=C	2.4		
	C≡C–**CH**$_3$	1.8	C≡C–**CH**$_2$–R	2.2	C≡C–**CH**R$_2$	2.6
	RCO–**CH**$_3$	2.2	RCO–**CH**$_2$–R	2.4	RCO–**CH**R$_2$	2.7
	ArCO–**CH**$_3$	2.6	ArCO–**CH**$_2$–R	2.9	ArCO–**CH**R$_2$	3.3
	ROOC–**CH**$_3$	2.0	ROOC–**CH**$_2$–R	2.2	ROOC–**CH**R$_2$	2.5
	ArOOC–**CH**$_3$	2.4	ArOOC–**CH**$_2$–R	2.6		
	N–CO–**CH**$_3$	2.0	N–CO–**CH**$_2$–R	2.2	N–CO–**CH**R$_2$	2.4
	N≡C–**CH**$_3$	2.0	N≡C–**CH**$_2$–R	2.3	N≡C–**CH**R$_2$	2.7
N	N–**CH**$_3$	2.3	N–**CH**$_2$–R	2.5	N–**CH**R$_2$	2.8
	ArN–**CH**$_3$	3.0	ArN–**CH**$_2$	3.5		
	RCON–**CH**$_3$	2.9	RCON–**CH**$_2$–R	3.2	RCO–N–**CH**R$_2$	4.0
	N$^+$–**CH**$_3$	3.3	N$^+$–**CH**$_2$–R	3.3		
	O$_2$N–**CH**$_3$	4.3	O$_2$N–**CH**$_2$–R	4.4	O$_2$N–**CH**R$_2$	4.7
O	HO–**CH**$_3$	3.4	HO–**CH**$_2$–R	3.6	HO–**CH**R$_2$	3.9
	RO–**CH**$_3$	3.3	RO–**CH**$_2$–R	3.4	RO–**CH**R$_2$	3.7
	C=CO–**CH**$_3$	3.8	C=CO–**CH**$_2$–R	3.7		
	ArO–**CH**$_3$	3.8	ArO–**CH**$_2$–R	4.3	ArO–**CH**R$_2$	4.5
	RCOO–**CH**$_3$	3.7	RCOO–**CH**$_2$–R	4.1	RCOO–**CH**R$_2$	4.8
			(RO)$_2$**CH**$_2$	4.8	(RO)$_3$**CH**	5.2
Hal	F–**CH**$_3$	4.3	F–**CH**$_2$–R	4.1	F–**CH**R$_2$	3.7
	Cl–**CH**$_3$	3.1	Cl–**CH**$_2$–R	3.6	Cl–**CH**R$_2$	4.2
	Br–**CH**$_3$	2.7	Br–**CH**$_2$–R	3.5	Br–**CH**R$_2$	4.3
	I–**CH**$_3$	2.1	I–**CH**$_2$–R	3.2	I–**CH**R$_2$	4.3
S	RS–**CH**$_3$	2.1	S–**CH**$_2$–R	2.4	S–**CH**R$_2$	3.2
	RSO–**CH**$_3$	2.5	RSO–**CH**$_2$–R	2.7		
	RSO$_2$–**CH**$_3$	2.8	RSO$_2$–**CH**$_2$–R	2.9		
			(RS)$_2$**CH**$_2$	4.2		
P	R$_2$P–**CH**$_3$	1.4	R$_2$P–**CH**$_2$–R	1.6	R$_2$P–**CH**R$_2$	1.8
Si	R$_3$Si–**CH**$_3$	0.0	R$_3$Si–**CH**$_2$–R	0.5	R$_3$Si–**CH**R$_2$	1.2
Se	RSe–**CH**$_3$	2.0				

R = alkyl group. These values will usually be within ±0.2 p.p.m. unless electronic or anisotropic effects from other groups are strong. An obsolete scale used τ values; these are related to δ values by the equation $\tau = 10 - \delta$

Estimation of ¹H Chemical Shifts in Alkanes

$$\text{For } R^1R^2R^3C\text{—H}, \quad \delta_H = 1.50 + \sum z_i \tag{4.23}$$

Table 4.24 Substituent constants *z* for Eq. (4.23)

R^i	*z*	R^i	*z*	R^i	*z*	R^i	*z*
H–	−0.3	CH_2=CH–	0.8	NC–	1.2	AcO–	2.7
Alkyl–	0.0	Ph–	1.3	H_2N–	1.0	Cl–	2.0
CH_2=CHCH_2–	0.2	HC≡C–	0.9	O_2N–	3.0	Br–	1.9
MeCOCH_2–	0.2	OHC–	1.2	HO–	1.7	I–	1.4
HOCH_2–	0.3	MeCO–	1.2	MeO–	1.5	MeS–	1.0
ClCH_2–	0.5	RO_2C–	0.8	PhO–	2.3	Me_3Si–	0.7

Table 4.25 ¹H Chemical shifts in some aliphatic compounds

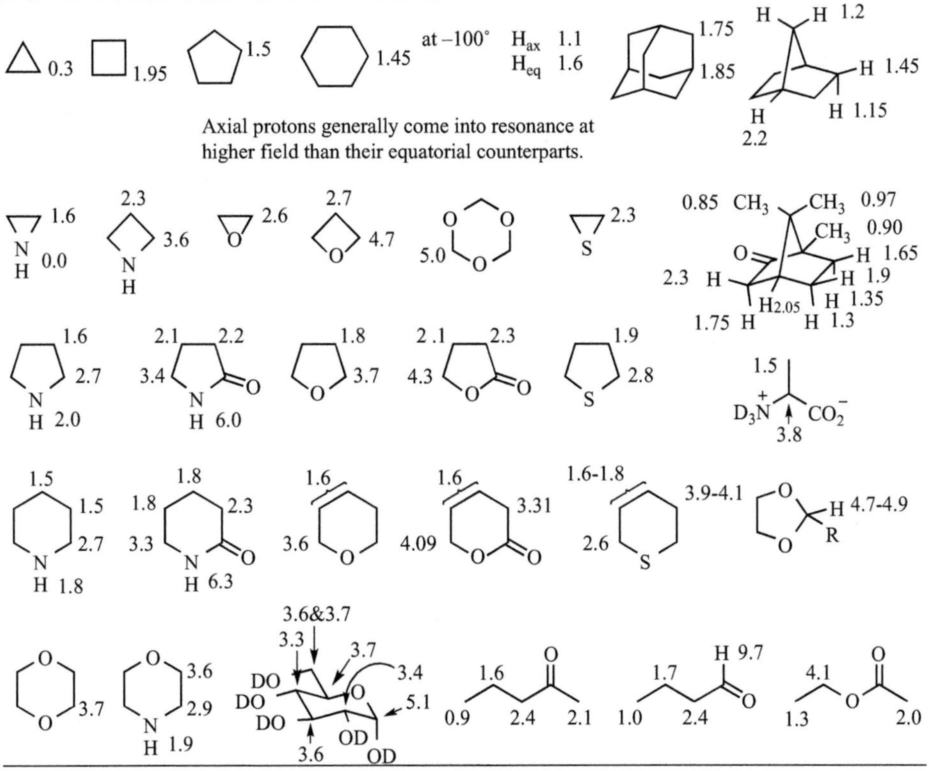

Axial protons generally come into resonance at higher field than their equatorial counterparts.

Estimation of [1]H Chemical Shifts in Alkenes

$$\delta_H = 5.25 + \Sigma z_{gem} + \Sigma z_{cis} + \Sigma z_{trans} \quad (4.24)$$

(Structure: R_{cis} and H on one carbon; R_{trans} and R_{gem} on the other carbon of the double bond.)

Table 4.26 Substituent constants z for Eq. (4.24)

	Substituent R	z_{gem}	z_{cis}	z_{trans}
H	H–	0	0	0
C	Alkyl–	0.45	–0.22	–0.28
	Ring-alkyl–[a]	0.69	–0.25	–0.28
	N≡CCH$_2$– or RCOCH$_2$–	0.69	–0.08	–0.06
	ArCH$_2$–	1.05	–0.29	–0.32
	R$_2$NCH$_2$–	0.58	–0.10	–0.08
	ROCH$_2$–	0.64	–0.10	–0.02
	HalCH$_2$–	0.70	0.11	–0.04
	RSCH$_2$–	0.71	–0.13	–0.22
	Isolated RCH=CH–	1.00	–0.09	–0.23
	Conjugated CH=CH–[b]	1.24	0.02	–0.05
	Ar–	1.38	0.36	–0.07
	OHC–	1.02	0.95	1.17
	Isolated RCO–	1.10	1.12	0.87
	Conjugated RCO–[b]	1.06	0.91	0.74
	Isolated HO$_2$C–	0.97	1.41	0.71
	Conjugated HO$_2$C–[b]	0.80	0.98	0.32
	Isolated RO$_2$C–	0.80	1.18	0.55
	Conjugated RO$_2$C–[b]	0.78	1.01	0.46
	R$_2$NCO–	1.37	0.98	0.46
	ClCO–	1.11	1.46	1.01
	RC≡C–	0.47	0.38	0.12
	N≡C–	0.27	0.75	0.55
N	(Alkyl)HN– or (Alkyl)$_2$N–	0.80	–1.26	–1.21
	(Conjugated alkyl or aryl)$_2$N–[b]	1.17	–0.53	–0.99
	AcNH–	2.08	–0.57	–0.72
	O$_2$N–	1.87	1.30	0.62
O	AlkylO–	1.22	–1.07	–1.21
	Conjugated alkyl or arylO–[b]	1.21	–0.60	–1.00
	AcO–	2.11	–0.35	–0.64
Hal	F–	1.54	–0.40	–1.02
	Cl–	1.08	0.18	0.13
	Br–	1.07	0.45	0.55
	I–	1.14	0.81	0.88
Si	R$_3$Si–	0.90	0.90	0.60
S	RS–	1.11	–0.29	–0.13
	RSO–	1.27	0.67	0.41
	RSO$_2$–	1.55	1.16	0.93

[a]Use the 'ring-alkyl' values when the double bond and the alkyl group are part of a five- or six-membered ring

[b]Use the 'conjugated' values when either the substituent or the double bond is further conjugated

Table 4.27 ^{1}H Chemical shifts of protons attached to multiple bonds

Structure	δ_H	Structure	δ_H
RC**HO**	9.4–10.0	R_2C=C**H**R	4.5–6.0
ArC**HO**	9.7–10.5	R_2C=C**H**–COR	5.8–6.7
ROC**HO**	8.0–8.2	R**H**C=CR–COR	6.5–8.0
R_2NC**HO**	8.0–8.2	R**H**C=CR–OR	4.0–5.0
RC≡C**H**	1.8–3.1	R_2C=C**H**–OR	6.0–8.1
R_2C=C=C**H**R	4.0–5.0	R**H**C=CR–NR_2	3.7–5.0
Ar**H**	6.0–9.0	R_2C=C**H**–NR_2	5.7–8.0

Table 4.28 ^{1}H Chemical shifts (largely attached to double bonds) in some unsaturated cyclic systems

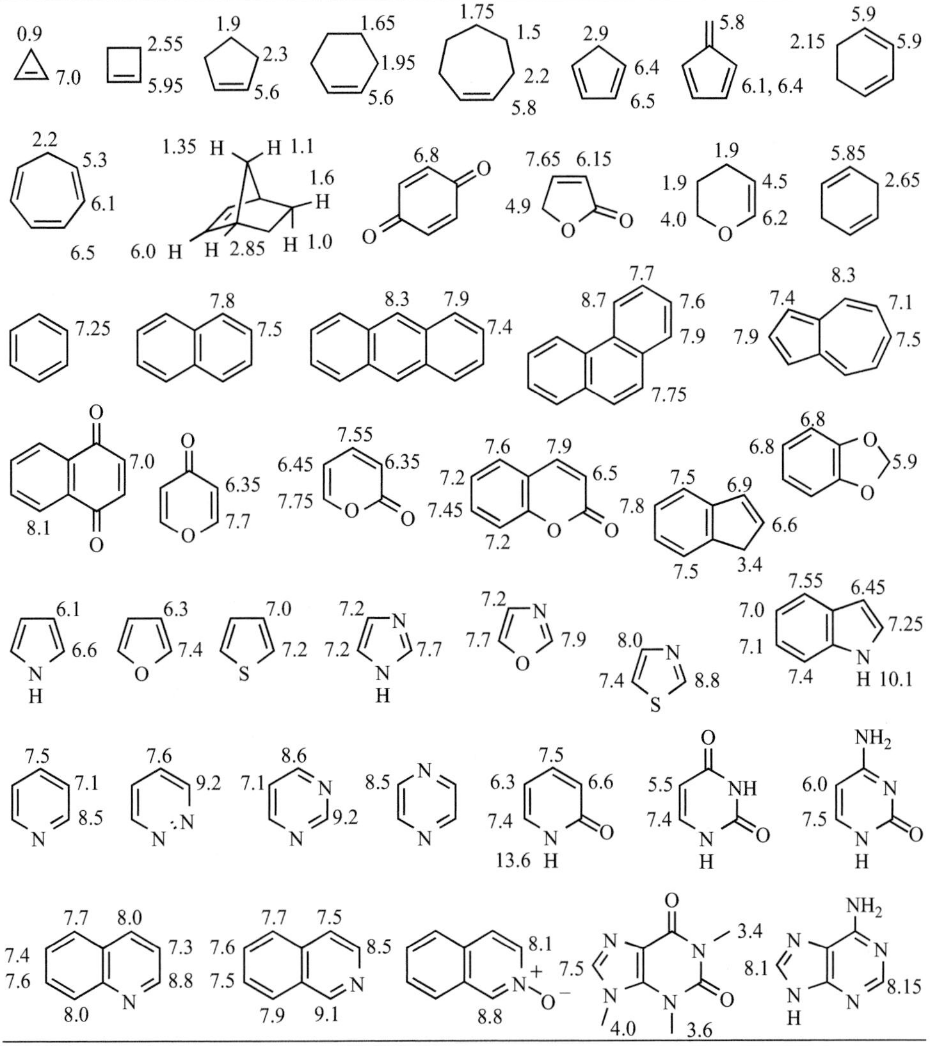

Estimation of ^{1}H chemical shifts in substituted benzenes

$$\delta_H = 7.27 + \Sigma z_i \quad (4.25)$$

(Structure: R–benzene ring with positions o, m, p)

Table 4.29 Substituent constants z for Eq. (4.25)

	Substituent R	z_{ortho}	z_{meta}	z_{para}
H	H–	0	0	0
C	Me–	–0.20	–0.12	–0.22
	Et–	–0.14	–0.06	–0.17
	Pr^i–	–0.13	–0.08	–0.18
	Bu^t–	–0.02	–0.08	–0.21
	H_2NCH_2– or $HOCH_2$–CH_2–	–0.07	–0.07	–0.07
	$ClCH_2$–	0.00	0.00	0.00
	F_3C–	0.32	0.14	0.20
	Cl_3C–	0.64	0.13	0.10
	CH_2=CH–	0.06	–0.03	–0.10
	Ph–	0.37	0.20	0.10
	OHC–	0.56	0.22	0.29
	MeCO–	0.62	0.14	0.21
	H_2NCO–	0.61	0.10	0.17
	HO_2C–	0.85	0.18	0.27
	MeO_2C–	0.71	0.10	0.21
	ClCO–	0.84	0.22	0.36
	N≡C–	0.36	0.18	0.28
N	H_2N–	–0.75	–0.25	–0.65
	Me_2N–	–0.66	–0.18	–0.67
	AcNH–	0.12	–0.07	–0.28
	O_2N–	0.95	0.26	0.38
O	HO–	–0.56	–0.12	–0.45
	MeO–	–0.48	–0.09	–0.44
	AcO–	–0.25	0.03	–0.13
Hal	F–	–0.26	0.00	–0.04
	Cl–	0.03	–0.02	–0.09
	Br–	0.18	–0.08	–0.04
	I–	0.39	–0.21	0.00
Si	Me_3Si–	0.22	–0.02	–0.02
P	$(MeO)_2P(=O)$–	0.48	0.16	0.24
S	MeS–	0.37	0.20	0.10

Errors are particularly likely to occur when substituents *ortho* to one another interfere with conjugation to the ring

Table 4.30 ^{1}H and ^{13}C chemical shifts in the common deuterated solvents

	Deuterated solvent				Undeuterated solvent
Solvent	δ_H [a]	[b]	δ_C	[b]	δ_C
Acetic acid	11.5[c]		179.0		178.1
	2.05		20.0	Septet	21.1
Acetone	2.05	Quintet	205.7		205.4
			29.8	Septet	30.5
Acetonitrile			118.2	m	118.2
	1.95	Quintet	1.3	Septet	1.7
Benzene	7.27		128.0	Triplet	128.5
t-Butanol	1.28[d]				
Carbon disulfide					192.8
Carbon tetrachloride					96.1
Chloroform	7.25		77.0	Triplet	77.2
Cyclohexane	1.40	Triplet	26.3	Quintet	27.6
1,2-Dichloroethane	3.72	br			
Dichloromethane (methylene chloride)	5.35	Triplet	53.1	Quintet	54.0
Dimethylformamide (DMF)	8.03		163.2	Septet	
	2.92	Quintet	34.9	Septet	
	2.75	Quintet	29.8	Septet	
Water in DMF	3.5[c]				
Dimethylsulfoxide (DMSO)	2.5	Quintet	39.7	Septet	40.6
Water in DMSO	3.3[c]				
Dioxan	3.55	Triplet	66.5	Quintet	67.6
Hexamethylphosphoricamide (HMPA)	2.60	Doublet[e]	35.8	Septet	36.9
Methanol	3.35	Quintet	49.0	Septet	49.9
	4.8[c]				
Nitromethane	4.33	Quintet	60.5	Septet	57.1
Pyridine	8.5		149.8	Triplet	150.3
	7.35		135.3	Triplet	135.9
	7.0		123.4	Triplet	123.9
Tetrahydrofuran (THF)	1.73	br	25.2	Quintet	26.5
	3.58	br	67.4	Quintet	68.4
Toluene	7.1	m	129.2	Triplet	
	7.00		128.3	Triplet	
	6.98	Quintet	125.5	Triplet	
	2.1	Quintet	20.4	Septet	

Table 4.30 (continued)

Solvent	Deuterated solvent				Undeuterated solvent
	δ_H[a]	[b]	δ_C	[b]	δ_C
Trifluoroacetic acid (TFA)	11.3[c]		164.2	Quartet[f]	163.8[f]
			116.6	Quartet[g]	115.7[g]
Water	4.7[c]				

[a]Residual protons in the deuterated solvent
[b]Singlet unless otherwise stated
[c]Variable, depends upon the solvent and its concentration
[d]$(CH_3)_3COD$ is usually used, not the fully deuterated solvent
[e]Coupled to P, $J = 9$ Hz
[f]Quartet from coupling to F 46 Hz
[g]Quartet from coupling to F 294 Hz

Table 4.31 ^{1}H Chemical shifts of protons attached to elements other than carbon[a]

	Structure	δ_H		Structure	δ_H
NH	RNH_2 and R_2NH	0.5–4.5	OH	Monomeric H_2O	~1.5
	$ArNH_2$ and ArNHR	3–6		Suspended H_2O	~4.7
	$RCONH_2$ and RCONHR	5–12		ROH	0.5–4.5
	Pyrrole NH	7–12		ArOH	4.5–10
				RCO_2H	9–15
SiH	Me_3SiH	4.0		$R_2C{=}NOH$	9–12
	Ar_3SiH	~5.5	Intramolecularly H-bonded OH		7–16
SnH	R_3SnH	~5.3	SH	RSH	1–2
PH	$(RO)_2P({=}O)H$	~6.8[b]		ArSH	3–4

[a]These values (except for SiH and SnH) are sensitive to temperature, solvent and concentration; the stronger the hydrogen bond, the lower field the chemical shift
[b]Doublet, $^1J_{PH}$ 140 Hz

Table 4.32 Geminal ($^2J_{HH}$) coupling constants in Hz[a]

Structure	$^2J_{HH}$
CH_4 (HH / H H)	−12.4
cyclopropane (HH)	−4.3
cyclobutane (HH)	−10.9
cyclohexane (HH)	−12.6
HH / H Ph	−14.3
HH / H COMe	−14.9
HH / H CN	−16.2
HH / Ph CN	−18.5
HH / NC CN	−20.3
HH / O= cyclopentene =O	−21.5
HH / H OH	−10.8
HH / H Cl	−10.8
HH / Cl Cl	−7.5
HH / R_2CH $SiMe_3$	−14.0
H H / C=C / H H	2.5
H H / C=C / H CH_3	2.1
H H / C=C / H Cl	−1.4
H H / C=C / H Ph	1.0
H H / C=C / H COMe	1.3
H H / C=C / H $SiMe_3$	3.8
H H / C=C=C (allene)	−9.0

[a]Signs are given only when they have been determined

Table 4.33 Vicinal ($^3J_{HH}$) coupling constants in Hz in some aliphatic compounds

Open-chain compounds

Structure	*$^3J_{HH}$ range*	*Typical value*
CH_3CH_2—	6-8	7
CH_3−CH<	5-7	6
—CH_2CH_2—	5-8	7
>CH−CH<	0-8	7
>=CH−CH<	4-11	6
>=CH−CH=<	6-13	11[a]
>CH CHO	0-3	2
>=CH−CHO	5-8	7
H H (cis alkene)	0-12	8
H / H (trans alkene)	12-18	15

Cyclic compounds

Structure	*Geometry*	*Ring size*	*$^3J_{HH}$ range*
R, H / H, R (cycloalkane)	cis	3	7-13
	tran	3	4.0-9.5
	cis	4	4.0-12.0
	tran	4	2.0-10.0
	cis	5	5.0-10.0
	trans	5	5.0-10.0
	cis	6	2.0-6.0
	trans	6	8.0-13.0[b]
H / H, R (cycloalkene, allylic)		3	1.8[c]
		4	−0.8[c]
		5	0.5[c]
		6	1.5[c]
		7	3.7[c]
		8	5.3[c]
H / H (cycloalkene, vinylic)		3	0.5-2
		4	2.5-4.0
		5	5-7
		6	8.5-10.5
		7	9-12.5
		8	10-13
norbornane (H^7, H^1, H^{2x}, H^{3x}, H^{2n}, H^{3n})		1-2x	3-4
		1-2n	0-2
		2x-3x	9-10
		2n-3n	6-7
		2x-3n	2-5
		1-7	0-3

[a]Found in dienes adopting the s-*trans* conformation
[b]J_{aa} = 8–13, J_{ee} = 2–5; note that J_{ee} is usually 1 Hz smaller than J_{ae}
[c]Value for the unsubstituted cycloalkene

Table 4.34 Vicinal ($^3J_{HH}$) coupling constants in Hz in some heterocyclic and aromatic compounds

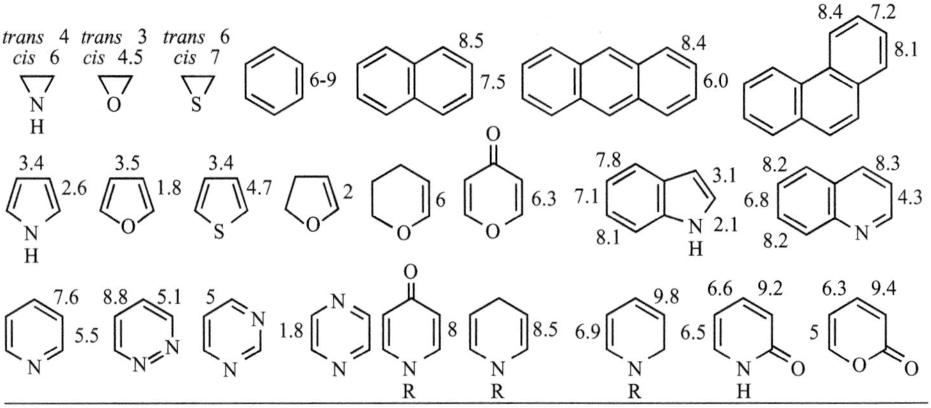

trans 4 cis 6 (N-H); trans 3 cis 4.5 (O); trans 6 cis 7 (S)

6-9; 8.5, 7.5; 8.4, 6.0; 8.4, 7.2, 8.1

3.4, 2.6; 3.5, 1.8; 3.4, 4.7; 2; 6; 6.3; 7.8, 7.1, 8.1, 3.1, 2.1; 8.2, 6.8, 8.2, 8.3, 4.3

7.6, 5.5; 8.8, 5.1; 5; 1.8; 8; 8.5; 9.8, 6.9; 6.6, 9.2, 6.5; 6.3, 9.4, 5

Table 4.35 Pascal's triangle giving the relative intensities of first-order multiplets for coupling to *n* equivalent nuclei of spin $I = 1/2$

n	*Relative intensity*
0	1
1	1 1
2	1 2 1
3	1 3 3 1
4	1 4 6 4 1
5	1 5 10 10 5 1
6	1 6 15 20 15 6 1
7	1 7 21 35 35 21 7 1
8	1 8 28 56 70 56 28 8 1

Table 4.36 $^4J_{HH}$ coupling constants in Hz with W features emphasised

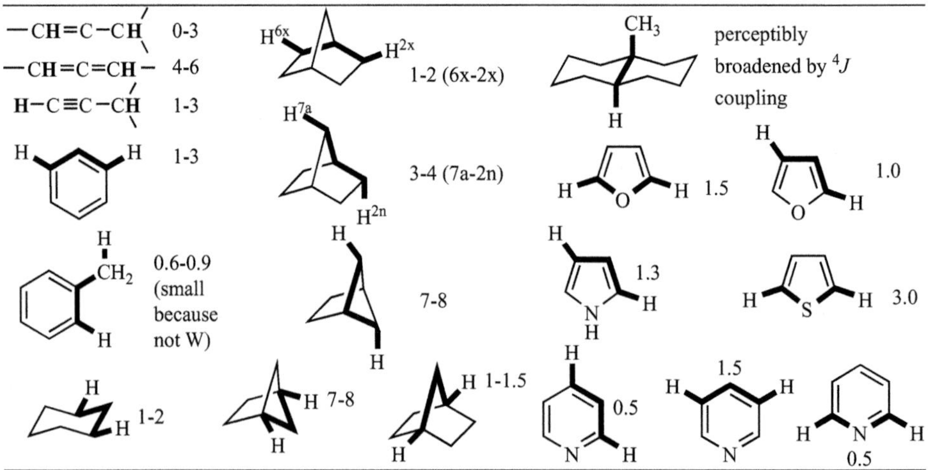

—CH=C—CH 0-3

—CH=C=CH— 4-6

H—C≡C—CH 1-3

H (benzene) H 1-3

H, CH₂, H 0.6-0.9 (small because not W)

H (cyclohexane) H 1-2

H6x, H2x 1-2 (6x-2x)

H7a, H2n 3-4 (7a-2n)

H, H 7-8

H 7-8, H

H 1-1.5, H

CH₃, H perceptibly broadened by ^4J coupling

H (furan) H 1.5; H (furan) H 1.0

H (pyrrole) 1.3 H

H (thiophene) H 3.0

H (pyridine) 0.5 H; H 1.5 H (pyridine); H (pyridine) H 0.5

Table 4.37 $^5J_{HH}$ coupling constants in Hz with zigzag features emphasised

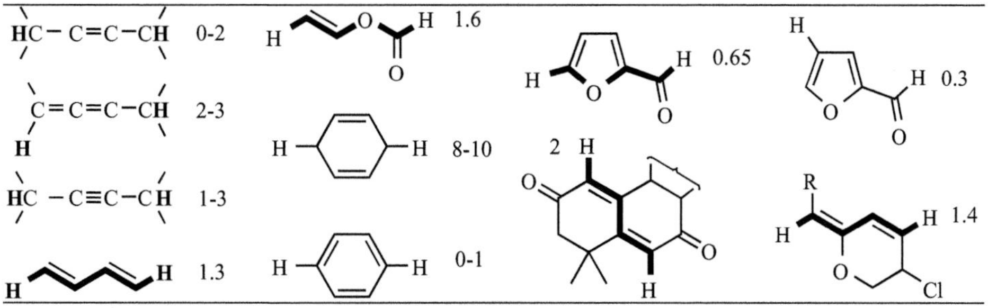

HC−C=C−CH	0-2
C=C=C−CH (H)	2-3
HC−C≡C−CH	1-3
H (zigzag diene) H	1.3
H (vinyl formate) H	1.6
H (cyclohexa-1,4-diene) H	8-10
H (benzene, para) H	0-1
H (furan-2-carbaldehyde) H	0.65
H (furan-3-carbaldehyde) H	0.3
2 H … H (bicyclic dienedione)	
R, H (dihydropyran, Cl) H	1.4

Table 4.38 Eu(dpm)$_3$-induced shifts of protons in some common environments[a]

Functional group	Shift p.p.m./mol of Eu(dpm)$_3$ per mol of substrate	Functional group	Shift p.p.m./mol of Eu(dpm)$_3$ per mol of substrate
$RCH_2N\mathit{H}_2$	~150	$RC\mathit{H}_2CHO$	11
$RCH_2O\mathit{H}$	~100	$RC\mathit{H}_2OC\mathit{H}_2R$	10
$RC\mathit{H}_2NH_2$	30–40	$RC\mathit{H}_2CO_2CH_3$	7
$RC\mathit{H}_2OH$	20–25	$RCH_2CO_2C\mathit{H}_3$	6.5
$RC\mathit{H}_2COR$	10–17	$RC\mathit{H}_2CN$	3–7
$RCH_2C\mathit{H}O$	19		

[a]The shifts refer to the protons in italics

Table 4.39 ^1H–^{19}F Coupling constants in Hz[a]

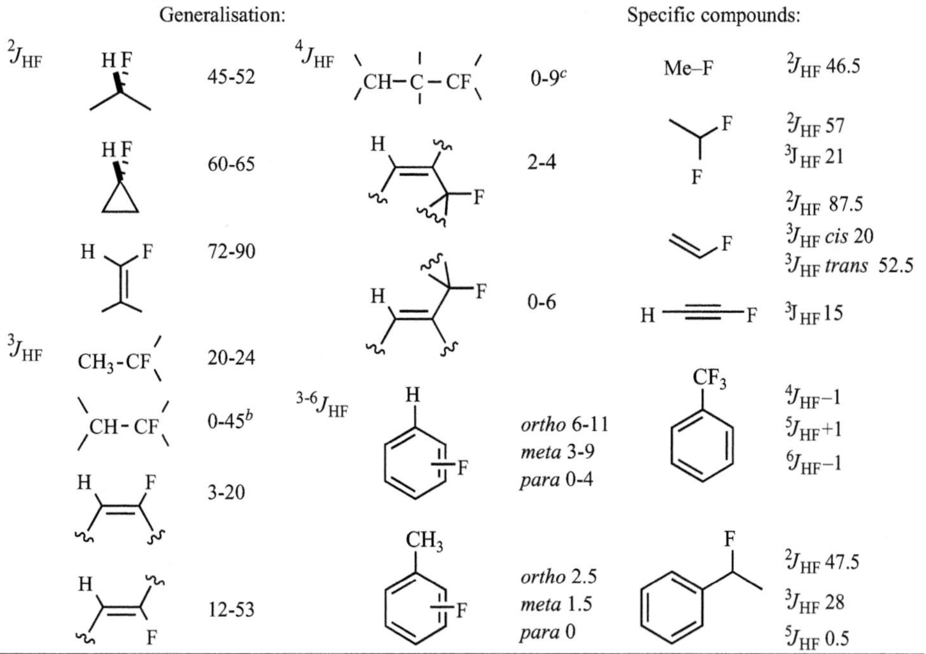

Generalisation:

$^2J_{HF}$	H F (sp³ carbon)	45-52
	H F (cyclopropane)	60-65
	H, F (alkene)	72-90
$^3J_{HF}$	CH_3-CF	20-24
	CH-CF	0-45[b]
	H, F (cis alkene)	3-20
	H, F (trans alkene)	12-53
$^4J_{HF}$	CH−C−CF	0-9[c]
	H, CF (alkene)	2-4
	H, CF (alkene)	0-6
$^{3\text{-}6}J_{HF}$	H, F (fluorobenzene)	*ortho* 6-11, *meta* 3-9, *para* 0-4
	CH_3, F (fluorotoluene)	*ortho* 2.5, *meta* 1.5, *para* 0

Specific compounds:

Me–F	$^2J_{HF}$ 46.5
F, F (1,2-difluoro)	$^2J_{HF}$ 57, $^3J_{HF}$ 21
vinyl F	$^2J_{HF}$ 87.5, $^3J_{HF}$ *cis* 20, $^3J_{HF}$ *trans* 52.5
H−C≡C−F	$^3J_{HF}$ 15
CF_3 (benzotrifluoride)	$^4J_{HF}$−1, $^5J_{HF}$+1, $^6J_{HF}$−1
F (1-fluoroethylbenzene)	$^2J_{HF}$ 47.5, $^3J_{HF}$ 28, $^5J_{HF}$ 0.5

[a]Signs are given only when they have been determined

[b]The higher end of the range (≥3.5) when the atoms are held in a W conformation

[c]0–12 when gauche and 10–45 when antiperiplanar

Table 4.40 ^{1}H–^{31}P Coupling constants in Hz[a]

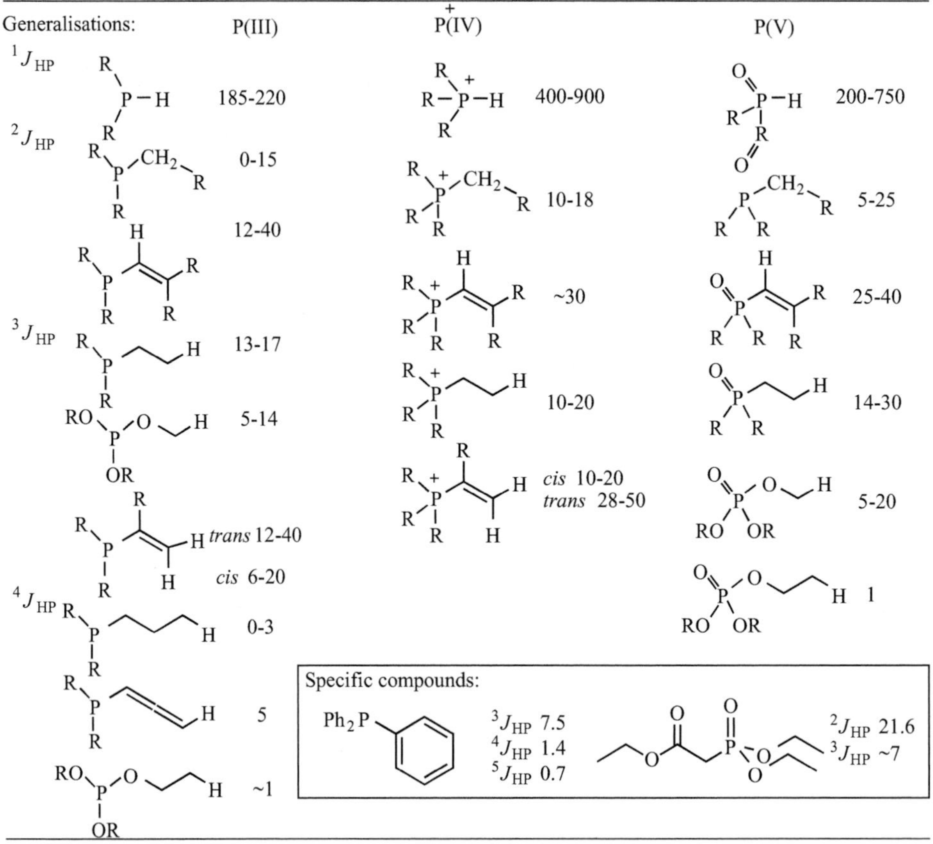

Generalisations:

	P(III)	$\overset{+}{P}$(IV)	P(V)
$^{1}J_{HP}$	185-220	400-900	200-750
$^{2}J_{HP}$	0-15	10-18	5-25
	12-40	~30	25-40
$^{3}J_{HP}$	13-17	10-20	14-30
	5-14	*cis* 10-20, *trans* 28-50	5-20
	trans 12-40, *cis* 6-20		1
$^{4}J_{HP}$	0-3		
	5		
	~1		

Specific compounds:

Ph_2P–phenyl: $^{3}J_{HP}$ 7.5, $^{4}J_{HP}$ 1.4, $^{5}J_{HP}$ 0.7

Ethyl (diethoxyphosphoryl)acetate: $^{2}J_{HP}$ 21.6, $^{3}J_{HP}$ ~7

[a]The coupling constants are often strongly dependent upon the groups attached to phosphorus, and therefore values outside the quoted range may be observed

Table 4.41 Some representative ^{11}B chemical shifts in p.p.m. relative to $Et_2O.BF_3$ (negative numbers are upfield and positive numbers downfield)

Structure	δ	Structure	δ	Structure	δ
Me_3B	87	$Me_2B(OMe)$	53	$Me_3B.NMe_3$	0.1
Me_2BF	60	$MeB(OMe)_2$	30	$Me_3B.PMe_3$	12
$MeBF_2$	8	$B(OMe)_3$	18	$H_3B.NMe_3$	−8
BF_3	10	$MeB(NMe_2)_2$	34	$H_3B.SMe_2$	−20
Ph_3B	68	BCl_3	47	BF_4^-	−2
$(CH_2{=}CH)_3B$	56	BBr_3	39		
B_2H_6	17				

For more about ^{11}B NMR, see [3, 4]

Table 4.42 Some representative one-bond ^{11}B coupling constants in Hz

Structure		J	Structure		J	Structure		J
Me_2BF	$^1J_{BF}$	119	BF_4^-	$^1J_{BF}$	1–5a	BH_4^-	$^1J_{BH}$	80
$MeBF_2$	$^1J_{BF}$	76	Me_3B	$^1J_{CB}$	47	Me_4B^-	$^1J_{CB}$	22
BF_3	$^1J_{BF}$	15						

[a]Solvent and temperature dependent

Table 4.43 Approximate ^{15}N chemical shifts in p.p.m. relative to $MeNO_2$ (negative numbers are upfield and positive numbers downfield)

Structure	δ	Structure	δ	Structure	δ
R_3N	−350	RN=C=NR	−250	Pyridines	50 to −50
R_4N^+	−350	RC≡N	−150	R_2C=NOH	0
$RNHNH_2$	−350	Pyrroles	−100 to −250	R_2C=NR	0 to −50
RNCO	−350	RCNO	−180	RN=NR	200
RN_3	−350, −190, −160	$(RCO)_2NR$	−180	R_2N–**N**=O	200
$R_3N^+–O^-$	−260	$RCONR_2$	−330	R–N=O	500
$RN^+(O^-)$=CHR	−100				

Table 4.44 Some representative ^{19}F chemical shifts in p.p.m. relative to $CFCl_3$ (negative numbers are upfield and positive numbers downfield)

Structure	δ	Structure	δ	Structure	δ
MeF	−272	$(CF_3)_3N$	−56	HF	40
CF_2H_2	−144	PhF	−116	F_2	429
CHF_3	−79	C_6F_6	−163	BF_3	−131
CF_4	−63	$PhCF_3$	−64	$BF_3.OEt_2$	−153
EtF	−213	$PhCH_2F$	−207	SiF_4	−163
c-C_6F_{12}	−133	$BrCF_3$	7	Et_2SiF_2	−143
CH_2=CHF	−114	CF_3CH_2OH	−78	Me_3SiF	−160
$CH_2=CF_2$	−81	CF_3CO_2Me	−74	PF_3	−34
$CF_2=CF_2$	−135	$(CF_3)_2CO$	−85	PF_5	−72
cis-CHF=CHF	−165	CH_3COF	+49	POF_3	−91
trans-CHF=CHF	−183	F–C≡N	−156	SF_6	57
CH_2=C=CHF	−176	F–C≡C–F	−95	SO_2F_2	33
$CF_3\mathbf{CF_2CF_2}CF_3$	−135	NF_3	145	$PhSO_2F$	65

For more about ^{19}F NMR, see [5–7]

Table 4.45 Some representative ^{19}F to ^{19}F coupling constants in Hz[a] (for coupling constants from ^{19}F to ^{13}C, ^{1}H and ^{11}B, see Tables 4.21, 4.39 and 4.42)

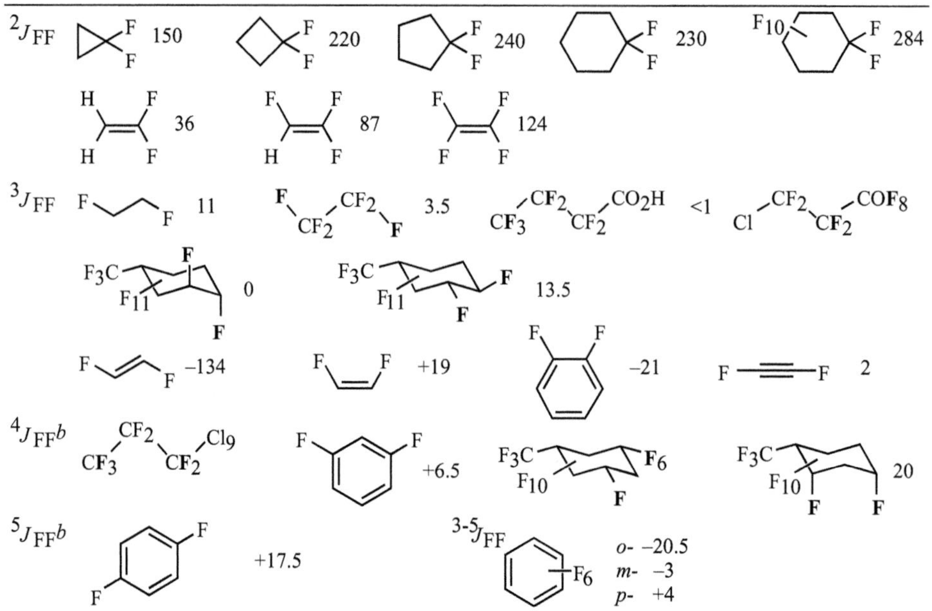

[a]Signs are given only when they have been determined
[b]$^{4}J_{FF}$ and $^{5}J_{FF}$ are often much larger than $^{3}J_{FF}$, especially when F is attached to trigonal carbon. They are very dependent upon the structure, including the spatial proximity of the F atoms

Table 4.46 Some representative ^{29}Si chemical shifts in p.p.m. relative to Me_4Si (negative numbers are upfield and positive numbers downfield)

Structure	δ	Structure	δ
Me_4Si	0	Cl_3SiH	−10
t-PhCH=CHSiMe$_3$	−7	$Me_3SiSiMe_3$	−20
Me_3SiH	−17	PhMe$_2$**Si**SiMe$_3$	−19
Me_2SiH_2	−37	PhMe$_2$Si**Si**Me$_3$	−22
$MeSiH_3$	−65	Me$_3$**Si**Me$_2$Si**Si**Me$_3$	−16
Me_3SiF	31	Me$_3$SiMe$_2$**Si**SiMe$_3$	−49
Me_2SiF_2	9	c-(Me$_2$Si)$_6$	−42
Me_3SiCl	30	Ph_3SiH	−21
Me_2SiCl_2	32	Ph_2SiH_2	−33
$MeSiCl_3$	12	$PhSiH_3$	−60
Me_3SiBr	26	$PhMe_2SiH$	−17
Me_3SiI	9	Ph_2SiCl_2	6
Me_3SiOMe	17	Ph_3SiLi	−9
(EtO)$_3$SiH	−59	$PhMe_2SiLi$	−29
(MeO)$_4$Si	−78	(PhMe$_2$Si)$_2$CuLi$_2$	−24
$Me_3SiOClO_3$	47	$AcNHSiMe_3$	6
(Me$_3$Si)$_2$O	7	Me(C=NH)OSiMe$_3$	18

Table 4.46 (continued)

Structure	δ	Structure	δ
c-$(Me_2SiO)_4$	–20	$AcN(SiMe_3)_2$	6
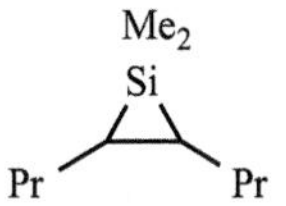 Me_2 Si Pr Pr	–60	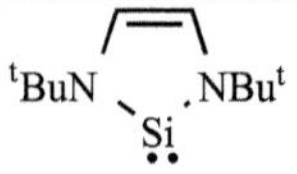 ^tBuN NBu^t Si	78
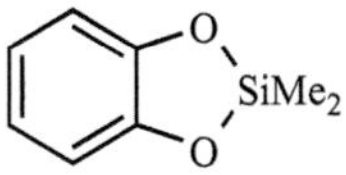 O SiMe$_2$ O	32	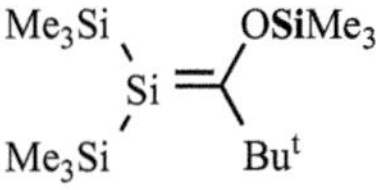 Me_3Si $O\mathbf{Si}Me_3$ Si Me_3Si Bu^t	13
$Ar_2Si{=}SiAr_2$ Ar = 2,4,6-$Me_3C_6H_2$	64	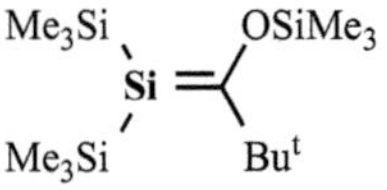 Me_3Si $OSiMe_3$ $\mathbf{Si}$ Me_3Si Bu^t	41.5
$Ar_3Si^+ (C_6F_5)_4B^-$ Ar = 2,4,6-$Me_3C_6H_2$	226	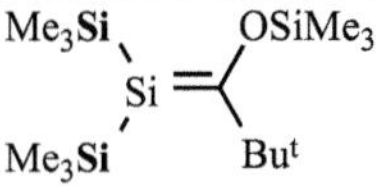 $Me_3\mathbf{Si}$ $OSiMe_3$ Si $Me_3\mathbf{Si}$ Bu^t	–12
5-coordinated Si	From –60 to –160	6-coordinated Si	From –120 to –220

For more about ^{29}Si NMR, see [8, 9]

Table 4.47 Some representative coupling constants from ^{29}Si to ^1H,[a] ^{13}C, ^{19}F, ^{29}Si and ^{31}P in Hz

Structure		Structure		Structure	
	$^1J_{SiH}$		$^1J_{SiH}$		$^2J_{SiH}$
SiH_4	202	$Me\mathbf{OSiH_3}$	216	$R_3\mathbf{SiCH_2}R$	~10
$Me_3\mathbf{SiH}$	184	$(MeO)_3\mathbf{SiH}$	298		
$Cl_2\mathbf{SiH_2}$	288				$^3J_{SiH}$
				$R_3\mathbf{Si}CH_2\mathbf{CH_2}R$	0–5
	$^1J_{SiF}$		$^1J_{SiF}$		
CCl_3SiF_3	264	$ClCH_2SiF_3$	267		
	$^1J_{CSi}$		$^1J_{SiSi}$		$^1J_{PSi}$
$(\mathbf{CH_3})_4\mathbf{Si}$	51	$Ph_3\mathbf{SiSi}Me_3$	87	$(Me_3\mathbf{Si})_3P$	27
$(\mathbf{CH_3})_3\mathbf{Si}Cl$	58	$(Me_3\mathbf{Si})_4\mathbf{Si}$	53		
$(CH_3)_3\mathbf{SiSi}(CH_3)_3$	44				

[a] $^1J_{SiH}$ is seen as two weak bands, one on each side of the main proton signal, similar to the ^{13}C side-bands in ^1H spectra but more intense (the natural abundance of ^{29}Si is 4.7%)

Table 4.48 Some representative $^{31}P(III)$ chemical shifts in p.p.m. relative to 85% H_3PO_4 (negative numbers are upfield and positive numbers downfield)

Structure	δ	Structure	δ	Structure	δ	Structure	δ
Me_3P	–62	i-Pr_3P	19	$MePH_2$	–164	$(RO)_3P$	125–145
Et_3P	–20	t-Bu_3P	63	Me_2PF	186	$PHal_3$	120–225
n-Pr_3P	–33	Me_2PH	–99	$MePF_2$	245		

For more about ^{31}P NMR, see [10]

Table 4.49 Some representative $^{31}P(V)$ chemical shifts in p.p.m. relative to 85% H_3PO_4 (negative numbers are upfield and positive numbers downfield)

Structure	δ	Structure	δ	Structure	δ	Structure	δ
$Me_3P{=}O$	36	Me_4P^+	24	$(RO)_3P{=}S$	60–75	PCl_4^+	86
$Et_3P{=}O$	48	$Hal_3P{=}O$	−80 to +5	$Ar_3P{=}CR_2$	5–25	PCl_6^-	−295
$Et_3P{=}S$	55	$(RO)_3P{=}O$	−20 to 0	PCl_5	−80		

Table 4.50 Some representative coupling constants from ^{31}P to ^{19}F in Hz

Structure	$^1J_{PF}$	Structure	$^1J_{PF}$	Structure	$^2J_{PF}$	Structure	$^3J_{PF}$
$Alkyl_2PF$	821–1450	Me_3PF_2	552	R_2PCFR_2	40–149	$CHF_2CH_2PH_2$	8
Ph_2PF	905	Ph_3PF_2	660	CF_3PF_2	87	$CHF_2CH_2PCl_2$	13
$Me_2P({=}O)F$	980	PF_6^-	706	$FCH_2\mathbf{CF}_2\mathbf{P}Cl_2$	99		

References

1. Ohtani I, Kusumi T, Kashman Y, Kakisawa H (1991). J Am Chem Soc 113:4092–4096
2. Reproduced with permission from Sanders JKM, Hunter BK (1993) Modern NMR spectroscopy 2nd edn. Oxford University Press, Oxford
3. Hermanek S (1992). Chem Rev 92:325–362
4. Nöth H, Wrackmeyer B (1978) NMR basic principles and progress, vol 14, p 1
5. Bruno TJ, Svoronos PDN (2003) Handbook of basic tables for chemical analysis 2nd edn. CRC Press, Boca Raton
6. Emsley JW, Phillips L (1971) Progress in NMR spectroscopy, vol 7, pp 1–520
7. Emsley JW, Phillips L, Wray V (1976) Progress in NMR spectroscopy, vol 10, pp 83–752
8. Williams EA (1989) NMR spectroscopy of organosilicon compounds. In: Patai S, Rappoport Z (eds) The chemistry of organosilicon compounds, Chap. 8. Wiley, New York
9. Brook MA (2000) Silicon in organic, organometallic, and polymer chemistry. Wiley, New York
10. Tebby JC (ed) (1991) CRC handbook of P-31 NMR data. CRC Press, Boca Raton

2D-Nuclear Magnetic Resonance Spectra

5

The conventional NMR spectrum is called a one-dimensional spectrum because it has one frequency dimension—the chemical shift and the coupling are displayed on the same axis with intensity plotted in the second dimension. In *two-dimensional NMR* spectra we have two frequency dimensions, displayed on orthogonal axes, with intensity as the third dimension. They are used to establish connections, coupling or spatial, between the nuclei giving the signals displayed on one axis and the nuclei giving the signals on the other axis. If both axes have spectra derived from the same element, the spectra are said to be *auto-correlated.* If the spectra on the two axes are from different elements they are said to be *cross-correlated.*

There are several multi-pulse sequences giving 2D-NMR spectra in common use, summarised in Sect. 5.11, greatly extending the range of information from that given by 1D-NMR spectra. 1D selective decoupling experiments (Sect. 4.11.1) and difference-NOE experiments (Sect. 4.13.2) are often perfectly adequate, but 2D-spectra can plot all the coupling information, or all the spatial information, in a single experiment. 2D-spectra often need long acquisition times and larger amounts of material, and are therefore taken only when they are needed. They are relatively costly in spectrometer time, but they add immeasurably to the power of NMR spectroscopy in structure determination.

5.1 The Basic Pulse Sequence

The basic pulse sequence for all 2D spectra is shown in Fig. 5.1. After the usual short delay, t_r, there is a *preparation* phase, typically a $\pi/2$ pulse, which tips the vector representing the bulk magnetisation into the *xy* plane. All the components making up this vector are then allowed to *evolve* as a function of the time t_1. A second $\pi/2$ pulse is a *mixing* phase preceding the *acquisition* phase, when the resulting magnetisation is detected by collecting the FID. The sequence is repeated from the preparation pulse to the end of the acquisition phase, with incrementally increasing values of t_1.

I. Fleming, D. Williams, *Spectroscopic Methods in Organic Chemistry*,
https://doi.org/10.1007/978-3-030-18252-6_5

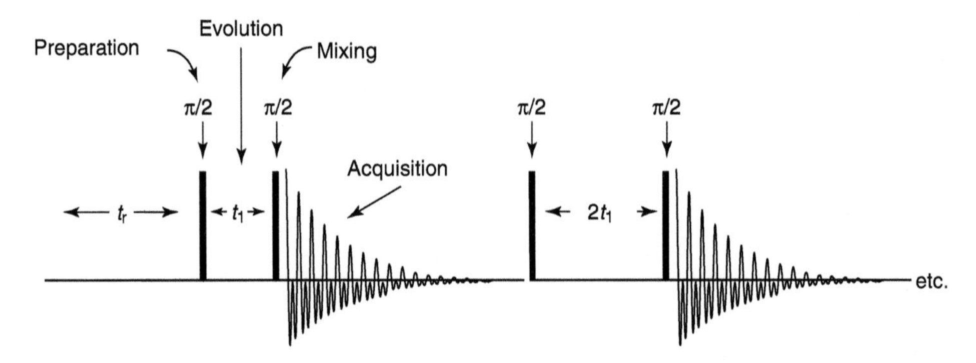

Fig. 5.1 The basic 2D pulse sequence

Figure 5.1 is not drawn to scale: the pulses are microseconds, the successive values of t_1 are multiples of a millisecond or two, and the acquisition phase is a second or two at each increment. To see how this can be used to create two dimensions of frequency, we follow what happens to the vectors in the rotating frame of reference (Sect. 4.14), as illustrated in Fig. 5.2.

The first π/2 pulse tips the vector representing the bulk magnetisation onto the y axis. The vectors then precess in the laboratory frame at their Larmor frequencies ν, and in the rotating frame at the frequencies ν away from whatever reference frequency we have chosen for the rotating frame. After the time t_1, they will have travelled to create angles given by $2\pi\nu t_1$. In Fig. 5.2, this is shown arbitrarily as a 60° clockwise rotation for one representative frequency. At this point the second π/2 pulse tips the component of the vector subtended onto the y axis, given by $\cos 2\pi\nu t_1$, onto the $-z$ axis, where it can make no contribution to the signal we detect. The intensity of the component left behind on the x axis, given by $\sin 2\pi\nu t_1$, is stored, and as the vector continues to precess in a clockwise direction, it is additionally recorded as an FID. The second pulse is given twice as long to precess before the second π/2 pulse, and all the vectors will have travelled twice as far as they did the first

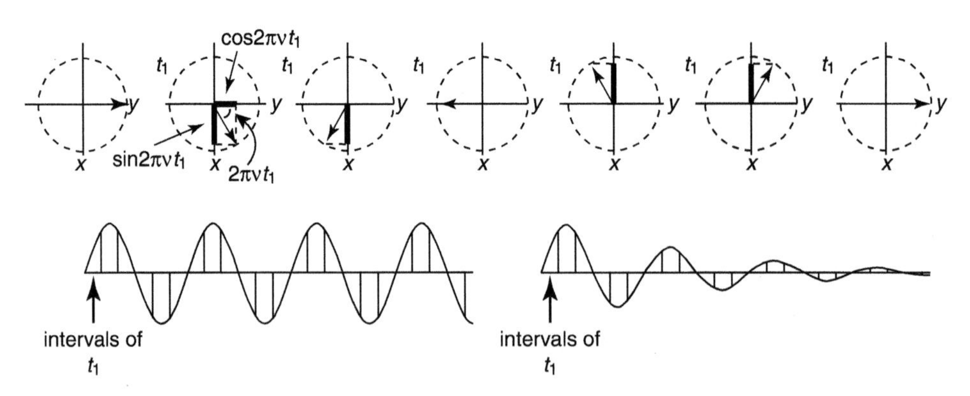

Fig. 5.2 The vectors in the rotating frame of reference for the basic 2D experiment

time. In the next iteration, they will have travelled three times as far, and so on. In Fig. 5.2, we see at the bottom left that each reading of the intensity on the x axis gradually creates a sine curve from each of the vectors, with points defined by the intensity after each successive interval of t_1. Add to this picture the fact that the signals are decaying, so that the curve produced is the decaying sine curve at the bottom right. The curves in Fig. 5.2 are those of a single frequency, but the sum of all the sine curves produced by all the vectors with all their different frequencies will in fact be stored as the intensity at the end of each t_1 increment, and all the frequencies will be extracted by Fourier transformation from the complex curve they define.

As in 1D-NMR, we need to know whether the frequencies in these decaying curves are higher or lower than the reference frequency, and the solution is to record the signal intensity on the x axis in one iteration and on the y axis in a second, alternating $\pi/2$ pulses for each t_1 unit, with the first on the x axis and the second, with the same value of t_1, on the y axis, with the second $\pi/2$ pulse alternating on the $+x$ and $-x$ axes. This creates the decaying sine curve in the first iteration, as illustrated in Fig. 5.2, and a decaying cosine curve in the second, as illustrated in Fig. 5.3. With both a sine and a cosine, we can determine whether the frequency is higher or lower than the reference frequency. The full sequence of phase cycling is more complicated than this, but this pair serves to demonstrate the idea. The next pair of pulses would use $2t_1$ intervals and the next $3t_1$, and so on.

At the end of the whole sequence, after many iterations, the decaying sine and cosine curves, defined by the successive values of t_1, will include the Larmor frequencies of all the spins in the sample. In addition, the FID collected at each increment records all the Larmor frequencies again, just as it does in 1D-NMR spectra with simultaneous readings

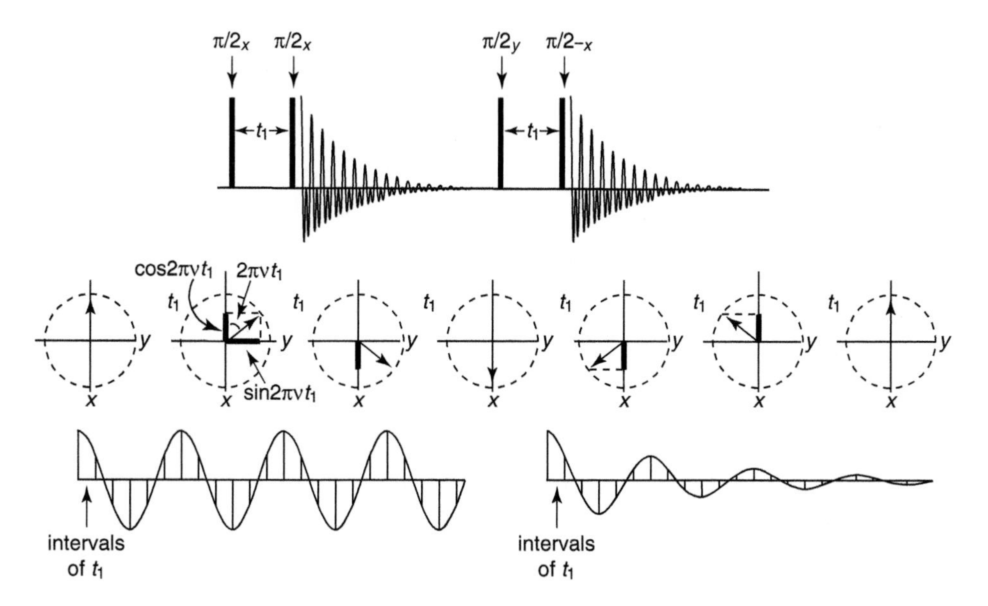

Fig. 5.3 Pulses and vectors identifying the sign of the frequency relative to a reference frequency

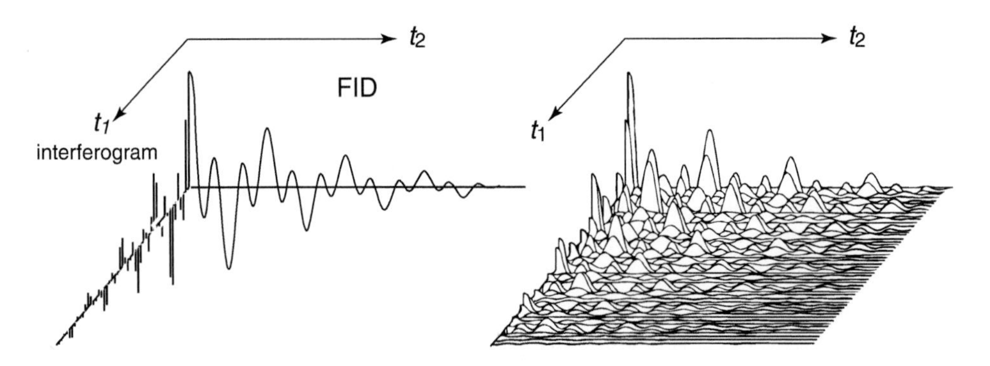

Fig. 5.4 Schematic illustration of a two dimensional surface with two frequencies

on the x and y axes. The FIDs are digitised, and the intervals in this time domain are called t_2.

Figure 5.4 illustrates on the left how the matrix of data points might look, showing only the first iteration in the t_2 dimension, for two frequencies from a pair of singlets with uncoupled spins with one frequency twice that of the other. The t_1 data are digital in nature and are called an interferogram. The t_2 values are digitised in the usual way for an FID with many data points, and are drawn as a smooth curve created from the sum of the two cosine curves. The intensity of the peaks in the interferogram on the t_1 axis follow the same curve with the same shape as that in the t_2 dimension, but it is visibly not as smooth or as well defined. Each FID takes seconds, and to acquire enough data points in the t_1 dimension to match the t_2 dimension would be intolerably time consuming. Typically the t_1 dimension might be defined by only a few hundred increments, compared to the thousands in the t_2 dimension. On top of all this, the whole experiment might be run a number of times, and the results added together to increase the signal-to-noise ratio, just as it is in 1D experiments.

On the right in Fig. 5.4 is a representation with the full set of FIDs added to that of the first iteration shown on the left; it creates a complex surface in which it is difficult but still possible to pick out the two frequencies in both dimensions, but easier to see on the projection onto the t_2 axis than onto the t_1.

In the next stage, illustrated in Fig. 5.5, two successive mathematical protocols similar to Fourier transformation extract the two frequencies from the time domain into the frequency domain. FT in the t_2 dimension creates a surface, illustrated schematically on the left in Fig. 5.5, in which the two frequencies are revealed as columns of peaks. Viewed from the front, we would see two Lorentzian peaks (with both positive and negative components), the higher frequency on the left and the lower on the right. Viewed from the side, each of the columns of peaks would be seen as decaying cosine curves. The FT in the t_1 dimension then extracts the same two frequencies, leading to two Lorentzian peaks, one at the bottom left identifying the higher frequency signal and the other at the top right identifying the lower frequency signal. The spectrum now looks the same whether viewed

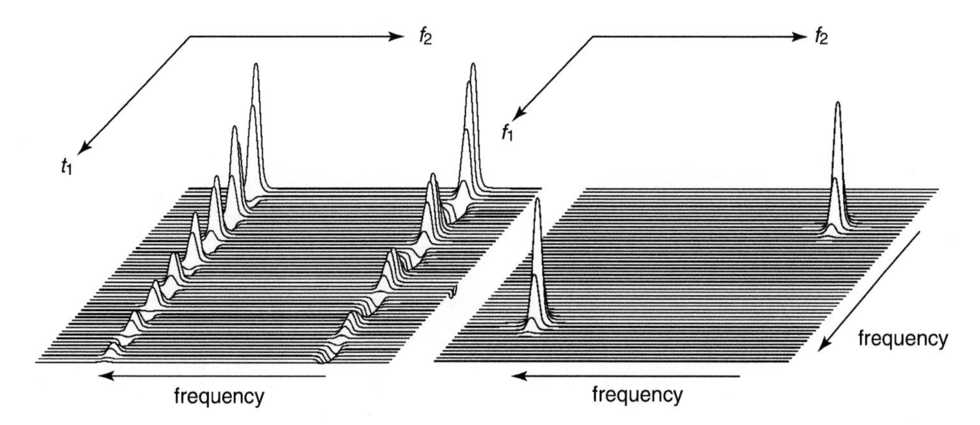

Fig. 5.5 Fourier transformation in two dimensions for a pair of uncoupled singlets

from the front or from the side, and the normal 1D spectrum also lies along the diagonal from the top right to the bottom left. The two frequency dimensions are given the labels f_1 and f_2, corresponding to t_1 and t_2, respectively. The former, derived from the successive readings at the end of each delay t_1, is called the *indirect dimension*, and the latter, derived in the usual way from the FIDs, is called the *direct dimension*.

5.2 COSY

5.2.1 Cross Peaks from Scalar Coupling

The pulse sequence in Fig. 5.3 is that of the most simple experiment—it creates a 2D spectrum known as a COSY spectrum (from COrrelation SpectroscopY). So far we have seen nothing new in the 2D spectrum with its two singlets. When spins interact by scalar coupling, useful information is added to the COSY spectrum. With two coupled spins, the t_1 and t_2 dimensions present a more complicated picture made from the sum of four decaying cosine curves. On the top left in Fig. 5.6, we can see a representation of the t_1 and t_2 data with four frequencies, modelled as having chemical shift frequencies of 10 and 20 Hz and a coupling constant between them of 4 Hz (ignoring the fact that such a system would be visibly roofing). The surface on the top right is more complicated than it was before, and the four frequencies can no longer be picked out by eye. FT in the t_2 dimension gives four columns of frequencies, illustrated at the bottom left of Fig. 5.6, corresponding to the four lines of the coupled AX system when viewed from the front. The difference from the picture in Fig. 5.5 is that each of these lines now *contains all four frequencies*, as revealed in the FT of the t_1 dimension, and illustrated in the bottom right of Fig. 5.6. These are grossly oversimplified pictures, to make the physics comprehensible at this stage of the explanation.

The 1D spectrum is still visible from the front, from the side, and along the diagonal, but the significant new feature is the presence of *cross peaks*, the group of four signals in

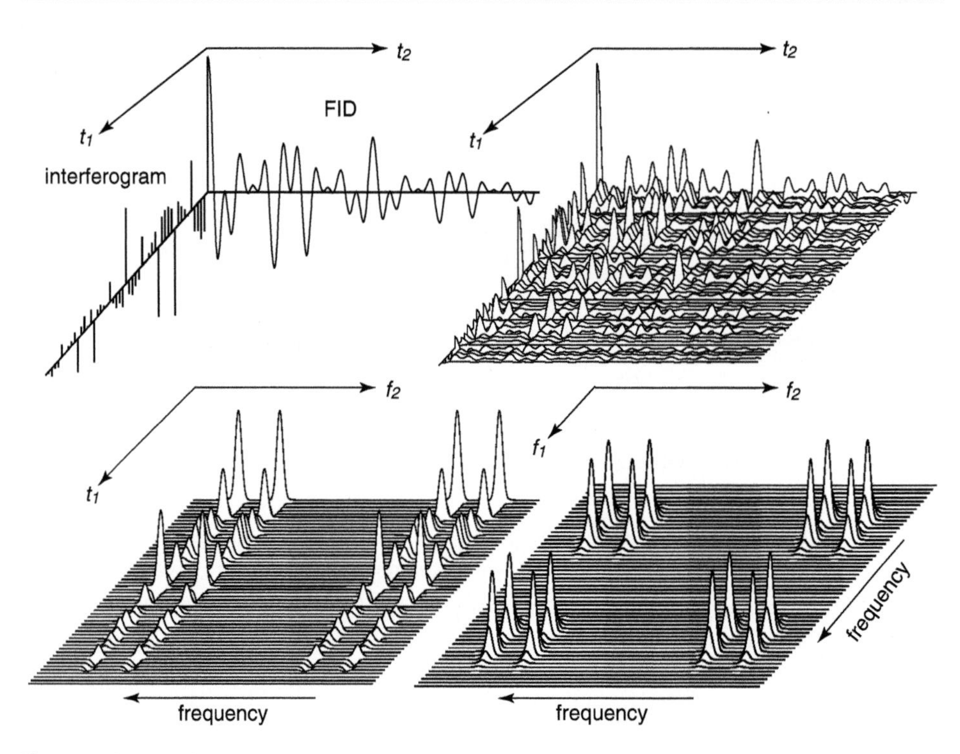

Fig. 5.6 Schematic representation of the surfaces for a coupled COSY spectrum

the top left and bottom right that establish that the A and X signals are coupled to each other.

Before we go on to see what the actual surface looks like, we need to learn why the two frequencies of one of the coupled partners are present in the columns of data for the other. To do that we must take a lengthy diversion, in order to understand a phenomenon called *polarisation transfer*. This is most simply understood with 1D spectra. We did not cover this topic in the previous chapter, because it is rarely applied in structure determination using only 1D spectra.

5.2.2 Polarisation Transfer

Staying with a coupled AX system, but looking only at a 1D spectrum (Fig. 5.7), the four lines are produced by the four transitions A_1, A_2, X_1 and X_2. The Boltzmann distribution of spins in excess of the bulk will be in the proportion four in the αα level, two each in the αβ and βα levels and zero in the ββ level. As we saw in the section on the nuclear Overhauser experiment (Sect. 4.13), the intensities of the signals generated by the A and X transitions are in proportion to the difference between the populations giving rise to them. Thus the straightforward AX system has each transition of equal intensity, because the difference in populations is two units for each.

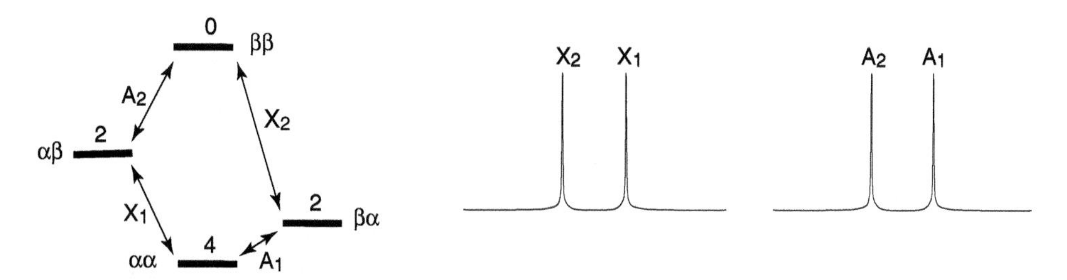

Fig. 5.7 Distribution of the excess spins for an unperturbed AX system

Polarisation transfer is achieved by irradiating precisely one signal of one of the doublets. If that irradiation is precisely at the frequency of, for example, the A_1 transition, and applied for just long enough to equalise the populations of the αα and βα levels, the populations will change to those shown in Fig. 5.8a, and the intensities of the four peaks will change from 1:1:1:1 to 1.5:0.5:1:0. In a pulse sequence this could be achieved by a soft π/2 pulse at the A_1 frequency. If instead the irradiation is continued for twice as long, or with twice as much power, or as a π pulse, the populations of the αα and βα levels will be inverted instead of equalised, and the four peaks will have the relative intensities 2:0:1:–1 shown in Fig. 5.8b.

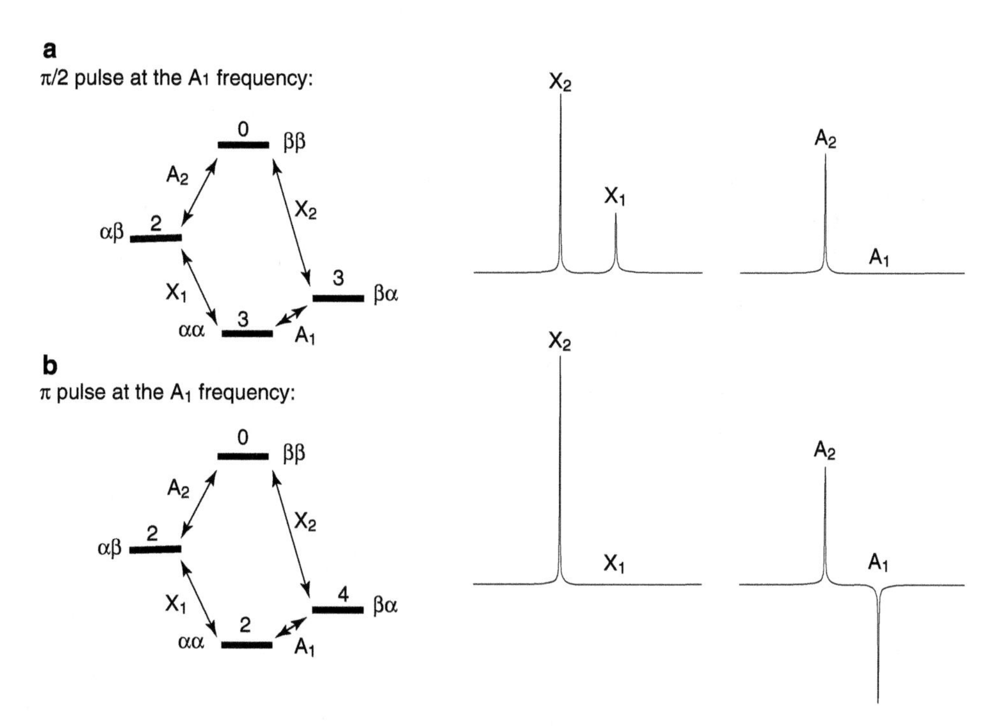

Fig. 5.8 Distribution of the spins in an AX system perturbed by irradiation at the A_1 frequency. (**a**) π/2 pulse at the A_1 frequency. (**b**) π pulse at the A_1 frequency

The overall intensity has not changed, but the individual lines have. These are called *selective polarisation transfer* and *selective polarisation inversion*, respectively. Either can be used to identify a coupling connection, because only the coupled spins would be changed. In practice it rarely is, because there are other ways to identify the connection, such as selective decoupling (Sect. 4.11.1) or examination of the COSY spectrum that we are working our way to understand.

The effect is greater if two different nuclei, such as ^{13}C and ^{1}H, are involved. The Boltzmann distribution is different as a consequence of the magnetogyric ratio of the proton being four times that of ^{13}C. The levels αα, αβ, βα and ββ have an excess population at equilibrium in the proportions of 10:2:8:0 (Fig. 5.9a). The proton peaks (the ^{13}C satellites in a ^{1}H spectrum) and the carbon peaks have intensities in the ratios of 4:4:1:1. A selective π/2 pulse at the frequency of the upfield ^{13}C satellite in the proton spectrum, labelled $^{1}H_1$, equivalent to the X_1 transition, equalizes the αα and αβ populations, and leads the carbon signals to have a 3:–1 ratio of intensities (not illustrated). A selective π pulse, illustrated in Fig. 5.9b, on the same upfield ^{13}C satellite, inverting the αα and αβ populations, leads the carbon signals to have a 5:–3 ratio of intensities.

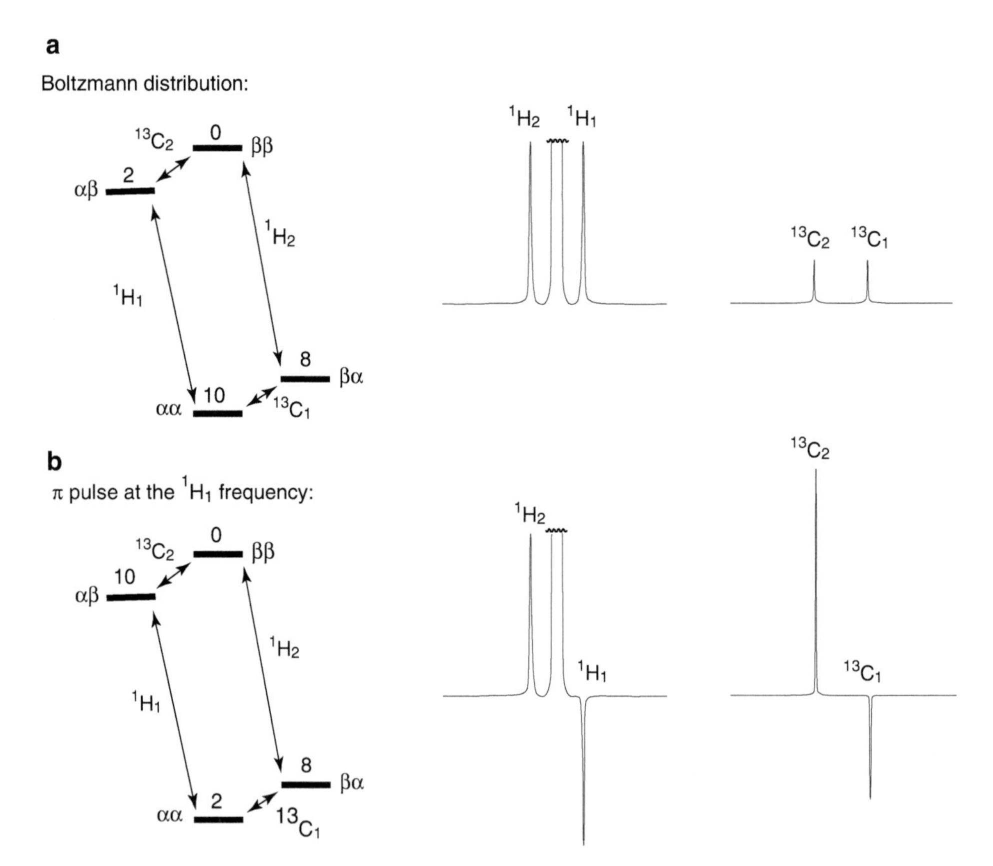

Fig. 5.9 Polarisation transfer in a C—H system. (**a**) Boltzmann distribution. (**b**) π pulse at the $^{1}H_1$ frequency

A more elaborate pulse sequence, called INEPT (Insensitive Nuclei Enhanced by Polarisation Transfer), based on the same principle, but made to apply simultaneously to all the carbons directly bonded to hydrogen atoms, can be used to intensify carbon signals by a factor of 8, somewhat better than that achieved by the NOE from proton decoupling. It involves spin echoes to refocus signals (Sect. 4.9.2), together with selective phase changes to invert the negative signals, and decoupling applied during the acquisition to combine them. The effect is even more pronounced with those nuclei having even smaller magnetogyric ratios, such as ^{29}Si and ^{15}N. These techniques are still used to enhance the intensity of the signals from these and other insensitive elements, effectively transferring the sensitivity of the protons to the signals of the less sensitive nuclei.

5.2.3 The Origin of Cross Peaks

In the COSY experiment, polarisation transfer has somehow been applied to all the signals simultaneously. To see, in outline only, how this is achieved, we return to the vectors of a doublet in the rotating frame of reference. Using the chemical shift of the doublet as the reference frequency, the vectors for the two lines rotate in opposite directions away from the reference frequency, as illustrated in Fig. 5.10, which uses the picture derived from a $\pi/2$ pulse on the x axis on the top line and a $\pi/2$ pulse on the y axis on the bottom line, giving the sine and cosine curves, respectively, to the accumulated readings in the t_1 dimension. After the first interval t_1, the vectors have magnitudes, projected onto the x axis in the top line and the y axis in the bottom, given by $\pm$-$\sin 2\pi \nu t_1$, where ν is $J/2$. At this point, the second $\pi/2$ pulse tips these components of the magnetisation onto the z and $-z$ axes. The component on the $-z$ axis is equivalent to giving *one* of the lines of the doublet, the low frequency line in this illustration, a π pulse. Since this happens to one line from each of the multiplets in the sample, it induces polarisation transfer to all of their coupling partners. In effect, since it continues in the same way at each successive interval of t_1, it codes into the successive readings of intensity in the t_1 dimension the frequencies of all the coupling partners. These in turn are therefore present in the columns of data after the first FT has created the f_2 dimension, and they give rise to the cross peaks. In this context, where all the

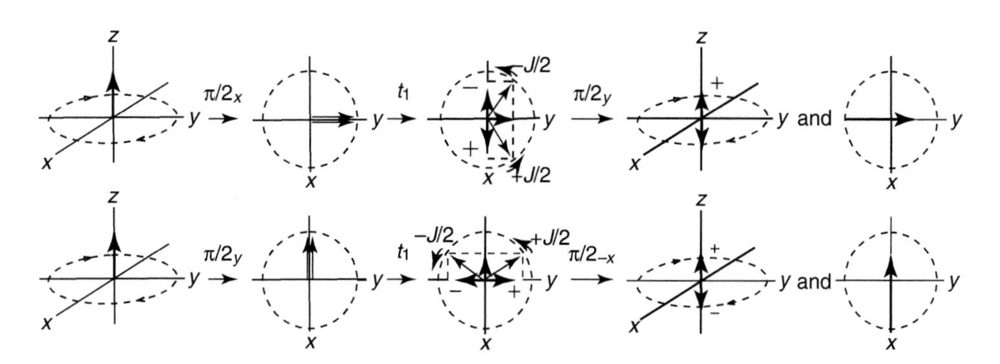

Fig. 5.10 The rotating frame of reference for a doublet after one interval of t_1

multiplets are affected, the phenomenon is called *coherence transfer*—an extension of the idea of polarisation transfer. The components of the vectors unaffected by the second $\pi/2$ pulse, given by $\cos 2\pi\nu t_1$, are left on the y axis in the top row and $-x$ axis in the bottom row, where they continue to precess. They are illustrated on the right in Fig. 5.10. They give rise to the interferogram and the FID, and, after FT, to the peaks on the diagonal.

Fourier transform, in more detail, gives two spectra, the real and the imaginary. Furthermore coherence transfer leads to negative as well as positive signals, as shown in the pictures of polarisation transfer in Figs. 5.8 and 5.9. The mountain view picture in Fig. 5.6 of the surface with all the peaks equally pointing upward is therefore a gross simplification of what is actually seen in a stacked plot. The effect of coherence transfer, together with the more detailed processing that is given to a COSY spectrum, leads to a complicated but orderly pattern of positive and negative signals, both to the signals on the diagonal and to the cross peaks. The cross peaks arise from out-of-phase vectors (labelled + and – in Fig. 5.10), which are processed to give absorption mode signals (Fig. 4.9), but anti-phase from one of the peaks of a doublet with respect to the other, just like the peaks A_1 and A_2 in the lower half of Fig. 5.8 illustrating the effect of a π pulse. The two vectors giving rise to the peaks on the diagonal are those on the right in Fig. 5.10, which are in-phase. They are processed to give dispersion mode signals (Fig. 4.9), which are broad, with positive and negative components, but in phase from one peak of the doublet to the next. In its raw state the simple COSY spectrum of a coupled AX system, processed in this way, is shown in Fig. 5.11, with the complicated, broad, diagonal peaks on the right and left, and the positive and negative, but sharp, cross peaks in the other two quadrants.

5.2.4 Displaying COSY Spectra

The aerial views of mountains and valleys are vivid and beautiful pictures, but inconvenient, and so we use a contour map instead, simultaneously taking slices of positive and negative peaks to reveal phase information, which is often colour-coded in the processing. It is important to choose levels that actually cut the peaks, but leave the noise on the floor. In a COSY spectrum, the two identical chemical shift axes are plotted orthogonally, with a 1D spectrum

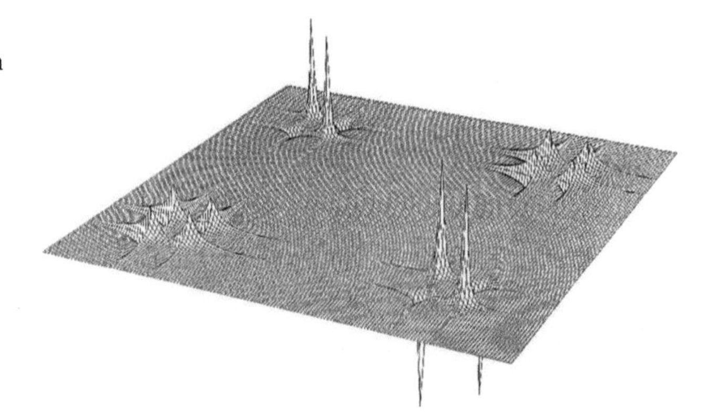

Fig. 5.11 The computed surface of the COSY spectrum of a coupled AX system. Reproduced with permission from http://www.ray-freeman.org/gallery-diagrams.html

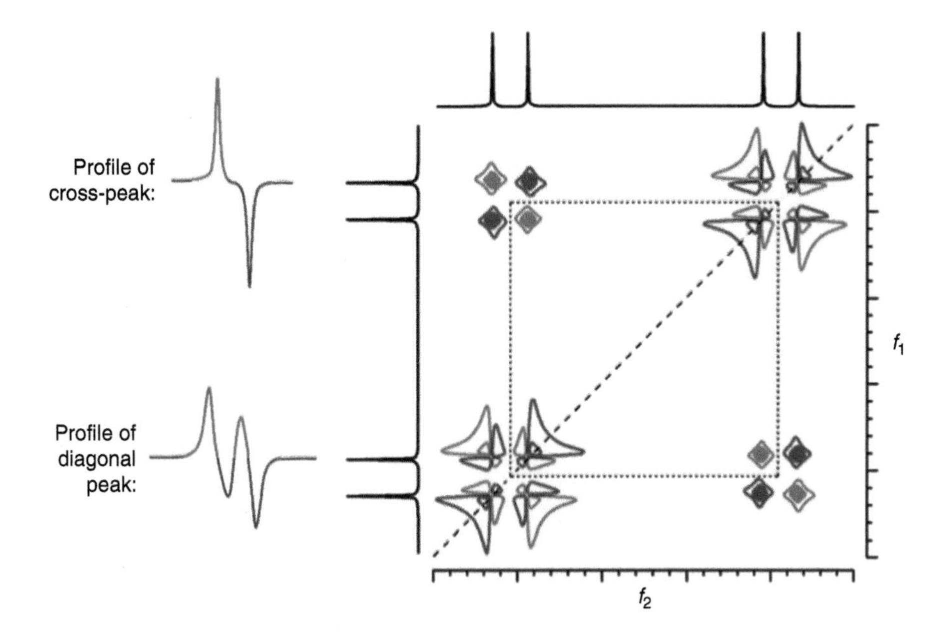

Fig. 5.12 Contour plot of the COSY spectrum of a coupled AX system

printed on each of the orthogonal chemical shift axes. The one-dimensional spectrum, viewed from above, also appears on the diagonal from the top right to the bottom left of the resulting square. In addition, all the peaks that are mutually spin-spin coupled are identified by cross peaks, which are symmetrically placed about the diagonal when both axes use the same scale for the chemical shift. All 2D spectra use this basic plan. The four-peak spectrum of an AX system would now look like Fig. 5.12, with the phases of the peaks coded in colour.

The diagonal with the coarsely dashed line gives the 1D spectrum, and the square with the finely dashed line leading into the cross peaks identifies that the two signals on the diagonal are from nuclei coupled to each other. This spectrum is highly idealised, and computed rather than taken using a spectrometer. It is time now to look at real COSY spectra and see how the cross peaks can be used to identify all the coupling connections in a molecule. The COSY experiment is the most often used 2D technique, routine in many laboratories today.

5.2.5 Interpreting COSY Spectra

To begin with a simple example, Fig. 5.13 shows the COSY spectrum of ibuprofen **1**. The spectrum on the left shows the projections of the 2D spectrum onto the top and side axes, which are noticeably less well resolved than the 1D spectra we are used to seeing, especially in the f_1 dimension. In the spectrum on the right, the spectra on the axes are the standard 1D spectra, taken separately, and transferred to the picture for the 2D spectrum. This is normal practice with 2D NMR spectra, because we can see more clearly which

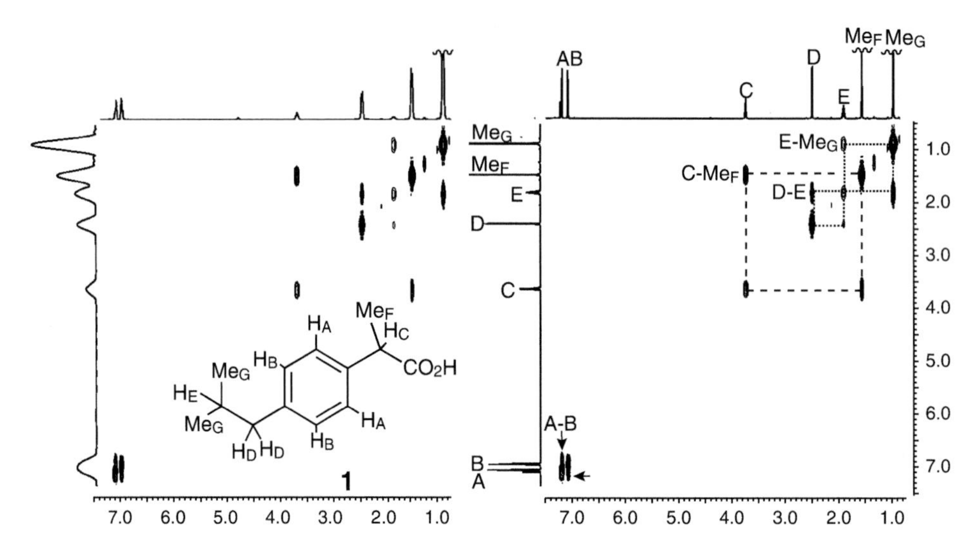

Fig. 5.13 600 MHz g-COSY spectrum, printed in magnitude mode, of ibuprofen **1** in $CDCl_3$

peaks are correlated. The contours of the diagonal and cross peaks in both spectra in Fig. 5.13 are identical, but they are not like their idealised versions in Fig. 5.12.

In the first place, all the signals have been processed to be positive in sign, with the intensity of both absorption mode and dispersion mode signals squared and the square root taken to give a spectrum called a *magnitude mode* spectrum. Secondly, there is t_1 noise, a string of extra peaks stretching in a line parallel to the f_1 dimension at the frequency of the strongest signals. This noise is largely caused by instabilities between one cycle ($\pi/2$-nt_1–$\pi/2$-FID) and the next [$\pi/2$-$(n + 1)t_1$–$\pi/2$-FID], which would actually be made worse by spinning the sample; the problem is minimised if everything (temperature, uniformity of the magnetic field, a solid floor, nothing magnetic moving near the spectrometer) is stable and the spectrometer in top condition. In the third place, COSY spectra are almost never run with the simple pulse sequence that would produce a spectrum looking like Fig. 5.12. There are several more elaborate pulse sequences in use, and the version used for the ibuprofen spectrum is called a gradient-COSY, abbreviated to g-COSY. It still uses the basic pulses to create the two dimensions, but it uses a gradient pulse with several advantages, the most important of which is that it takes less spectrometer time to run. We shall return later in Sect. 5.2.7 to learn how the g-COSY pulse sequence works, and how some of the other problems when taking 2D spectra are solved.

Looking at the spectrum on the right, the square defined by the coarsely dashed lines identifies that the proton H_C is coupled to the protons of the methyl group Me_F. The finely dashed squares identify that the protons H_D are coupled to the proton H_E, which in turn is coupled to the two methyl groups Me_G. And the cross peaks between H_A and H_B are just clear enough of the diagonal to identify the unexceptionable presence of coupling between the two pairs of protons on the benzene ring. It would have been possible to identify these connections using selective decoupling, or from a careful measurement of the coupling constants in a 1D spectrum. Equally, a set of TOCSY spectra would have separated the three spin systems, but the COSY spectrum reveals it all quickly and easily. COSY spectra have another advantage over 1D-decoupling. Irradiation of one of the signals in a 1D experiment inevitably hits other protons with similar chemical shifts at the same time, and we are then unable to tell which of them is involved in the coupling we are trying to identify. In COSY spectra the cross peaks can be lined up with the signals in the 1D spectrum on the axes with much greater precision.

Moving on to a more complicated example, the COSY spectrum of the α,β-unsaturated ester **2** is reproduced in Fig. 5.14. It is a DQF-COSY, with a different pulse sequence (discussed in Sect. 5.2.8), which removes the complicatedly broad peaks to be seen on the diagonal in simple COSY spectra. The ester **2** is not pure, as we can see from several peaks that are manifestly too small to be significant components of the mixture. We have to learn to ignore them, and the cross peaks they are apt to show, just as we ignore the peak at δ 7.26 from the benzene solvent.

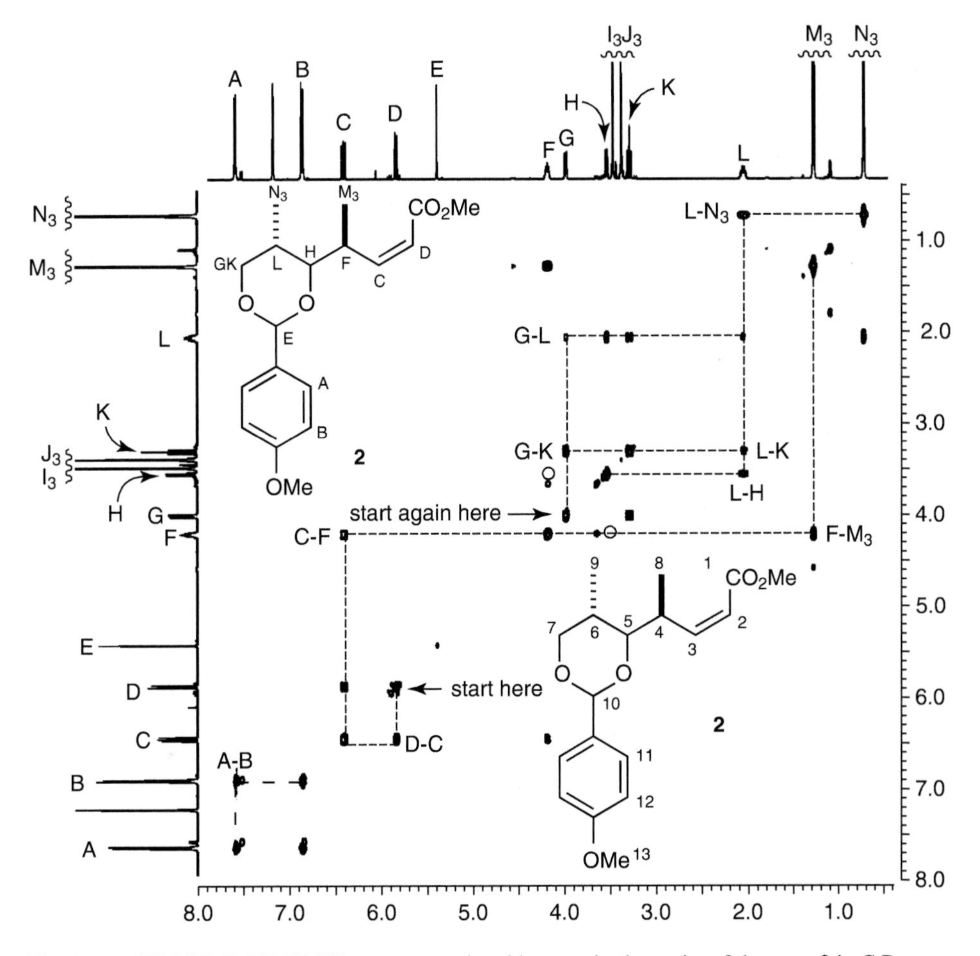

Fig. 5.14 500 MHz DQF-COSY spectrum, printed in magnitude mode, of the ester **2** in C_6D_6

This COSY spectrum establishes connections in three spin systems. The protons on the benzene ring giving rise to the signals A and B are coupled to each other. They have no other cross peaks, and are therefore not coupled to anything else. Starting on the diagonal with the signal labelled D, we can follow a chain of connections D-C-F-M_3. And in the third spin system, starting again on the diagonal with the signal labelled G, using the finely spaced dashed line, we can follow the chain of connections G-K&L, which branches from L to K, H and N_3. If now we look at the structure of this compound at the top left, we can see how the *p*-disubstituted benzene ring gives rise to the first set, A-B. The chain from C-2 to C-4 and the methyl group numbered 8 in the structure on the bottom right accounts for the second set D-C-F-M_3, and the sequence GK-L(N3)-H identifies the sequence from C-5 to the diastereotopic protons G and K on C-7, with the methyl group numbered 9 branching off from C-6. The methoxy groups are clearly responsible for the peaks labelled I_3 and J_3, with no cross peaks, and no coupling connections to anything, but not yet identified as to which is which. Finally the peak labelled E is from the methine proton on C-10, with no coupling connections either.

It is well to check what you can identify in the COSY spectrum with what the signals look like in the 1D spectrum, and to measure the coupling constants. In particular, this is needed in this case, because the connections in the COSY spectrum are interrupted by the absence of a cross peak between the signals for the protons F and H on C-4 and C-5. There ought to be cross peaks where the small circles are drawn in Fig. 5.14 between the signals labelled F and H, and there is nothing. (There is actually a weak signal near the circles, but it is easy to see that it is not directly on the lines that would link these two signals—it must be from an impurity.)

We can hope to find the missing coupling as an otherwise unaccounted for splitting between the 1D signals for these two protons. Figure 5.15 shows the multiplets for this compound, with the small coupling constant between the protons on C-4 and C-5 identified as the 2.3 Hz from H-5 (H) to H-4 (F), visible in the signal labelled H. It is low, because of the gauche dihedral angle between them and because H-4 is antiperiplanar to an oxygen atom, and a low coupling constant leads to an absent COSY cross peak. If we

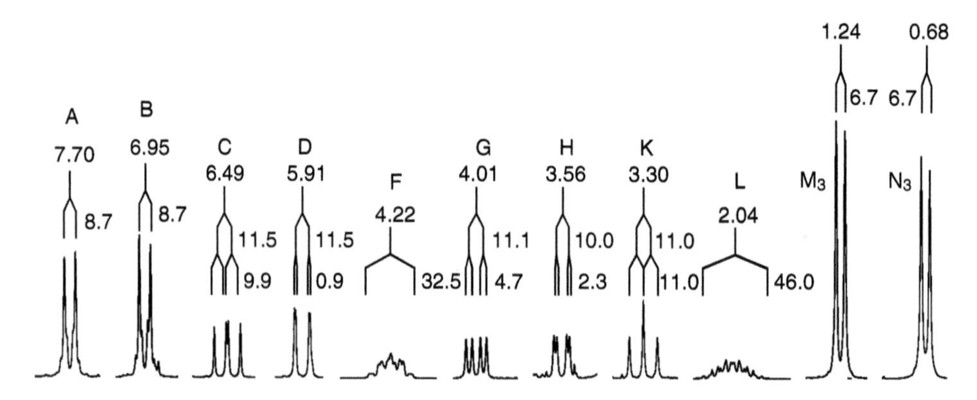

Fig. 5.15 The 1D multiplets A–D, F–H and K–N_3 of the ester **2**

had not known what the structure of this compound was, we would not know where to look in order to connect the two spin systems.

The signals are unambiguously assignable from the combination of the COSY spectrum and the 1D spectrum. Bearing in mind the Karplus equation, we can see that the coupling constants are consistent with the conformations shown in the drawings **3** and **4**. The methine multiplets F and L at δ 4.22 and 2.04, respectively, are not well enough resolved to pick out the coupling constants within them, but we can identify them with certainty from the COSY spectrum as the signals from H-4 and H-6, respectively. The sum of all the splittings shown on the drawing **3** for the multiplet L at δ 2.04 is 45.9 Hz, which is, as it should be, close to the total width of this signal. The sum of all the splittings identified on the drawing **4** for the multiplet F at δ 4.22 is 30.0 Hz, to which we should add the 2.3 Hz for the coupling between H-4 and H-5, and the allylic coupling of 0.9 Hz between the protons on C-2 and C-4, making the total 33.2 Hz close to the measured value of 32.5 Hz. The allylic coupling is just resolved in the signal D for the C-2 proton, but there is no cross peak for this either.

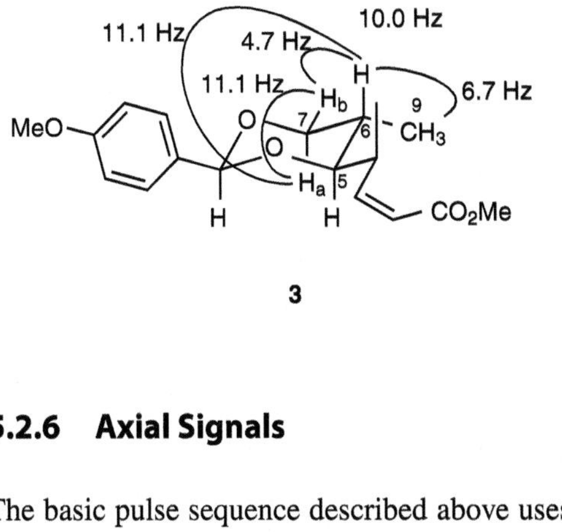

5.2.6 Axial Signals

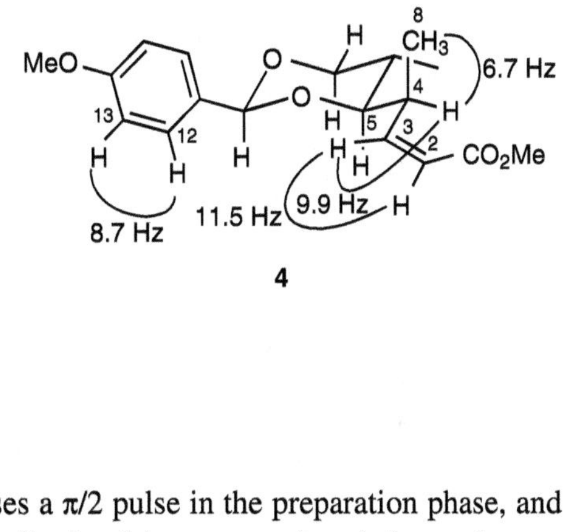

The basic pulse sequence described above uses a $\pi/2$ pulse in the preparation phase, and then another before taking a reading of the amplitude of the vector subtended onto the x or y axis at the end of each t_1 period. If the first $\pi/2$ pulse is not perfect, some magnetisation is left on the z axis, and the second $\pi/2$ pulse will tip the residual magnetisation onto the xy plane, where it will contribute to the FID. The residual magnetisation has no consequence in the most simple 1D experiments, because it stays on the z axis, but it does matter in multipulse sequences like the spin-echo (Figs. 4.73 and 4.74) and in all 2D experiments, because they all have a second $\pi/2$ pulse. The pulses we have been calling $\pi/2$ pulses are almost never precisely $\pi/2$, partly because it would be difficult to arrange for the intensity and time to be exactly right, and partly because it would take too long for the magnetisation to return to the Boltzmann distribution for an efficient use of time between pulses, especially in 2D experiments, which use a large number of them. In simple 1D experiments, the incomplete pulse is no problem, because there is always a component of the vector in the xy plane, and that is all that we need.

In a 2D experiment, the residual magnetisation on the z axis does not evolve during the times t_1, it makes no contribution to the spread of frequencies in the f_1 dimension, but does appear as a signal with zero frequency in the f_2 dimension—a horizontal row of signals, from each of the resonances, displayed parallel to the f_2 axis. These are called axial signals. Where the row appears is a matter of how the signals are processed, but one presentation puts the row in the precise centre of the spectral width. The residual vector on the z axis poses a problem with any multi-pulse experiment using incomplete $\pi/2$ pulses followed by another $\pi/2$ pulse. There are several solutions, one of which alternates the phase of the receiving signal between one cycle and the next, and then subtracts the two results, as in Fig. 4.74. It is also incorporated into the phase cycling we saw earlier in Fig. 4.93. Since the vector giving rise to the axial peaks doesn't move, subtraction cancels it, but the important signals, subtracted but having been received with opposite phases, effectively add to each other.

5.2.7 Gradient Pulses

Another method that cancels the axial peaks has more profound uses, and is incorporated into the pulse sequences of many 2D experiments, including the g-COSY spectrum that we have already seen. The g stands for gradient, which is a gradient in the magnetic field created by a pulse B_g applied on the z axis, using a dedicated built-in coil reaching from below the sample to above it for a fixed time t_g. This creates a gradient in the magnetic field, uniformly stronger at the top (say) than the bottom during the time t_g. The vectors at the top in the stronger field precess further, and those at the bottom less far, eventually leading all the vectors to be out of phase with each other, and the net magnetisation to drop to zero. This appears to be deeply counterproductive, but if the gradient in the magnetic field is reversed for the same period t_g by a pulse with the opposite polarity $-B_g$, all the spins return to being in phase. Those in the region of highest field will have precessed most in the first stage, and equally the most in the second, but in the opposite direction, back to where they started from. The pulse sequence and the vectors are illustrated in Fig. 5.16.

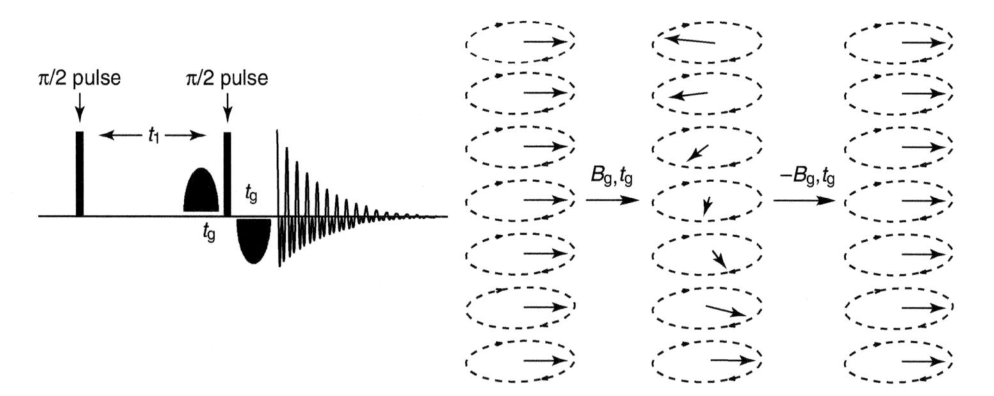

Fig. 5.16 Pulse sequence and vectors for the basic g-COSY experiment

In the first segment, before the second π/2 pulse, the vectors precess in the usual way during t_1. The gradient field takes them out of phase with each other in an orderly way during t_g. Then, after the second π/2 pulse, the same vectors are refocused. All other spins continue to precess out of phase with each other, including those which had been in the residual magnetisation on the *z* axis, which only come into the *xy* plane with the second π/2 pulse. This is an astonishingly effective device for selecting only the spins of interest, which are the only ones brought into phase with each other before the acquisition. It is efficient in spectrometer and computer time, and removes the need for, and the time consumed by, both phase cycling and the separate pulses for axial peak suppression. It is also much more complicated in detail, but we can leave it here. All the times involved are so short that diffusion plays little part in dephasing the signals. g-COSY spectra are relatively strong, and take a relatively short time to run, typically about 20 min. They are often processed without phase information coded into the cross peaks, which are narrow in the f_2 dimension, but broad in the f_1 dimension. A simple g-COSY spectrum is often the first to be taken, and the more time-consuming COSY spectra are taken only when there is a need for the greater detail they provide.

5.2.8 DQF-COSY

Greater detail is needed when two (or more) signals have very similar chemical shifts and we want to know if the protons are coupled to each other. The peaks on the diagonal in simple g-COSY spectra are so strong that they obscure cross peaks close to the diagonal. The commonly used COSY experiment to solve this problem is called DQF-COSY (Double Quantum Filtered-COSY). Figure 5.17 shows the basic pulse sequence, which includes two spin-echoes, for each increment of t_1, a third π/2 pulse, and the gradient filter. An alternative uses a sequence of phase cycling in place of the gradient filter but takes longer to run. This is too complicated to explain, and the explanations using a rotating frame and vectors are inadequate anyway. Double quantum transitions are those between energy levels that have two spins inverting (between the ββ and the αα levels, for example). This has not complicated the story so far, because the coherence they induced was lost, having been left as vectors pointing diagonally opposite to each other, cancelling each other out as they precess in the *xy* plane during the acquisition phase, and they are inherently unobservable when they are on the *z* axis.

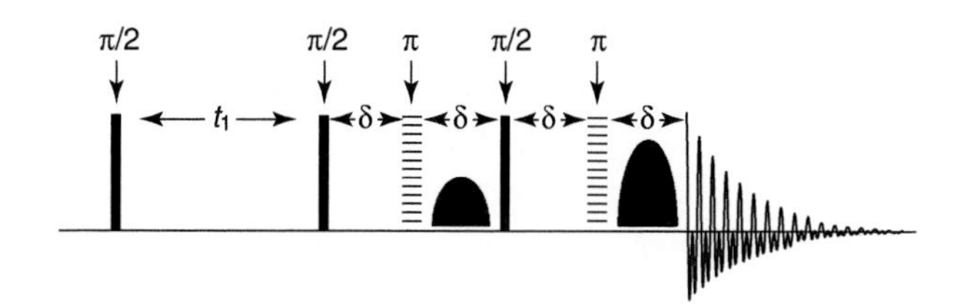

Fig. 5.17 The basic pulse sequence for a DQF-COSY spectrum

The sequence in the diagram with the delays 2δ provides the spin-echoes, and the third $\pi/2$ pulse in between them brings the double quantum coherences back as single-quantum coherences into the *xy* plane, where they can be detected. The gradient filter removes everything that did not involve the double quantum coherence. The result is that the dispersion mode peaks that were in phase on the diagonal in the standard COSY experiment are now in absorption mode and out of phase, and therefore largely cancel each other out. Singlets, which cannot participate in double quantum coherence (and in any case are of no interest in COSY experiments) disappear (or almost disappear) from the diagonal, along with the t_1 noise often associated with them. Figure 5.14 shows a DQF-COSY, where we can see the exceedingly weak signals on the diagonal for the proton labelled E and for the protons of the methoxy groups. Some of the signal intensity is lost in a DQF-COSY spectrum, but it is so much cleaner and better resolved, that it is worth the sacrifice when cross peak detail is needed, and especially when coupling between signals near the diagonal is important. Figures 5.13 and 5.14 give some idea of the difference in appearance between a simple g-COSY and a DQF-COSY spectrum.

5.2.9 Phase Structure in COSY Spectra

DQF-COSY spectra are often processed to leave the phase structure in the cross peaks, giving spectra said to be in *phase-sensitive mode*. The fine structure in the cross peaks when the spectrum is processed in the phase-sensitive mode can be useful, but the relative intensities and the phases of cross peak multiplets are not straightforwardly easy to recognise. In the cross peaks, a doublet is produced with two antiphase lines, as in the schematic picture in Fig. 5.12. A triplet appears as two antiphase lines separated by a null, and a quartet by four lines, the higher frequency in phase in one sense and the lower frequency pair in phase in the opposite sense. This phase structure appears in the cross peaks for the coupling identified in that cross peak, called the active coupling. Any other coupling visible in a cross peak but not involving the protons connected by that cross peak is called passive coupling. In passive coupling the lines appear in phase with each other. Analysis of the phases in cross peaks can be useful for identifying coupling relationships and for measuring coupling constants. Figure 5.18 is an example, showing a detail of the DQF-COSY spectrum, processed in phase-sensitive mode, of a fragment **5** of quinine, in which the positive peaks have red contours and the negative peaks blue.

The cross peak between H_A and H_C (at the top left) has peaks, reading from left to right, in the order red-blue-red-blue, indicating that the first pair (and the second pair) are the active couplings, evidently having the smaller of the two coupling constants in the double-doublet. The cross peak between H_A and H_B in contrast has peaks in the order red-red-blue-blue, indicating that the first and third peaks (and the second and fourth) are the active coupling, with the larger of the two coupling constants. Although it seems likely that this is correct from their general appearance in the 1D spectrum, it is not possible to be sure that this is the right way round just from looking at the signals in the 1D spectrum.

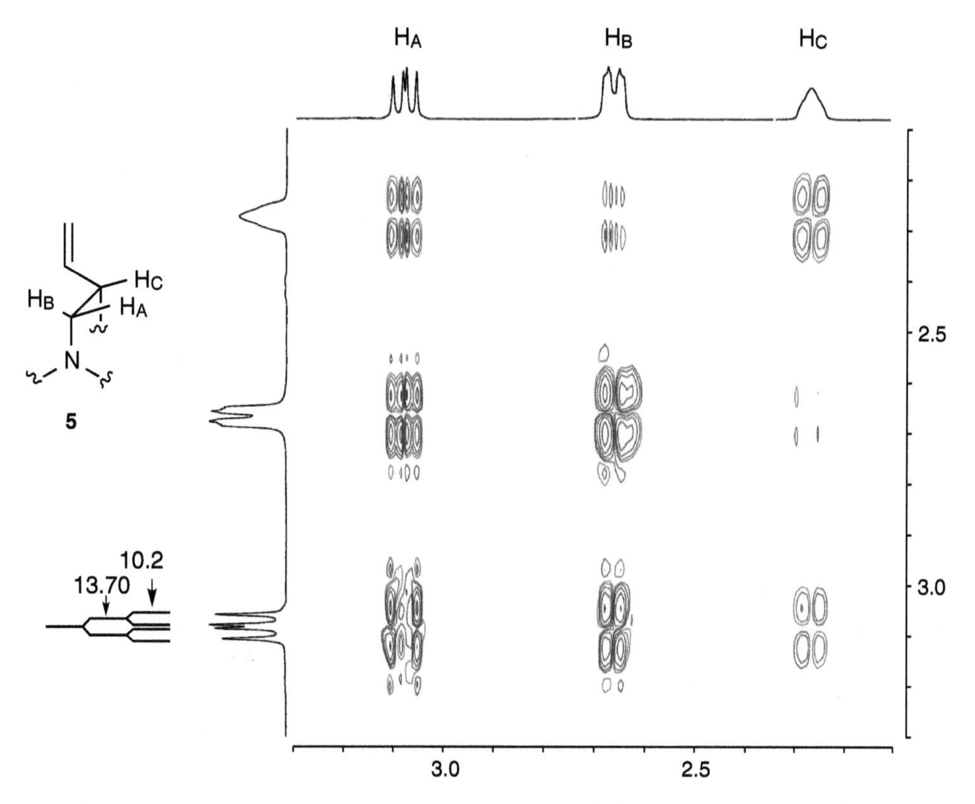

Fig. 5.18 Detail from a 500 MHz DQF-COSY spectrum of quinine, printed in phase-sensitive mode. (Irrelevant peaks in the full quinine spectrum have been removed for clarity)

The coupling constants can be measured on the cross peaks (10 and 16 Hz), but they are not particularly accurate (the lines in the 1D spectrum for H_A give more reliable data: 10.2 and 13.7 Hz). On the other hand the coupling constant between H_B and H_C is not measurable at all on the 1D spectrum, but it can be estimated to be about 6.5 Hz from the space between the red and blue contours on the weak cross peak in the middle of the top line.

For all the power of COSY spectra, they do have two limitations. When the coupling constant is small, the active cross peaks start to overlap causing them to cancel because they are anti-phase. Cross peaks are intense for the larger coupling constants, but often weak or invisible for small coupling constants, as we already saw in Fig. 5.15, where the connection between the protons on C-4 and C-5 did not give rise to a COSY cross peak. Secondly, bear in mind that COSY spectra are not always better than 1D difference decoupling, since difference decoupling reveals the residual multiplicity in a multiplet, which COSY does not, unless you have available and can analyse the phase structure within the cross peaks.

Although magnitude mode g-COSY and either magnitude mode or phase-sensitive mode DQF-COSY, with several variants within these designations, are the most common procedures used in structure determination, other refinements also have their uses. COSY-β, also called COSY-45 (with the normal COSY called COSY-90), has cross peaks with a different appearance if they are geminal rather than vicinal. This follows from the former usually having a negative J value and the latter a positive J. Another called Delayed-COSY, changing the pulse sequence to include a delay, which needs to be of the order of $1/2J$, makes long range coupling more easily detected than in a normal COSY. This can make the picture complicated, but if a normal COSY is run first it is clear what the Delayed-COSY has added.

5.3 2D-TOCSY

We saw in Sect. 4.12 how a 1D-TOCSY spectrum identified the components of a spin system, and separated them from all other signals, both from within the molecule and from impurities. In a larger molecule with a larger number of separate spin systems, it can be tiresome, although frequently worth the effort, to go through all the possible connections, irradiating each of the signals in turn to collect a 1D spectrum for each. It would also be difficult to irradiate only one signal if a large number were close together in chemical shift. For larger molecules, a 2D version of TOCSY solves these problems, and reveals all the spin systems in one experiment. One of the possible pulse sequences used is shown in Fig. 5.19, using a phased sequence of pulses called DIPSI-2 between the two $\pi/2$ pulses that follow the t_1 period. They give the z magnetisation time to mix all the couplings within a spin system, while holding all the chemical shifts steady before the final pulse puts them into the xy plane. By holding the chemical shift steady, but allowing the coupling to develop, they make all the couplings strong, since the chemical shifts are essentially all equal during the DIPSI-2 sequence. This makes all the transfers of coupling from spin to spin efficient. The δ delays are there to give the pulses in DIPSI-2 time to develop and decay. A gradient pulse pair replaces the other complications like phase cycling that could otherwise be used.

The 2D-TOCSY spectrum looks just like a COSY spectrum, but shows all the coupling connections in a spin system instead of only those with substantial coupling constants. The 2D-TOCSY spectrum in Fig. 5.20 is the high-field part of the spectrum of naloxone **6**.

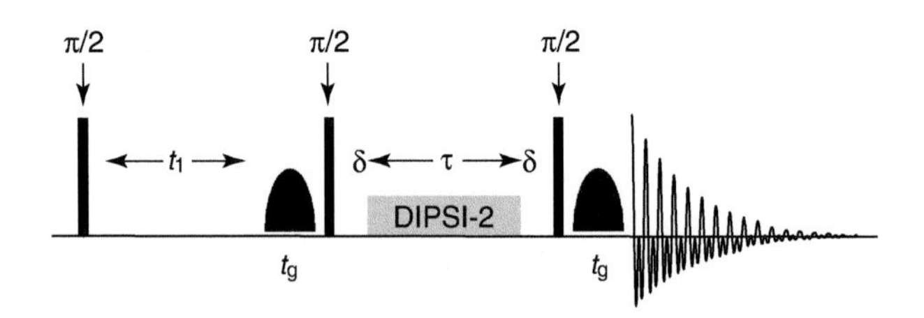

Fig. 5.19 A pulse sequence for a 2D-TOCSY experiment

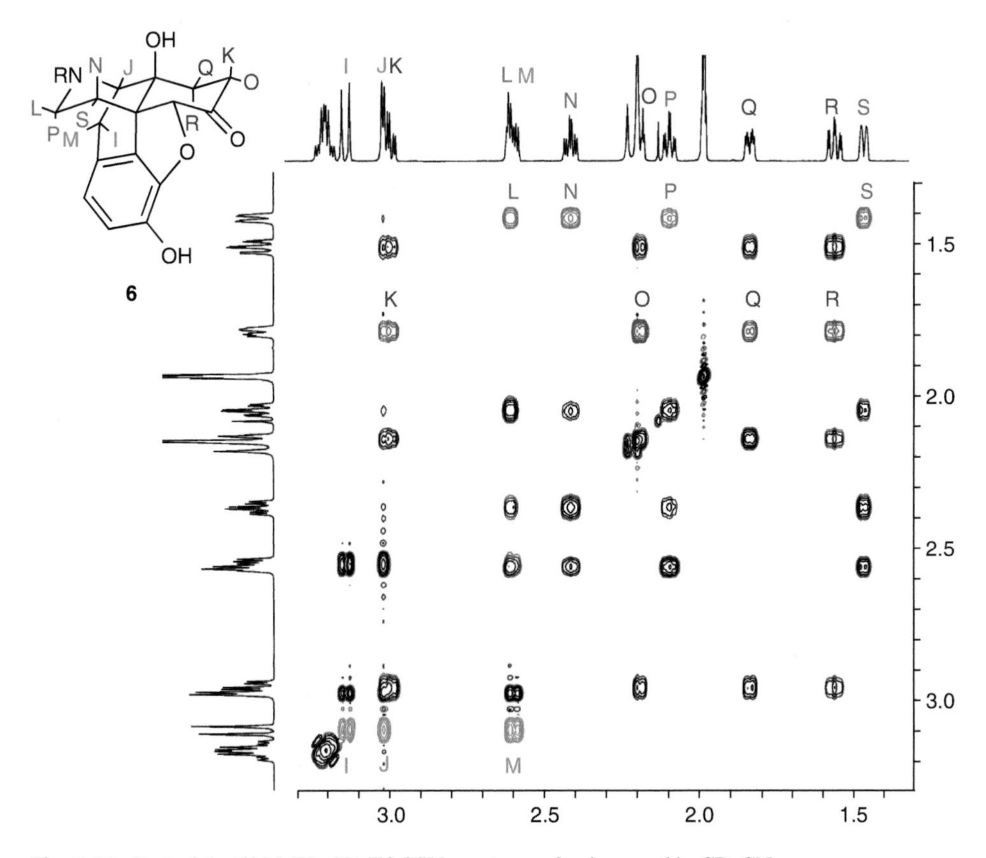

Fig. 5.20 Part of the 500 MHz 2D-TOCSY spectrum of naloxone **6** in CD_3CN

As usual the normal spectrum is on the diagonal, and cross peaks are found for protons that are part of the same spin system. Naloxone was the subject of the NOE-difference experiment in Chap. 4 (Fig. 4.86), which was used to separate the overlapping signals of the pair of protons labelled J and K, and of the pair of protons labelled L and M. In the 2D-TOCSY spectrum in Fig. 5.20 we can see in the set of red peaks at the top that the proton labelled L is in the same spin system as the protons labelled N, P and S. In the blue set of peaks below them, we can see that the proton labelled K is in the same spin system as the protons labelled O, Q and R, and finally in the green set at the bottom, we see that the proton labelled J is in the same spin system as the protons I and M. The cross peaks for the signals J and K have clearly different structures, as do the cross peaks for the signals L and M, making it possible to identify which signal of the overlapping pair belongs in which spin system. A COSY spectrum for this molecule might have revealed much the same information, but some of the low coupling constants might have hidden some of the connections. In addition, the TOCSY spectrum identifies that each of the groups of signals is self-contained, and there are no other protons in any of these three spin systems.

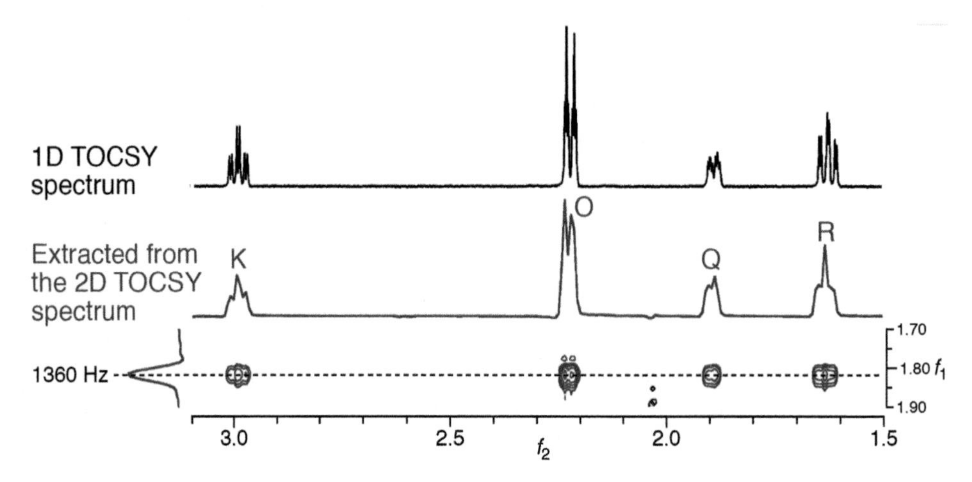

Fig. 5.21 1D spectrum extracted from a 500 MHz 2D-TOCSY spectrum of naloxone **6** in CD_3CN

It is possible to extract a 1D-TOCSY signal from a 2D-TOCSY spectrum using an NMR processing program. Choosing the appropriate tool in the computer program and placing it on a signal in the ill-resolved f_1 axis of a 2D-TOCSY spectrum brings up the 1D-spectrum of the spin system on the f_2 axis. The act of selecting dismisses the 1D spectra from the axes, and replaces them with the projection of the spectra on the f_1 and f_2 axes at the site of selection. Both projections are less well resolved than the 1D spectra usually placed on the axes, as we saw in Fig. 5.13 for a COSY spectrum, but the projection on the f_2 axis is often good enough to give us useful information, although rarely as good as the same spectrum in a 1D-TOCSY experiment. Figure 5.21 shows the result of extracting the blue KOQR line of signals in the naloxone spectrum, together with the spectrum of the same spin system taken as a 1D-TOCSY spectrum. The signals are just well enough resolved in the 2D extract, coloured blue, to recognise the larger couplings as triplet, doublet, doublet and triplet, but the finer doublings visible in the 1D spectrum, in black, are not resolved in the extract.

The NH protons in amides, unlike the NH protons of amines, frequently show coupling to the adjacent α-CH, and therefore appear as part of the spin system of each component amino acid. The NH signals are often well spaced, because many of them indulge in more or less hydrogen bonding, which spreads them out on the chemical shift axis. They can often be identified from the TOCSY spectra, first by taking a 2D-TOCSY to get the whole picture, and then by carrying out selective 1D-TOCSY experiments, like those we saw earlier, irradiating at the frequency of the signal from each NH proton, to reveal the coupling pattern within the individual amino acids. Peptides and proteins are water soluble, and it is necessary to remove most or all of the signal from the water. It would not help to use D_2O, because this would exchange with the NH protons that are so useful. The unwanted resonance of H_2O can be largely suppressed by presaturation of its resonance before the acquisition of each TOCSY spectrum.

5.4 NOESY

A two-dimensional spectrum which records all the proton-proton NOEs occurring in a molecule in a single experiment is called a NOESY spectrum. Superficially, it looks like a COSY spectrum—the normal spectrum appears on the diagonal and both orthogonal axes usually have the ^{1}H NMR spectrum printed on them. However, it differs crucially in that the cross peaks now indicate those protons that are close in space, just as 1D-NOE difference experiments detect through-space interactions. Thus, a NOESY spectrum provides information about the geometry of molecules.

The pulse sequence is another one too complicated to explain properly, but in outline it looks like that in Fig. 5.22. It begins the same as the COSY, with a sequence of regularly increasing t_1 times to create the second dimension, but this is followed by a fixed time τ before the acquisition. As with COSY, the second of the $\pi/2$ pulses tipped some of the magnetisation onto the $-z$ axis, which we abandoned in the COSY experiment. In the COSY experiment we only processed the magnetisation that was left in the xy plane, accepting the influence of the z magnetisation for the coherence that gave us the cross peaks. As in the COSY experiment, the magnetisation in the z axis corresponds to an inversion of the population with more spins in the β than in the α state. The population inversion gives rise to an NOE, which builds during the delay τ. In the NOESY sequence the vector on the $-z$ axis is tipped back to the xy plane by the third $\pi/2$ pulse and recorded. In the meantime, the COSY peaks, and everything else in the xy plane, have been dephased by the pulsed field gradient (PFG) making phase cycling unnecessary.

The positive NOEs from small molecules (ca. 100–400 Da) are rather weak, and spatial connections are often more reliably identified by NOE difference spectroscopy. The NOESY spectrum in Fig. 5.23, which has been taken using the same α,β-unsaturated ester **2** as in the DQF-COSY spectrum in Fig. 5.14, shows that the diagonal peaks are positive

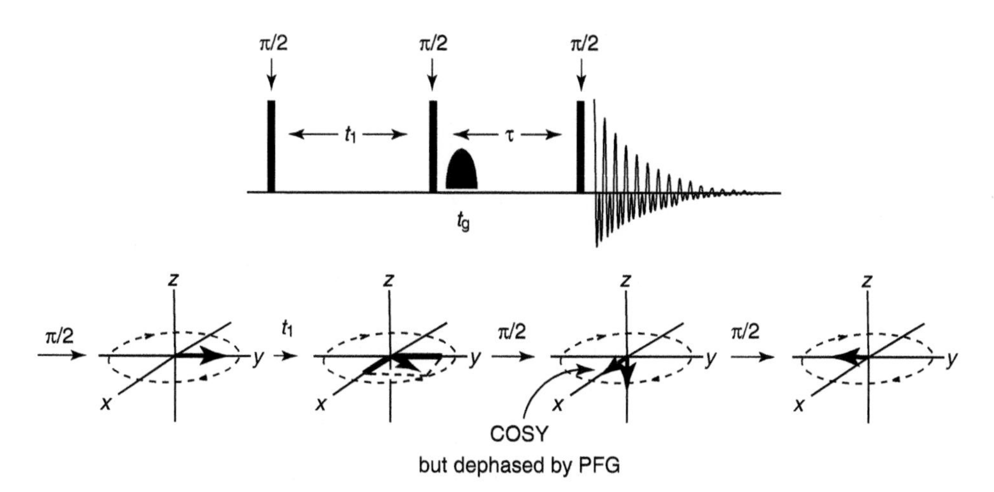

Fig. 5.22 Pulse sequence for a NOESY spectrum

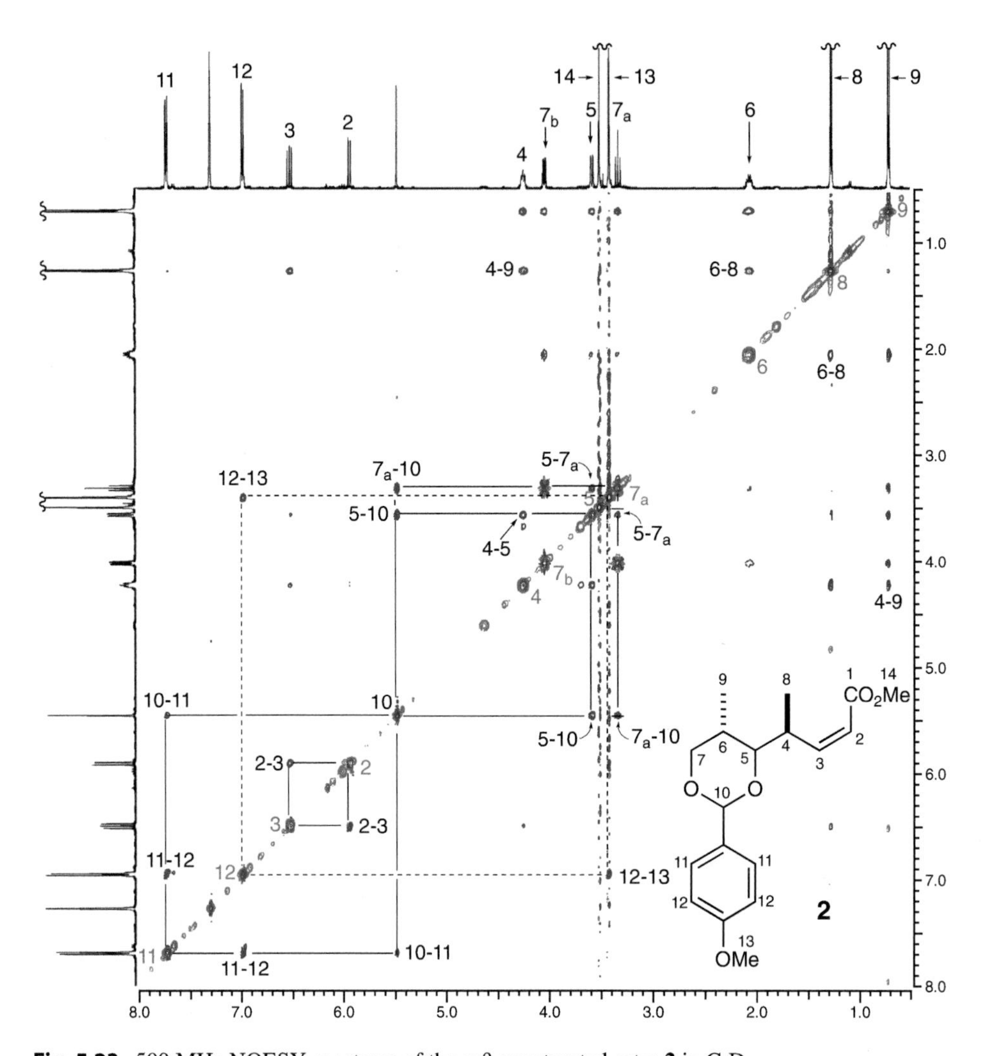

Fig. 5.23 500 MHz NOESY spectrum of the α,β-unsaturated ester **2** in C_6D_6

and the cross peaks negative, revealed by the different colours. It is exceptionally clean, containing relatively little 'noise,' which can be identified whenever there is no corresponding peak in the 1D spectrum, and/or when the cross peak is not accompanied by a partner which is its reflection point about the diagonal running from top right to bottom left. For example, the signals on the diagonal at 4.4 and 4.7, among others, and the t_1 noise below the intense methyl signals appearing at δ 0.70, 1.25, 3.38 and 3.50, can easily be ignored in the interpretation.

The NOESY spectrum allows us to confirm all the stereochemistry, especially that at C-10, which was not established by the COSY spectrum. All protons showing geminal, and most showing vicinal coupling, also spend time close enough in space to give cross peaks to each other in the NOESY spectrum. It is the *additional* cross peaks, labelled in Fig. 5.23, that give us new information, including the cross peak from 4 to 5, which is strong in the NOESY, but was absent in the COSY because of the small coupling constant. Furthermore, the cross peaks labelled 5–10, 7_α–10 and 5–7_α show that these three protons are close in space because they are all axial. The 4–9 and 6–8 cross peaks are strong, the 3–4 cross peaks are weak, but there are no 4–6 cross peaks, indicating that the molecule spends most of its time in the conformation **7**, which illustrates the NOEs discussed so far.

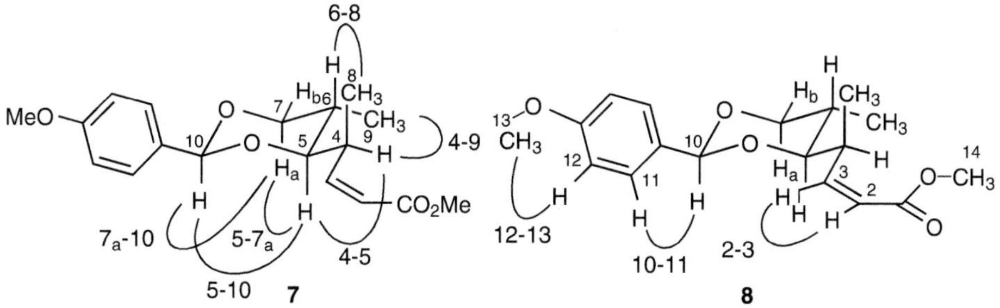

7

8

The other informative NOEs allow us to assign the methoxy peaks at δ 3.50 and 3.38 to the ester and ether groups, respectively **8**, because of the presence of the cross peaks 12–13 connected by the dashed lines in Fig. 5.23. The ester methyl group will be oriented away from the rest of the molecule in the lower energy s-*trans* conformation about the carbonyl-oxygen single bond, and does not show any NOE. There is little doubt about the assignment of the two aromatic protons, since their chemical shift values are definitive, but the assignment is supported by the 10–11 cross peak **8**. Among the cross peaks that are informative, even though they can also be seen in the COSY spectrum, is the cross peak 2–3, which confirms that the olefinic protons are *cis*-related.

Nuclear Overhauser effects are inherently weak for molecules with molecular weights between about 600 and 1500. This problem is overcome to some extent with a different pulse sequence called ROESY (Rotating frame Overhauser Enhancement SpectroscopY). The pulse sequence (Fig. 5.24) is closely similar to the sequence for a NOESY spectrum (Fig. 5.22), but the magnetisation is held after the second $\pi/2$ pulse to give the NOE time to develop by a continuous irradiation on the y axis, called a spin lock (because it keeps the magnetisation there). The field induced by the spin lock B_1 is so much weaker than the B_0 field that the tumbling frequency that will match it is brought down from megahertz to kilohertz. The FID gives all the signals of the cross peaks positive in sign and all the diagonal peaks negative. It has not depended upon a high tumbling rate, so it works no matter what the molecular weight. ROESY spectra look similar to, and are interpreted in the same way, as NOESY spectra.

NOESY and ROESY cross peaks are seen when the distance apart of the irradiated and receptive nuclei is small, governed by the same r^{-6} constraint that 1D-NOE suffers. A useful guide for larger molecules is: large cross peaks, r = 2.0–2.5 Å; medium-sized cross peaks, r = 2.0–3.0 Å; small cross peaks, r = 2.0–5.0 Å.

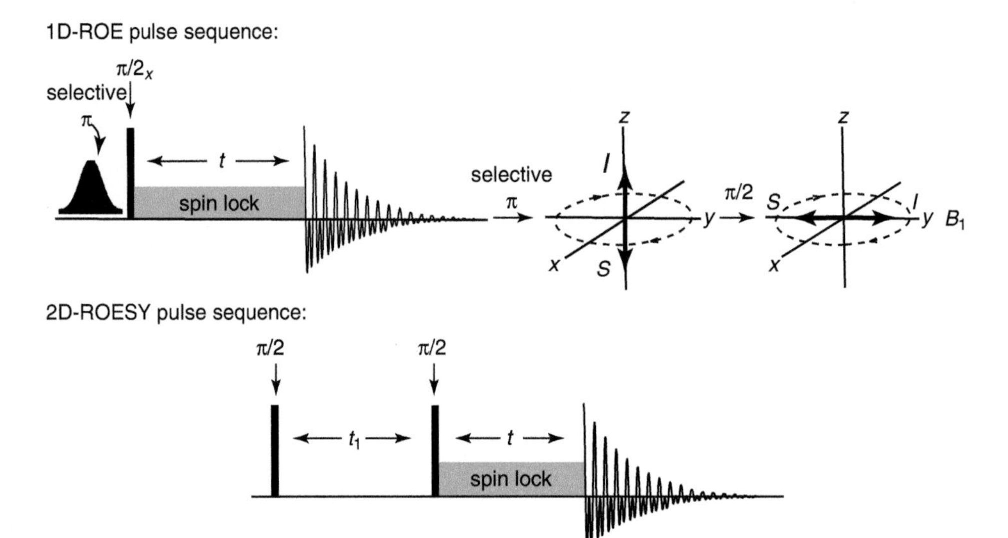

Fig. 5.24 2D-ROESY pulse sequence

5.5 Cross-Correlated 2D Spectra Identifying 1-Bond Connections

The 2D spectroscopic techniques that we have seen so far have given auto-correlated spectra, identifying connections—scalar coupling or proximity in space—from protons to protons. There are equivalent techniques for fluorine to fluorine. It is also possible to examine the connections from one nucleus to the nucleus of a different element. If the proton spectrum can be correlated with the carbon spectrum, for example, then the complete assignment of both is easier. These spectra are all COSY spectra in principle, but pulses now have to be delivered to both nuclei independently, and the various techniques are given new names. The first cross-correlated techniques identify directly bonded pairs of elements, most commonly ^{1}H to ^{13}C.

5.5.1 Heteronuclear Multiple Quantum Coherence (HMQC) Spectra

The one-bond, ^{1}H–^{13}C cross-correlated experiment is called a Heteronuclear Multiple Quantum Coherence (HMQC) spectrum. In outline it uses the pulse sequence in Fig. 5.25.

The delay $1/2J$ is set to allow the ^{1}H vectors of the proton doublet to precess 180° in the xy plane. At this point, the $\pi/2$ pulse in the ^{13}C channel tips its magnetisation onto the y axis in the usual way. At this stage the N_α and N_β energy levels of each pair are equally occupied, and there is no single quantum contribution to coherence. The protons and the carbons can achieve multiple quantum coherence by two-quantum changes between the ββ-to-αα and zero-quantum αβ-to-βα changes, but this cannot be described adequately with the vector model. The successive intervals of t_1 generate the second dimension in the usual way, with both ^{13}C and ^{1}H vectors evolving in the xy plane during this period. The

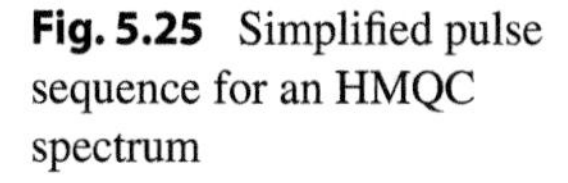

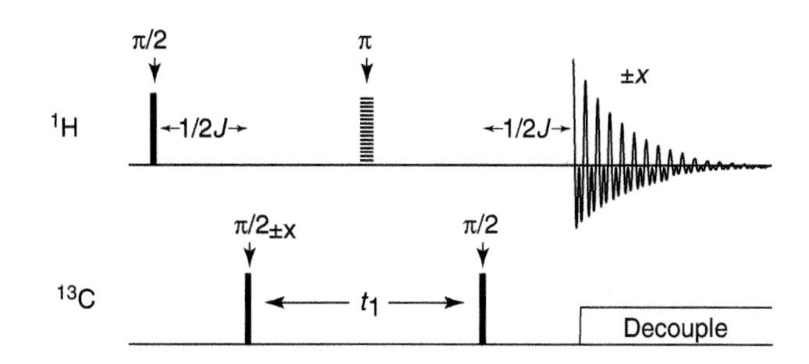

Fig. 5.25 Simplified pulse sequence for an HMQC spectrum

^{1}H chemical shifts are refocused with a spin-echo π pulse half way through, so that they have no influence on the f_1 dimension, and the second π/2 pulse in the carbon channel brings any further multiple quantum coherence from the proton signal to a halt. The second 1/2*J* delay then brings the proton vectors back into phase so that their frequencies can be collected as they evolve to give chemical shift information (i.e. the 1D ^{1}H NMR spectrum) in the f_2 dimension. Finally, the proton coupling in the carbon spectrum may be removed during acquisition using the usual decoupling frequency, in order to keep the cross peaks relatively narrow in the f_1 dimension. In practice the decoupling is often omitted, because it avoids damage that the decoupling currents can do to the receiver coils. The spreading of the signal in the f_1 dimension is not severe, and it does not affect the spectrum printed on the f_1 axis, which is derived independently from a 1D experiment. Critically, in the HMQC experiment, the 1/2 *J* value is chosen as representative of the *direct* ^{1}H–^{13}C coupling constant, a value near the middle of the 120–170 Hz range for $^1J_{CH}$ coupling (making 1/2*J* close to 3.3 ms), in order to achieve a correlation between each ^{13}C atom and the proton to which it is directly attached.

On top of all this, quadrature detection and phase cycling, or, more likely, a gradient pulse is used to suppress the usual unwanted signals. In this case, phase cycling has to include an alternation in the phase of the ^{13}C π/2 pulses followed by subtracting the two signals, in order to cancel the strong signals that would otherwise be collected from all the protons attached to ^{12}C.

5.5.2 Heteronuclear Single Quantum Coherence (HSQC) Spectra

Before we look at examples of these spectra, there is a closely related sequence, giving us essentially the same information in the same form, that is called an HSQC (Heteronuclear Single Quantum Coherence) spectrum, which has the advantage that it removes ^{1}H–^{1}H couplings that broaden the ^{13}C signals in the f_1 dimension in HMQC spectra. They are small and fairly inoffensive there, because ^{1}H–^{1}H coupling constants are so much smaller than the ^{1}H–^{13}C couplings we are using and then suppressing. The pulse sequence for HSQC in Fig. 5.26 is as complicated (and schematic) as we are going to see, and we shall leave it unexplained, except to point out that the second π/2 pulse to the protons means that the proton magnetisation is no longer distributed with equal occupancy of the N_α and N_β energy levels at the same time as that situation is created in the carbon spins, and *single* quantum coherences can develop. Also, the gradient pulses are in the same direction, because, by the

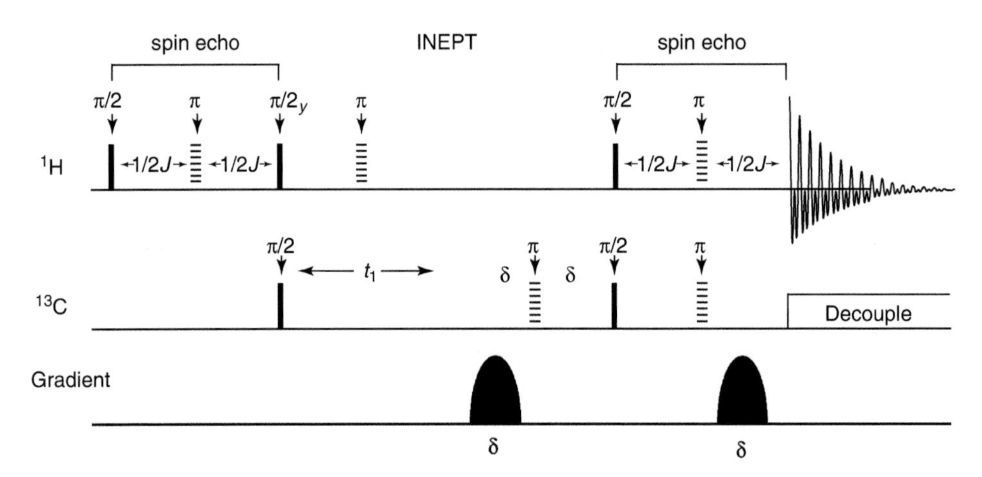

Fig. 5.26 Simplified pulse sequence for an HSQC spectrum

time the second gradient pulse is applied, the vectors are precessing in the opposite direction to the direction they were precessing in at the time of the first gradient pulse.

You will come across both HMQC and HSQC spectra, both of which are in use. They look very much alike, there is not that much to choose between them. HSQC spectra take longer to run, but they are probably what you will see most often.

5.5.3 Examples of HSQC Spectra

The HSQC spectrum for *o*-dibromobenzene **9** is given in Fig. 5.27, to which we have attached the 1D proton spectrum on the f_2 axis and the 1D carbon spectrum on the f_1 axis. In Chap. 4, *o*-dibromobenzene (**36** in Chap. 4) was a substrate for ^{13}C-^{13}C coupling at natural abundance (Fig. 4.26), an example of an AA′BB′ system in a proton spectrum (Fig. 4.56), and a substrate for a T_1 measurement (Sect. 4.9.1). In the HSQC spectrum in Fig. 5.27, the cross peak identified by the dashed line confirms that the proton giving the downfield signal at δ 7.6 is attached to the carbon atom giving the downfield signal at δ 133.8.

Previously we had identified the downfield proton signal H_a as coming from the protons attached to C-3 and C-6 by their larger T_1 values, and we had identified the downfield carbon signal as coming from C-3 and C-6 from its having two different ^{13}C-^{13}C couplings in the sidebands. Now we see that they are definitely connected to each other. Likewise, the upfield proton signal H_b has a cross peak with the upfield carbon signal we assigned to C-4 and C-5. Equally usefully, the HSQC spectrum confirms that the ^{13}C signal at δ 124.9 is from C-1 and C-2, because it does not have a cross peak with any proton signal. Although it was fairly safe to do so, we had merely assumed this was the correct assignment on the basis that it was a low-intensity signal.

In general, there is no diagonal in an HSQC spectrum. Although there is an approximate correlation between the chemical shifts of protons and carbons, giving rise to more signals in the top right and bottom left than in the other two quadrants, there is no straight

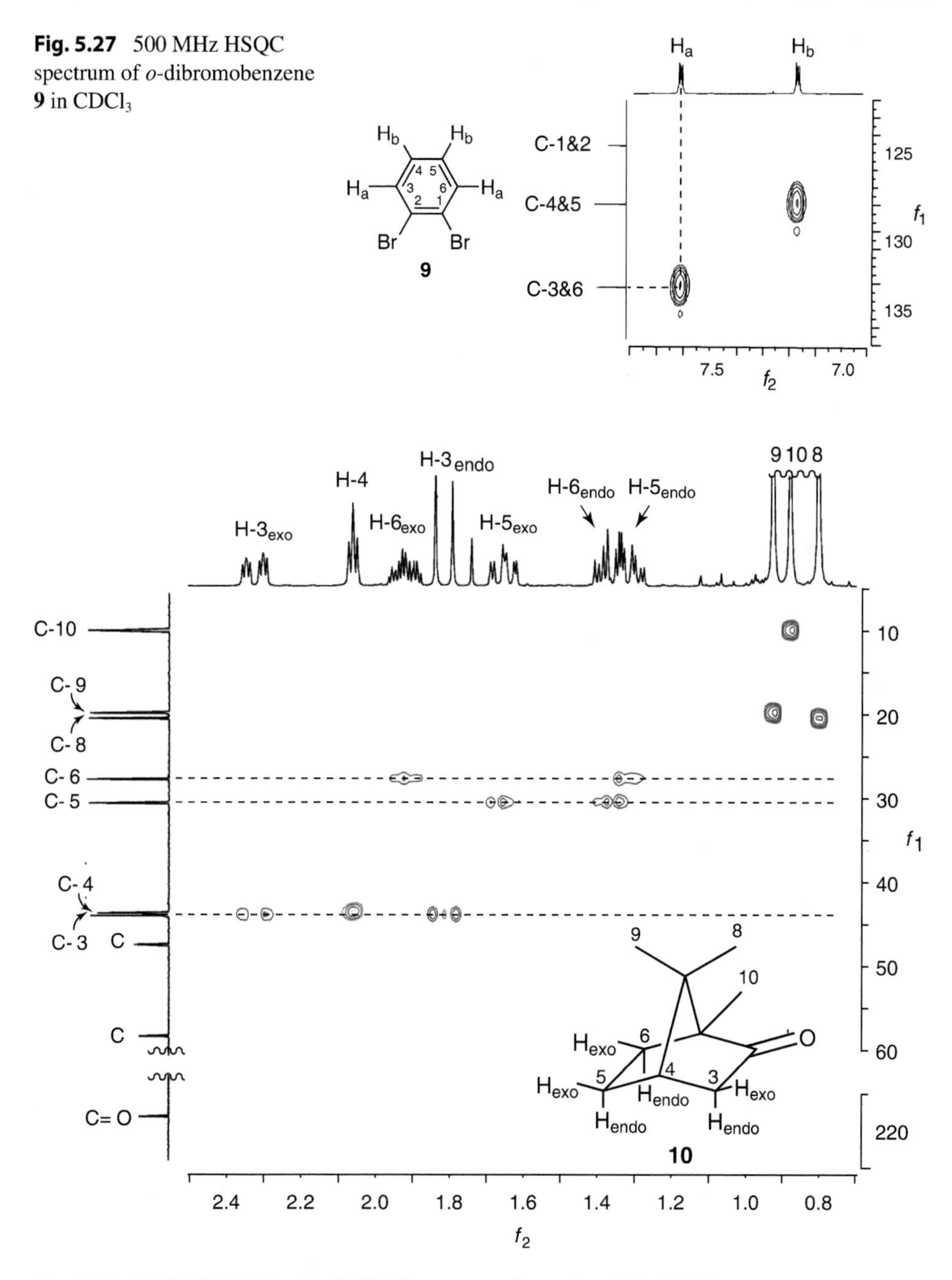

Fig. 5.27 500 MHz HSQC spectrum of *o*-dibromobenzene **9** in $CDCl_3$

Fig. 5.28 500 MHz DEPT-edited HSQC spectrum of camphor **10** in $CDCl_3$

line. In the case of *o*-dibromobenzene in Fig. 5.27, the proton giving the upfield signal and the carbon giving the upfield signal are connected, but this is not always the case. For example, in the HSQC spectrum of camphor **10** (**141** in Chap. 4) in Fig. 5.28, the protons at highest field are from the methyl group C-8, but they correlate with the carbon signal at δ

19.8 at the lowest field of the three methyl carbons. Evidently, the protons and the carbon atom to which they are attached have different spatial arrangements in the field of the anisotropic carbonyl group. Since the three methyl singlets in the proton spectrum had been assigned by their NOEs (Fig. 4.85), it is now clear which methyl carbon signal is which. It is essential in assigning carbon signals to take an HMQC or HSQC spectrum beforehand, and not to rely upon an approximate correlation of chemical shifts.

Assigning which proton is directly connected to which carbon is a prerequisite before moving on to an HMBC spectrum (Sect. 5.6.1), but HSQC spectra supply other useful information. Thus we learn from Fig. 5.28 that the ^{13}C signals at δ 57.7 and 46.8, as well as the carbonyl carbon signal at δ 219.8, are from quaternary carbons, because they have no cross peaks at all. We also learn that the three carbon signals at δ 27.0, 29.8 and 43.3 have cross peaks (on the dashed lines) to two proton signals each, identifying them as methylene carbons. The cross peaks are frequently processed to give HSQC spectra with colour coding, achieved using a procedure called DEPT editing, as reproduced in Fig. 5.28. The methine and methyl cross peaks are printed in red and the methylene cross peaks printed in blue, making it unnecessary to take a DEPT-135 spectrum if you have a DEPT-edited HSQC spectrum.

Furthermore, it is now clear how the diastereotopic protons on the three methylene groups C-3, C-5 and C-6, are paired up: the H-3_{exo} and H-3_{endo} signals on one of the dashed lines; the H-5_{exo} and H-5_{endo} signals on another; and the H-6_{exo} and H-6_{endo} signals on the third. This makes the analysis of proton-proton couplings easier, because it is now clear which coupled pairs of protons are related as geminal and which as vicinal. The methine carbon at δ 43.0 is almost coincident with the signal at δ 43.3 from the protons on the methylene group at C-3, but even drawn as small as it is in Fig. 5.28, the spectrum makes it clear that the red cross peak between the signals from C-4 and H-4 is further upfield in the carbon dimension than the cross peaks from C-3 to the two methylene protons.

5.5.4 Non-Uniform Sampling (NUS)

All 2D NMR spectra take a relatively long time to run. They make big demands on the stability of the spectrometer, and on the willingness of one's colleagues to wait their turn. The intervals of t_1 that create the f_1 dimension are not themselves time consuming, but the time taken to acquire the t_2 data at each t_1 data point, and give time for the FID to decay, adds up to a long time. A COSY or HSQC spectrum routinely takes 10–20 min, and some of the more elaborate experiments take much longer. A major contribution to cutting down the time they take is to use a deeply counter-intuitive pulse sequence that records *fewer* t_1 data points, saving the accumulated times of the attendant t_2 data not sampled. It is called non-uniform sampling (NUS).

The Nyquist theorem (Sect. 4.2) states that the minimum rate of sampling must be twice the highest frequency you are trying to detect. This imposes limits on 2D spectra, because of the number of data points in the t_1 dimension that have to be taken. The Nyquist theorem is valid only when the wave is being sampled at *regular* intervals. It is possible to take readings at irregular intervals instead of at regular intervals, and the curious consequence is that we are no longer obliged to collect as much data. It is also necessary mathematically to process the readings in a different way; it is not strictly a Fourier transform, but it achieves the same effect of extracting the frequency information out of the time domain data, and is often referred to as FT.

Non-uniform sampling (NUS) dips into the matrix of data points, typically taking one-quarter of the number randomly distributed in the time dimension. Thus, instead of taking each of the data points in the t_1 dimension, and collecting all the t_2 data points for each, we choose a random selection of the data points in the t_1 dimension and only collect in the t_2 dimension the FID from those. One method of randomising is to use a radial distribution in the matrix, which weights the sampling in favour of the points generated early in the t_2 dimension, where the signal is strongest.

The processing, although more complicated than the regular FT, gives the same result—a 2D plot of frequency information, with intensity in the third dimension. The key advantage is that the whole spectrum can be taken in shorter times overall with essentially no loss of definition even down to about 12.5% sampling. The HSQC spectrum of camphor in Fig. 5.28 was a conventional HSQC spectrum taken with 256 data points in the t_1 dimension, and taking 8.5 min. The spectrum in Fig. 5.29 was taken with the

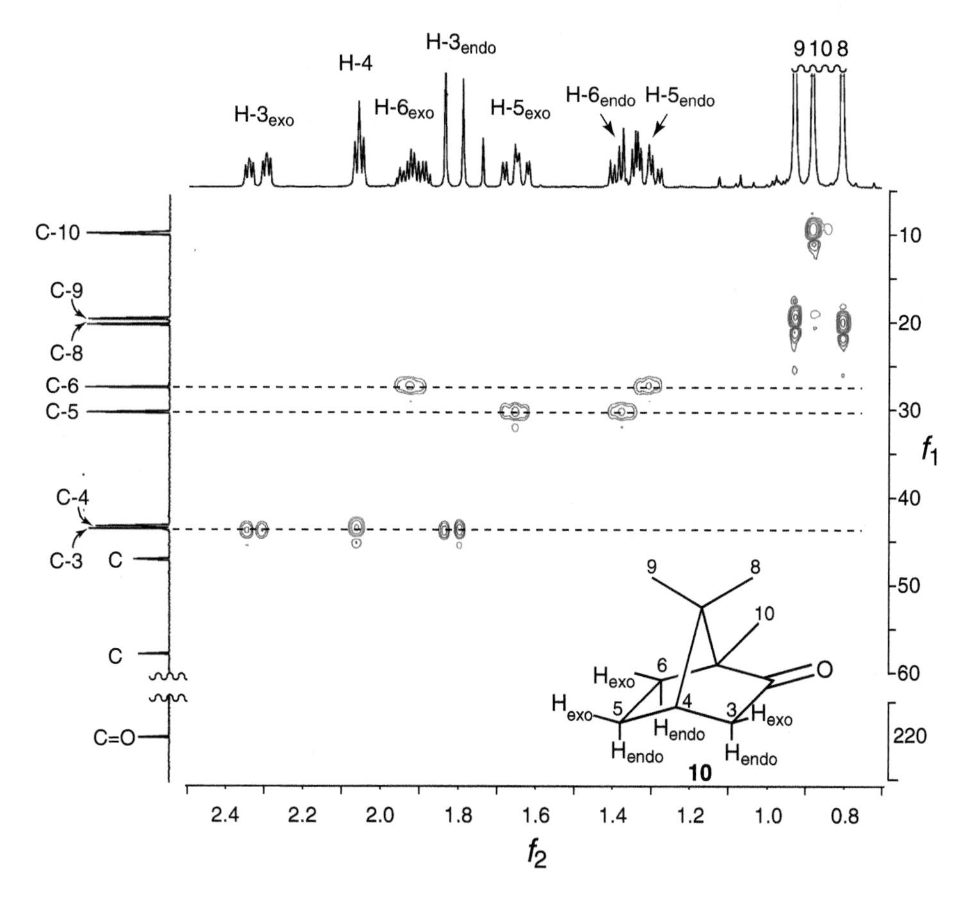

Fig. 5.29 500 MHz DEPT-edited HSQC spectrum of camphor **10** taken with 12.5% NUS in $CDCl_3$

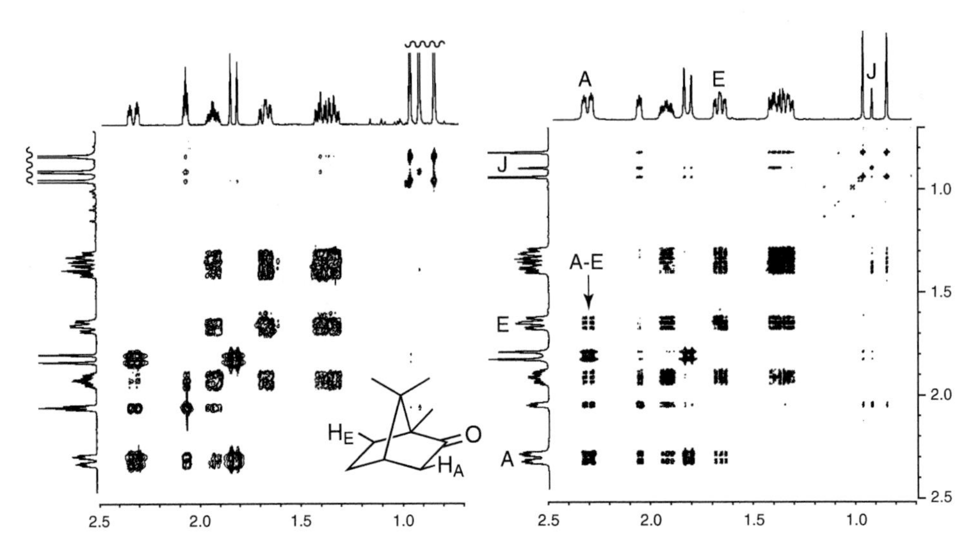

Fig. 5.30 500 MHz magnitude mode COSY spectra of camphor **10** taken normally and with 25% NUS in $CDCl_3$

same sample and the same number of data points in t_1, but in a little over 1 min using NUS to select only 12.5% of them, showing astonishingly little loss of detail, and only some extra t_1 noise.

Alternatively, if the times are not too long, we can set many more data points in t_1, to improve the resolution, but then select randomly only a fraction of them using NUS, to gain better resolution in the same time as a conventional spectrum. Here is an example: the DQF-COSY spectrum of camphor **10** on the left in Fig. 5.30, was processed in magnitude mode using a standard set of 256 data points in t_1 over 23 min. It is good enough for most purposes. With the aim of improving even this spectrum, the one on the right was taken with eight times as many points in t_1, but sampling only 25% of them, only doubling the time of the experiment, but improving the detail in the cross peaks. The spectrum on the right is so good that it has been plotted using the projections of the f_1 and f_2 axes, instead of the usual 1D spectrum, with little loss of resolution in either. You can tell that the axes are projections of a DQF spectrum by how small the peaks for the methyl groups have become relative to the other signals, especially small for the signal J of the bridge-head methyl group, because it has so little coupling to anything. One improvement visible in the NUS spectrum is the arrival of the cross peak between the signals A and E, which was not present in the spectrum on the left—it is a $^5J_{HH}$ coupling with an extended version of W-coupling.

Shortening the time in which a 2D spectrum can be taken by using non-uniform sampling, or improving its quality by using more points in the same time, is gaining importance. It is even more important, indeed essential, when it comes to the 3D (and 4D) spectra that are so important in determining the three-dimensional structure of proteins. Most of the experiments described from here on either could have been run with non-uniform sampling, or actually were run with non-uniform sampling.

5.5.5 Cross-Peak Detail: Determining the Sign of Coupling Constants

Figure 5.31 shows part of the HSQC spectrum of F-tenofovir **11**. The proton axis shows the two double-double-doublets for the isolated spin system of the methylene group with the protons coupling to each other and to the fluorine and phosphorus atoms. The carbon axis shows the signal from the carbon atom as a double doublet from its coupling to the fluorine and phosphorus atoms.

In the cross peak at the top left, the fine structure shows that the peaks labelled 1, 2, 4 and 6 are aligned with the first, second, fourth and sixth lines in the H_A multiplet. The pairs 1–2 and 4–6 are for the coupling of 6.7 Hz, as are the peaks labelled 3–5 and 7–8 in the lower left cross peak. The pair 2 and 6 line up with the lowest frequency line of the carbon spectrum, and the pair 1 and 4 line up with the next lowest frequency line of the carbon

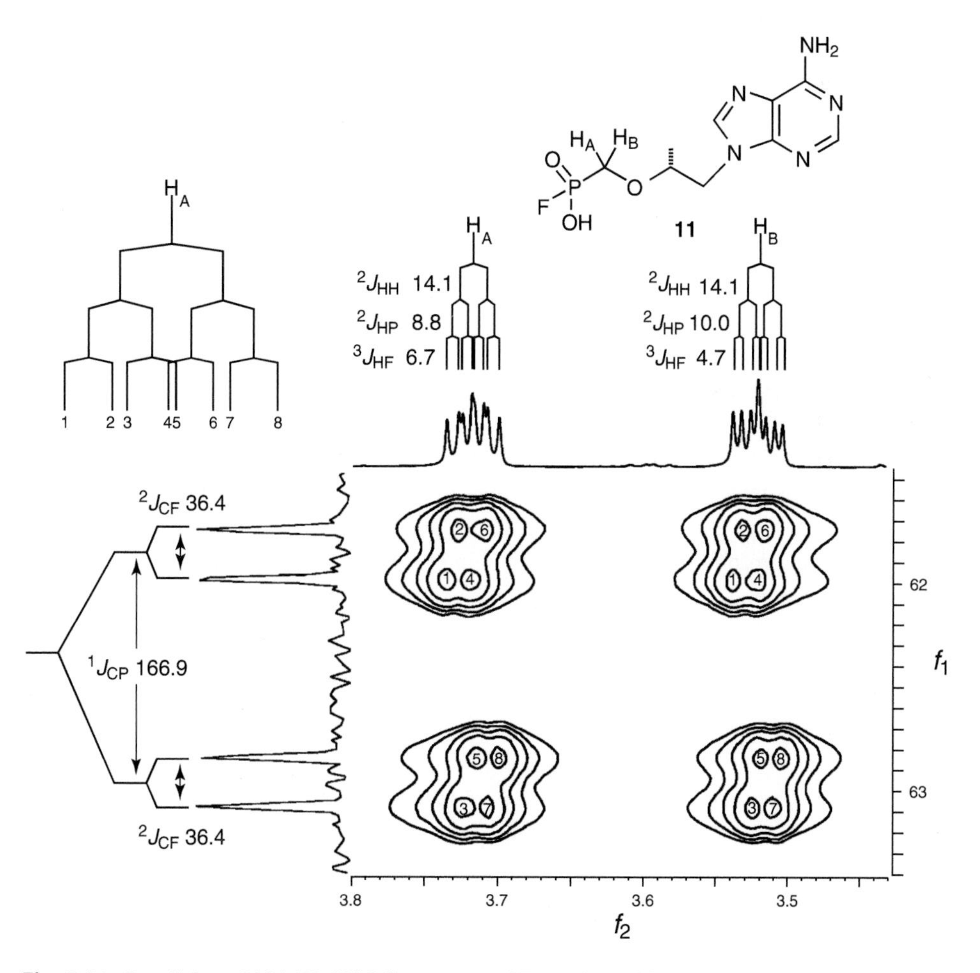

Fig. 5.31 Detail from 800 MHz HSQC spectrum of F-tenofovir **11** in $CDCl_3$

spectrum; similarly the vertical separation of the pair 3 and 7 from 5 and 8 lines up with the two higher-frequency lines of the carbon spectrum. This shows that the coupling of 6.7 Hz is from the proton H_A to the fluorine atom, since the doublets in the carbon spectrum with a coupling constant of 36.4 Hz are from coupling from the carbon to the fluorine. Similarly, if we look at the two right hand cross peaks, we see the same feature, and can deduce that the smaller coupling there, 4.7 Hz, is also to the same atom, presumably the fluorine.

Looking at the vertical separation of the two sets of lines, 1, 2, 4 and 6 from lines 3, 5, 7 and 8 in the left hand pair of cross peaks. The vertical separation of lines 1 and 3, for example matches the separation of 166.9 Hz of the two doublets in the carbon spectrum from the coupling to phosphorus. Therefore, the horizontal separation of 8.8 Hz for lines 1 and 3 in the proton spectrum corresponds to the coupling to phosphorus. Similarly for each of the other pairings, 2 and 5, 4 and 7, and 6 and 8. And again on the right hand cross peak we see that same vertical separation of 166.9 Hz and the horizontal separation matching the 10 Hz coupling to phosphorus for H_B.

But the most important piece of information in these cross peaks is not the reassuring matching up of all the couplings. In the top left cross peak, the peaks 2 and 6 are above (to lower frequency) the peaks 1 and 4, lining up with the low and high frequency lines of the carbon doublet, respectively. Likewise, peaks 5 and 8 are at lower frequency than 3 and 7, lining up with the low and high frequency lines of the carbon doublet. The high frequency lines in the proton spectrum match the high frequency lines of the fluorine-induced doublet. *This tells us that the* $^{3}J_{HF}$ *coupling of 6.7 Hz, is positive in sign.* Similarly, in the right hand cross peaks, the 4.7 Hz coupling is positive in sign. In contrast, the four peaks 1, 2, 4 and 6 are in the upper cross peak, and the peaks 3, 5, 7 and 8 are in the lower. Thus the four higher frequency lines 1, 2, 4 and 6 in the proton spectrum are aligned with the two lower frequency lines in the carbon spectrum. *This tells us that the* $^{2}J_{HP}$ *coupling of 8.8 Hz, is negative in sign.* With the 6.7 Hz positive and the 8.8 Hz negative, we can be sure that these are the couplings to fluorine (3-bond) and phosphorus (2-bond), respectively. The only snag is that couplings to these and other elements outside the carbon and proton range are not as reliable in sign as we have been assuming, but in fact it is known that $^{3}J_{HF}$ and $^{2}J_{HP}$ couplings are usually positive and negative in sign, respectively.

5.5.6 Clip-HSQC

A variant of the HSQC spectrum is called a CLIP-HSQC, in the pulse sequence for which the decoupler is not used. By not using the decoupler much longer acquisition times are possible, and much better resolution is achieved in consequence. The cross peaks display the proton signals in the f_2 dimension without proton decoupling as *double* multiplets centred on the proton chemical shift, and separated by a frequency difference corresponding to the $^{1}J_{CH}$ coupling constant. The spectrum on the left in Fig. 5.32 is a normal HSQC for

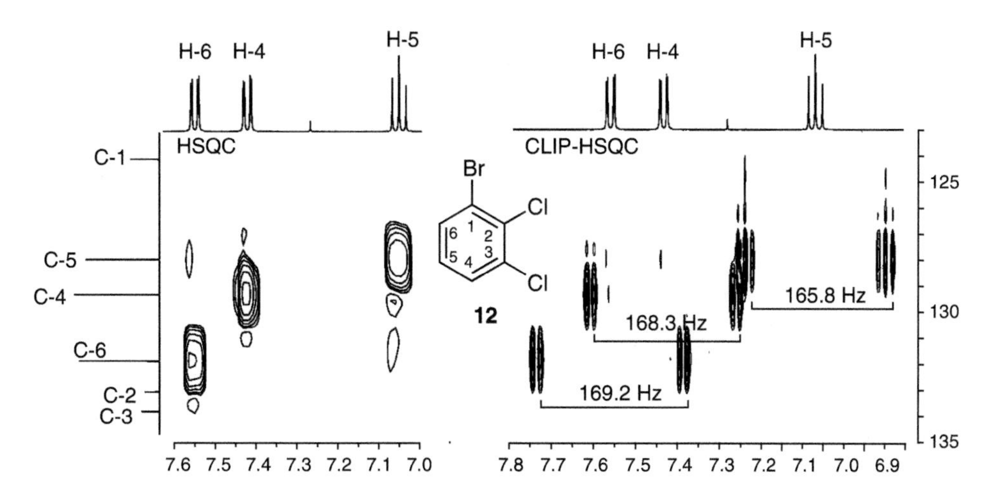

Fig. 5.32 500 MHz HSQC and CLIP-HSQC spectra of 2,3-dichlorobromobenzene in $CDCl_3$

2,3-dichlorobromobenzene **12**, and the spectrum on the right is the corresponding CLIP-HSQC. The latter allows us to measure the three coupling constants, H–C-6 as 169.2 Hz, H–C-4 as 168.3 Hz and H–C-5 as 165.8 Hz, accurate to ±0.5 Hz. It supplements, and largely replaces, the method using the ^{13}C-satellites in proton spectra (Sect. 4.5.2), which are often not easy to pick out.

5.5.7 Deconvoluting a ^1H-Spectrum Using the HSQC Spectrum

We have seen how 1D- and 2D-TOCSY, and difference-NOE, can pull overlapping multiplets apart, but the HSQC spectrum is even more versatile for this purpose. Selecting one of the carbon lines in an HSQC spectrum brings up the ^1H spectrum of the protons attached to that carbon atom, free of all the other proton signals. Since the carbon signals are usually well enough resolved from each other, this technique is available for many more compounds. It does not suffer from the limitation of TOCSY, which only works if a signal in the spin system is well enough isolated to be irradiated selectively, and if the overlapping signals are in different spin systems (or are far enough apart in the spin system). Nor does it have the limitation of NOE, that there must be a resolved signal from a proton held in space close to one of the protons in an overlapping multiplet, and not to the others. Unfortunately, as we have already seen with 2D-TOCSY spectra (Sect. 5.3), the deconvoluted spectra displayed on the f_2 axis by selection from a routine HSQC spectrum do not usually have the same degree of resolution that a 1D spectrum has, and multiplets are only resolved if the coupling constants are quite large.

Figure 5.33 shows an example using naloxone **6**, chosen because we are familiar with its two overlapping multiplets (Figs. 4.86 and 5.21). We see at the top of the figure the result of selecting in turn each of four carbon signals, revealing one methine proton, J, and

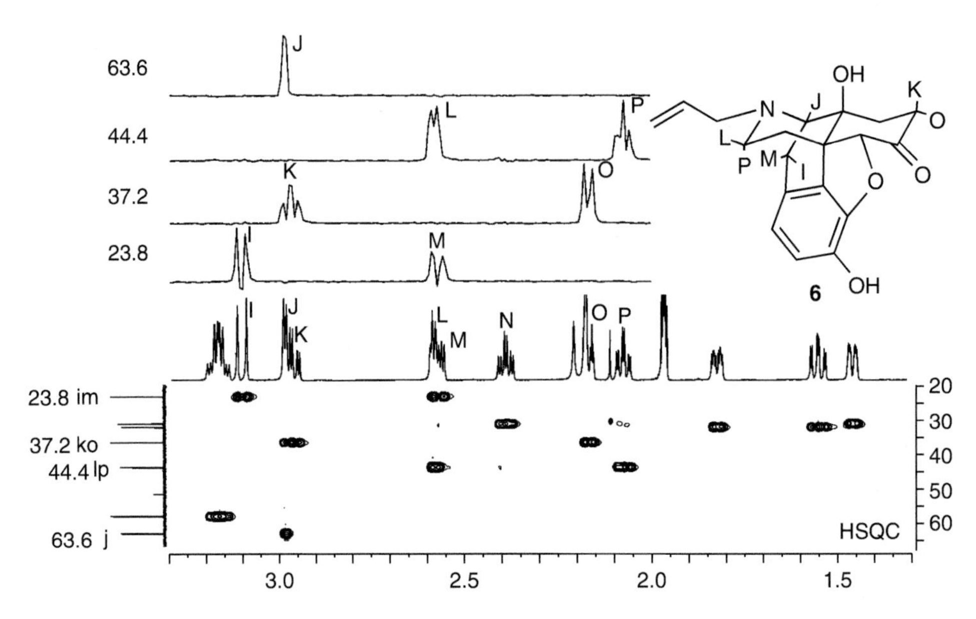

Fig. 5.33 Deconvoluting ^{1}H signals using the attached carbon signals in a 750 MHz HSQC spectrum of naloxone **6** in CD_3CN

three pairs of methylene protons, LP, KO and IM. The doublet and triplet structure is clear but nothing better than that.

The even greater loss of resolution, relative to that which we have already seen being lost in 2D spectra, is caused by problems from the proton-decoupling. With the decoupling signal on during the acquisition, the steady power of the signal unacceptably warms up the cryoprobe ("fries" it in the vernacular). If we turn it off, as we do in a CLIP-HSQC spectrum, the FID can be collected for longer periods. But all the proton signals are doubled, potentially complicating the peaks, and making them harder to interpret. The compromise reached in a standard HSQC spectrum is to have the decoupling signal on at the beginning of the acquisition as usual, but to turn it off, and stop recording the FID, soon after, before the decoupler does any harm. The signal quality is correspondingly low, since a smaller fraction of the FID has been taken during each increment of t_1. It is already a short acquisition time in every 2D spectrum; now it is even shorter. Typically, an HSQC spectrum would acquire the signal for as little as 0.15 s, and then allow the FID to decay over only another 1 s.

When a CLIP-HSQC would not make the spectrum too complicated, it can be used to overcome this problem. The signals from a diastereotopic pair of methylene protons in the substructure **13** give a 1D spectrum (at the top in Fig. 5.34) in which the upfield multiplet is incompletely exposed. The signal from H_L looks like part of a quartet but is it? The CLIP-HSQC spectrum below it in Fig. 5.34, splits the signal into two quartets separated by the $^{1}J_{CH}$ coupling constant, but extracts it with much better resolution as shown by its projection on the f_2 axis. The only penalty is that the downfield partner of the H_L signal now overlaps with the upfield double-triplet from the H_I signal. Fortunately this does not

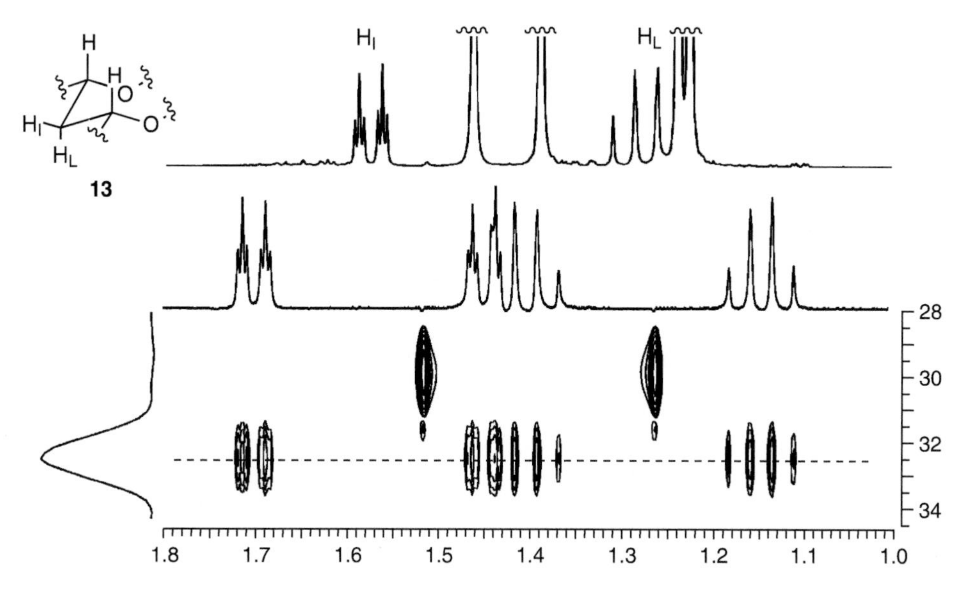

Fig. 5.34 500 MHz CLIP-HSQC spectrum in $CDCl_3$ for deconvoluting signals

interfere, and the downfield double-triplet from the H_I signal is also well resolved. This spectrum incidentally was taken using 25% NUS.

5.5.8 HSQC-TOCSY

A related HSQC experiment is called HSQC-TOCSY, being an extension of the deconvolution experiment, but bringing up the whole spin system connected to the attached protons instead of just the attached proton signals seen in an HSQC spectrum. Figure 5.35 shows an example from a 1,2-disubstituted benzene ring **14** embedded in a larger molecule. The four protons labelled B, F, G and H in the 1D spectrum overlap in the δ 7–8 region with other aromatic protons C, D and E in the same molecule. It might be that there was no clear signal to irradiate for a conventional TOCSY, but in an HSQC-TOCSY, we have the opportunity to use instead one of the signals from one of the carbon atoms carrying a proton in this spin system. Selecting the signal g (the signal from the carbon atom attached to the proton G), displayed in Fig. 5.35 from the 1D ^{13}C spectrum, brings onto the f_2 axis the ^1H spin system, revealing the expected pair of doublets and the pair of triplets with the usual low resolution. The beauty of this technique is that we get access to the ^1H spectrum from a carbon spectrum, where the signals are usually better resolved.

5.5.9 HETCOR

An older technique achieving the same correlation of ^1H directly bonded to ^{13}C is called HETCOR, and you may come across it. Whereas HMQC and HSQC use the protons for detection, somewhat limiting the resolution on the carbon axis, where it rarely matters,

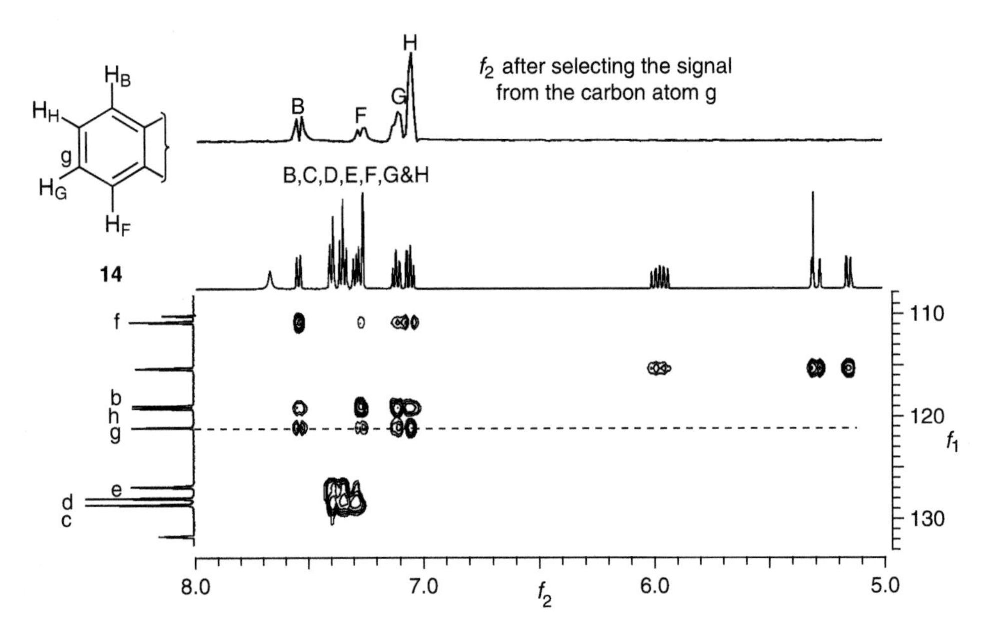

Fig. 5.35 Selection of a spin system using the signal of a ^{13}C attached to one of the protons in a 500 MHz HSQC-TOCSY spectrum taken in $CDCl_3$

HETCOR uses the weaker FID of the ^{13}C signal for detection in the f_2 dimension, and could still be useful if a very crowded carbon spectrum has to be analysed. You may come across HETCOR spectra, and all you need to be aware of is that they will usually have the axes transposed—the f_2 axis, the direct spectrum collected in the acquisition stage, is plotted as usual on the abscissa, and this is now the carbon axis, and the proton spectrum is the indirect spectrum on the ordinate. Otherwise the interpretation of HETCOR spectra is the same as for HMQC and HSQC spectra.

5.6 Cross-Correlated 2D Spectra Identifying 2- and 3-Bond Connections

5.6.1 The HMBC Pulse Sequence

The pulse sequence giving an HMBC spectrum, is essentially the same as that which gives an HMQC or HSQC spectrum shown in Fig. 5.25. The difference is that the delay $1/2\,J$ is changed, typically, to about 50 ms to match coupling constants in the region of 10 Hz. Since many $^2J_{CH}$ and $^3J_{CH}$ coupling constants are in the range 2–20 Hz (Sect. 4.7), the ^{13}C chemical shifts are now correlated with the chemical shifts of those protons *separated from them by two and three bonds*. It is an unfortunate limitation that the absolute magnitudes of the two- and three-bond couplings are so similar, and that it is not possible to identify whether a connection, revealed in the HMBC spectrum, is through two or three bonds. The technique is nevertheless powerful, not least because it is possible to identify the connection between one spin system and another, even when there is an insulating group like an oxygen atom, a quaternary carbon or a carbonyl group in between. In this way, complete connectivity in complex skeletons can often be deduced. Before an HMBC spectrum is run, it is essential to take an HMQC or HSQC spectrum, in order to identify which carbon atoms are directly attached to which protons.

5.6.2 HMBC Spectra

Looking at a simple example, methyl salicylate **15**, we would know from its HSQC spectrum how the protons and the carbons correlate, but neither the ^{1}H nor the ^{13}C spectrum would tell us whether the methyl group was attached to the carboxylic oxygen or the phenolic oxygen. In its HMBC spectrum in Fig. 5.36, the 'quaternary' carbon signals have been labeled with the letter Q, and the signal labelled Q_1 must, from its chemical shift, be the signal from the carbonyl carbon. It shows a strong HMBC connection to the protons F of the methyl group. This tells us that there are two or three bonds between them, and that places the methoxy group firmly in a methyl ester where it has a $^{3}J_{CH}$ connection, shown bold for one of the three protons of the methyl group in the drawing **15**. Had the molecule been the methyl ether, there would have been five bonds between the carbonyl carbon and the methyl protons.

There is also a strong HMBC connection from the carbonyl carbon to the downfield proton H_B, which identifies H_B as the signal from the proton *ortho* to the carbonyl group. Similarly, we can see connections from Q_2 to H_B and H_C and from Q_3 to H_D and H_E. All of these are 3-bond connections, a *meta* arrangement within the benzene ring, giving strong HMBC peaks, because the nuclei are antiperiplanar to one another. Also showing *meta* connections are the d-H_E, b-H_C, e-H_D and c-H_B cross peaks. The two-bond *ortho* cross peaks are weaker, as usual in aromatic rings: b-H_E, c-H_E and Q_2-H_D. Even the four-bond *para* cross peaks Q_3-H_F and Q_1-H_D are nearly as strong as the two-bond cross peaks in this case. Figure 5.36 also shows a feature common in HMBC spectra, picked out with the course dashed line at the top of the spectrum, which we learn to discount as an undesirable artefact frequently seen, especially in the stronger signals. It is an HSQC cross peak, like a CLIP-HSQC cross peak, connecting H_F and the carbon f to which the F protons are directly attached.

Moving on to a more elaborate example, Fig. 5.37 shows the HMBC spectrum of the ketone **16**. The carbon signals had been assigned to the appropriate proton signals from an HSQC spectrum, and the several separate spin systems, including two with the sequence–$CH_2CH(CH_3)$–, had been identified with a set of 1D-TOCSY spectra. The problem was to join the fragments up.

The more informative two- and three-bond connections are labelled. The C8 signal at δ 73.1 has a cross peak labelled C8-H10 establishing the three-bond connection from C8 to the protons *ortho* to it on the benzene ring. Equally, there is a strong three-bond cross peak C10-H8, confirming the connection. Key peaks, subtended by the solid horizontal and

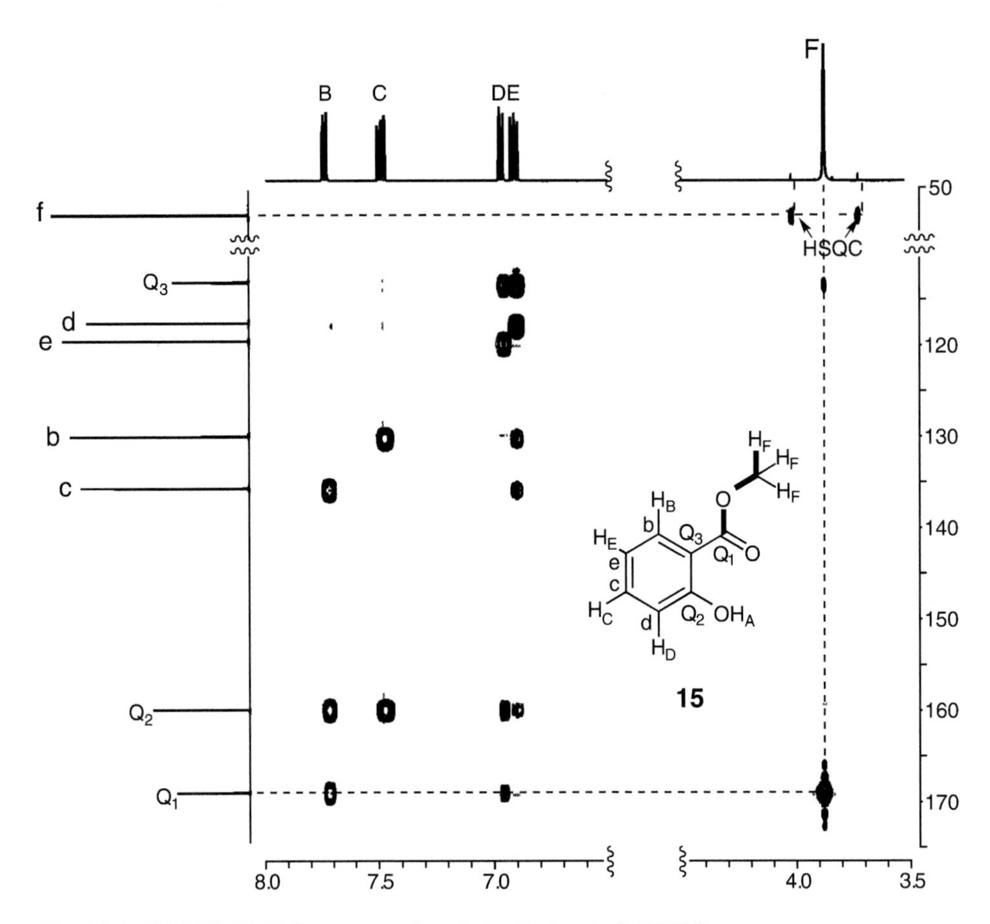

Fig. 5.36 500 MHz HMBC spectrum of methyl salicylate in d_6-DMSO

vertical lines, are those labelled C8-H5. They show that C8 in its turn, is two or three bonds from the protons on C5, which give the diastereotopic methylene pair of signals at δ 3.78 and 3.69. This information is transmitted through three bonds, and permits a structural connection of the previously established spin system associated with the aromatic ring to the spin system $-CH_2CH(CH_3)-$ extending to the right of the oxygen atom. Thus, there is communication across the ether linkage that allows a connection to be made between fragments identified from the ^1H and TOCSY spectra. The same connection in the other direction is provided by the cross peak labelled C5-H8.

The HMBC spectrum also tells us how the fully substituted carbons C3, C9 and C12 are connected to the rest of the molecule. The weak carbon signal, at δ 159.3 from C12 shows HMBC correlations to the signals from the protons on C10 and C11, and, most tellingly, to the methoxy signal at δ 3.79. The other aromatic carbon with no attached protons, C9 at δ 129.8, has no cross peak with the methoxy signal, establishing the full connectivity in the benzene ring. The peaks labelled C3-H1$_A$ and C3-H5$_B$ and those labelled C3-H4 and C3-

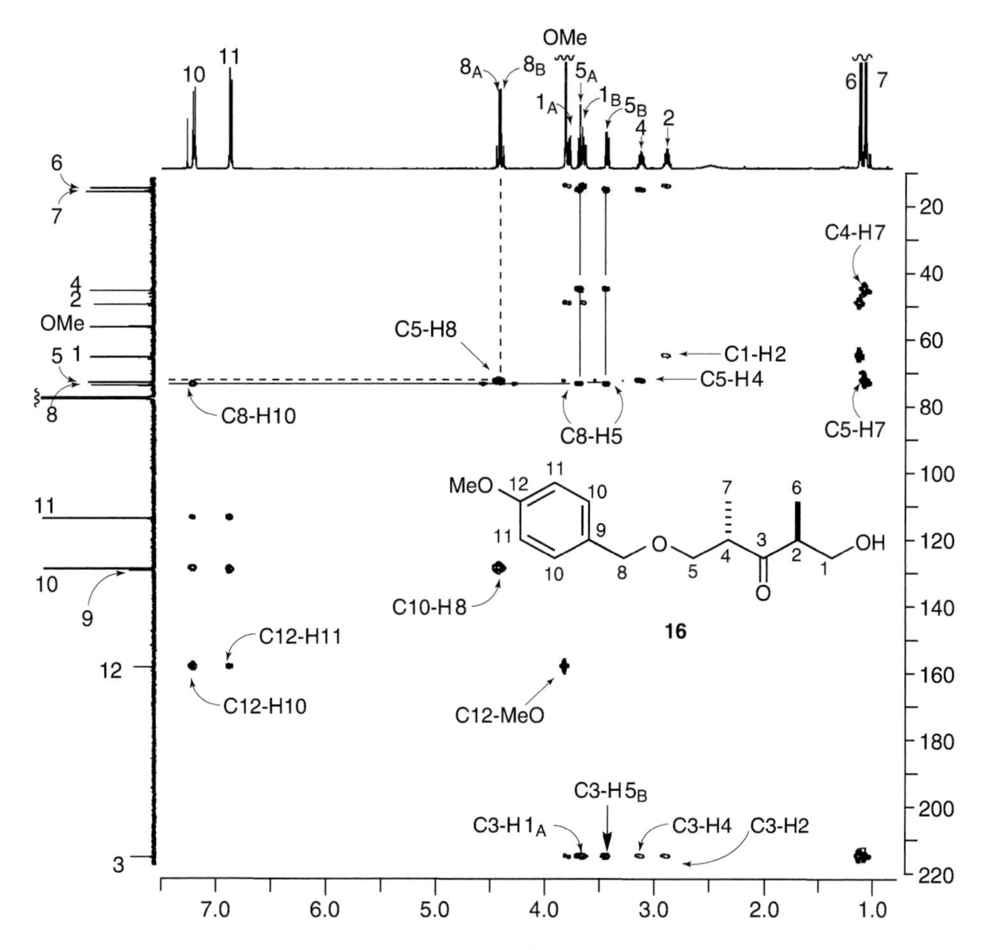

Fig. 5.37 500 MHz HMBC spectrum of the ketone **16** in $CDCl_3$

H2 connect the carbonyl carbon to the spin systems $–CH_2CH(CH_3)–$ that are attached to each side of it. Thus, HMBC allows the extension of skeletal information not only across heteroatoms, but also across carbonyl groups.

This spectrum is unusually free of CLIP-HSQC peaks, but it does show what look like HSQC cross peaks for the connection between the protons labelled 10 and the carbons labelled 10, and between the protons labelled 11 and the carbons labelled 11. This is a consequence of the fact that these are from two identical atoms of each, so that there is both a 1-bond connection, giving a cross peak in the HSQC spectrum, and a 3-bond (meta) connection giving a cross peak in the HMBC spectrum.

COSY, TOCSY, NOESY (or ROESY), HSQC and HMBC spectra are the workhorses of 2D-NMR spectroscopy, one or more of which would be called for following the acquisition and careful interpretation of the standard 1D spectra. Many problems of structure determination can be solved—even *most* problems if the research has been designed, by looking ahead, to take advantage of the information these standard NMR methods can be expected to provide. Several other 2D techniques are available for special circumstances. Section 5.7 describes three of them, but there are many more for the expert to turn to. In general, they consume more spectrometer time, sometimes much more, and are deployed only when they are needed.

5.7 Some Specialised NMR Techniques

5.7.1 ADEQUATE: Identifying ^{13}C–^{13}C Connections

ADEQUATE (Adequate sensitivity DoublE-QUAnTum coherencE) spectra use the coupling of one ^{13}C atom to another. The problem is one of sensitivity: the natural abundance of ^{13}C is close to 1%, making ^{13}C signals inherently weak, but, in addition, the probability of finding one ^{13}C attached to another is also close to 1%. In some cases, as we have seen for DDQ (**35** in Chap. 4), one-bond connections can be deduced from their direct coupling constants measurable as side-bands in high quality ^{13}C spectra. But this technique cannot be relied upon for a structure determination, because the J values have to be different enough, in order to pair up which carbon is bonded to which carbon. The side-bands also have to be detected and resolved, which is not always possible.

The ADEQUATE pulse sequence is complicated, but more general. It detects double-quantum coherence from one ^{13}C directly bonded to another, when at least one of them has a proton attached. The proton, with its relatively strong signal, is used to report the connection. The parameters that are fed to the spectrometer when taking the spectrum include an estimate of the ^{1}H–^{13}C coupling constant and an estimate of the ^{13}C–^{13}C coupling constant. If the latter is a number in the middle of the range for $^{1}J_{CC}$ values, the spectrum will detect ^{1}H–^{13}C–^{13}C connectivity. Typically a value close to 60 Hz is used, but other values might be more appropriate if a better estimate can be made. The choice of parameters allows fine tuning for molecules with trigonal-trigonal bonds (Sect. 4.5.3: $^{1}J_{CC}$ ~65 Hz, $^{1}J_{CH}$ 150–170) or tetrahedral-tetrahedral bonds ($^{1}J_{CC}$ ~30 Hz, $^{1}J_{CH}$ 120–150), for example. The ADEQUATE spectrum so generated is called a 1,1-ADEQUATE, because it picks out the directly bonded carbon atoms.

Figure 5.38 shows the HMBC spectrum of 2-acetylthiophen **17** on the left and the 1,1-ADEQUATE spectrum on the right. There is no doubt about the structure, which is completely determinable from the ^{1}H spectrum alone. But there is ambiguity about the assignment of the protons on C-3 and C-5 to the signals labelled A and B, which are so close in chemical shift that they could be either way round. The HMBC spectrum shows cross peaks from the carbonyl carbon to both A and B, with that to B noticeably weaker. The coupling from the carbonyl carbon to the proton on C-5, is $^{4}J_{CH}$, and to the proton on C-3 is $^{3}J_{CH}$, which indicates that the signal A is from the proton on C-3, since we expect a three-bond connection to give more intense cross peaks than a four-bond connection. C-2 also has cross peaks to both A and B, this time equally. The one to the proton on C-5, is

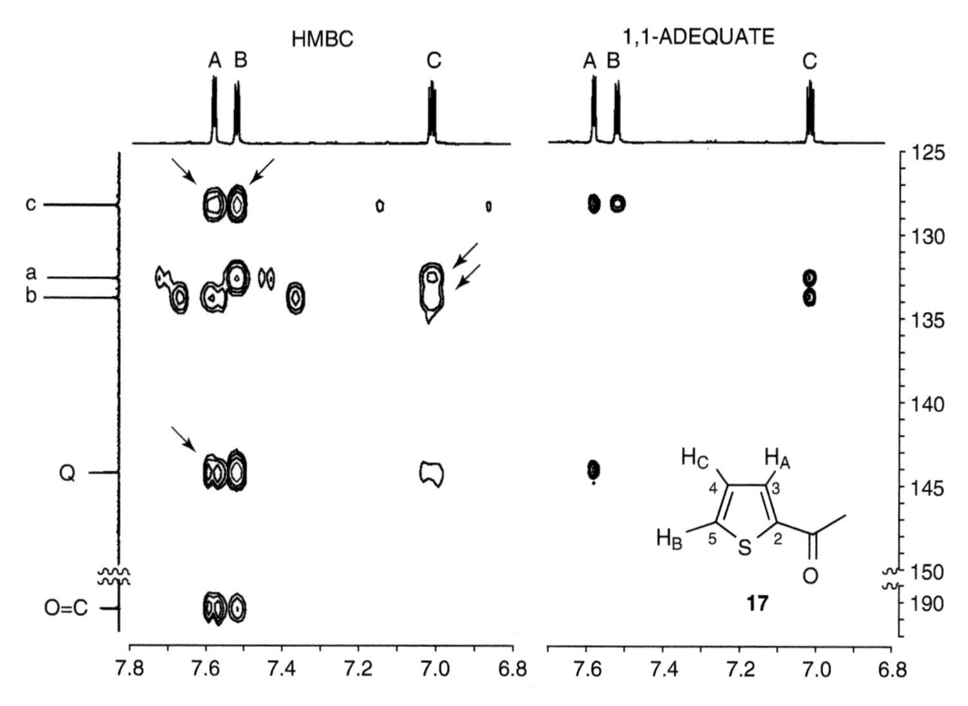

Fig. 5.38 500 MHz HMBC and 1,1-ADEQUATE spectra of 2-acetylthiophen in $CDCl_3$

$^3J_{CH}$, and the other, to the proton on C-3 is $^2J_{CH}$, both of which might reasonably be strong. The 1,1-ADEQUATE spectrum solves the problem by identifying which of the two carbons giving the signals labelled a and b has the 2-bond coupling to C-2, and it sometimes adds cross peaks that are too weak in the HMBC spectrum to show up. The spectrum on the right in Fig. 5.38 was taken using 180 Hz as the trial value for $^1J_{CH}$ and 65 Hz for $^1J_{CC}$, and, with a relatively concentrated solution, took 2 h to run.

The ADEQUATE spectrum can be read from the carbon scale or from the proton scale. Start with the carbon signal labelled c, which is unambiguously from C-4 (chemical shift and two couplings for the attached proton C): it has cross peaks below the proton signals for A and B. This shows that both carbons a and b, to which protons A and B are attached, are directly connected to the carbon c. Now look at the carbons labelled a and b: they have cross peaks below the proton signal C. This shows that both carbon atoms a and b are bonded to the carbon atom c, to which the proton labelled C is attached. None of this is in any doubt, but it shows how to read the spectrum. The significant peak is that opposite the signal Q from the carbon C-2. It has a cross peak under the signal labelled A. This shows that the carbon a is directly bonded to C-2, proving that the proton A is indeed on C-3. The carbonyl carbon signal has no cross peaks visible within the range of this spectrum, because it has no direct connection to any of the ring carbons carrying a hydrogen atom. It does have a strong cross peak to the signal from the methyl carbon much further upfield, not shown in Fig. 5.38.

The other way of reading the ADEQUATE spectrum is to look at the signals below the proton signals. The proton signal A has below it cross peaks to the carbons c and Q, showing that the carbon a, which carries the proton A, is connected to carbons c and Q, in other words it sits between C-2 and C-4. Similarly proton B has a cross peak to carbon c, showing that the carbon b, which carries the proton B, is connected directly to carbon c, identifying the connection between C-4 and C-5. Likewise the cross peaks to both carbons a and b below the signal to proton C, represent the connections from C-4 to C-3 and C-5.

If you mentally lift the ADEQUATE spectrum and lower it onto the HMBC spectrum, you will see that the coincident peaks identify the five $^2J_{CH}$ cross peaks in the HMBC spectrum, picked out in Fig. 5.38 with arrows, and the absence of coincident cross peaks in the ADEQUATE spectrum identifies all the other HMBC cross peaks as $^3J_{CH}$ and $^4J_{CH}$ connections, solving the ambiguity inherent in all HMBC spectra. It is not uncommon for $^2J_{CH}$ connections to be missing from HMBC spectra when they are weak; their appearance in ADEQUATE spectra is useful in filling in the gaps. The cover picture on this book shows the ADEQUATE spectrum in red superimposed on the HMBC spectrum in yellow (and orange for the HSQC peaks).

It is also possible to use an alternative parameter set for the ^{13}C–^{13}C coupling constant to match 2-bond and 3-bond values, say 10 Hz instead of the 1-bond value of 65 Hz used in the experiment described above. This creates a 1,n-ADEQUATE spectrum that picks up protons on one of the carbons and identifies the carbon atoms that are 2- and 3-bonds away from the carbon that the proton is attached to. The spectra are read in the same way as 1,1-ADEQUATE spectra, and they help to identify the 3- and 4-bond connections in HMBC spectra, although not distinguishing between them.

5.7.2 INADEQUATE: Identifying ^{13}C–^{13}C Connections

Another experiment called INADEQUATE (Incredible Natural Abundance DoublE QUAnTum coherencE) also detects the direct connection from one ^{13}C to another, but it uses only the ^{13}C signals, and is correspondingly even less sensitive than the ADEQUATE experiment. Unless the molecule is enriched in the ^{13}C isotope, an INADEQUATE spectrum can only be taken in concentrated solution, and even so taking many hours. It is not a routine experiment, but as instruments grow ever more sensitive, it may well increase in importance, since it is an extraordinarily powerful technique, detecting directly bonded pairs of carbon atoms, even when neither of them has an attached proton.

Figure 5.39 shows the INADEQUATE spectrum of 2-acetylthiophen **17**, which also, like the ADEQUATE spectrum above, allows us to assign which carbon signal is which, but this time by detecting the continuous connections from one carbon to the next. Whereas COSY spectra identify coupling from the cross peaks reflected across a diagonal, the INADEQUATE spectrum is displayed in a different way. It has the carbon spectrum on a diagonal in Fig. 5.39, but the cross peaks do not line up with the carbon atoms that are connected (and hence coupled to each other).

To find a route through an INADEQUATE spectrum presented like this, we start with an unambiguous signal—the far-upfield carbon signal labelled d from the methyl group. We see a lot of t_1 noise vertically below it. To find the peak we are looking for, we have to enlarge the spectrum and look for doublets within that line of t_1 noise. All the signals we are interested in are doublets, because we are looking for directly bonded ^{13}C atoms, and the probability of one ^{13}C atom being bonded to more than one ^{13}C is vanishingly small. Doublets can be recognised, and distinguished from the large amount of noise, by the phase change, represented here as the colour change from red (positive) to blue (negative). Vertically below the signal d we find a doublet, enlarged in the box on the right of Fig. 5.39. Now we change direction and look along the horizontal line looking for another doublet, and find it in the enlarged spectrum directly below the signal that must, from its chemical shift, be that from the carbonyl carbon. This horizontal line establishes that the methyl and carbonyl carbons are joined together. The horizontal line connecting these two signals appears on the vertical axis at a level corresponding to the sum of the frequencies of the two signals connected by that line. In this 125.85 MHz spectrum, the frequencies are

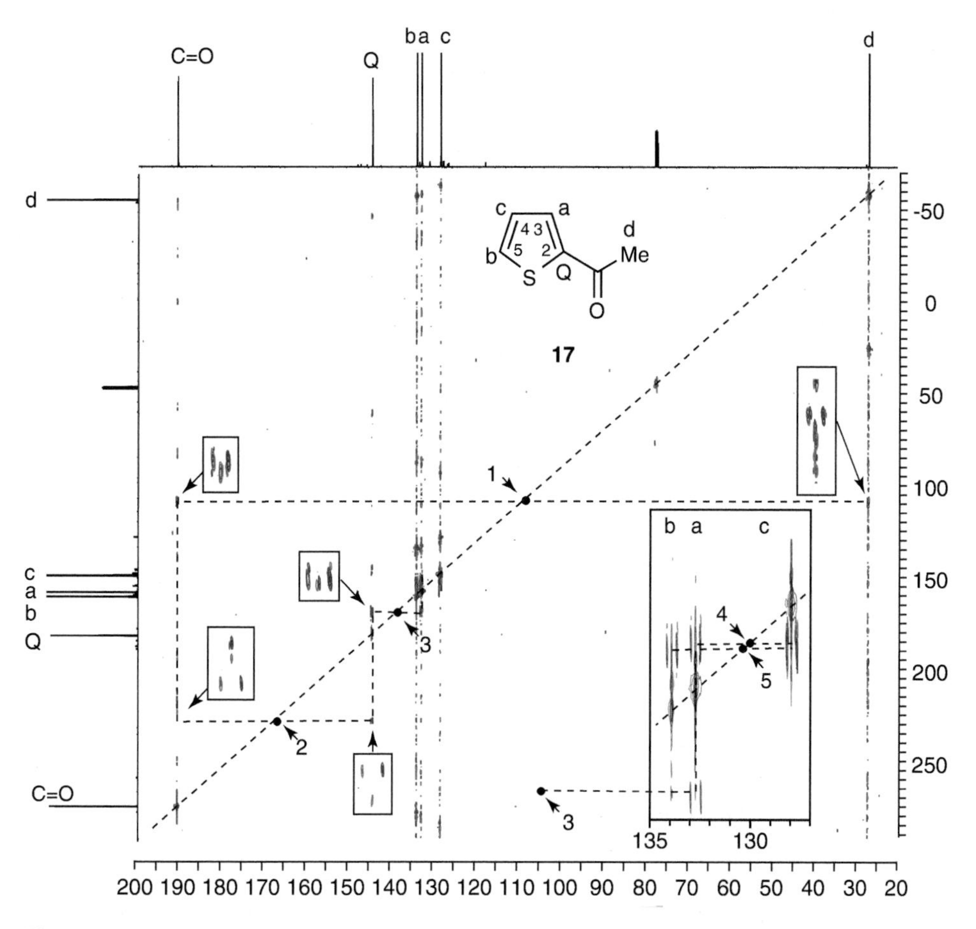

Fig. 5.39 125 MHz INADEQUATE spectrum of 2-acetylthiophen **17** in $CDCl_3$

3356.4 Hz for signal d, and 23978.6 Hz for the C=O signal. The sum is 27,335 Hz, which is divided by 2 and converted to δ units for the scale drawn on the right. This geometry places the mid-point of the horizontal dashed line on the diagonal, with a frequency half way between the frequencies of the connected carbons. The mid-point is emphasised with a black dot and labelled 1 in Fig. 5.39. Thus searching for a connected pair of doublets is made easier, because they are always equally spaced horizontally on either side of the diagonal.

The next step is to change direction and look straight up or down until we find another doublet. Having found it, in this case below, we change direction again and look horizontally along the line, with the mid-point labelled 2, for another, and find it vertically below the carbon signal labelled Q. This horizontal line establishes that the carbonyl carbon is directly bonded to the ring carbon with no hydrogen atoms attached to it, C-2 in the structure **17**. Turning through right angles again, we see vertically above this signal a doublet on the same horizontal level, with the mid-point labelled 3, as a doublet directly below that from the carbon atom labelled a. This is the key connection that tells us that the carbon signal Q from C-2 is directly bonded to the carbon atom giving the signal labelled a, which must therefore be from the carbon numbered C-3 in the drawing **17**. Looking vertically again, more clearly by going to the expanded inset in the bottom right of Fig. 5.39, we find a peak on another horizontal line, with the mid-point labelled 4, as the peak below the carbon signal labelled c. The horizontal line 4, connecting signals from carbons a and c, identifies the next carbon in the chain, which must be from C-4. Turning to the vertical axis we find nothing, and must therefore deduce that this signal is an elongated one, because we do see a doublet, with the mid-point labelled 5, leading from the lower half of the signal below c to the signal below the one labelled b. We have now completed the set of connections, in which each C—C bond is identified by a horizontal hashed line between pairs of doublets, and the order in which they come, 1–5, is established by the vertical dashed lines that connect them. In a weak spectrum, the diagonal helps us to locate the connecting doublets, and the doublet structure helps us to distinguish the peaks from the noise. There is a singlet corresponding to each signal on the f_2 axis on the diagonal in this spectrum, but it is sometimes too weak to register completely in other spectra. This spectrum used 150 mg in a run taking 7 h that was just good enough to find all the connections.

Figure 5.40 shows a more ambitious INADEQUATE spectrum, processed and displayed in a different way, more like a COSY spectrum. In this spectrum of the ester **18**, signals from only three of the carbon atoms show up on the diagonal, namely the peaks labelled α, β and γ, which are the signals from C-9, C-1 and C-2′, respectively. Each of these atoms is at the terminus of a chain, and that appears to be why they register on the diagonal.

As before, we start with a signal that can be assigned unambiguously, and we turn through 90° for each connection. The signal β, which appears, typically downfield, at δ 170.8, must be that given by the ester carbonyl group C-1. Working from C-1, the cross peak labelled *a* is horizontally aligned with the signal β from C-1 on the vertical scale at the left-hand side of Fig. 5.40 and it is directly below the signal at δ 41.1 on the horizontal scale at the top. This tells us that the signal at δ 41.1 is produced by C-2. The cross peak *a* is, in turn, on the same vertical line (dashed) as the cross peak *b*, which is horizontally

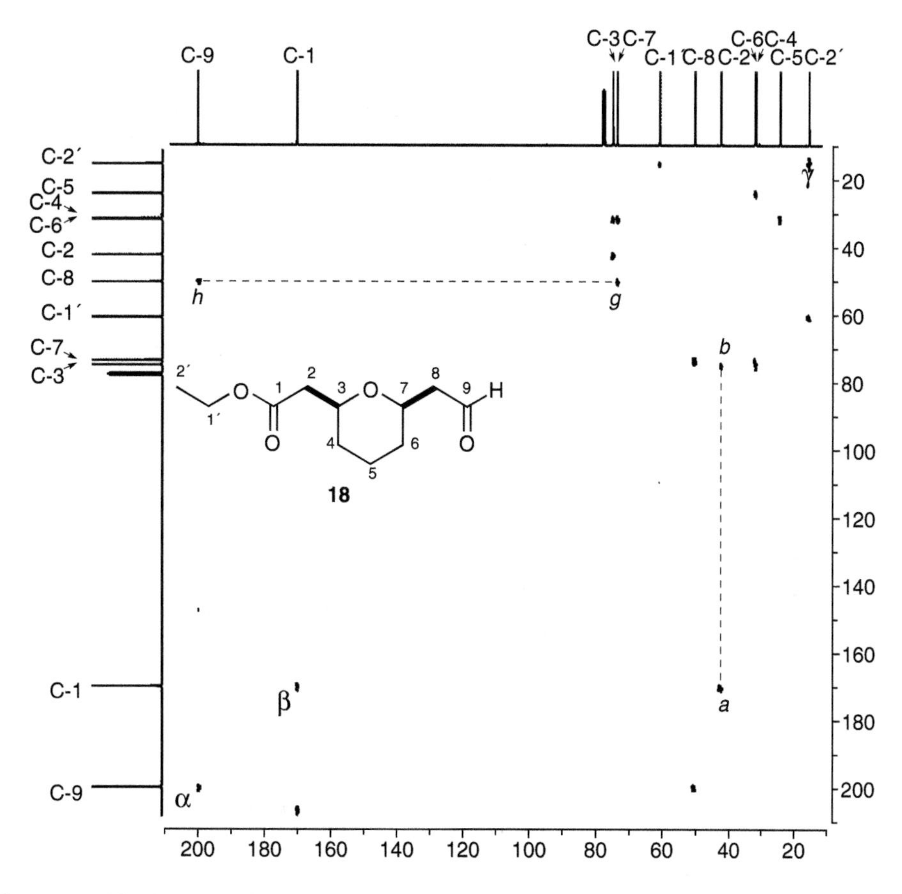

Fig. 5.40 125 MHz INADEQUATE spectrum of the ester-aldehyde **18** in $CDCl_3$

aligned with the signal at δ 74.3 on the vertical axis, showing that the signal at δ 74.3 must be from C-3. To continue the assignment, it is helpful to look at the enlargement of the high-field end of the spectrum shown in Fig. 5.41.

Continuing the analysis using Fig. 5.41, we have reached the cross peak labelled *b*, which is connected by the horizontal line to the cross peak *c*. This cross peak is directly below the signal at δ 30.1, which must be from C-4. Following the vertical dashed line from *c* to the cross peak *d* then identifies the connection from C-4 to C-5, which we can see gives rise to the peak at δ 22.9. From *d*, there is no obvious connection, and so we deduce that *d* must connect to the signal *e*, which is so close to *d* that we cannot easily see its fine structure. The cross peak *e* is vertically below the signal at δ 30.7, which must be the signal from C-6, not surprisingly very close in chemical shift to the signal from C-4. The cross peak *e* is in turn

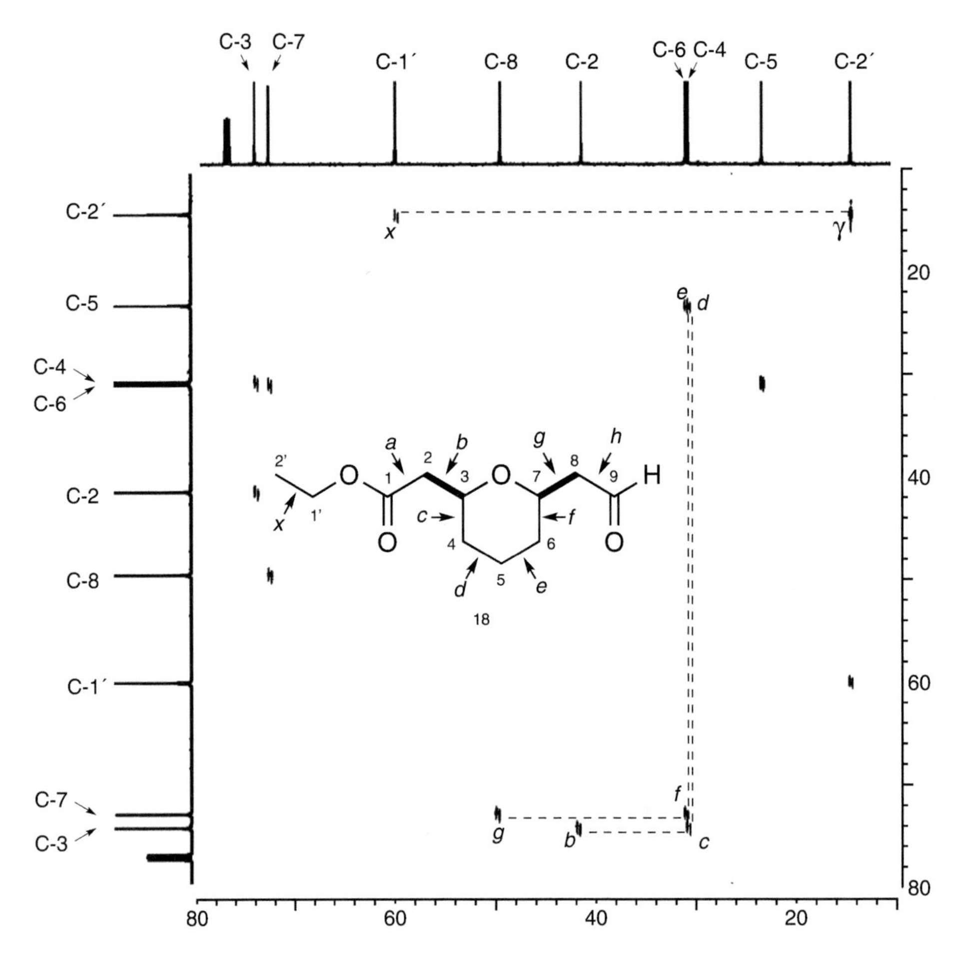

Fig. 5.41 Detail from the 125 MHz INADEQUATE spectrum of the ester-aldehyde **18** in $CDCl_3$

vertically above the cross peak *f*, which is aligned with the signal at δ 73.1 on the vertical axis, and which must therefore be from C-7. The cross peak *f* is connected horizontally to the cross peak *g*, vertically below the signal at δ 49.5 on the horizontal axis, which is the signal from C-8. Finally, returning to the full spectrum in Fig. 5.40, we can see that the cross peak *g* is connected to the cross peak *h*, which is vertically below the signal at δ 201.1, the signal from the aldehyde carbonyl carbon, C-9. The cross peak *x* (Fig. 5.41), which does not connect on to any other cross peaks, identifies the connection between C-1′ and C-2′ in the ethyl group. In summary, the cross peaks *a-h* and *x* identify the bonds shown in the drawing **18** in Fig. 5.41.

Note that, as we make each connection, we change direction through 90° as we look for the next cross peak. We could equally easily have started from the aldehyde carbonyl group, and followed the chain of connections through to C-1, using the cross peaks not labelled in Figs. 5.40 and 5.41. As before, all the cross peaks are doublets, because the signals detected are from one ^{13}C attached to only one ^{13}C. The separation of the doublets measures the coupling constant, $^1J_{CC}$. In this case, fairly typically for a chain of tetrahedral carbons, all the coupling constants are in the range 30–40 Hz (Sect. 4.5.3).

This INADEQUATE spectrum, free enough from noise and from the need to identify the doublets by the phase change, required 250 mg of sample, a cryoprobe, and 48 h of acquisition time. Clearly it is an immensely powerful technique for identifying connections through a carbon skeleton, but it is a costly one, often limited by low solubility.

5.7.3 HSQC-HECADE: Measuring the Sign and Magnitude of ^{13}C–^{1}H Coupling Constants

The HSQC-HECADE experiment measures the sign and magnitude of coupling constants from carbon to hydrogen. The sign enables us to distinguish $^2J_{CH}$, $^3J_{CH}$ and $^4J_{CH}$ coupling constants, because they are more or less reliably, negative, positive and negative, respectively. The magnitude gives us information about conformation, since $^3J_{CH}$ coupling constants are dependent upon dihedral angles, just as $^3J_{HH}$ coupling constants are. We saw in Sect. 4.5.2 that $^1J_{CH}$, $^2J_{CH}$ and $^3J_{CH}$ can be measured in the ^{13}C spectrum when proton decoupling is turned off, but, with more complicated compounds than the ester **1** in Chap. 4, it is not so easy to measure these coupling constants, and it can be impossible to identify the protons to which the ^2J and ^3J coupling occurs. Determining relative stereochemistry using NMR spectroscopy is a particularly difficult task in open-chain compounds, where conformation may be unpredictable and NOEs unreliable.

As an example to illustrate how HSQC-HECADE spectra are read, we shall use the diol **20**, which was prepared by the addition of propynyl lithium to the meso dialdehyde **19**, followed by separating it from the mixture of the four diastereoisomeric products. Two of the products were identifiably centrosymmetric by their having only one set of NMR signals for each half. The problem was to determine which of these two was which.

OBn OHC CHO OBn **19** Li —≡— → OH OBn 6 4 5 OBn OH **20** + 3 other diastereo-isomers

The diol **20** turned out to be the anti isomer with respect to the relative configuration between C-4 and C-6. COSY and HMQC spectra had allowed all the signals to be assigned for both centrosymmetric isomers. As we shall see, the HSQC-HECADE spectrum in

Fig. 5.42 allows signs to be attached to the proton-proton coupling constants, confirming which were vicinal (positive) and which geminal (negative). These in turn allowed the relative stereochemistry between C-4 and C-6 to be determined. Only the relevant part of the HSQC-HECADE spectrum for the anti isomer **20** is shown; it is an HSQC spectrum, but with cross-peak structure providing information about the $^{2}J_{CH}$ and $^{3}J_{CH}$ coupling constants.

Let us first look at the cross peaks between C-6 (at δ 81.05) and the double-doublet for H-6 (at δ 3.96). In an ordinary HSQC spectrum this would be a single peak identifying only that these two atoms are connected, but in the HSQC-HECADE spectrum in Fig. 5.42 it is a pair of peaks, separated in both dimensions by 142 Hz. This is the $^{1}J_{CH}$ coupling constant between these two atoms. This particular cross peak does not provide useful

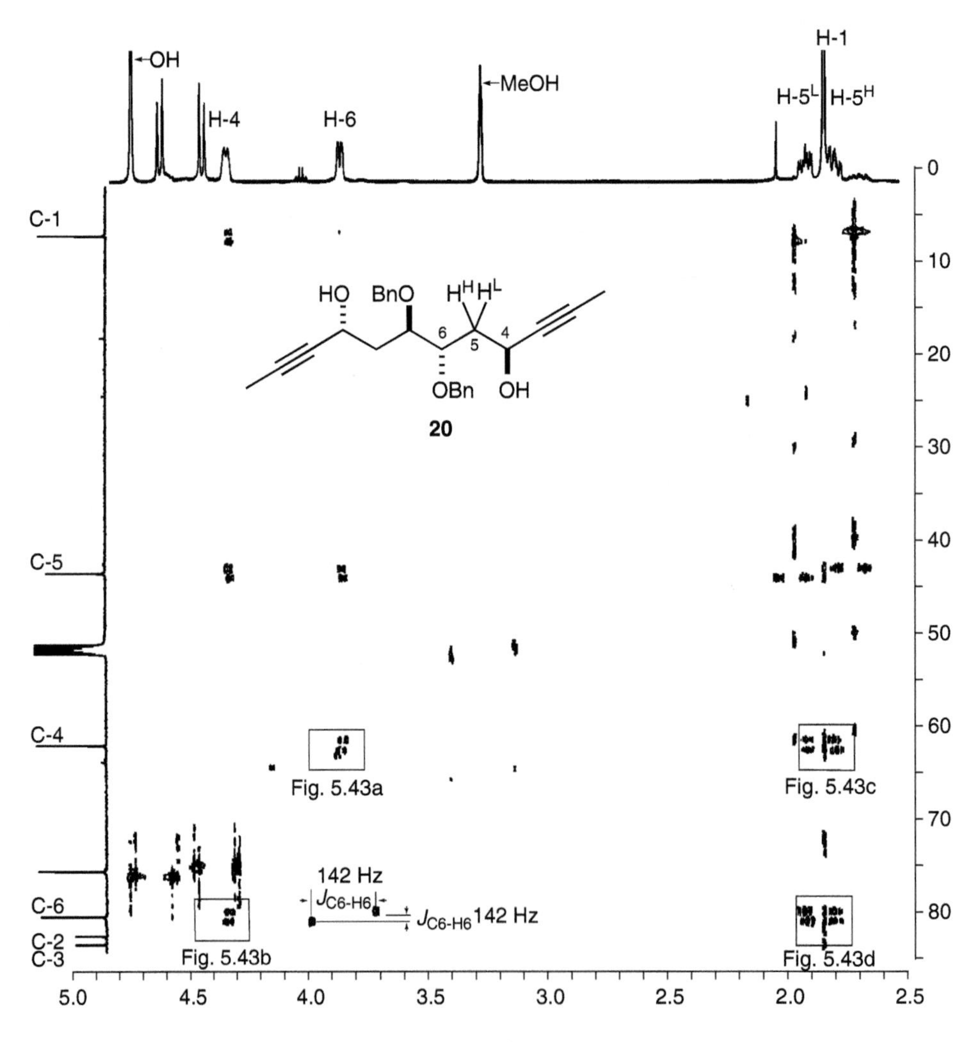

Fig. 5.42 Part of the 500 MHz HSQC-HECADE spectrum of the diol **20** in $CDCl_3$

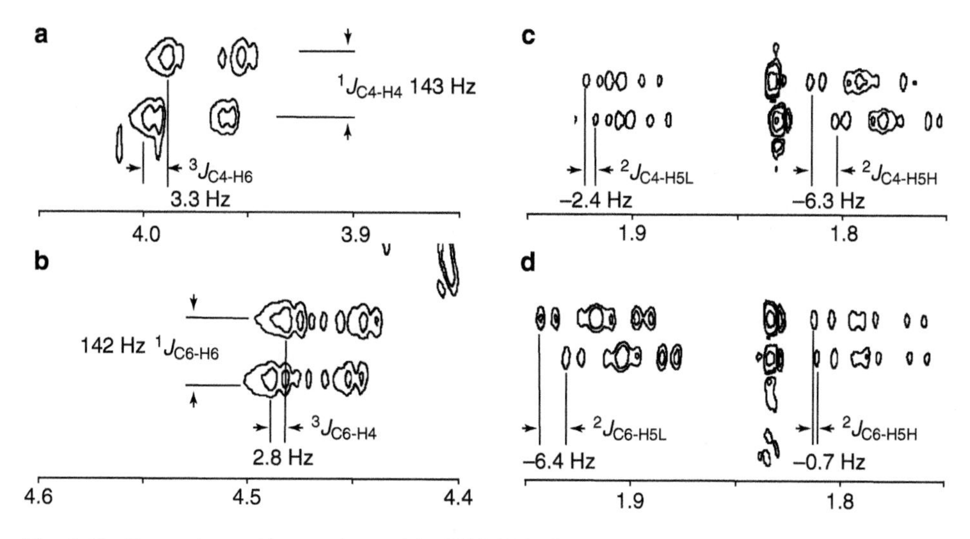

Fig. 5.43 Expansions of four regions of the HSQC-HECADE spectrum from Fig. 5.42

information for structure determination, but it illustrates how HSQC-HECADE spectra are read. For more detailed use of the HSQC-HECADE spectrum, we look at the more telling expansions in Fig. 5.43.

The $^3J_{CH}$ values are measurable in the same way as the $^1J_{CH}$ values—they can be seen more clearly in Figs. 5.43a and b—with the $^3J_{CH}$ values of 3.3 Hz for $^3J_{C4\text{–}H6}$ and 2.8 Hz for $^3J_{C6\text{–}H4}$ represented by the horizontal displacement. The sign of the coupling constant is given by the direction of the horizontal displacement—if the upper (lower frequency) pair is to the right (to lower frequency) of the lower it is positive. (We have seen this pattern already in Fig. 5.31.) The values measured are probably only accurate to the nearest whole number. Being at the lower end of the range (0–8) for $^3J_{CH}$ values, they suggest that the carbon atoms and the hydrogens are gauche rather than antiperiplanar. At this stage it is not known whether this isomer has the anti or the syn relationship between the stereocentres at C-6 and C-4. Therefore, the conformation about the C4-C5 bond could be a consequence of any relative weighting of the conformations **21** and **22** (anti isomer), or of **23** and **24** (syn isomer). Similarly, the conformation about the C5-C6 bond could be a consequence of any relative weighting of the conformations **25** and **26**. (We are assuming that the conformations with the three hydrogen atoms all together are little populated.) The $^1J_{CH}$ values for C-4 and C-6 of 143 and 142 Hz, respectively, are easily measurable by the vertical displacements in Fig. 5.43a and b, but they are not needed in this structure determination.

21	**22**	**23**	**24**	**25**	**26**
C-6 (front); C, H-4; H, H (rear); OH	C-6 (front); H-4, OH; H, H (rear); C	C-6 (front); HO, H-4; H, H (rear); C	C-6 (front); H-4, C; H, H (rear); OH	C (front); H, H; H-6, OBn (rear); C-4	OBn (front); H, H; C, H-6 (rear); C-4

The next step in the analysis is to assign which is which of the two methylene protons on C-5, labelled H-5^L for the low-field signal at δ 1.91 and H-5^H for the high-field signal at δ 1.79. The vicinal proton-proton coupling constants are 3.3 Hz between H-6 and H-5^H, 9.3 Hz between H-6 and H-5^L, 9.3 Hz between H-4 and H-5^H, and 4.0 Hz between H-4 and H-5^L. These are governed by a Karplus relationship in the usual way, and indicate that H-5^L is predominantly anti to H-6 and gauche to H-4, whereas H-5^H is predominantly gauche to H-6 and anti to H-4, indicating that the conformations **21**-**26** can be relabelled with a little more detail as **21a**-**26a**.

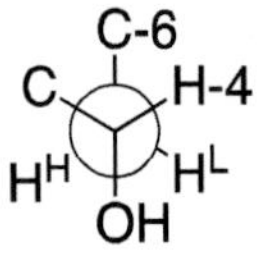

21a 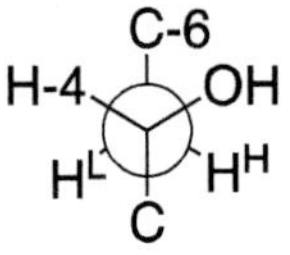 **22a** 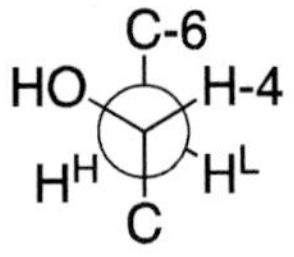 **23a** 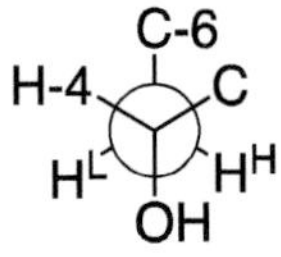 **24a** 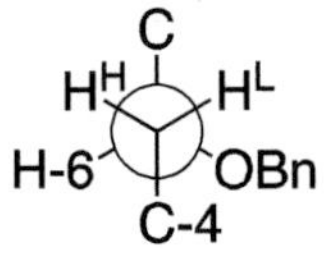 **25a** 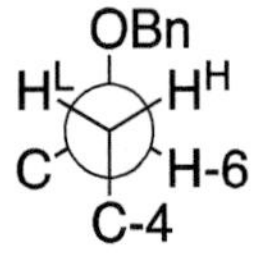 **26a**

To decide between the various possibilities, we need the $^2J_{CH}$ coupling constants which are found in the detail of Figs. 5.43c and d. The $^2J_{CH}$ displacements are in the opposite direction from the $^1J_{CH}$ and $^3J_{CH}$ displacements, because $^2J_{CH}$ coupling constants are negative. Figure 5.43c shows that H-5^L and H-5^H have $^2J_{CH}$ coupling constants to C-4 of −2 and −6 Hz, respectively, and Fig. 5.43d shows that H-5^L and H-5^H have $^2J_{CH}$ coupling constants to C-6 of −6 and −1 Hz, respectively. These numbers are not governed by a Karplus equation. They are governed by the electronic relationships of the carbon and proton to the other substituents, especially when one of them is electronegative, as here. Assigning the conformational relationships from them uses a large amount of empirical data, which suggests that a proton held antiperiplanar to an electronegative element will have a smaller (more positive, Sect. 4.7.1) coupling constant to the carbon carrying the electronegative substituent. In this case, the relationships in **22a**, **23a** and **25a** have the oxygen substituent anti to the proton with the smaller $^2J_{CH}$ coupling constant, suggesting that these conformations explain the differences in the coupling constants, which the others do not. The full protocol used to identify which of the possibilities like those shown as **21a**-**26a** is true is given for the general case in [1]. The result in this case is the unambiguous assignment that the conformation **25a** is the more populated of the pair **25a** and **26a**. This shows that H^H is the hydrogen atom which would project towards the reader in the drawing **20** and H^L is the hydrogen atom which would project away from the reader. This matches the configuration in **22a** and not that in **23a**. Putting together the two drawings **25a** and **22a** defines the preferred configurational relationship between C-4 and C-6 as anti. Furthermore, it defines the most populated conformation of this half of the molecule as that shown in Fig. 5.44, which includes the coupling information that was used to arrive at it. The other half of the centrosymmetric molecule, of course, exactly mirrors this half.

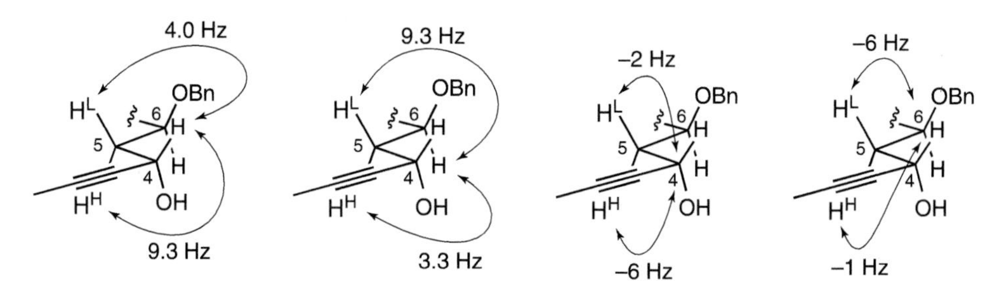

Fig. 5.44 Conformation of the diol **20** deduced from the HSQC-HECADE spectrum

5.8 Three- and Four-Dimensional NMR

When the signals of an NMR spectrum are spread out into a second dimension, the dispersion of the spectroscopic information is thereby improved. It would clearly be an advantage to extend this concept by obtaining three- (and even four-) dimensional NMR spectra. Such spectra are in fact available, with the HSQC-TOCSY (Sect. 5.5.8) an example which we have already seen. That spectrum was conveniently easy to display in two dimensions, and so did not need a separate section to explain it. Although only a very limited treatment of this rather specialised subject is justified here, 3D and 4D spectra are powerful in determining the three-dimensional structures of relatively large and important molecules such as polypeptides, proteins and lengths of DNA duplexes (DNA double helices). This area is important in the realm of structural biology, but is covered only in very brief outline here.

Consider the problem of determining the three-dimensional structure of a small protein, which typically might consist of a sequence of 100 amino acids, where the segment **27** shown has arbitrarily an alanine and a serine residue, which may or may not be adjacent.

27

The first step would be to identify the proton spin systems of each amino acid, which can in principle be achieved using 2D TOCSY spectra (Sect. 5.3), but in practice too many of the signals are likely to overlap in a large molecule. For example, if the N*H* of the alanine residue and the N*H* of the serine residue in the protein, whether they are adjacent or not, had the same proton chemical shift, the proton spin systems of the two amino acids

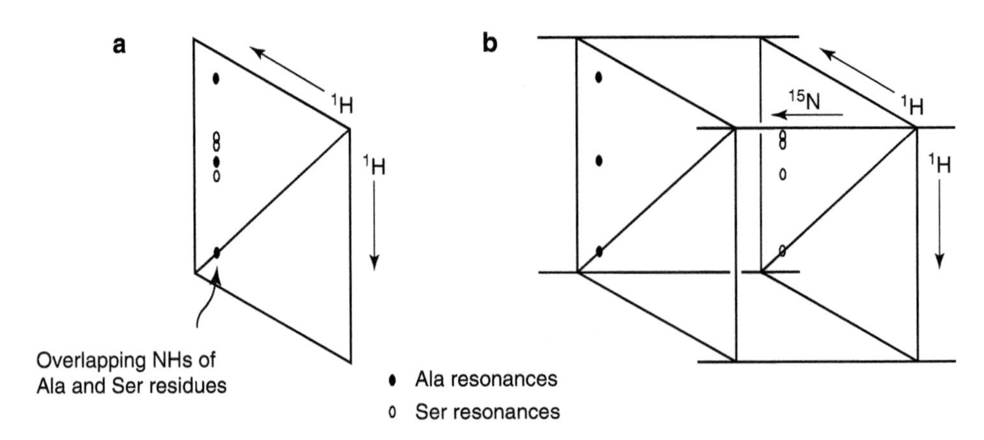

Fig. 5.45 Scheme for a 3D TOCSY-^{15}N-HMQC spectrum

would both have cross peaks on a line defined by the N*H* chemical shift in the TOCSY spectrum, which might look like Fig. 5.45a. The problem arising because the two NHs have the same proton chemical shift can be overcome by carrying out the biosynthesis of the protein in an environment where it is given ^{15}N-enriched ammonium nitrate as its nitrogen source. The protein is then produced as its ^{15}N-enriched analogue, and it is possible to take a 3D TOCSY spectrum so that each of the 2D TOCSY spectra are separated into a third dimension. Each TOCSY spectrum is found in a plane defined by the ^{1}H axes in the usual way, but the planes are separated according to the ^{15}N chemical shift of each amino acid. If the alanine N and the serine N have different chemical shifts, the two planes might look something like Fig. 5.45b, where the TOCSY signals from the alanine and the valine are separated onto different planes. As the name for this experiment implies, 3D TOCSY-^{15}N-HMQC, these spectra can be regarded as the combination of a 2D TOCSY, with an HMQC in the third dimension, except that in this case the heteronuclear component is ^{15}N and not ^{13}C.

Once the spin systems of the amino acids of the protein are assigned, all the NOEs which are available between the assigned proton resonances must be analysed. The 3D experiment for this assignment is the 3D NOESY-^{15}N-HMQC spectrum. In this spectrum the planes that show the TOCSY cross peaks in Fig. 5.45b now have the NOESY cross peaks. Thus, as long as the N*H* proton resonance of each amino acid is defined by a unique combination of ^{15}*N*H and N*H* chemical shifts, all the protons near in space to the NH proton of a specified amino acid can be identified. This is particularly useful in defining which parts of the protein form α-helices and which form β-sheets. In an α-helix **28**, the NH of the *i*th amino acid will give an NOE to the NH of the (i + 1)th, whereas in a β-sheet **29** it is the α-CH of the *i*th amino acid which gives an NOE to the NH of the (i + 1)th amino acid. Using these NOEs, and others which will not be detailed here, further detail of the secondary structure in solution can often be assigned.

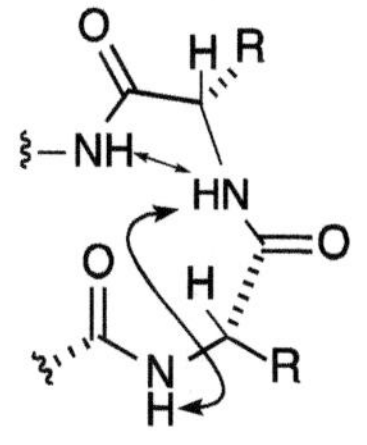

28

Two amino acids in the conformation of an α-helix: NOEs from NH_i to NH^{i+1} are seen.

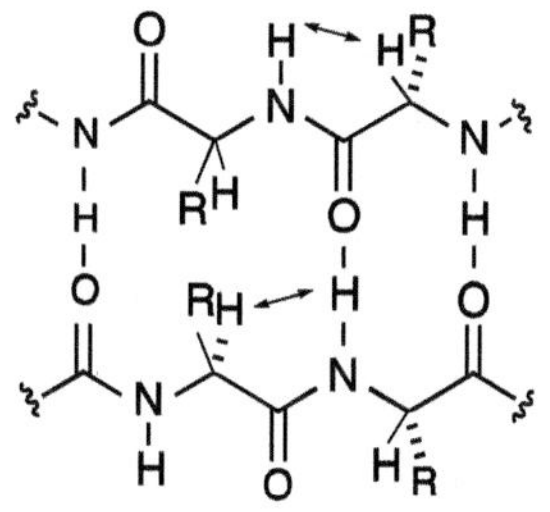

29

The conformation of a β-sheet: hydrogen bonds are indicated by dashed lines. NOEs are seen from the α-CH of the *i*th to the NH of the (*i*+1)th amino acid

Finally, other through-space interactions can be picked up using a third 3D experiment giving a 3D TOCSY-NOESY spectrum, in which a series of 2D TOCSY spectra are created with each collecting all the NOESY connections to a single identifiable proton. The diastereotopic methyl groups of valine and leucine, for example, are apt to give rise to recognisable sharp doublets, and to be held in space in a position to interact through space with other residues. Thus, a series of 3D spectra can identify the signals from the component amino acids, and then establish how the amino acid residues, which may be distant from each other in the linear sequence, can be close in the folded structure. Given enough NOEs and a good enough separation of the ^{15}NH and N*H* signals, the full folded structure of a protein can be described with a precision approaching that found by X-ray crystallography. Such information has the further advantage over most X-ray structures that it gives details of where the hydrogen atoms are, and it applies to the solution structure.

5.9 Hints for Structure Determination Using 2D-NMR

1. After studying all the 1D spectra, identify any still uncertain spin connections from the COSY spectrum and, if still necessary, from a 2D TOCSY spectrum.
2. At some stage, if necessary, use one or more of the methods for deconvoluting overlapping multiplets. The methods discussed in this and the previous chapter are difference-decoupling, 1D- or 2D-TOCSY, difference-NOE or NOESY, CLIP-HSQC, HSQC itself or HSQC-TOCSY, any one of which might offer you a site of attack.
3. Correlate the carbon and proton assignments using an HSQC spectrum, using it also to identify the pairs of diastereotopic methylene protons. It is helpful to put labels (a, b_2, c_2, d, e…x_3, for example) on the carbon signal attached to the protons already labelled A, B_2, C_2, D, E…X_3, for example. Use a code of your own devising, if you don't like this one.

4. Use an HMBC spectrum to work out the 2- and 3-bond connections from protons to carbon atoms, either from one spin system across a barrier to another, or even within the spin system if you have not already made those connections. This is the most challenging stage of an analysis, largely because of the difficulty of not knowing, in the 1D proton spectrum, in the COSY spectrum and in the HMBC spectrum, whether a coupling or a cross peak identifies a 2- or 3-bond connection (or even an unusually prominent 4-bond connection). It is often necessary to hold several possibilities in play before you find good arguments to narrow them down.
5. Look back at the 1D spectra, to see if there are any features there that add detail, or clarify ambiguities, in the light of the more advanced analysis you will have reached with the 2D spectra. In particular, and if appropriate, use the coupling constants from the ^1H spectrum and NOE connections to assign conformation and relative configuration.
6. Use the computer programs within ChemDraw, or those within the NMR processing programs, to predict the ^{13}C chemical shifts, and compare them with the values on the spectrum. These programs are fairly reliable to within a few p.p.m.; a serious discrepancy from the predicted value should give you pause, and make you think again.
7. As a last resort, rarely necessary and only if you have enough compound and spectrometer time, take one or both of the ADEQUATE or INADEQUATE spectra to identify the connectivity of the carbon atoms. Consider whether an ADEQUATE or HECADE spectrum will clear up the 2- or 3-bond ambiguity. It may.

5.10 Further Reading

See list in Chap. 4.

5.11 Table of Information

Table 5.1 A checklist of techniques and conditions, typical for an organic compound of MW ~300 using a 500 MHz instrument (less material may be used, with a corresponding increase in acquisition times)

Name		Description	Typical sample size	Typical run time
APT	1D ^{13}C	Plots ^{13}C signals for C & CH_2 on one side of a horizontal line, and CH and CH_3 on the other	30 mg	15 min
DEPT	1D ^{13}C	Plots ^{13}C signals separately for C, CH, CH_2 and CH_3	30 mg	15 min
1D TOCSY	1D ^1H–^1H	Plots ^1H signals separately for each ^1H-to-^1H spin system	10 mg	2 min

Table 5.1 (continued)

Name		Description	Typical sample size	Typical run time
1D-NOE-difference	1D ^1H–^1H	Plots ^1H signals for each ^1H in close proximity to another	10 mg	10 min
NUS	2D	A method for reducing the time taken by 2D and 3D experiments	–	25% of normal
COSY	2D ^1H–^1H	Cross peaks identify ^1H-to-^1H scalar coupling	10 mg	20 min
2D TOCSY	2D ^1H–^1H	Cross peaks identify members of the same ^1H-to-^1H spin system	10 mg	2 h
NOESY and ROESY	2D ^1H–^1H	Cross peaks identify ^1H-to-^1H proximity in space; NOESY for low and high MW compounds, ROESY for any compound	10 mg	8 h[a]
HMQC and HSQC	2D ^1H–^{13}C	Cross peaks identify ^1H-to-^{13}C one-bond connections	10 mg	40 min
HMBC	2D ^1H–^{13}C	Cross peaks identify ^1H-to-^{13}C two- and three-bond connections	10 mg	2 h
HSQC-HECADE	2D $^{2\&3}J_{CH}$	Cross peaks measure sign and magnitude of $^{2\&3}J_{CH}$ values	30 mg	8 h[a]
INADEQUATE	2D ^{13}C–^{13}C	Cross peaks identify ^{13}C-to-^{13}C one-bond connections	500 mg	48 h[a,b]
1,1-ADEQUATE	2D ^{13}C–^{13}C–^1H	Cross peaks identify ^{13}C–^{13}C one-bond connections when one of the C atoms carries an H	100 mg	4 h
3D TOCSY	3D ^1H–^1H	Cross peaks identify members of the same ^1H-to-^1H spin system, with each spin system in a different plane created from a different pulse sequence	30 mg	24 h[a]

[a]Acquisition time is much reduced with a cryoprobe

[b]To obtain a high-quality spectrum, a cryoprobe is nearly essential

Reference

1. Matsumori N, Kaneno D, Murata M, Nakamura H, Tachibana K (1999). J Am Chem Soc 64:866–876

Worked Examples 6

6.1 General Approach

There is no fixed way of tackling a problem in structure determination using the four spectroscopic methods. Each problem has its own unique features, and some knowledge of the provenance makes a big difference to how one starts. At one extreme, the product of a reaction with known starting materials is often predictable, and the purpose of the spectroscopic investigation is largely to check that the compound actually isolated does have the expected structure. On the other hand, a complete unknown, extracted from a plant for example, will have little information to be drawn on. In the former case one must guard against a too easy assumption that the predicted structure fits the spectra; in the latter one can only draw on one's knowledge of the known structures of natural products—often of limited use since these are astonishingly diverse. In the discussion that follows, we work through some representative examples in which provenance is missing. This is not all that realistic, since compounds handled in research rarely have so little information attached to them, but these examples will serve to show how one can go from one spectrum to the next to draw out all the information needed in order to assign a structure. The sequence we use here to put together the pieces of spectroscopic information in order to deduce the structures is by no means the only one that would work, or even necessarily the shortest. We hope that what we do serves to show how expeditiously the various leads can be connected.

It is almost always best to begin with the molecular weight of the unknown derived from its mass spectrum, and ideally to obtain the molecular formula and hence (Eq. (1.4)) the number of double bond equivalents (DBE) from a high-resolution measurement or a combustion analysis. Fragmentation in EI-MS is rarely crucial except when there is too little material for any other spectra to be taken, or when an oligopeptide is to be sequenced; if necessary, it can be called upon to see if something obvious shows up there—such as an easily recognised fragment like benzoyl (105), benzyl (91), phenyl (77) or methoxycarbonyl

I. Fleming, D. Williams, *Spectroscopic Methods in Organic Chemistry*,
https://doi.org/10.1007/978-3-030-18252-6_6

(59), for example. A striking absence of fragmentation or an even-numbered McLafferty rearrangement product can also be helpful. Sometimes, UV provides critical information on the extent of conjugation to hold in the back of your mind as you put together a structure. The IR spectrum giving information about the presence or absence of functional groups and, perhaps critically, the size of ring in which a carbonyl group is incorporated, is almost always worth keeping in mind from the beginning. The spectra of greatest power in structure determination, however, are ^{13}C and ^{1}H NMR. We begin with some exceptionally simple examples to set the scene.

6.2 Simple Worked Examples Using ^{13}C NMR Alone

The ^{13}C NMR spectrum gives a wealth of information, often making it the first place to look after a formula has been deduced. It gives the number of differently situated carbon atoms in the molecule, approximately subdivided by chemical shift into digonal, trigonal and tetrahedral carbons, and it gives information about symmetry elements, which cause two or more carbons to be in identical environments. An APT or DEPT spectrum helpfully gives the numbers of fully substituted, CH, CH_2 and CH_3 carbons. Furthermore, some functional groups, such as carbonyl groups, have distinctive signals. Carbon spectra are often presented simply as a list of chemical shifts, with each shift being followed by (s), (d), (t) or (q) for the fully substituted, CH, CH_2 and CH_3 carbons, derived from an off-resonance decoupled spectrum (Sect. 4.5.2) or, more likely today, followed by (+) or (−) from an APT or DEPT spectrum (Sect. 4.15). Longer-range C–H coupling, $^{2}J_{CH}$ and $^{3}J_{CH}$, although examined at length in Sect. 4.5.2, is rarely used at this stage, and when it is it is taken from 2D HSQC and HMBC spectra (Sects. 5.5 and 5.6), rather than from the 1D ^{13}C spectrum.

You will find it helpful to consider the ^{13}C spectrum as divided into three regions, starting at the low-field end on the left:

1. The region from 220 to 160 p.p.m., in which the various kinds of carbonyl carbons will give signals (Table 4.15): aldehydes and ketones in the range 220–190 p.p.m., and carboxylic acids, esters and amides at higher field, that is lower p.p.m. values, in the 190–160 p.p.m. range.
2. The region from 160 to 100 p.p.m., in which the various kinds of trigonal carbons other than carbonyl carbons come into resonance.
3. The region from 100 to 0 p.p.m., in which the various kinds of tetrahedral carbons come into resonance. Note that there is some overlap in all these regions, but even the anomeric carbons of sugars, with two attached oxygen atoms, are deshielded to only *ca.* 100 p.p.m. By counting the carbonyl and trigonal carbons, the numbers of double bonds in the unknown can be estimated, but bear in mind that nitrile carbons (–C≡N) occur in the 120–105 p.p.m. region and acetylenic carbons (–C≡C–) in the 90–65 p.p.m. region.

Example 1 Two isomeric hydrocarbons C_5H_{10}, **1** and **2**, were separated by GC and gave the following ^{13}C data. **1**: δ 132(s), 118(d), 26(q), 17(q) and δ 13(q). **2**: δ 147(s), 108(t), 31(t), 22(q) and 13(q). Deduce structures for the olefins.

The hydrocarbons contain one DBE, which is present as a single double bond (each gives two signals in the 100–160 p.p.m. range). Each has no equivalent carbons, and the multiplicities indicate the ten required hydrogens: in **1** as three methyl groups and one trigonal CH, and in **2** as two methyl groups, one tetrahedral CH_2, and one trigonal CH_2. The structures are therefore:

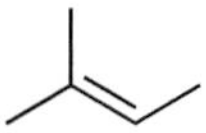

1 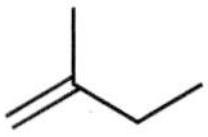 **2**

Example 2 A compound $C_5H_6N_2$ gives a molecular ion in its EI spectrum at *m/z* 94, unchanged when the compound is introduced into the mass spectrometer in the presence of D_2O. The compound gives signals in its ^{13}C NMR spectrum at δ 119(s), 22(t) and 16(t). Deduce its structure.

The compound has four DBEs. Since the molecular weight is unchanged in the presence of D_2O, the six hydrogens must be attached to carbon. The carbon atoms giving rise to the triplet signals at δ 22 and 16 carry two protons each, and therefore, to give six hydrogens in total, one of these signals must correspond to two carbons. Hence, to give five carbons in total, the signal at 119 must also correspond to two carbons, neither of which has any hydrogen atoms attached to it.

If the signal at 119 were from two trigonal carbons, then we could account for only one DBE in this way—clearly not enough. Also, acetylenic carbons (δ 90–65) are precluded, but the two carbons at 119 are consistent with the presence of two nitriles (Table 4.11). This leaves the signals at 16 and 22 to account for three tetrahedral carbons in an acyclic structure **3**, which also accommodates the symmetry required by the data.

NC CN

3

6.3 Simple Worked Examples Using ^{1}H NMR Alone

In simple ^{1}H NMR spectra, singlets, doublets, triplets and quartets are represented as (s), (d), (t) and (q); quintets and sextets (and beyond) are spelled out to avoid ambiguity; unresolved or uninterpretable multiplet signals are indicated by (m), and broad signals by (br). It is again not uncommon to work from left to right (from low to high field). Given the molecular formula of an unknown, the nature of the functional groups can often be inferred, and even complete structures derived solely from these spectra.

Examples 3–5

Three isomeric compounds $C_4H_8O_3$, each of which must therefore contain one DBE, give the following ^{1}H NMR spectra. Deduce plausible structures for each of them.

Example 3 δ 12.1 (1H, s), 4.15 (2H, s), 3.6 (2H, q, $J = 7$ Hz) and 1.3 (3H, t, $J = 7$ Hz).

The quartet signal at δ 3.6 in the ^{1}H NMR spectrum immediately suggests the presence of an OCH_2CH_3 group—its position of resonance is appropriate for an OCH_2 group and the multiplicity tells us that a methyl group is joined to it. The OCH_2CH_3 signal is also present as a triplet at δ 1.3. In more detail, the precise chemical shift of the lower field signal suggests an ethyl ether (OC_2H_5) rather than an ethyl ester ($RCOOC_2H_5$) (Table 4.23). The very low-field proton at δ 12.1 is consistent with the presence of a carboxylic acid (Table 4.31), plausibly accounting for the one DBE. Thus, all that remains is a CH_2 group, which must be placed so as not to be coupled, and its chemical shift at δ 4.15 indicates that it is adjacent to an electronegative element. Therefore the structure is $CH_3CH_2OCH_2COOH$.

Example 4 δ 4.15 (1H, 1:5:10:10:5:1 sextet, $J = 7$ Hz), 2.35 (2H, d, $J = 7$ Hz) and 1.2 (3H, d, $J = 7$ Hz). Spectrum run in D_2O.

Only six of the eight protons are observed in the spectrum, indicating that two protons are probably bound to oxygen, and therefore after exchange with D_2O become OD groups. The integrated areas of the signals, in conjunction with their multiplicities, indicate the structural element CH_3CHCH_2, to which the atoms CH_2O_3 must be attached. The observed deuterium exchange behaviour and the water solubility of the compound can therefore be satisfied by OH and CO_2H groups so that two structures might fit: $CH_3CH(OH)CH_2CO_2H$ or $CH_3CH(CO_2H)CH_2OH$. The methine proton observed in the spectrum (δ 4.15) is in good accord with the expected chemical shift (δ 3.9 ± 0.2 in Table 4.23) if the former structure is correct, with the methine proton's being β to the carboxyl group taking it just downfield of the normal range.

Example 5 δ 4.05 (2H, s), 3.8 (3H, s) and 3.5 (3H, s).

The two three-proton singlets suggest the presence of two methoxyl groups. Perhaps one of these is part of a methyl ester (δ 3.8) and the other part of a methyl ether (δ 3.5), a proposal which takes care of the three oxygen atoms, leaving only a CH_2 group to satisfy the molecular formula. The two-proton uncoupled signal at δ 4.05 must come from that methylene group, which is downfield because it is next to an ether oxygen atom and next to a carbonyl group. The data are therefore consistent with the structure $CH_3OCH_2CO_2CH_3$.

6.4 Simple Worked Examples Using the Combined Application of MS, UV, IR and 1D-NMR Spectroscopic Methods

We suggest that you see how much of the structure you can deduce for yourself from the spectra, and only then read our analysis.

Example 6 The formula is $C_7H_{14}O$, and there is therefore one double-bond equivalent. The UV spectrum has λ_{max} 295 nm, but before doing anything with this information, it is important always to work out the ε value from Eq. (2.2) in order to find out how intense the absorption is. It is usefully rewritten as Eq. (6.1).

$$\varepsilon = \frac{\text{absorbance} \times \text{molecular weight} \times 100}{\text{weight of compound in mg in 100 ml} \times \text{path length in cm}} \tag{6.1}$$

In this case, the numbers are:

$$\varepsilon = \frac{0.52 \times 114 \times 100}{186 \times 1} = 32$$

The absorption is therefore very weak, and typical of the $n \rightarrow \pi*$ absorption of a saturated ketone or aldehyde (Sect. 2.13). The absorption could easily be a trace of strongly absorbing impurity, but in this case, the presence of a ketone group is immediately apparent in the IR spectrum with its very strong carbonyl band at 1710 cm^{-1} (Table 3.7), with no aldehyde absorption in the δ 9–10 region of the ^1H NMR spectrum. With only one double bond equivalent and one heteroatom to account for, we have now found them both. There can be no further functionality, so all we have to do is arrange the carbon skeleton.

The ^{13}C NMR has a distinctive signal at δ 214.3, which is clearly the ketone carbon (Table 4.15). It also shows only six signals, telling us that one of the carbons must be duplicated. In the ^1H NMR spectrum, we see a six-proton doublet at δ 1.09, which must be from two methyl groups, accounting for the doubling of the carbon signals. Since they are a doublet, they must have one proton as a neighbour, and thus be the two methyl groups of an isopropyl group. We also see a three-proton triplet (δ 0.90), which must be a methyl group attached to a methylene group—an ethyl group. From the ^{13}C NMR spectrum we know that there are two methylene groups [labelled (−)], one of which is in the ethyl group and the other could be attached to the ethyl group or to the isopropyl group.

Example 6 $C_7H_{14}O$

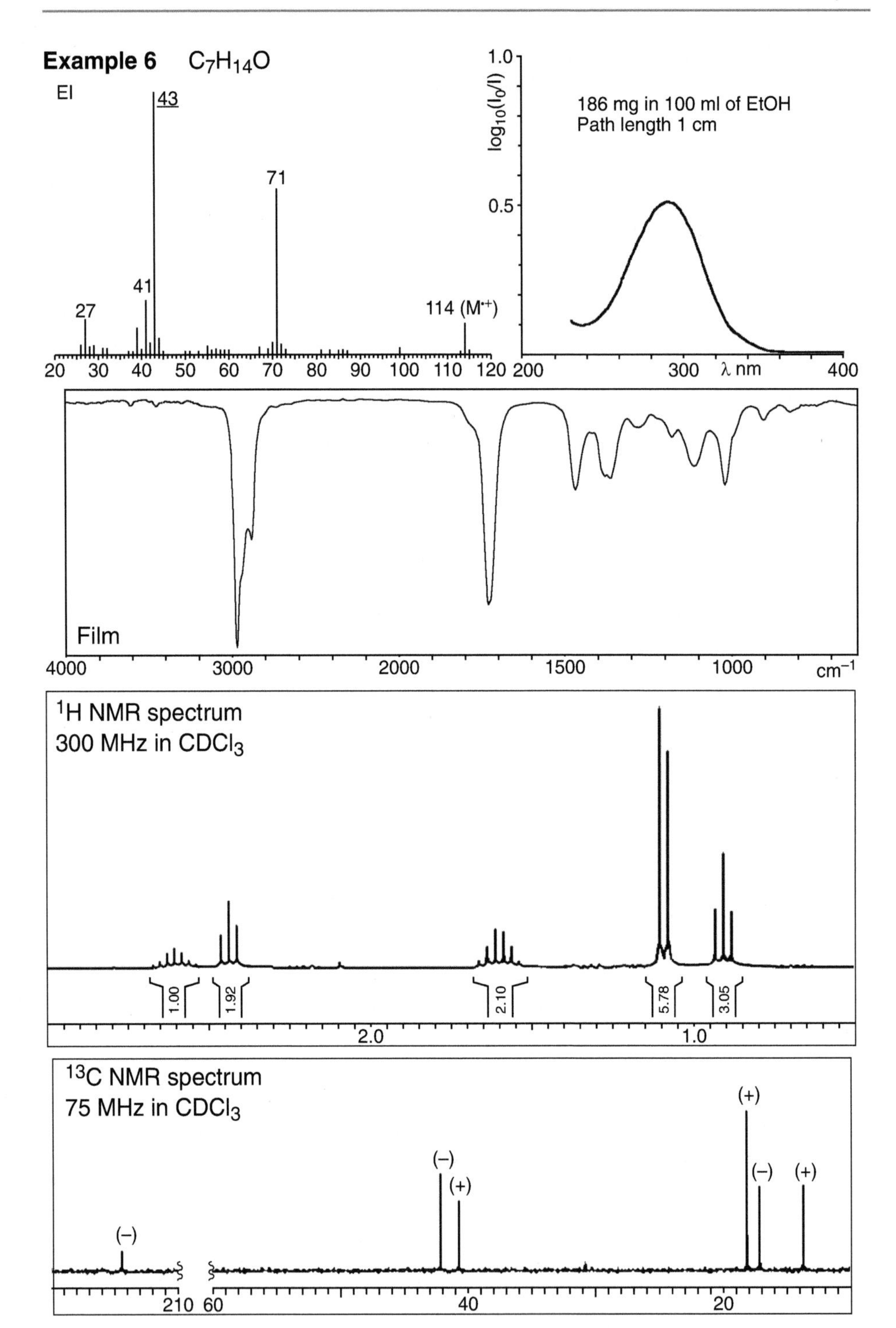

EI
43
71
41
27
114 (M•+)
20 30 40 50 60 70 80 90 100 110 120
1.0
0.5
log10(I0/I)
186 mg in 100 ml of EtOH
Path length 1 cm
200 300 λ nm 400
Film
4000 3000 2000 1500 1000 cm−1
1H NMR spectrum
300 MHz in CDCl3
1.00
1.92
2.10
5.78
3.05
2.0
1.0
13C NMR spectrum
75 MHz in CDCl3
(–)
(+)
(–)
(+)
(+)
(–)
(–)
210 60
40
20

The ^{1}H NMR spectrum has a one-proton septet and a two-proton triplet at lower field, and a two-proton sextet at higher field. This shows that the ethyl group, which will have a two proton signal that must be at least a quartet, is in fact attached to a second methylene group to make a n-propyl group $CH_3CH_2CH_2$ with the methylene group between the methyl and the second methylene a sextet (five-neighbouring protons). This leaves the one-proton signal with only the two methyl groups to couple to, and hence a septet. The structure is 2-methylhexan-3-one $(CH_3)_2CHCOCH_2CH_2CH_3$. The second methylene group is next to the carbonyl group, and so is the methine, accounting for why these two signals are at lower field than that of the middle methylene group. Note how the outer lines of the multiplets get smaller and smaller as the multiplicity increases (Table 3.35).

We could have deduced something from the mass spectrum, which shows α fragmentation (Sect. 1.4.5) giving only a propyl (43) and a propionyl group (71), but this does not tell us that they are different on each side of the ketone group.

Example 7 The molecular formula from the combustion analysis figures is $C_{11}H_{20}O_4$ showing that there are two DBEs. The absence of UV absorption shows that these are not from two conjugated double bonds. The IR spectrum shows a strong carbonyl band at 1740 cm^{-1}, which could be a five-ring ketone or a saturated ester. There is no C=C double-bond absorption in the IR, so the two double-bond equivalents are either two carbonyl groups or one carbonyl group and a ring. There is no OH absorption in the IR, so the four oxygen atoms must be in ketone, ester or ether groups.

The ^{13}C NMR spectrum shows only eight different kinds of carbon atom in a molecule having 11 in all (ignoring the $CDCl_3$ solvent). Some carbon atoms must be repeated in identical structures symmetrically disposed in the molecule. The 1:3:3:1 quartet at δ 4.2, and the 1:2:1 triplet at δ 1.27 in the ^{1}H NMR spectrum suggest the presence of an OCH_2CH_3 group, as we saw in Example 3. In more detail, the precise position of the OCH_2 resonance (δ 4.2) suggests (Table 4.23) that the OCH_2CH_3 group is actually an ester, $CO_2CH_2CH_3$, and not an ether. The intensity of the OCH_2 signal corresponds to *four* hydrogens, which means that there are two (likely identical) $CO_2CH_2CH_3$ groups, thus accounting for the presence of only eight different carbon atoms in the molecule, with three of them duplicated in the CO_2Et groups.

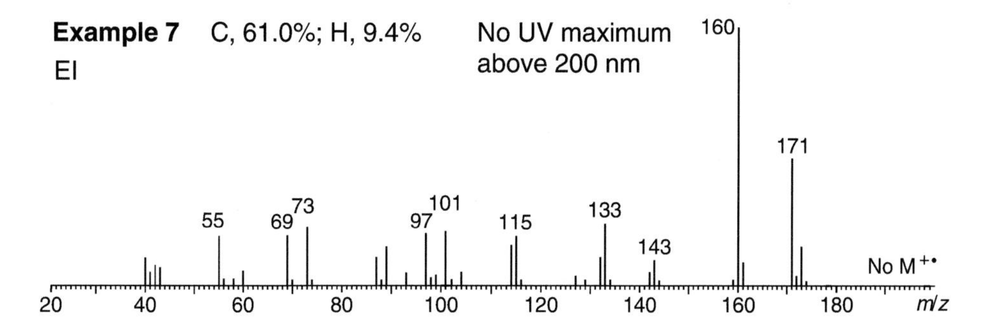

Example 7 continued

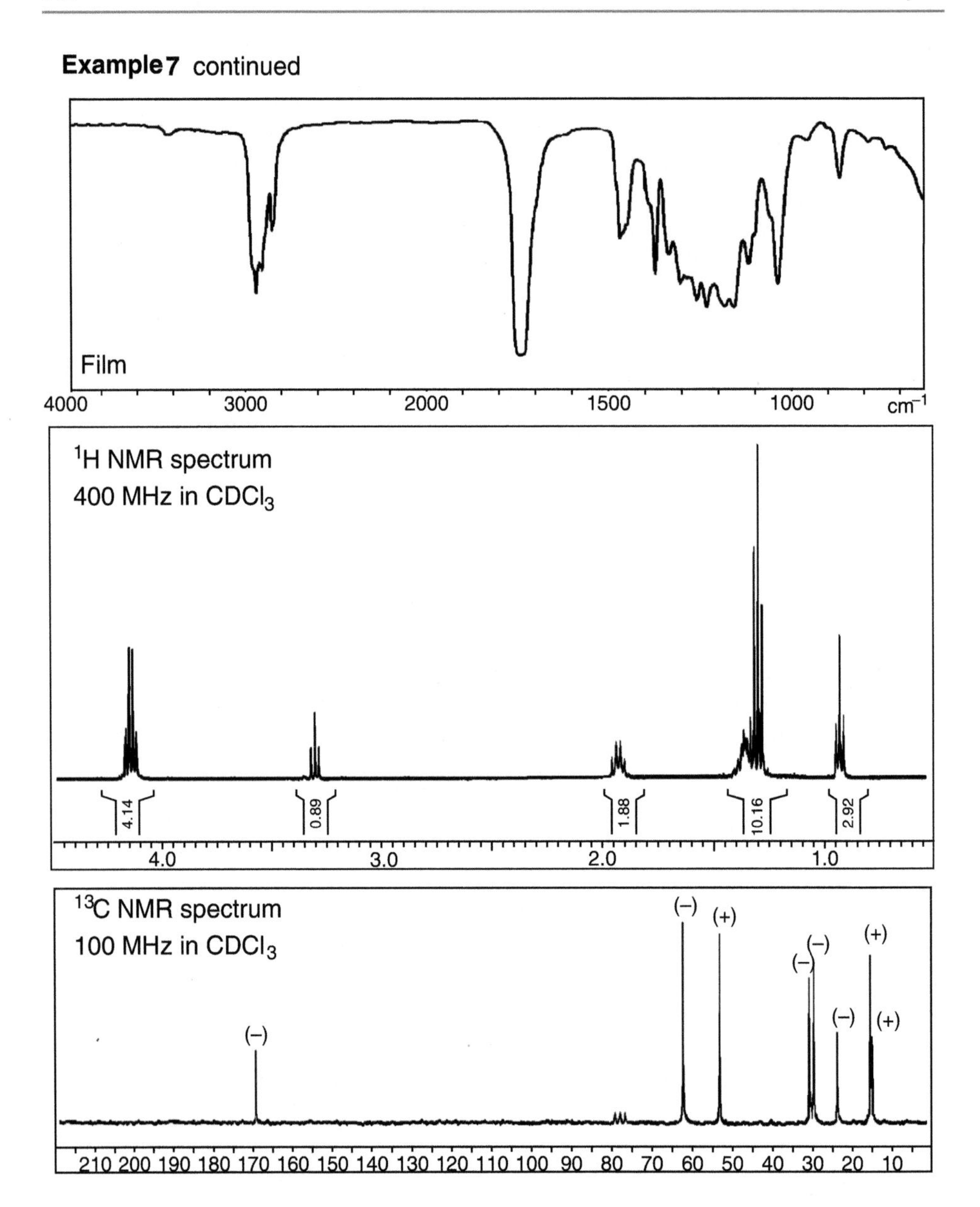

The next informative signal is the one-proton triplet at δ 3.3; we have to account for the chemical shift at which it resonates. Since there is no other functionality in the molecule (the CO_2Et groups account for the two double-bond equivalents and all the heteroatoms), a one-proton triplet at this position can only be produced by the grouping $CH_2C\mathit{H}(CO_2Et)_2$, the carbethoxy groups causing the downfield shift of the hydrogen adjacent to them. Furthermore, the two-proton quartet at δ 1.9 is likely to be the signal from the $C\mathit{H}_2CH(CO_2Et)_2$

hydrogens, and, since it is a quartet, it too is bonded to a CH_2 group (to make the total number of vicinal hydrogens three). Thus, we have the fragment $CH_2CH_2CH(CO_2Et)_2$. The three-proton triplet at δ 0.9 can only be produced by a CH_3CH_2 group; the CH_2 signal of this group, and of the $CH_2CH_2CH(CO_2Et)_2$ group, give the unresolved multiplet at δ 1.3–1.4. The two fragments we have now identified, CH_3CH_2– and –$CH_2CH_2CH(CO_2Et)_2$, account for all the atoms of the molecular formula, and the structure must be made by joining them together. Note how we have been able to put the structure together without having been able to analyse the detail within the methylene envelope at δ 1.3–1.4.

This structure also fits the ^{13}C NMR spectrum: (a) There are two different kinds of methyl group (at δ 14.10 and 13.81), with the signal at a lower field about twice as intense (because it comes from two identical methyl groups). We know that they are methyl groups, not only because of their chemical shift but also because they are displayed on opposite sides of the base line from the $CDCl_3$ signal in the APT spectrum (Sect. 4.15.1). (b) There are three C–CH_2–C groups labelled (−) at δ 29.5, 28.5 and 22.4. (We discussed the assignment of these signals in Sect. 4.18 in the example of the application of Eq. (4.19).) (c) There is a methine group labelled (+) at δ 52.0. (d) There is a CH_2O carbon at δ 61.1 and the carbonyl carbon at δ 169.3, appropriate for an ester carbonyl group.

The mass spectrum also confirms the structure, and the base peak, notable in being an even number (*m*/*z* 160), is the result of β-cleavage with γ-hydrogen (McLafferty) rearrangement (Sect. 1.4.5 and Table 1.11).

Example 8 The mass spectrum in this example is unhelpful—we can tell that the peak at *m*/*z* 56 is not a molecular ion by the presence of a large number of peaks immediately to lower molecular mass than it. The combustion analysis gives a formula, $C_5H_{11}NO_4$. There is therefore one double-bond equivalent, which is clearly not a carbonyl group because there is no absorption in the IR spectrum between 1900 and 1600 cm^{-1} and no signal in the ^{13}C NMR at appropriately low field. Instead, there is exceptionally strong absorption at 1545 cm^{-1}, which is likely to be from a nitro group (Table 3.11), a formulation supported in the first place by the UV spectrum with a weak $n \rightarrow \pi*$ band at 275 nm (ε 24), and in the second by the absence of the molecular ion in the mass spectrum, which is common with aliphatic nitro compounds, because of the ease with which the NO_2 radical is released. The IR spectrum, with strong absorption at 3350 cm^{-1}, also shows the unmistakable presence of a hydroxyl group.

The ^{13}C NMR spectrum shows that there are only four different kinds of carbon atom, which means that two carbons are in the same magnetic environment and are most likely to be two identical groups symmetrically disposed in the molecule. The integration in the ^1H NMR spectrum distributes the 11 hydrogens thus: two hydrogens in each of the signals at δ 4.2 and 3.9, two hydrogens to the multiplet (actually an ill-resolved triplet) at δ 2.95, two hydrogens to the quartet at δ 1.9, and three hydrogens to the triplet at δ 0.9. These last two signals (the two-proton quartet and the three-proton triplet) are obviously an ethyl group, and the chemical shift of the quartet means it must be bound to a carbon atom. The fact that it is only a quartet means that the carbon atom is fully substituted.

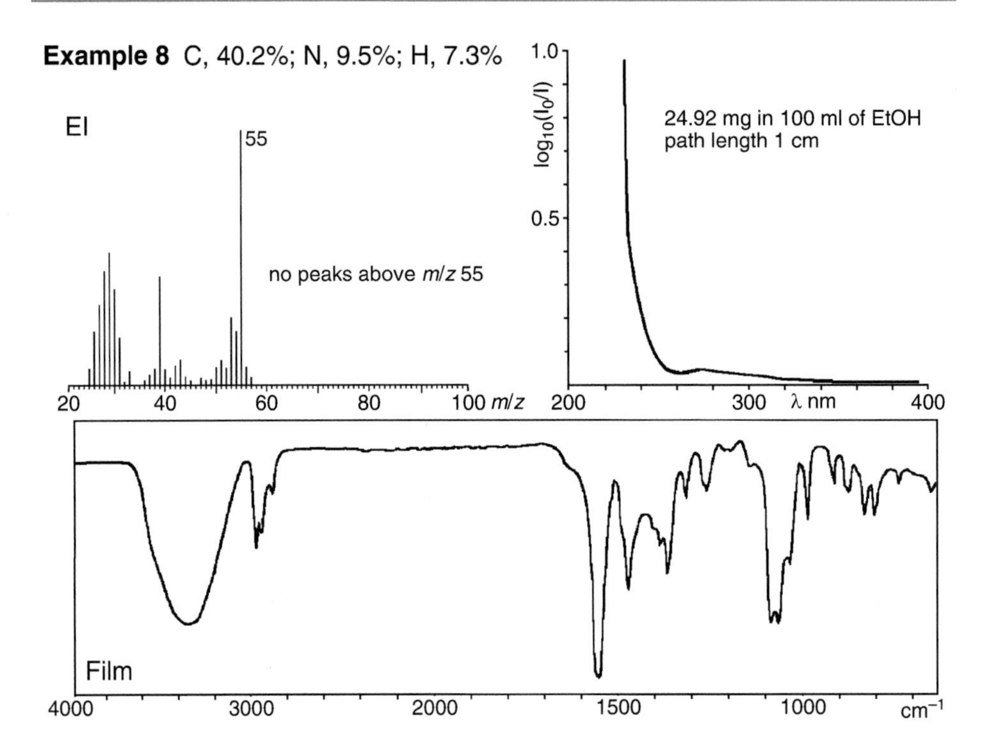

Going back to the ^{13}C NMR spectrum, the *C*-ethyl group is responsible for the signal (+) at δ 7.7 (CH_2CH_3) and either of the (−) signals at δ 63.5 and δ 26.8 (CH_2CH_3), with the latter more likely. The fully substituted carbon atom must be the weak signal (−) at δ 94.2 (fully substituted carbons are usually weak, see Sect. 4.3). Thus, the two identical carbon atoms must be responsible for whichever of the two signals δ 63.5 and 26.8 is not produced by the CH_2 group of the ethyl group. The one at δ 63.5 is nearly twice as intense as the other, so it is likely that it comes from the two identical groups, which are CH_2 groups, labelled (−) because they are on the same side of the base line as the $CDCl_3$ signal in the APT spectrum. They are at a comparatively low field, suggesting that they are attached to an electronegative heteroatom (Table 4.8).

When the ^{1}H NMR spectrum is taken after a shake with D_2O (Sect. 4.4.7), the two-proton multiplet at δ 2.95 and the doubling of the doublets at δ 4.2 and 3.9 disappears, showing that there are two CH_2OH groups in which the hydrogen atoms of the CH_2 group are not equivalent. We have found the fragments **4**, **5** and **6**, which account for all the atoms, and there is only one way of putting them together, namely **7**.

Example 8 continued

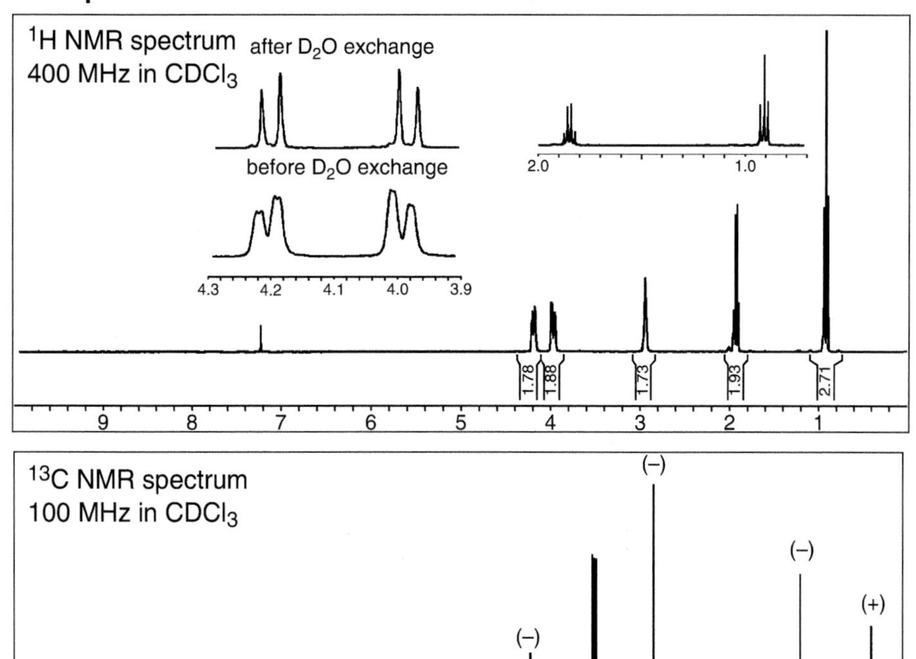

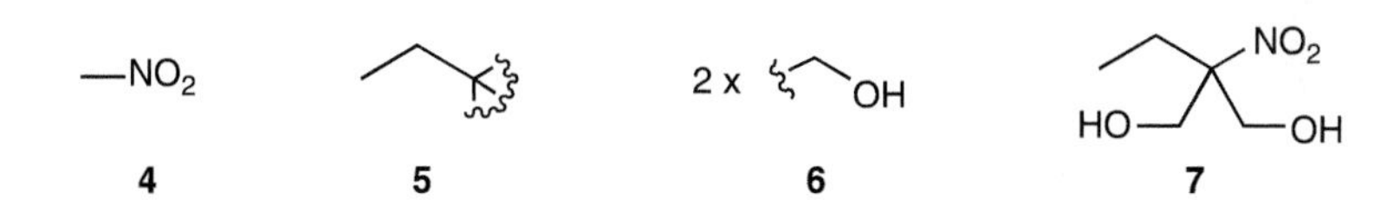

A remarkable feature of the ^{1}H NMR spectrum is the AB system (δ_A 4.21, δ_B 3.98, J_{AB} = 12 Hz) from the CH_2OH hydrogens after the D_2O shake. The CH_2OH groups are bonded to a prochiral centre, not a stereogenic centre, but the A and B hydrogens of each of the CH_AH_BOH groups are diastereotopic in a similar way to the signal from the diastereotopic methylene groups in Fig. 4.47, but in this case giving rise to a simple AB system.

Example 9 C, 70.7%; H, 5.9%

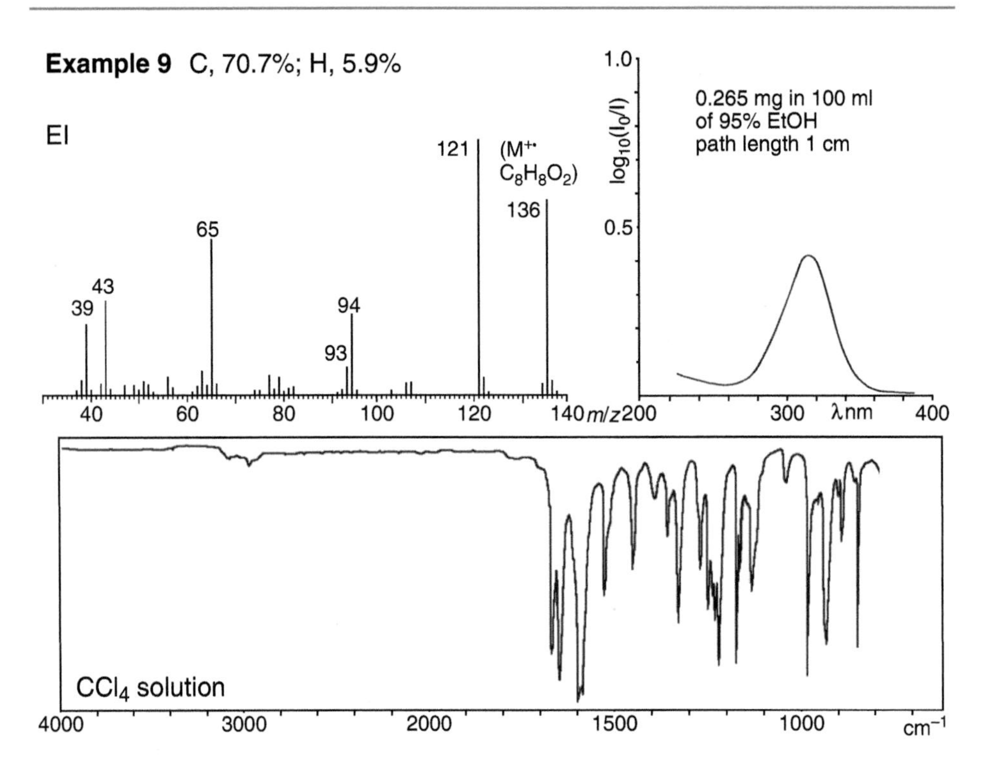

Example 9 The formula $C_8H_8O_2$ shows that five double-bond equivalents are present, and the UV, with λ_{max} at 316 nm (ε of 22,000), suggests that four or five double bonds are conjugated to each other. The carbonyl region in the IR spectrum with two bands at 1695 cm^{-1} and 1675 cm^{-1} seem likely to be from an α,β-unsaturated ketone, aldehyde or acid. An acid is eliminated by the absence of the H-bonded OH absorption in the 3000–2500 cm^{-1} region, and an aldehyde by the absence of absorption in the δ 9–10 region of the ^{1}H NMR spectrum. Bands at 1600, 1685 and 1480 cm^{-1} suggest an aromatic type of compound, which, conjugated to the unsaturated ketone, would account for the UV absorption.

The ^{13}C NMR shows eight signals—all, with the exception of a methyl group (+) at δ 27.8, in the trigonal ($\delta > 100$) region, with a carbonyl carbon at δ 197.4 consistent with an unsaturated ketone. In the ^{1}H NMR spectrum, the sharp singlet at δ 2.3 suggests that the methyl group is part of a methyl ketone **8**, supported by the mass spectrum with a base peak at M – 15 (*m/z* 121) and peaks at *m/z* 94 (*M* – CH_2=C=O), 93 (*M* – Ac) and 43 (Ac).

The ^{1}H NMR spectrum shows that the remaining five hydrogen atoms are attached to double bonds, and the ^{13}C NMR, with five signals (+) in the APT spectrum, shows that each is attached to a different carbon atom. There are 12 lines in this region of the ^{1}H NMR spectrum, numbered 1–12 in the expanded traces. Lines 3 and 4 are a one-proton doublet with a coupling constant of 16 Hz, matched by the separation of lines 7 and 8. Since they are doublets, with no further coupling, lines 3, 4, 7 and 8 must be an AX system, and the coupling constant of 16 Hz shows that they are *trans* **9**. This, incidentally, is supported by the presence in the IR spectrum of a strong band at 970 cm^{-1}, typical of this feature

Example 9 continued

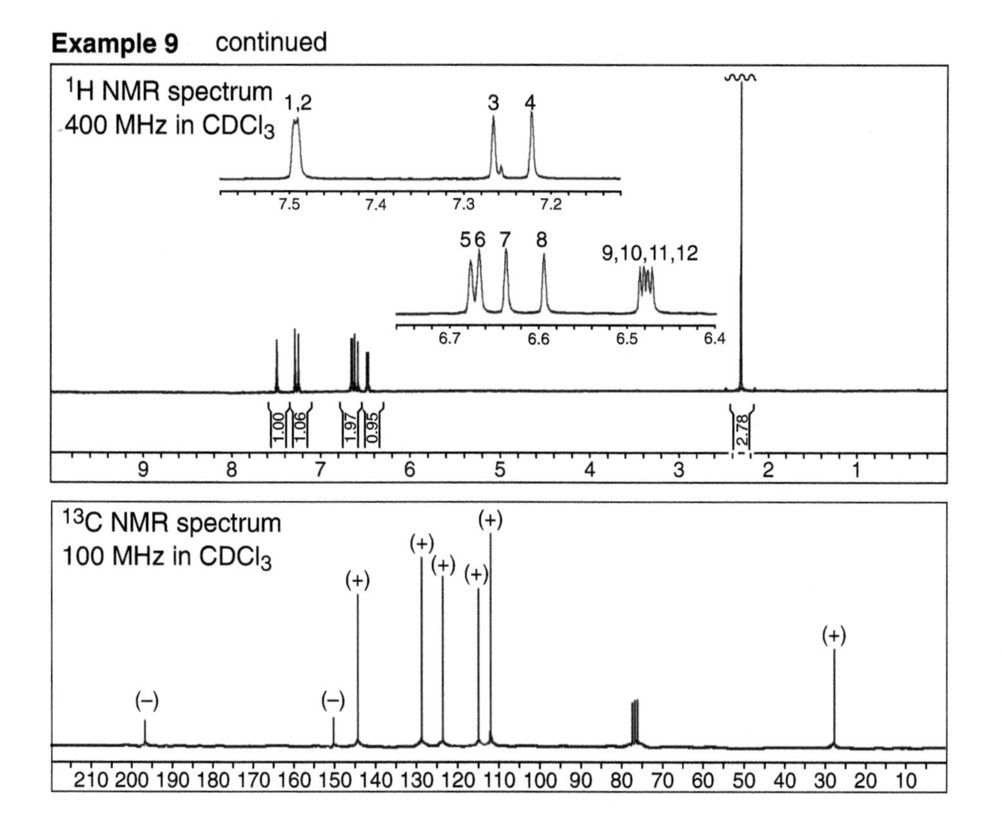

(Table 3.2). We now have fragments **8** and **9** (in which X and Y carry no protons, since the AX system is not coupled to anything else).

8 **9** **10** **11**

The remainder of the molecule is a C_4H_3O unit, which does not have a carbonyl group (^{13}C NMR), does not have an OH (IR) and has each of the three hydrogens on a different carbon. With a little thought, the only reasonable possibility is a furan ring with one substituent **10**, and all we have left to deduce is whether that substituent is on C-2 or C-3. The chemical shift of the furan carbon with no attached proton at δ 150.5 suggests that the attachment is at C-2 and we have a 2-substituted furan **11**. The splittings of the three furan protons confirm this assignment: H-5 is the fine doublet (lines 1 and 2) with J_{45} = 1.5 Hz, H-3 is also a doublet (lines 5 and 6) with J_{34} = 3.5 Hz, and H-4 is a double doublet (lines 9–12). The low coupling constants are typical of a furan ring (Table 4.34).

6.5 Worked Examples Using the Combined Application of MS, UV, IR, 1D-NMR and 2D-NMR Spectroscopic Methods

Example 10 The mass spectrum gave MH^+ (ESI) 195.1489, corresponding to a formula of $C_{11}H_{18}N_2O$, and a weak fragment for the loss of an ethyl group ($M - 29$). The UV spectrum indicates that there is some conjugation, but its weakness (ε 3000–4000) is indicative of an aromatic system. The IR spectrum, with peaks near 3300 cm^{-1}, and the ^1H spectrum, with a broad singlet at δ 13.15 indicate a hydrogen-bonded H, but the molecular formula excludes the possibility of its being a carboxylic acid. The ^{13}C NMR spectrum has ten lines, indicating that one carbon is duplicated, and the DEPT spectrum indicates that there is one methylene group (−) and three carbon atoms with no protons attached, two of which are in the carbonyl region near 160 p.p.m. Together with the peak at 1645 cm^{-1} in the IR spectrum, this suggests that there is an amide group –CONH–.

The ^1H NMR spectrum is complicated with the coupling constants not matching well enough to pair up the coupling partners. The first step is to label each of the proton signals—devise a system of your own if you like, but here we use a letter code A–F starting at the low-field end of the spectrum, with G_3–I_3 for the three methyl signals, one of which G_3 integrates for twice the intensity of each of the other two. The enlarged signals from the three methyl groups are on the same horizontal scale as for the signals C–F, but are scaled down in height to fit into the picture. The six one-proton signals A–F and the 12 methyl protons account for the 18 hydrogen atoms we know are present.

It helps to look briefly at the HSQC spectrum to find which pair of the five protons C–F is in the methylene group, and we find them, on the dashed line, as the signals attached to the carbon atom labelled ef, using as a code the lower-case letters for the carbon atom bonded directly to the proton (or protons) with the corresponding upper-case letters. If

Example 10 $C_{11}H_{18}N_2O$

HR-ESI-MS gave M^+ 195.1489 ($C_{11}H_{18}N_2O + H^+$).
The optical rotation was $[\alpha]_D$ +1.99 (c. 0.13 in $CHCl_3$).

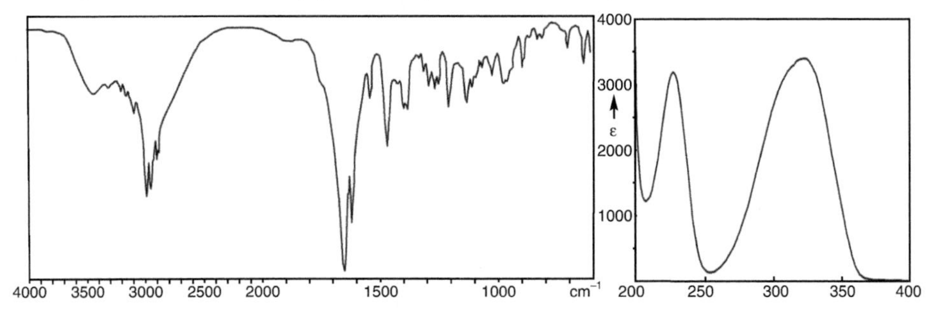

Example 10 continued

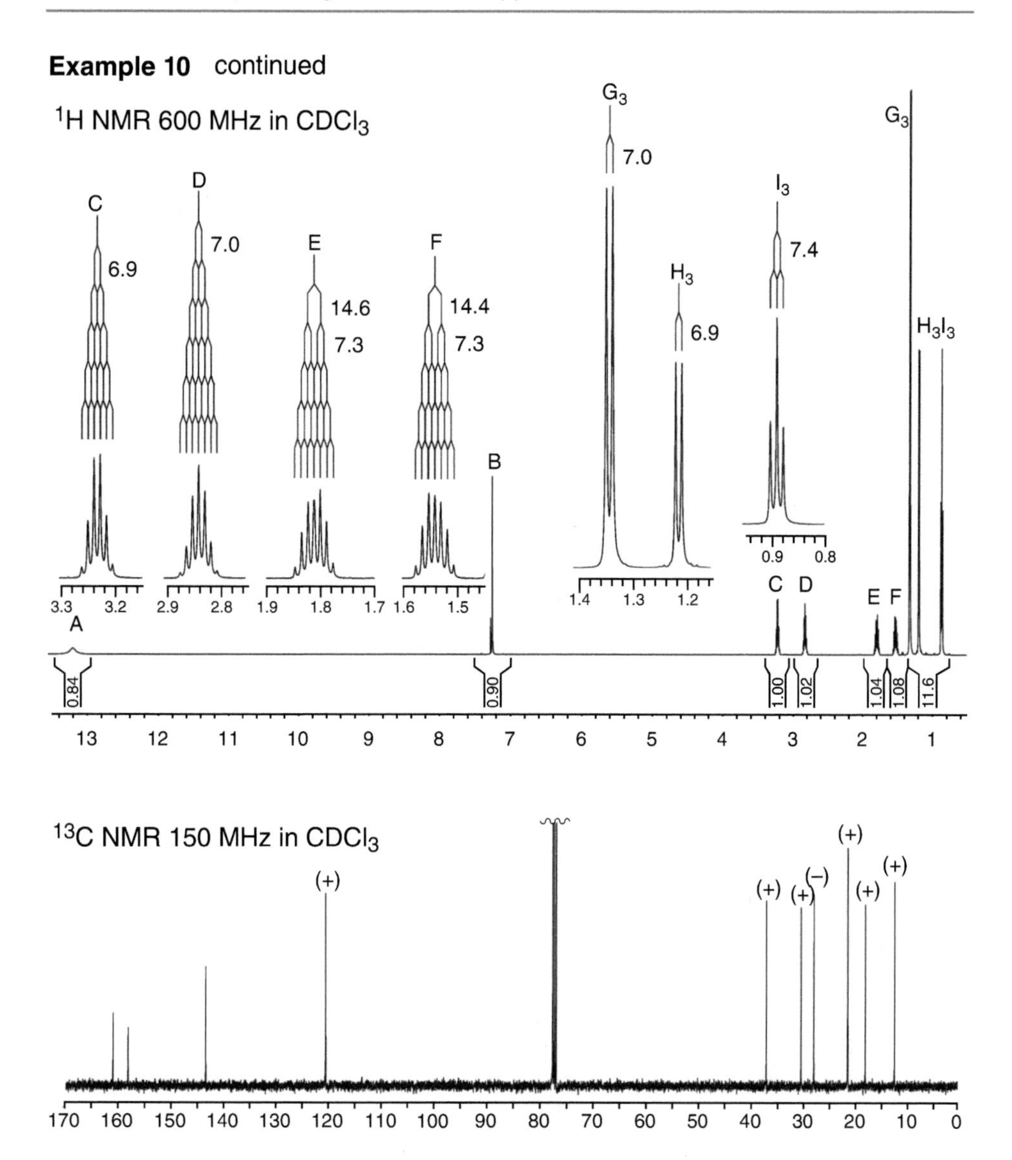

now we look at the signals from E and F, we see the mutual coupling of 14.5 Hz, which is appropriate for a geminal coupling. The other coupling constant in these two multiplets is 7.3 Hz, and this matches the coupling constant for the triplet I_3, indicating that we have found the ethyl group, with each of the methylene protons E and F having at least a quartet of signals from coupling to the methyl group I_3. Both signals E and F appear to be double quintets, indicating that they are further coupled with a similar coupling constant to another proton. We have the fragment $–CHCH_2CH_3$.

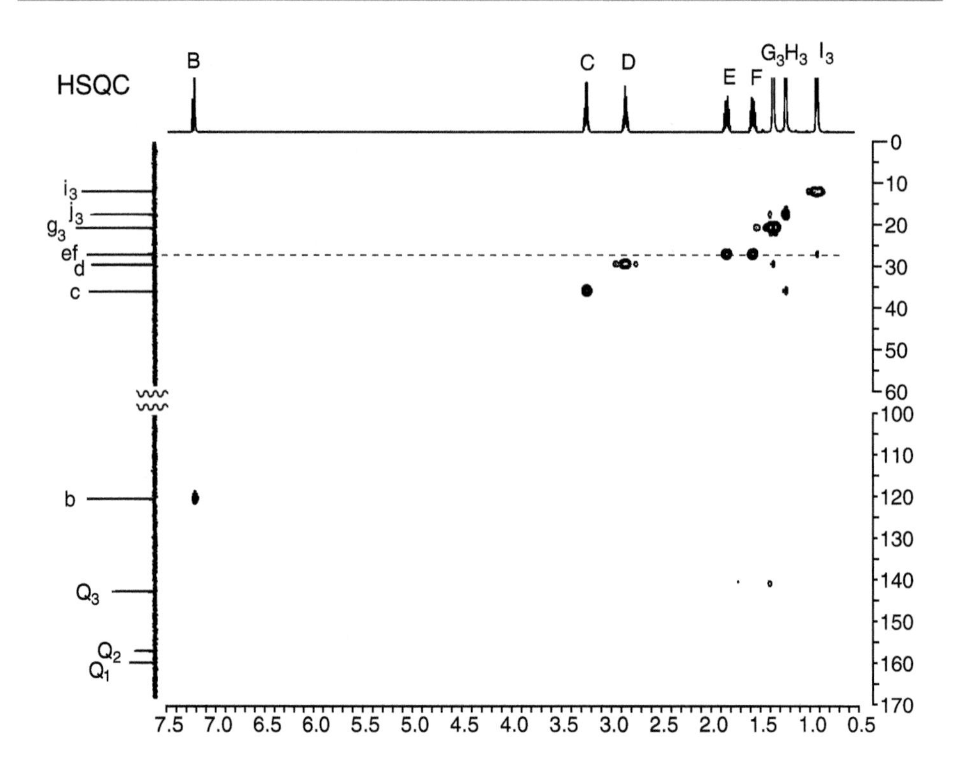

The methyl signal G_3 integrating for six protons is a doublet with a coupling constant of 7.0 Hz. This matches the septet D, which has seven lines in the right proportions (Table 4.35) for first-order coupling, and indicates the presence of an isopropyl group $(CH_3)_2CH$– with no further connections to any protons. Among the proton signals on tetrahedral carbons, this only leaves C and H_3; the former is a sextet and the latter a doublet, which can be accommodated by joining them to the fragment to make a secondary butyl group $-CH(CH_3)CH_2CH_3$. The signal A is a broad singlet, and the signal B a sharp singlet. These seem likely to be —NH— and =CH— signals, respectively. We now have the fragments **12–15**, which account for all the hydrogen atoms, and a carbon and a nitrogen atom with which to complete the empirical formula. They must be an imine **16**.

12 **13** **14** **15** **16**

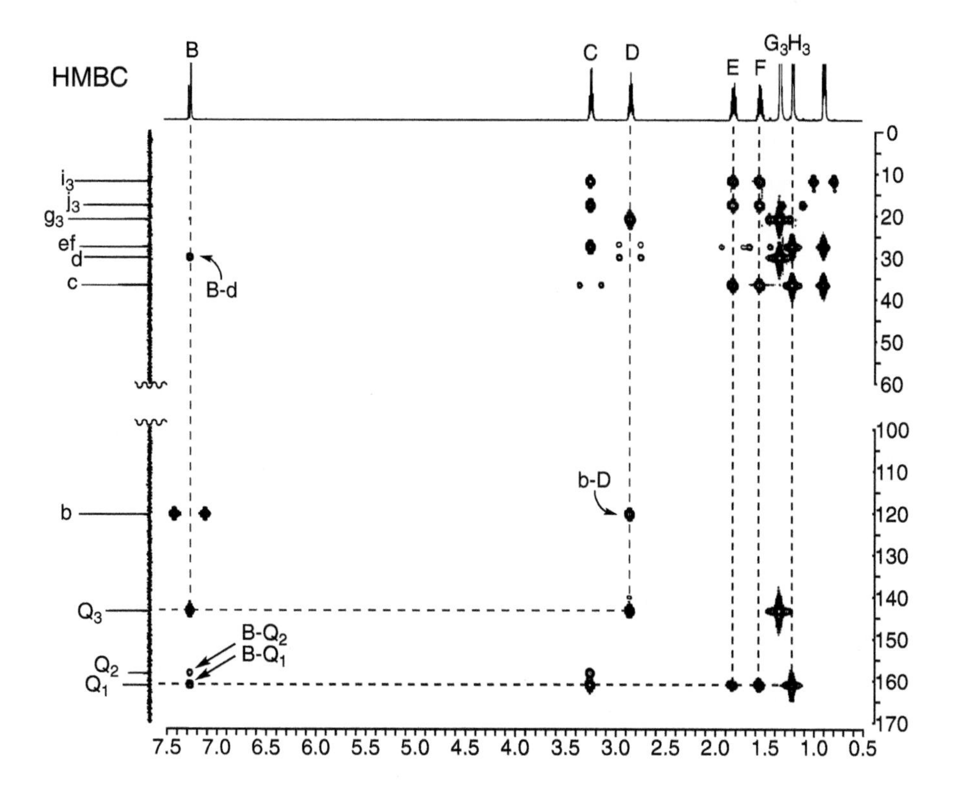

To put the pieces together we turn to the HMBC spectrum. Having assigned in the HSQC spectrum which carbon atoms carry which protons, we can start to look for connections over two or three bonds. For example, there are cross peaks identified by the coarse dashed lines between the methine proton D of the isopropyl group and the carbons labelled Q_3 and b. Equally the proton B also has a cross peak with the carbon labelled Q_3. In spite of the ambiguity in all HMBC spectra, that we cannot tell if these are two- or three-bond connections, the only way to accommodate this information is to have the isopropyl group and the proton B attached one at each end of the C=C double bond **17**.

H_B b d Q_3 H_D

17

N Q_1 c N H Q_2 O

18

H_B N Q_1 N H Q_2 O

19

Looking at the fine dashed line from features within the s-butyl group, both the methyl group labelled H_3 and the methylene protons labelled E and F have strong cross peaks to the signal labelled Q_1, which must therefore be attached to the s-butyl group. The methine proton labelled C has cross peaks to both of the signals labelled Q_1 and Q_2. This means that Q_1 must be bonded to Q_2—one of the cross peaks from the proton C is a two-bond and the other a three-bond connection. One of Q_1 and Q_2 must be the carbonyl group and the other a signal from a structural unit likely to appear at comparably low field. The only sensible candidate is the C=N imine, but which is which. If Q_1 were to be the carbonyl group, it would be a ketone, since it is certainly bonded to the carbon atom Q_2, in which case it would have a chemical shift much further downfield, in the region δ 195–220 p.p.m. Since it does not, it must be Q_2 that is the amide carbonyl group, and we have extended the structure to **18**.

The cross peak between the B proton on the double bond and both signals Q_1 and Q_2, allows us to put the two pieces together. The carbon atom labelled b cannot be directly bonded to either Q_1 or Q_2, because they would not then give signals so far downfield. The nitrogen atom of the imine must be in between, and we can complete a structure **19**. The cross peak between the proton B and Q_1 is noticeably stronger than the one to Q_2, which identifies H_B as having a three-bond coupling to Q_1, and the weak cross peak to Q_2 must be a not unusual four-bond coupling. This example illustrates just one route from the spectra to a structure. You should look at the HMBC spectrum again, and see that there are several other connections confirming the conclusion which might equally easily have given us the structure. It can help to have some chemical knowledge in a problem like this. The left and right hand halves of the structure are readily identified as having come from condensation between valine and isoleucine with additional oxidation level changes.

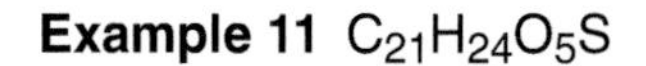

Example 11 $C_{21}H_{24}O_5S$

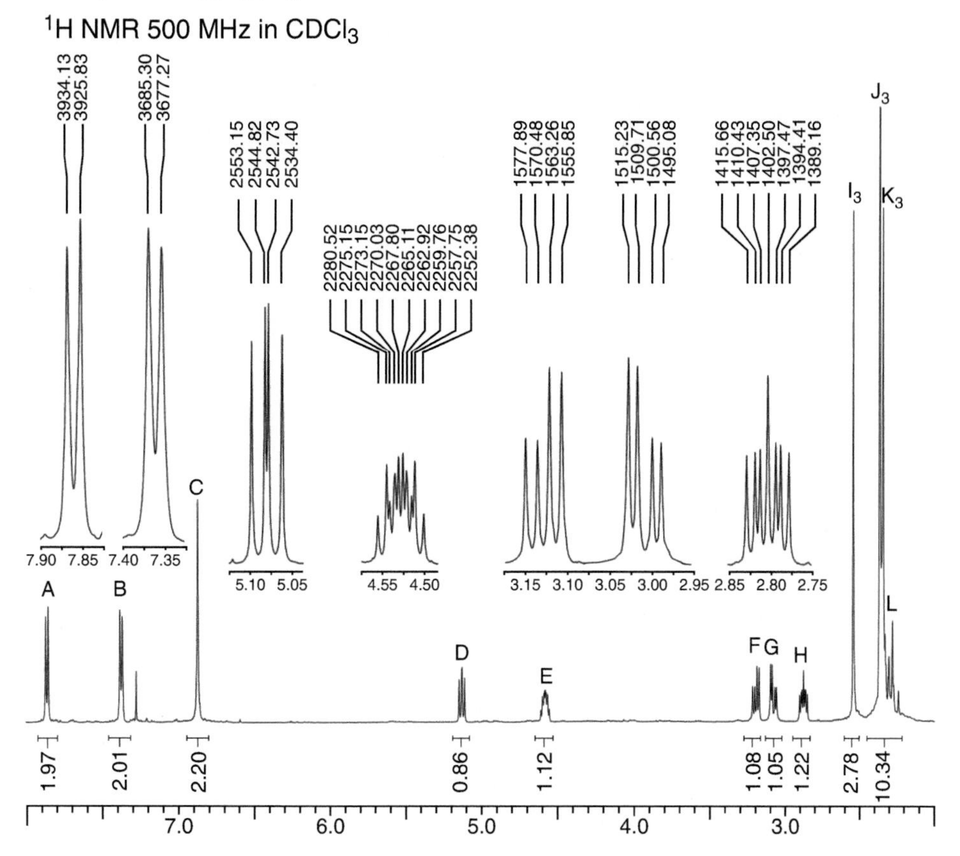

Example 11 It is not unusual to carry out a chemical reaction and find yourself with a product that you did not expect. When the triketone **20** was reduced with sodium borohydride, the product was not the expected triol, nor was it any of the straightforward intermediates of incomplete reduction. The product was clearly an alcohol, but its spectra were not definitive. The toluene-*p*-sulfonate of the alcohol gave the spectra shown on the following pages, from which the structure can be determined. This is a common situation—to know the structures of the reagents that went into the reaction, but not be able to guess easily what the unexpected structure actually is.

20 $\xrightarrow{NaBH_4}$ $C_{14}H_{18}O_3$ (ν_{max} 1768 cm^{-1}) $\xrightarrow[Et_3N]{p\text{-}CH_3C_6H_4SO_2Cl}$ $C_{21}H_{24}O_5S$ (1788 cm^{-1})

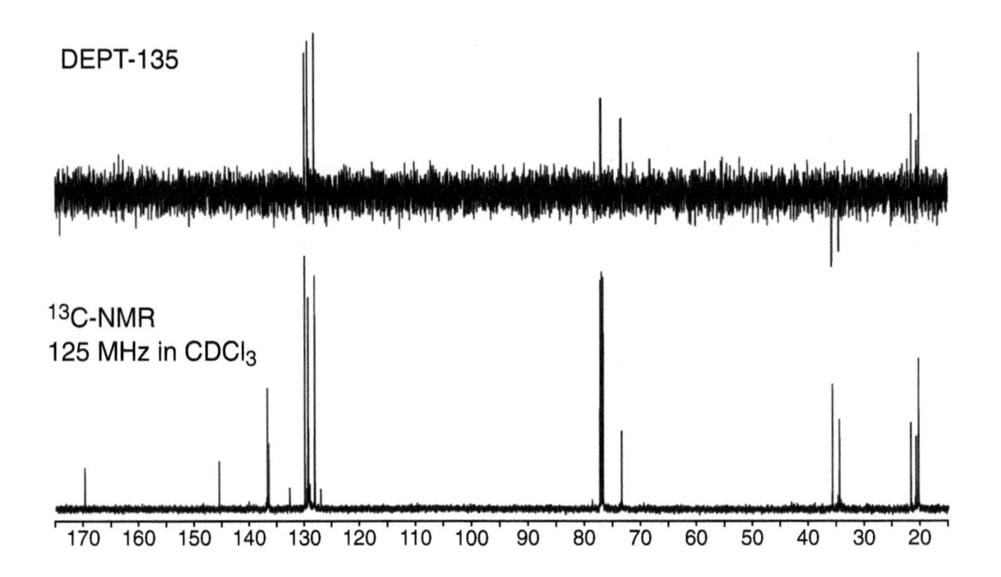

Knowing the reagents helps. The mesityl ring is most unlikely to have been disturbed by borohydride or toluenesulfonyl chloride, and the toluenesulfonate is sure to be attached to an oxygen atom in the final compound—we can look for the signals from these components in order to discount them. In the ^{1}H NMR spectrum, the AA´BB´ system, labelled A and B, of the *p*-disubstituted benzene ring, and one of the methyl groups I_3 or K_3 will be from the *p*-toluenesulfonyl group, and the aromatic singlet C, the other methyl group of the pair I_3 or K_3 and the stronger methyl signal J_3 will be from the mesityl group. We can be reasonably confident that we have both aromatic rings **21** and **22.**

21 **22** **23** **24**

We are left with the signals from six protons labelled D–H and L. Equally, in the ^{13}C NMR spectrum we see a number of signals in the aromatic region, with three strong peaks from the carbons in the benzene rings carrying protons, and three methyl signals. We also

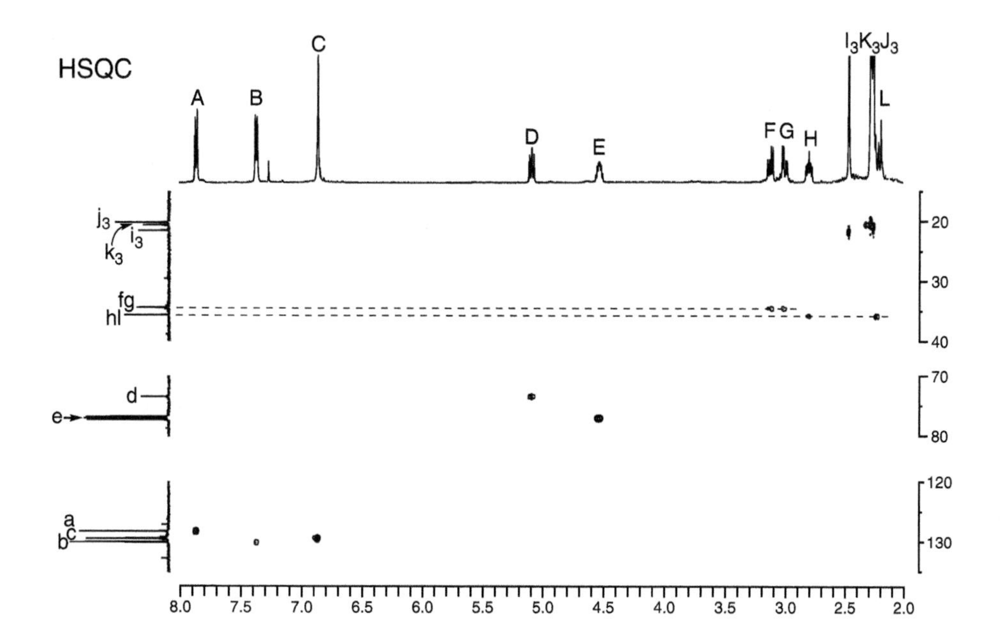

see in the DEPT spectrum that there are two methylene signals in the δ 30–40 region, and two methine signals significantly further downfield, one of which was hidden under the $CDCl_3$ signal in the ^{13}C NMR spectrum. These are the carbon atoms that carry the six protons D–F and L.

The other significant signal in the ^{13}C NMR spectrum is from a carbonyl carbon at δ 169.7, which must be from the carbonyl carbon of an ester (Table 4.15). The carbonyl band in the IR spectrum indicates that it is most probably a five-membered ring lactone with an unusually high frequency. Before the sulfonate group was introduced the frequency 1768 cm^{-1} was normal for a γ-lactone; the sulfonation raised that frequency to 1788 cm^{-1}, and we might guess that this could be explained if the sulfonyloxy group is α to the carbonyl, since electronegative substituents α to carbonyl groups raise the C=O stretching frequency (Table 3.7). We can be reasonably confident that we have the group **23**, but we have to do more work before we place the sulfonyloxy group with confidence.

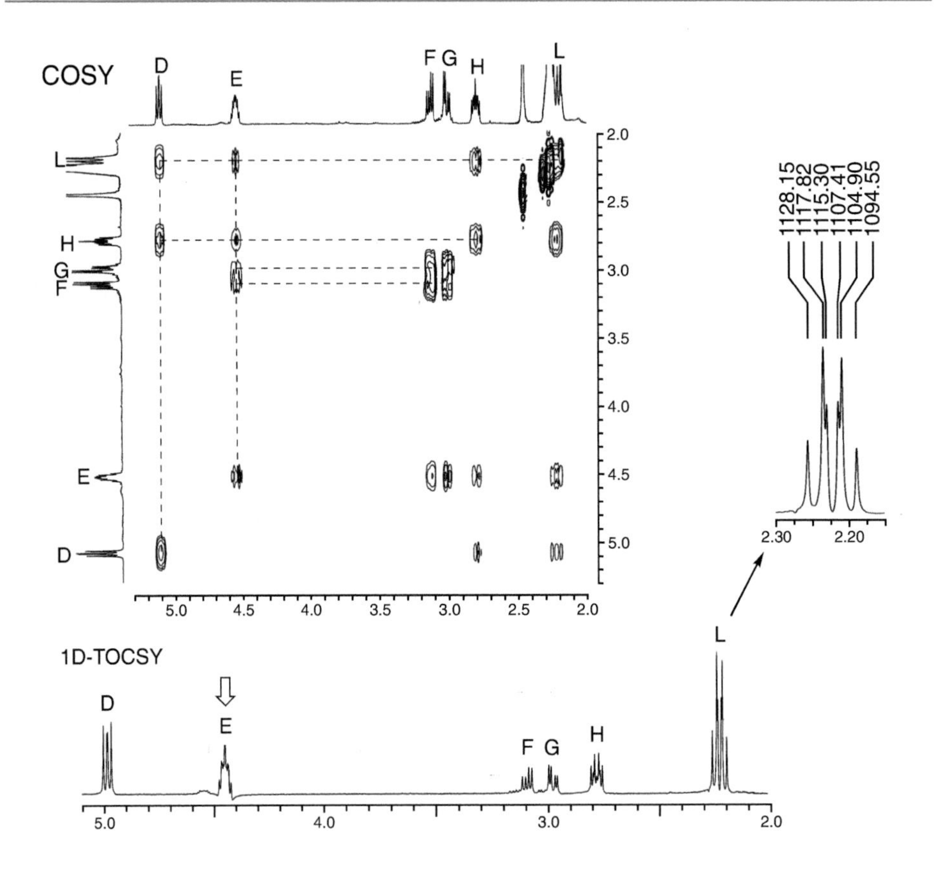

Turning to the HSQC spectrum we quickly identify the methylene protons as those giving the signals F and G on one carbon, and H and L on another, and the methine protons are D and E. The COSY spectrum then reveals the spin connections. Starting at D, there are cross peaks to both H and L, which in turn have cross peaks to the methine signal E, which has cross peaks to both methylene signals F and G. The six protons are arranged as in **24**, which can be confirmed by using this information to match up the coupling constants measured in the ^{1}H NMR spectrum. The signal L, partly obscured by the methyl signals, is revealed in the 1D-TOCSY as a double triplet, which helps to identify all the coupling constants. As usual the experimentally measured values do not match each other perfectly, but the drawing **24**, with the numbers rationalised to ±0.05, pairs them up well enough to be convincing.

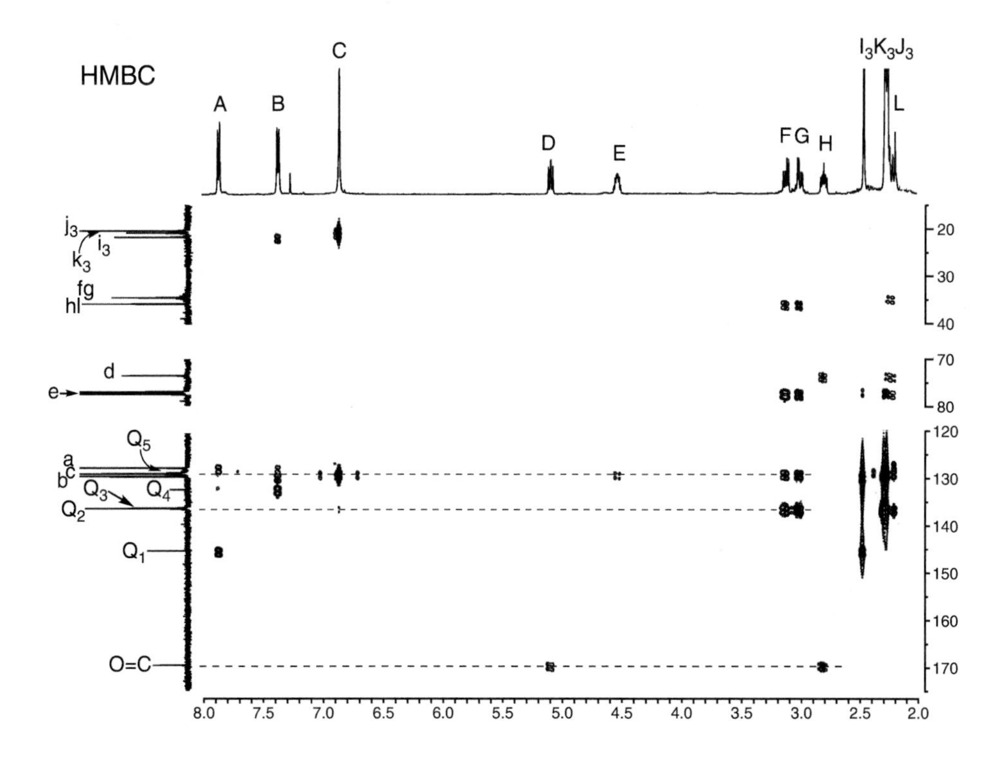

The signals D and E, and the carbon atoms labelled d and e to which they are attached, are downfield in the region that indicates that they have an oxygen atom attached to the carbon atoms. One of these oxygens must be from the sulfonyloxy group, and the other from the lactone oxygen. To make the connections between these fragments we turn to the HMBC spectrum.

The lactone carbonyl signal has cross peaks to D and H. The carbonyl group must be attached to carbon d, with proton D having a two-bond connection and proton H a three-bond connection. The lactone oxygen must therefore be attached to carbon e, and the sulfonyloxy group to carbon d, confirming the explanation for the high carbonyl stretching frequency. Further support comes from the connections shown with the coarse dashed lines from the protons F and G to signals from the aromatic rings. The two carbon signals labelled Q_2 and Q_3 are barely resolved, but enlarged they can be seen to be in a 2:1 ratio. They have cross peaks to the methyl signals J_3 and K_3, and are from the carbon atoms carrying the methyl groups in the mesityl ring, with two carbons for Q_2 and one for Q_3. The cross peaks from the protons F and G to one of these shows that the mesityl group is attached to carbon fg. The

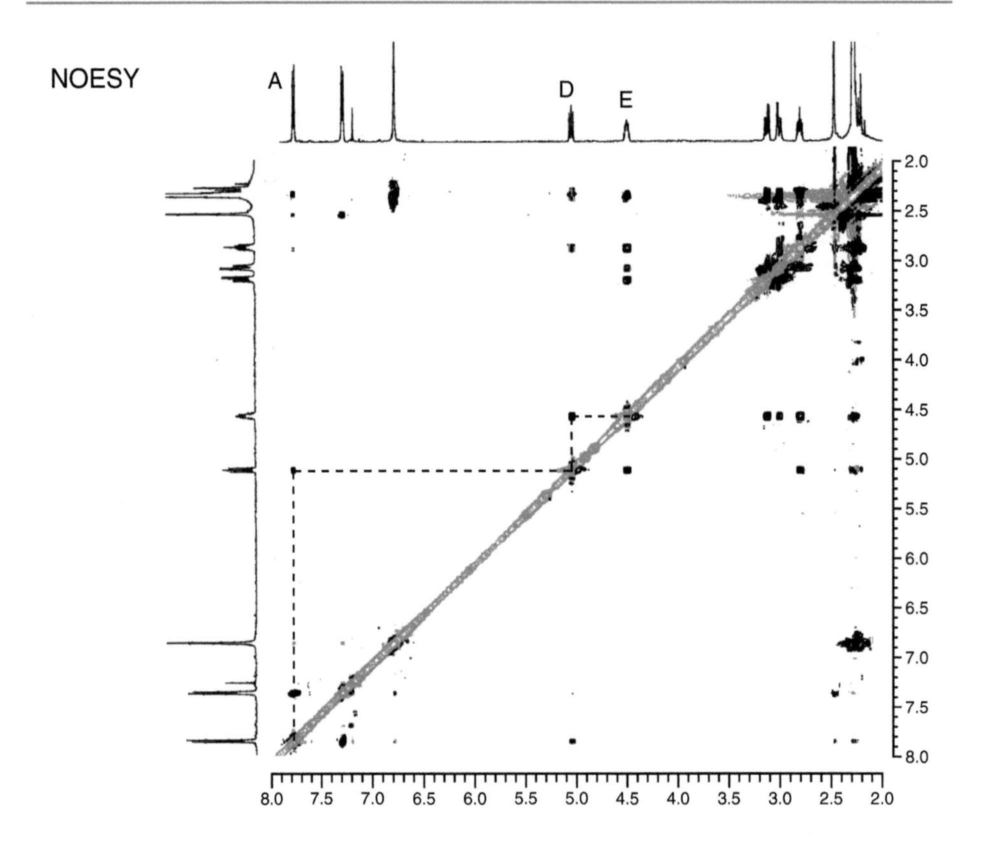

upper dashed line from the F and G signals leads into a busy region on the ^{13}C axis, with three strong methine signals, and a buried signal Q_5, which is from the carbon atom at the point of attachment of the mesityl group, with a further indication that this is correct from the cross peak with the proton E. The cross peaks with the A and B signals, show that the signals Q_1 and Q_4 are the fully substituted carbons in the toluenesulfonyl ring, and the connection from Q_4 to the signal I_3 identifies that as the signal from the methyl group in that ring.

We now have a complete structure **25**, and all we have left to do is try to identify whether the two substituents on the lactone ring are *cis* or *trans*.

NOE H_E mesitiyl H_D H_G H_H H_F $TolSO_2O$ H_L

25 **26**

We have two ways to tackle this: find cross peaks from protons close to each other in the NOESY spectrum or analyse the information in the coupling constants to identify plausible dihedral angles (Sect. 4.7.2). With both methods we are on weak ground because this is a five-membered ring with an uncertain conformation, which might allow vicinally related protons on opposite sides of the ring to be closer to each other than protons on the same side of the ring (Sect. 4.7.2). For this reason, we can ignore the cross peaks from protons on adjacent atoms in the NOESY spectrum, and look only at those from more distant pairs. The ones that are significant are firstly the one between H_A and H_D, which confirms that the toluenesulfonyl group is attached to the carbon α to the carbonyl group, a detail we could not use the HMBC spectrum for, because the benzene ring and the lactone ring are separated by too many bonds from each other. Secondly the cross peak between the signals H_D and H_E, between protons that have a 1,3-relationship, shows that they are on the same side of the ring—making the substituents *cis*. The conformation drawn in the picture **26** is a plausible approximation: it would account for the large coupling constant (10.4 Hz) between H_L and both H_D and H_E. There is even a hint that H_G, but not H_F, has an NOE to H_H, allowing us tentatively to assign the signals in the methylene group. The large geminal coupling constant between H_F and H_G (14.6 Hz), incidentally, might have been a way of deciding that the aryl ring was attached to the carbon atom labelled fg, if we had not found it by another route.

Problem Sets

7

The following problems, graded from relatively simple to quite challenging, will give you some practice in interpreting spectra. The APT or DEPT data, when not illustrated, are reported on the ^{13}C spectrum with a + representing signals that have three or one hydrogen atoms attached, and a – representing signals that have two or zero.

7.1 Chemical Shift Problems

1. Three compounds of molecular formula C_4H_8O gave the following ^{13}C and IR data.
   Compound **1**:
   - IR: 1730 cm^{-1}
   - ^{13}CNMR: δ 201.6, 45.7, 15.7 and 13.3

   Compound **2**:
   - IR: 3200 (broad) cm^{-1}
   - ^{13}CNMR: δ 134.7, 117.2, 61.3 and 36.9

   Compound **3**:
   - IR: no peaks except CH and fingerprint
   - ^{13}C NMR: δ 67.9 and 25.8

   Suggest a structure for each compound, and then see whether your suggestions are compatible with the following information. Compound **1** reacts with $NaBH_4$ to give compound **4**, $C_4H_{10}O$, IR 3200 (broad) cm^{-1} and ^{13}C NMR δ 62.9, 36.0, 20.3 and 15.2. Compound **2** reacts with hydrogen over a palladium catalyst to give the same product **4**, while compound **3** reacts with neither reagent.

Electronic Supplementary Material The online version of this chapter (https://doi.org/10.1007/978-3-030-18252-6_7) contains supplementary material, which is available to authorized users.

I. Fleming, D. Williams, *Spectroscopic Methods in Organic Chemistry*,
https://doi.org/10.1007/978-3-030-18252-6_7

2. Given the spectroscopic data, deduce structures for the compounds **5–7**.

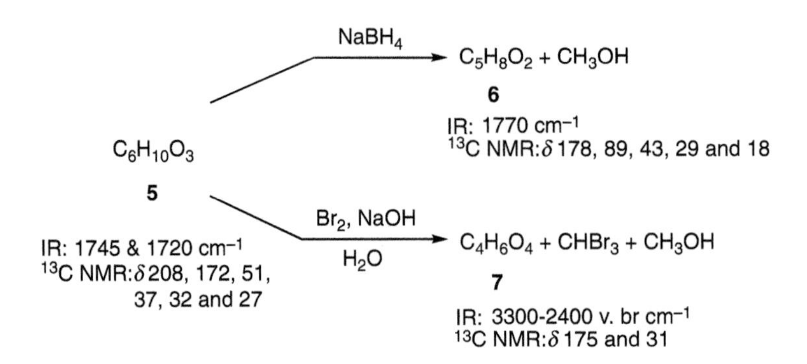

3. Two isomeric compounds C_4H_8O gave the following ^{1}H NMR data, in which J values <1.5 Hz were not resolved. What are their structures?
   Compound **8**: δ 3.8 (2H, s), 3.5 (3H, s) and 1.8 (3H, s)
   Compound **9**: δ 4.95 (1H, broad s), 4.8 (1H, broad s), 4.0 (2H, s), 2.2 (1H, s, removed on shaking with D_2O) and 1.7 (3H, s)
4. IR and simple NMR spectra are shown for six compounds. What are their structures?

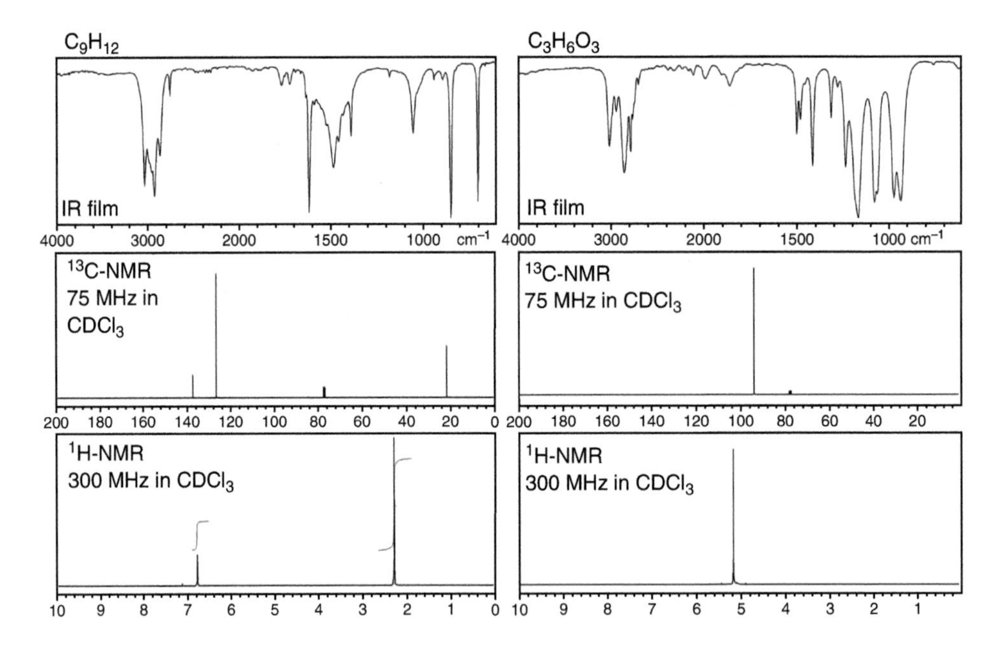

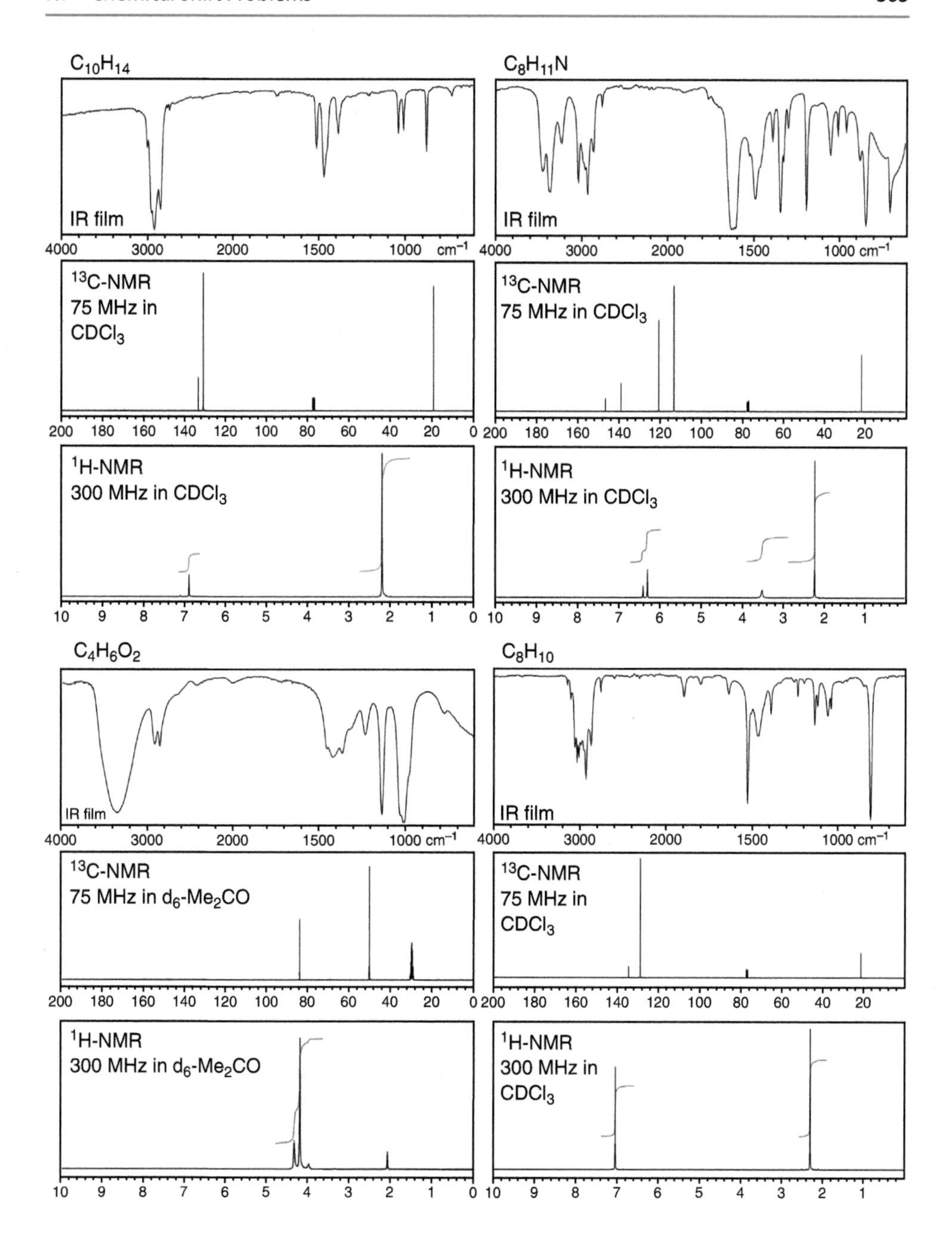

$C_{10}H_{14}$
IR film
4000 3000 2000 1500 1000 cm⁻¹
13C-NMR
75 MHz in
CDCl3
200 180 160 140 120 100 80 60 40 20 0
1H-NMR
300 MHz in CDCl3
10 9 8 7 6 5 4 3 2 1 0
$C_8H_{11}N$
IR film
4000 3000 2000 1500 1000 cm⁻¹
13C-NMR
75 MHz in CDCl3
200 180 160 140 120 100 80 60 40 20
1H-NMR
300 MHz in CDCl3
10 9 8 7 6 5 4 3 2 1
$C_4H_6O_2$
IR film
4000 3000 2000 1500 1000 cm⁻¹
13C-NMR
75 MHz in d6-Me2CO
200 180 160 140 120 100 80 60 40 20 0
1H-NMR
300 MHz in d6-Me2CO
10 9 8 7 6 5 4 3 2 1 0
C_8H_{10}
IR film
4000 3000 2000 1500 1000 cm⁻¹
13C-NMR
75 MHz in
CDCl3
200 180 160 140 120 100 80 60 40 20
1H-NMR
300 MHz in
CDCl3
10 9 8 7 6 5 4 3 2 1

5. The five carbonyl compounds **52–56** (phenylacetate, anisaldehyde, *p*-toluic acid, methyl benzoate and phenylacetic acid) from Question 3 in Chap. 2 gave the five ^{13}C-NMR spectra labelled **A-E** below, not respectively. Assign the structures to the spectra **A–E**.

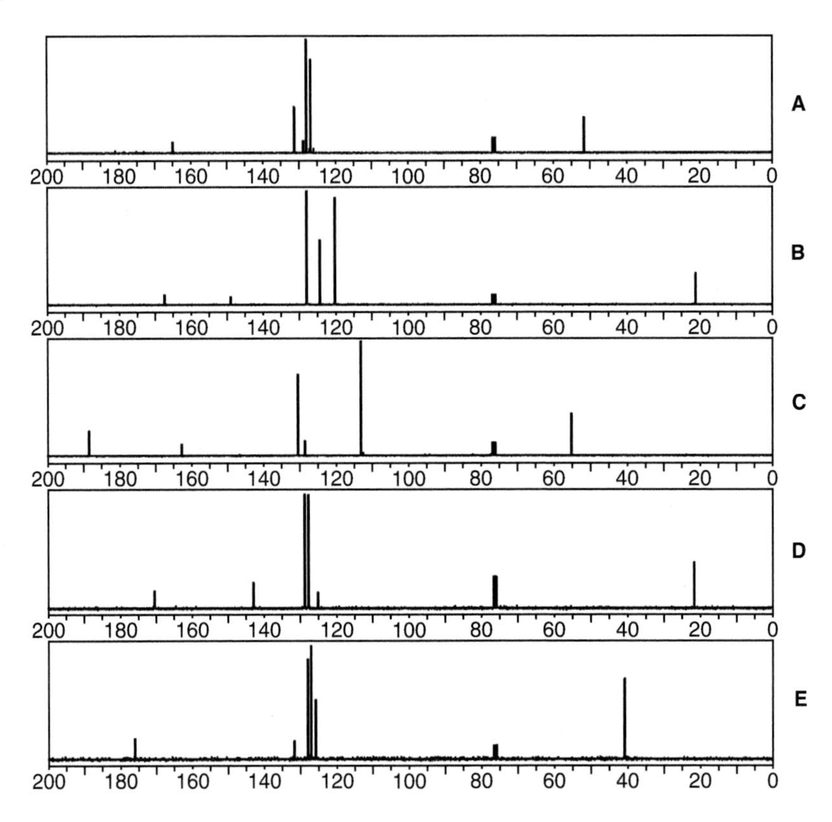

7.2 1D-NMR Chemical Shift and Coupling Problems

6. The five carbonyl compounds **52–56** (phenylacetate, anisaldehyde, *p*-toluic acid, methyl benzoate and phenylacetic acid) from Question 3 in Chap. 2 gave the five ^{1}H-NMR spectra labelled **A–E** below, not respectively. Assign the structures to the spectra **A–E**.

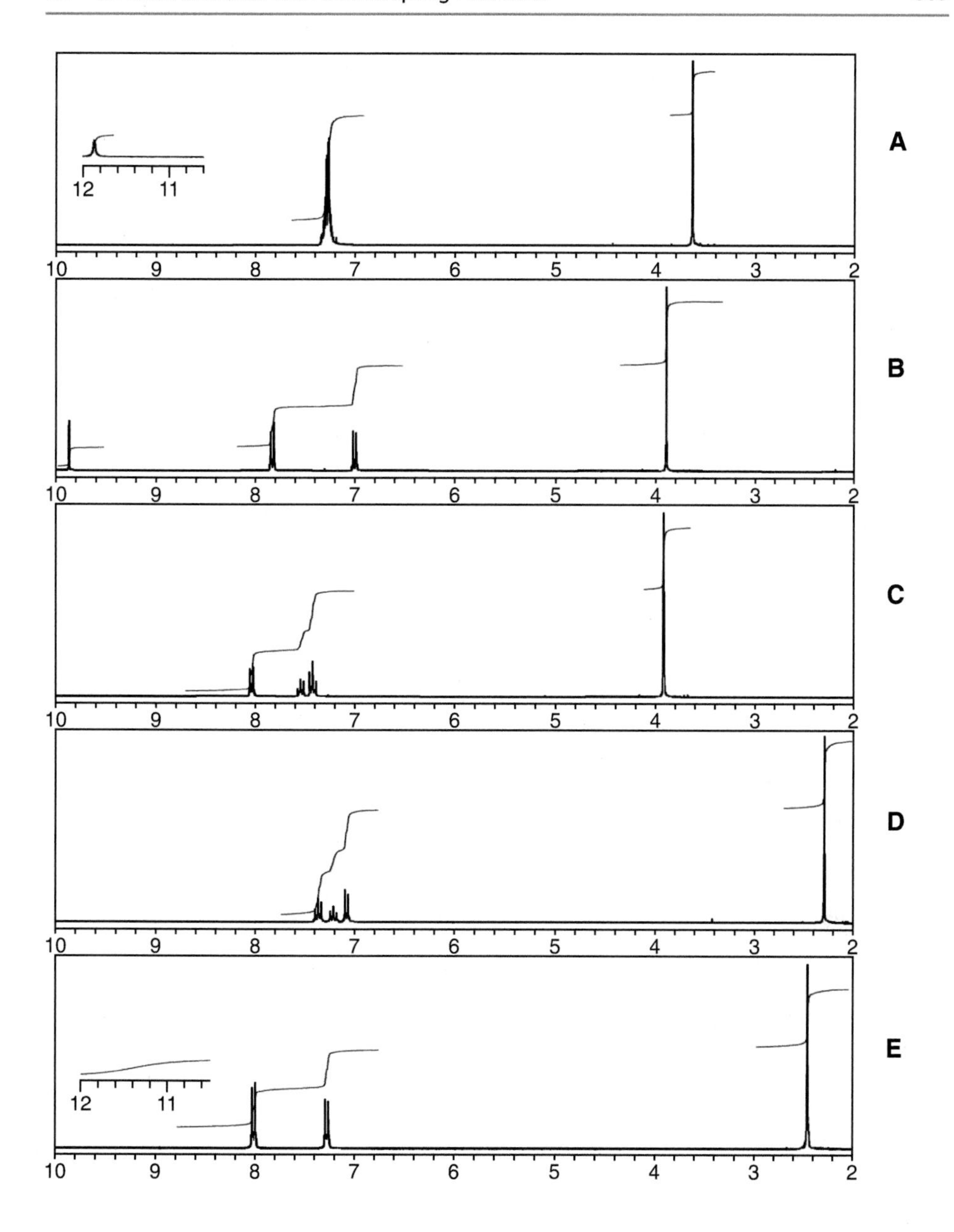
A
B
C
D
E
12
11
10
9
8
7
6
5
4
3
2

7. Three compounds, **10–12**, of formula C_4H_6 have been prepared by the routes shown below. Suggest structures for these isomers.

NaNH$_2$, Bu^{t}OH → **10** ^{1}H NMR: δ 5.35 (2H, s) and 1.0 (4H, s)
^{13}C NMR: 3 different signals

NaNH$_2$, THF → **11** ^{1}H NMR: δ 6.4 (1H, t, $J = 1$ Hz), 2.13 (3H, s) and 0.84 (2H, d, $J = 1$ Hz)
^{13}C NMR: 4 different signals

LiNH$_2$, dioxan → **12** ^{1}H NMR: δ 7.18 (2H, d, $J = 1$ Hz), 1.46 (1H, qt, $J = 5$ and 1 Hz) and 0.97 (3H, d, $J = 5$ Hz)
^{13}C NMR: 3 different signals

Compound **10** reacts with *m*-chloroperbenzoic acid to give compound **13**, which rearranges to compound **14** in the presence of lithium iodide. Suggest structures for compounds **13** and **14**.

10 —*m*-ClC$_6$H$_4$CO$_3$H→ **13** (C_4H_6O) IR: no strong peaks except in the fingerprint region

^{1}H NMR: δ 3.0 (2H, s) and 0.90-0.80 (4H, A_2B_2 system)

13 —LiI→ **14** (C_4H_6O) IR: ν_{max} 1770 cm^{-1}
^{1}H NMR: δ 3.02 (4H, t, $J = 5$ Hz) and 1.98 (2H, q, $J = 5$ Hz)

8. Deduce structures for the products **15–17** of the following reactions.

Cl Cl

O

H$_2$O, 100°C → **15** MS: M$^{\cdot+}$ 128 and 130
IR: ν_{max} 3500, 1600 and 1500 cm^{-1}
^{1}H NMR: δ 7.1 (2H, d, $J = 7$ Hz), 6.8 (2H, d, $J = 7$ Hz) and 5.4 (1H, br s)

O

NaOH → **16** MS: M$^{\cdot+}$ 306
IR: ν_{max} 1600 and 1500 cm^{-1}
^{1}H NMR: δ 8.05 (3H, s), 7.64 (6H, d, $J = 6$ Hz) and 7.5-7.3 (9H, m)
^{13}C NMR: δ 142, 141, 129, 127, 126 and 125 (142 and 141 are weak)

COCl

AlCl$_3$ → **17** MS: M$^{\cdot+}$ 110
IR: ν_{max} 1720 cm^{-1}
^{1}H NMR: δ 7.7 (1H, dt, J = 6 and 3 Hz), 6.2 (6H, dt, $J = 6$ and 2 Hz), 2.2 (1H, dd, $J = 3$ and 2 Hz) and 1.1 (6H, s)

9. Iodolactonisation of the acid **18** gives a single product **19**; its 400 MHz ^{1}H NMR spectrum shows the signals below in addition to others. What is the stereochemistry of compound **19**?

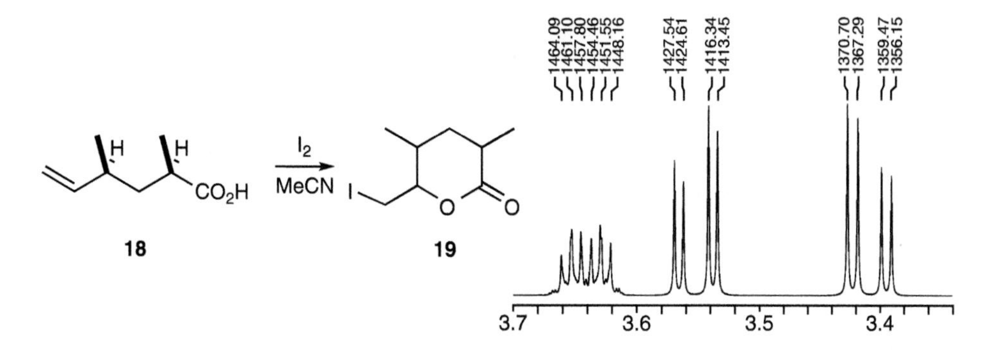

10. Compounds **20** and **21** are produced by the following reaction scheme. Deduce the structure of the final product **21**, including the relative stereochemistry and the conformation, from the ^{1}H-NMR data: δ 7.5–7.0 (5H, m), 3.5 (1H, d, $J = 12$ Hz), 3.3 (1H, dd, $J = 12$ and 2 Hz), 2.6 (1H, dqd, $J = 12$, 7 and 6 Hz), 2.1 (1H, ddd, $J = 13$, 6 and 2 Hz), 1.8 (1H, dd, $J = 13$ and 12 Hz), 1.4 (3H, s), 1.3 (3H, d, $J = 7$ Hz) and one exchangeable proton. Irradiation of either the signal at δ 3.3 or at 2.1 gives a small positive enhancement of the signal at δ 1.4.

CN, Ph + CO_2Me —NaOMe→ $C_{14}H_{17}NO_2$ **20** —Raney Ni, H_2→ $C_{13}H_{17}NO$ **21**

11. The bis(bromomethyl)benzil **22** reacts with sodium hydroxide to give a product **23**. Deduce its structure.

Br, O, O, Br, **22** —NaOH→ $C_{16}H_{12}O_3$ **23**

IR: ν_{max} 1700 cm^{-1}
^{1}H NMR: δ 8.0-7.0 (8H, m), 5.5 (1H, d, $J = 16$ Hz), 5.3 (1H, d, $J = 13$ Hz), 5.2 (1H, d, $J = 13$ Hz) and 4.9 (1H, d, $J = 16$ Hz)
^{13}C NMR: δ 189(–), 12 aromatic carbons (8+ and 4–), 109(–), 74(–) and 63(–)

12. Given the following spectroscopic and chemical data, deduce a structure for a toxic substance **24**, $C_{11}H_{16}N_2O_5$, isolated from the Colorado potato beetle.

MS (FIB): *m/z* 257

UV: λ_{max} (nm) 230 (ε 8000)

IR: ν_{max} (cm^{-1}) 3400–2500 (br), 1660 and 1590.

^{13}C NMR

δ	APT	δ	APT	δ	APT	δ	APT
178.6	–	135.1	+	122.4	–	33.4	–
175.6	–	133.2	+	56.5	+	28.0	–
175.2	–	127.7	+	54.9	+		

^{1}H NMR (in D_2O)

δ	Intensity	Multiplicity	*J* (Hz)
6.75	1H	dt	16.7 and 10.2
6.26	1H	t	10.2
5.43	1H	dd	10.2 and 9.6
5.38	1H	dd	16.7 and 1.8
5.29	1H	dd	10.2 and 1.8
5.06	1H	d	9.6
3.74	1H	t	6.0
2.45	2H	t	7.5
2.13	2H	m	

Hydrogenation of the toxin with H_2 on Pd/C gave a compound whose ammonia CI spectrum showed *m/z* 261 and fragment ions at 132 and 130. When CF_3CO_2H was added to a D_2O solution of this reduced product, a triplet in its proton NMR spectrum shifted from δ 3.8 (1H, $J = 6$ Hz) to δ 4.2. This same reduced product, on treatment with 2,4-dinitrofluorobenzene, followed by acid hydrolysis (6 *N* HC1, 373 K, 12 h) gave one molar equivalent of the amino acid **25**.

25 $\overset{+}{N}H_3$ CO_2^-

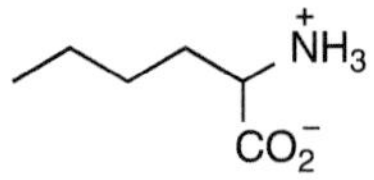

13. The multiplets labelled A-F below are all one-proton signals taken from six different compounds. They are first order with, at most, a little roofing. Identify the multiplicity and draw a tree diagram to account for the pattern of each signal.

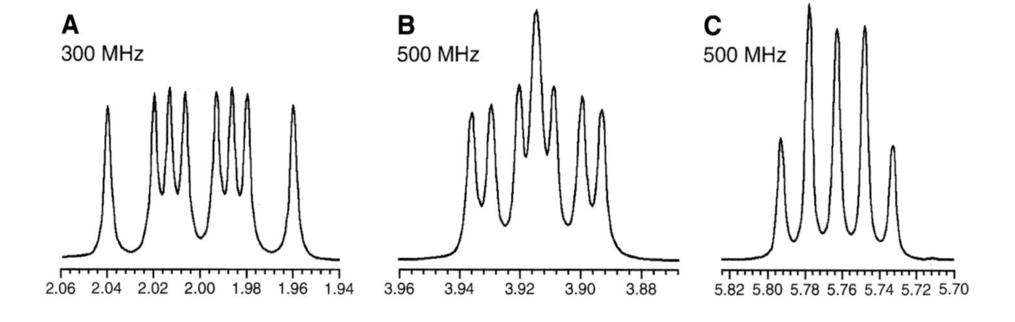
A
300 MHz
2.06 2.04 2.02 2.00 1.98 1.96 1.94
B
500 MHz
3.96 3.94 3.92 3.90 3.88
C
500 MHz
5.82 5.80 5.78 5.76 5.74 5.72 5.70

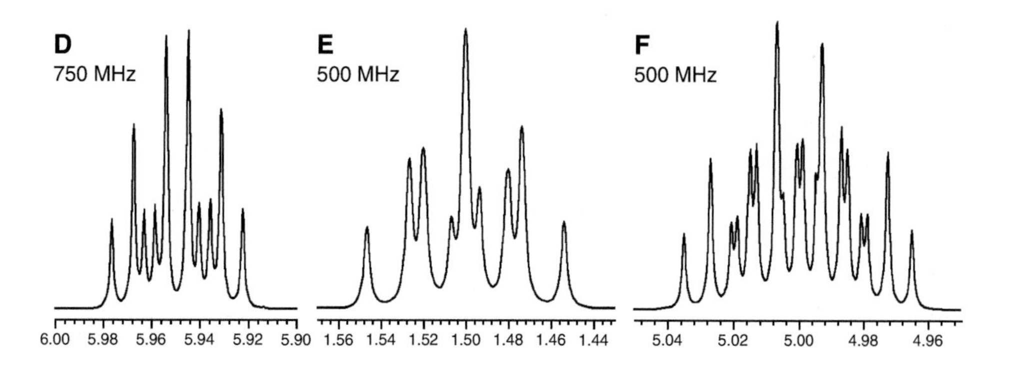
D
750 MHz
6.00 5.98 5.96 5.94 5.92 5.90
E
500 MHz
1.56 1.54 1.52 1.50 1.48 1.46 1.44
F
500 MHz
5.04 5.02 5.00 4.98 4.96

7.3 Problems Using all the Spectroscopic Methods

In each of the Problems 14–34, deduce the structure of the compound giving the set of spectra.

Problem 14 C, 49.4%; H, 9.8%; N, 19.1%

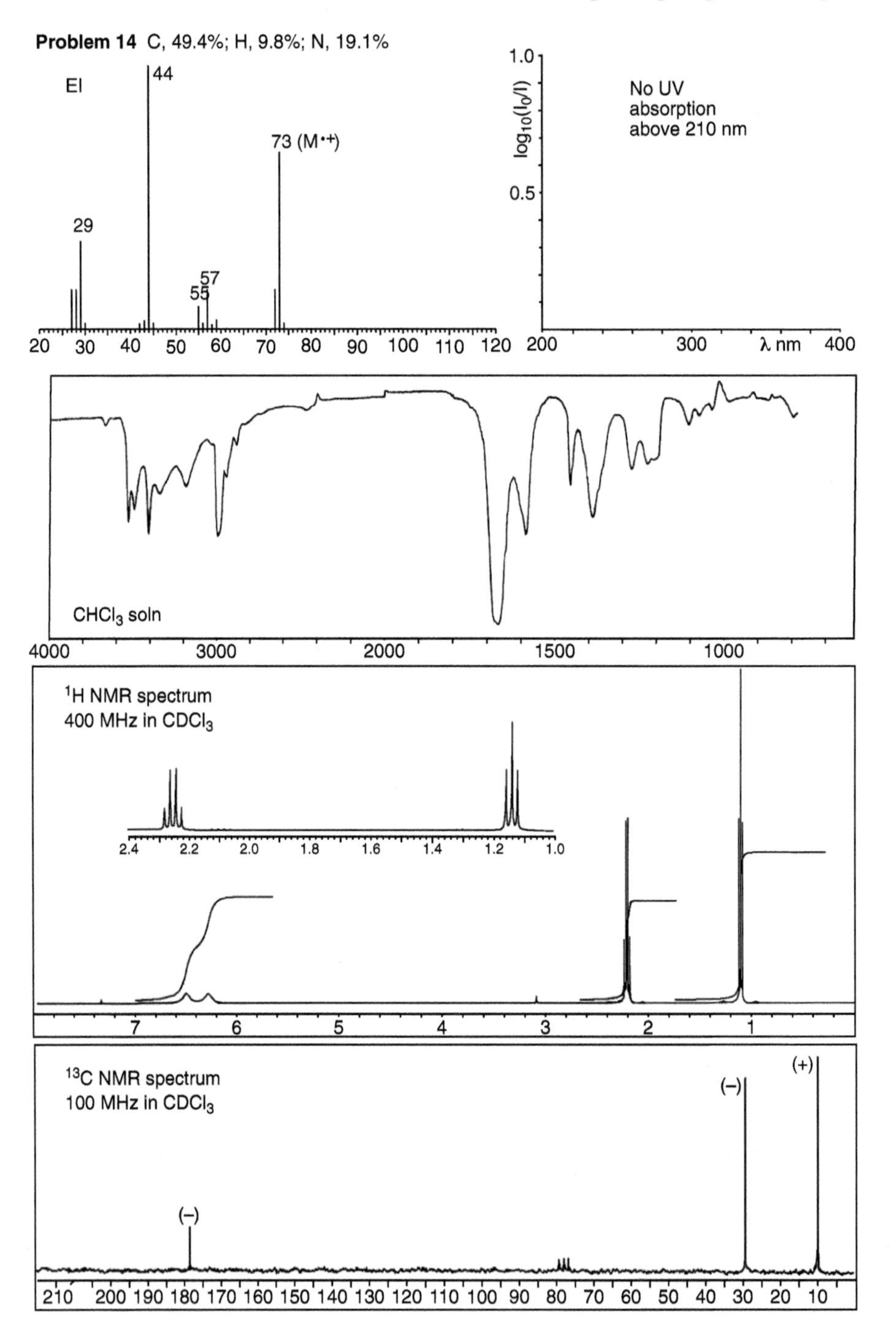

Problem 15 C, 80.0%; H, 4.8%

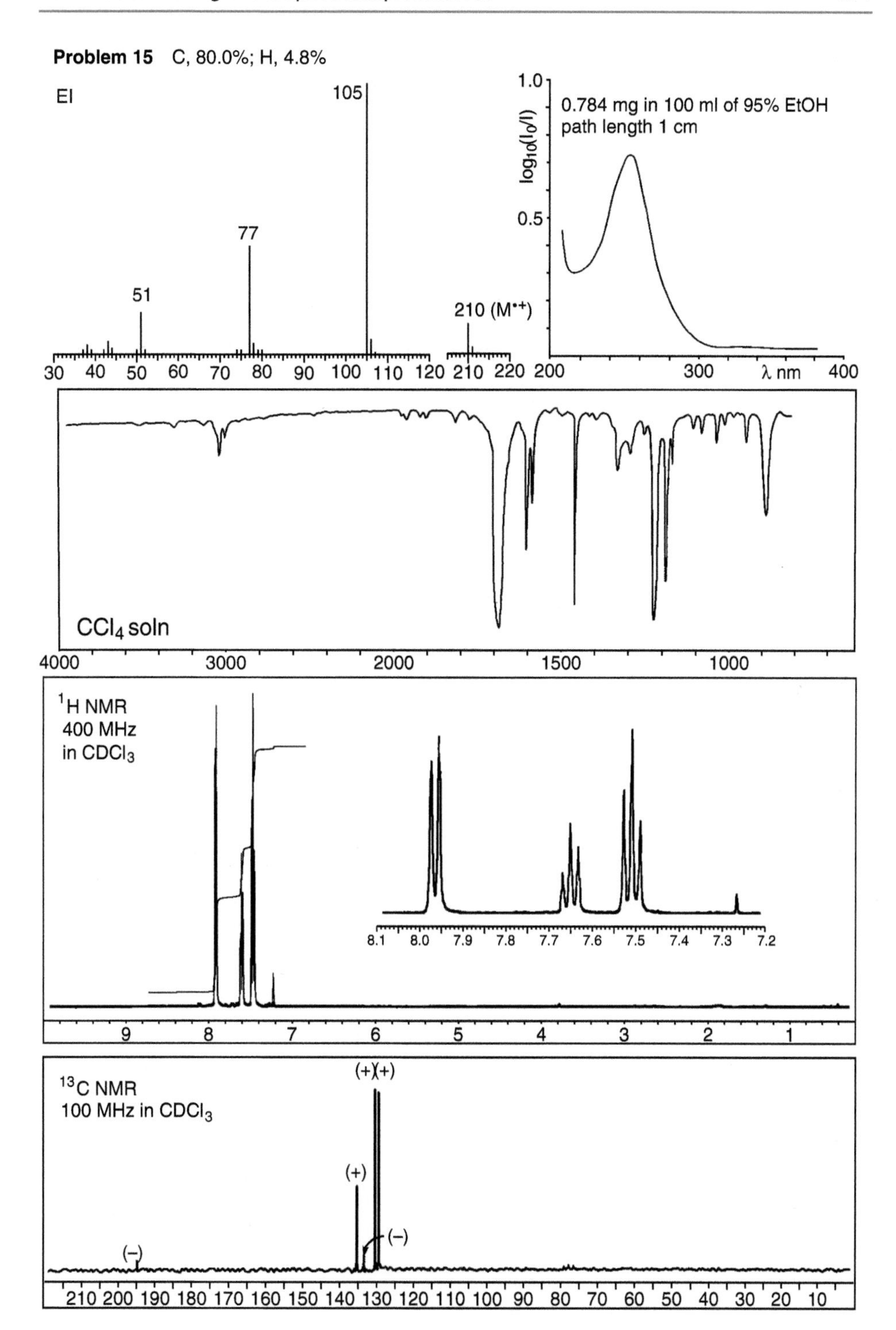

EI
105
77
51
210 (M•+)
30 40 50 60 70 80 90 100 110 120 210 220
1.0
0.5
log10(I0/I)
0.784 mg in 100 ml of 95% EtOH
path length 1 cm
200 300 λ nm 400
CCl4 soln
4000 3000 2000 1500 1000
1H NMR
400 MHz
in CDCl3
8.1 8.0 7.9 7.8 7.7 7.6 7.5 7.4 7.3 7.2
9 8 7 6 5 4 3 2 1
13C NMR
100 MHz in CDCl3
(+)(+)
(+)
(−)
(−)
210 200 190 180 170 160 150 140 130 120 110 100 90 80 70 60 50 40 30 20 10

Problem 16 C, 64.7%; H, 10.9%

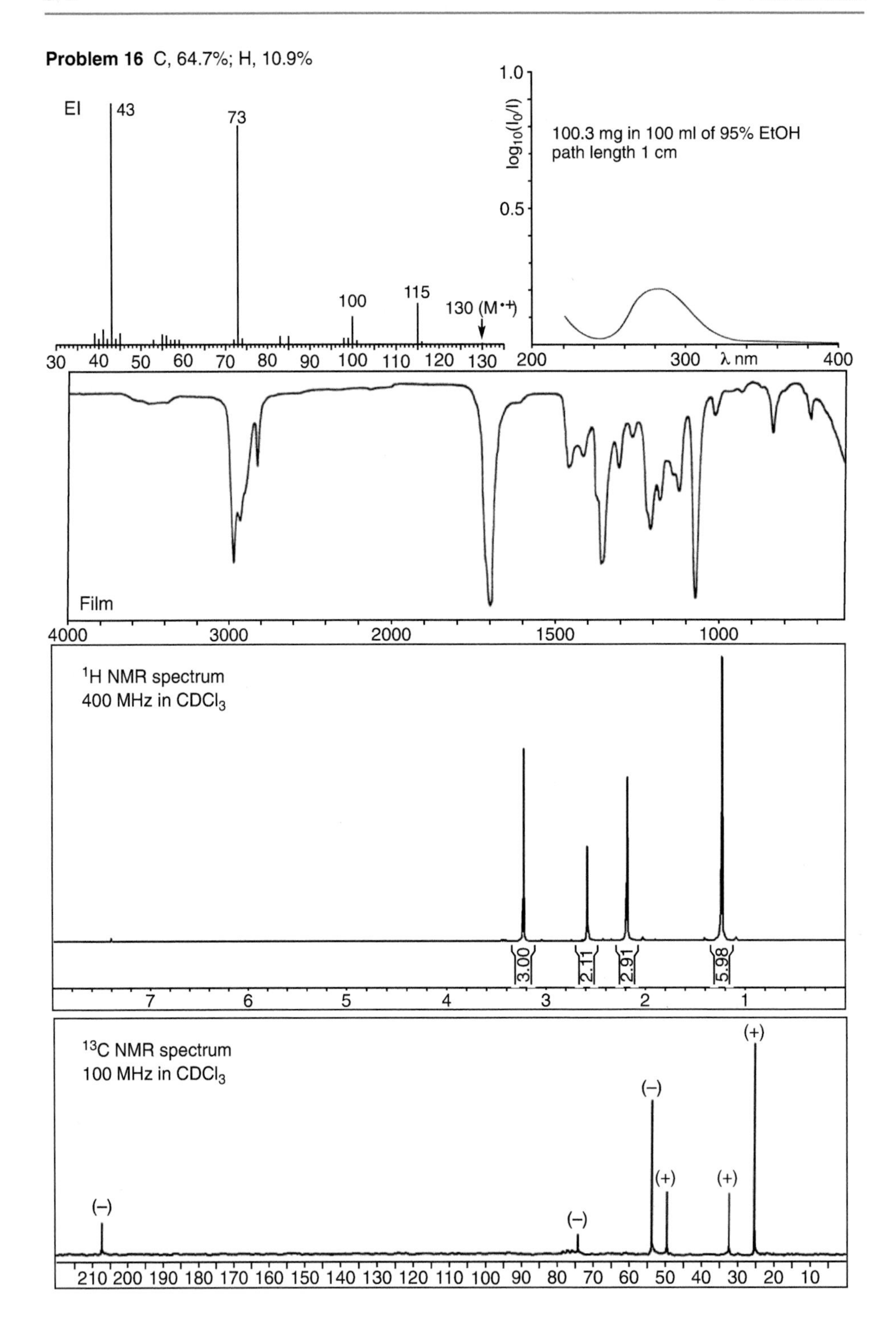
EI
43
73
100
115
130 (M•+)
30 40 50 60 70 80 90 100 110 120 130
1.0
log10(I0/I)
0.5
100.3 mg in 100 ml of 95% EtOH
path length 1 cm
200 300 λ nm 400
Film
4000 3000 2000 1500 1000
1H NMR spectrum
400 MHz in CDCl3
3.00
2.11
2.91
5.98
7 6 5 4 3 2 1
13C NMR spectrum
100 MHz in CDCl3
(–)
(–)
(+)
(+)
(–)
(+)
210 200 190 180 170 160 150 140 130 120 110 100 90 80 70 60 50 40 30 20 10

Problem 17 C, 58.2%; H, 8.5%

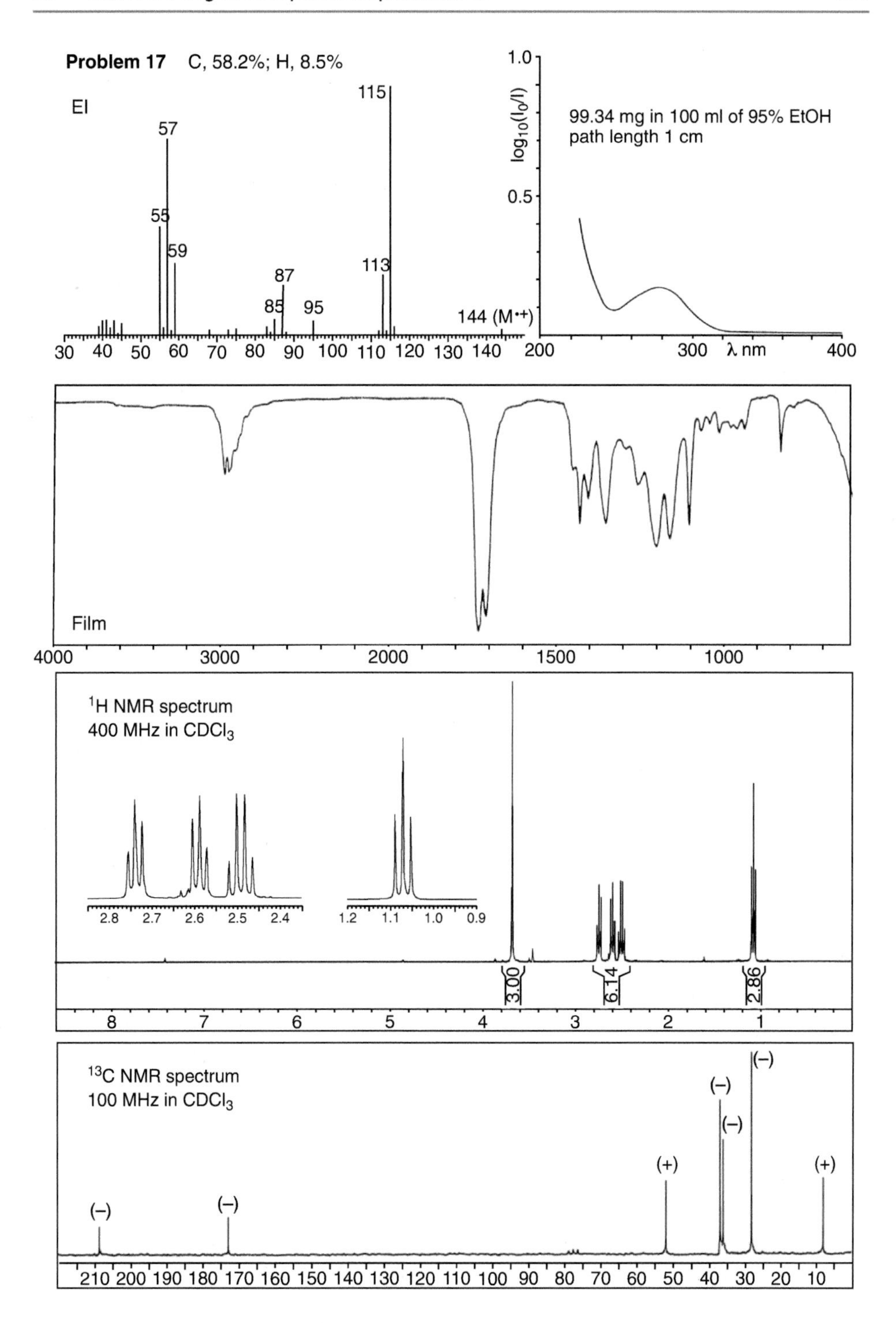

Problem 18 C, 36.5%; H, 10.0%

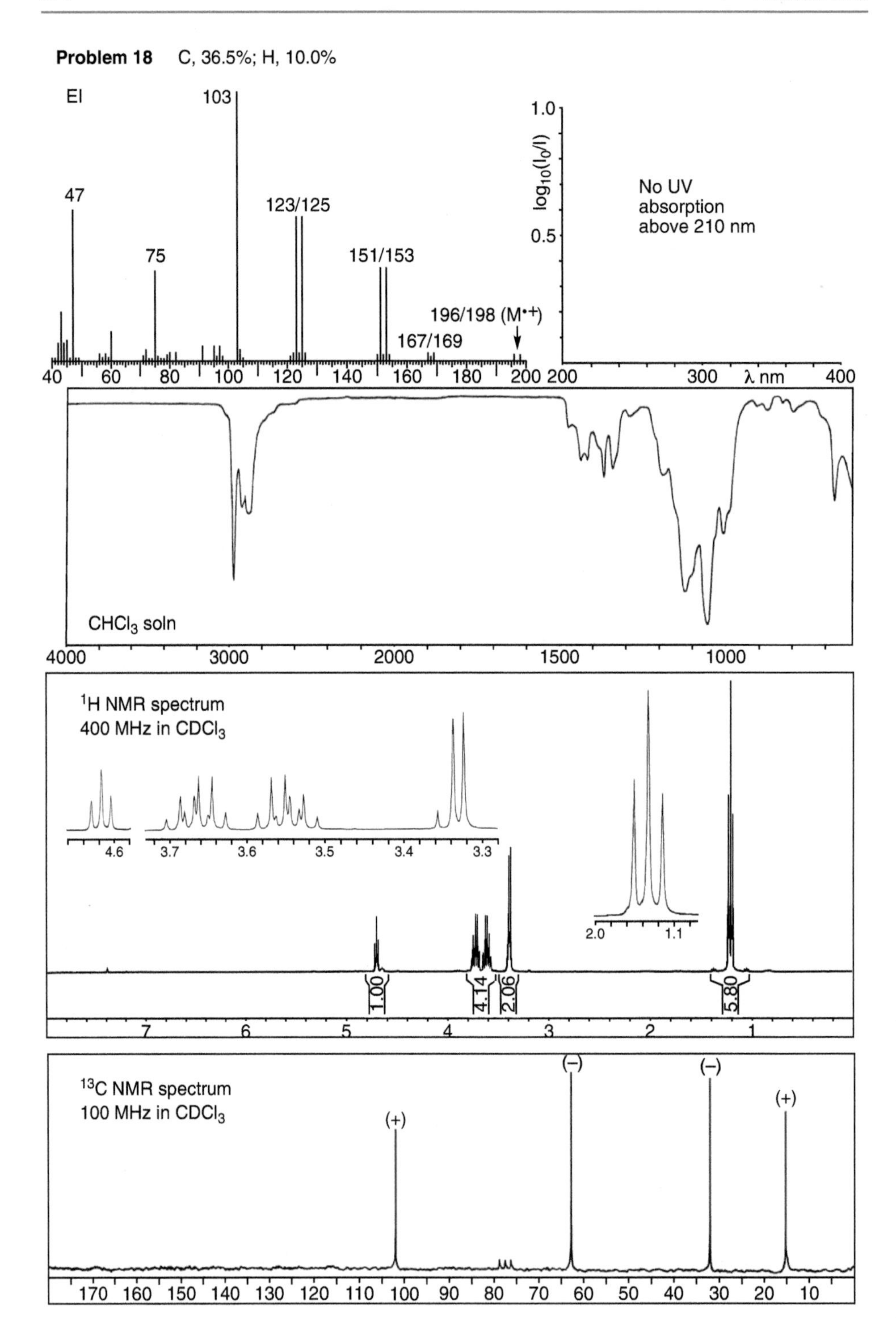

Problem 19 C, 39.8%; H, 7.3%

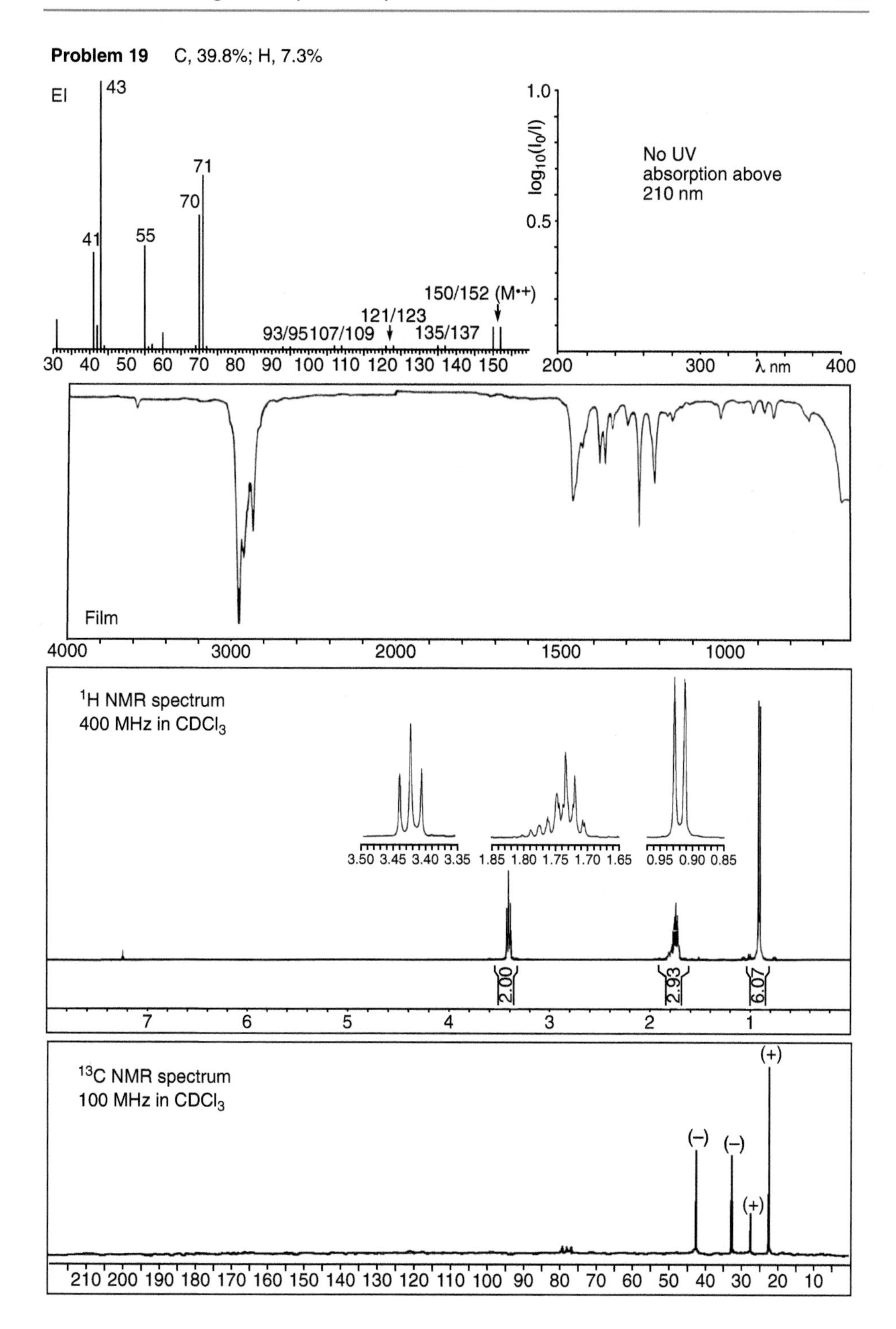

Problem 20 C, 75.5%; H, 7.5%; N, 8.1%

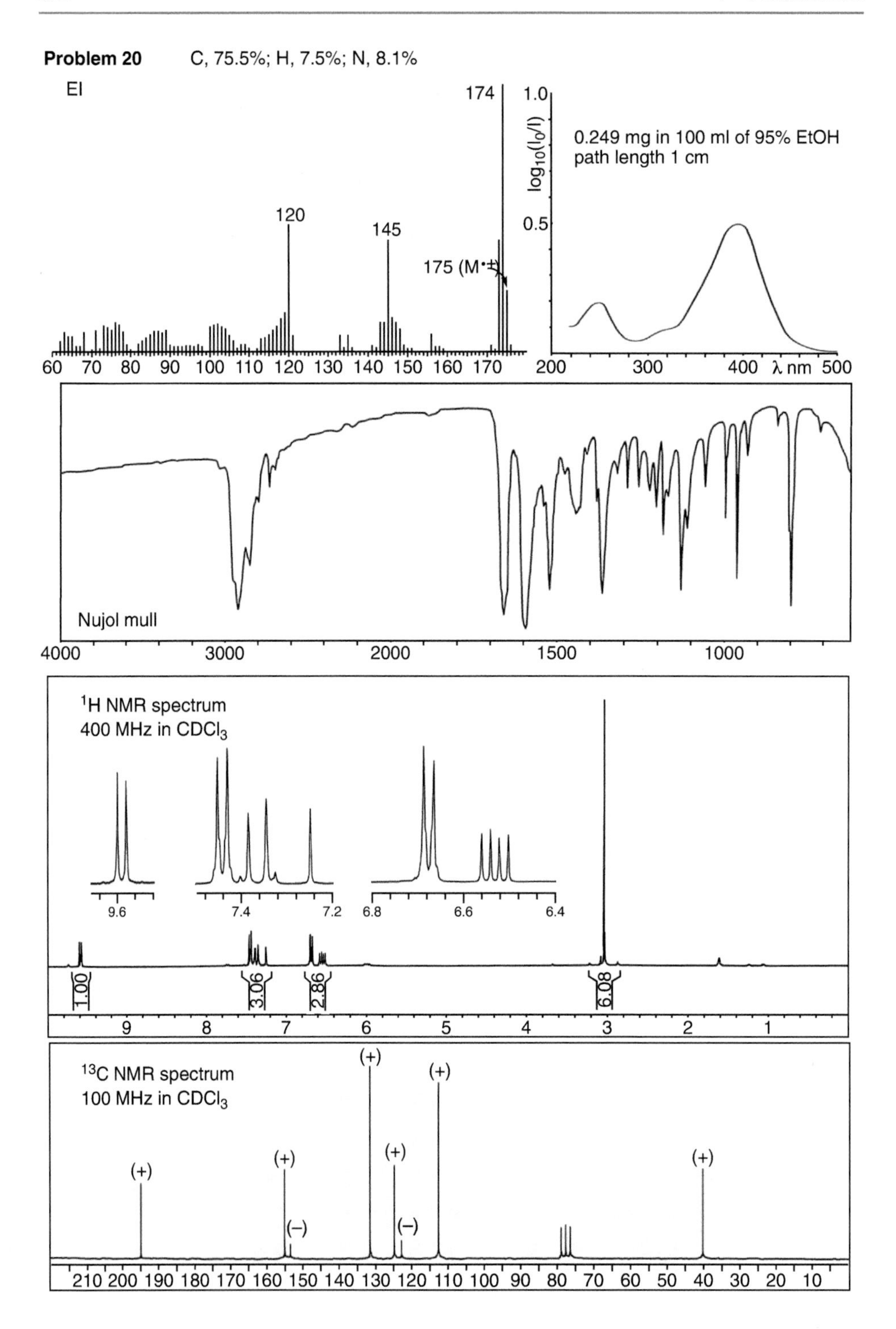
EI
174
120
145
175 (M•+)
60 70 80 90 100 110 120 130 140 150 160 170
1.0
log10(I0/I)
0.5
0.249 mg in 100 ml of 95% EtOH
path length 1 cm
200 300 400 λ nm 500
Nujol mull
4000 3000 2000 1500 1000
1H NMR spectrum
400 MHz in CDCl3
9.6
7.4 7.2
6.8 6.6 6.4
1.00
3.06
2.86
6.08
9 8 7 6 5 4 3 2 1
13C NMR spectrum
100 MHz in CDCl3
(+)
(+)
(–)
(+)
(+)
(–)
(+)
(+)
210 200 190 180 170 160 150 140 130 120 110 100 90 80 70 60 50 40 30 20 10

Problem 21 C, 51.1%; H, 2.8%; N, 10.0%

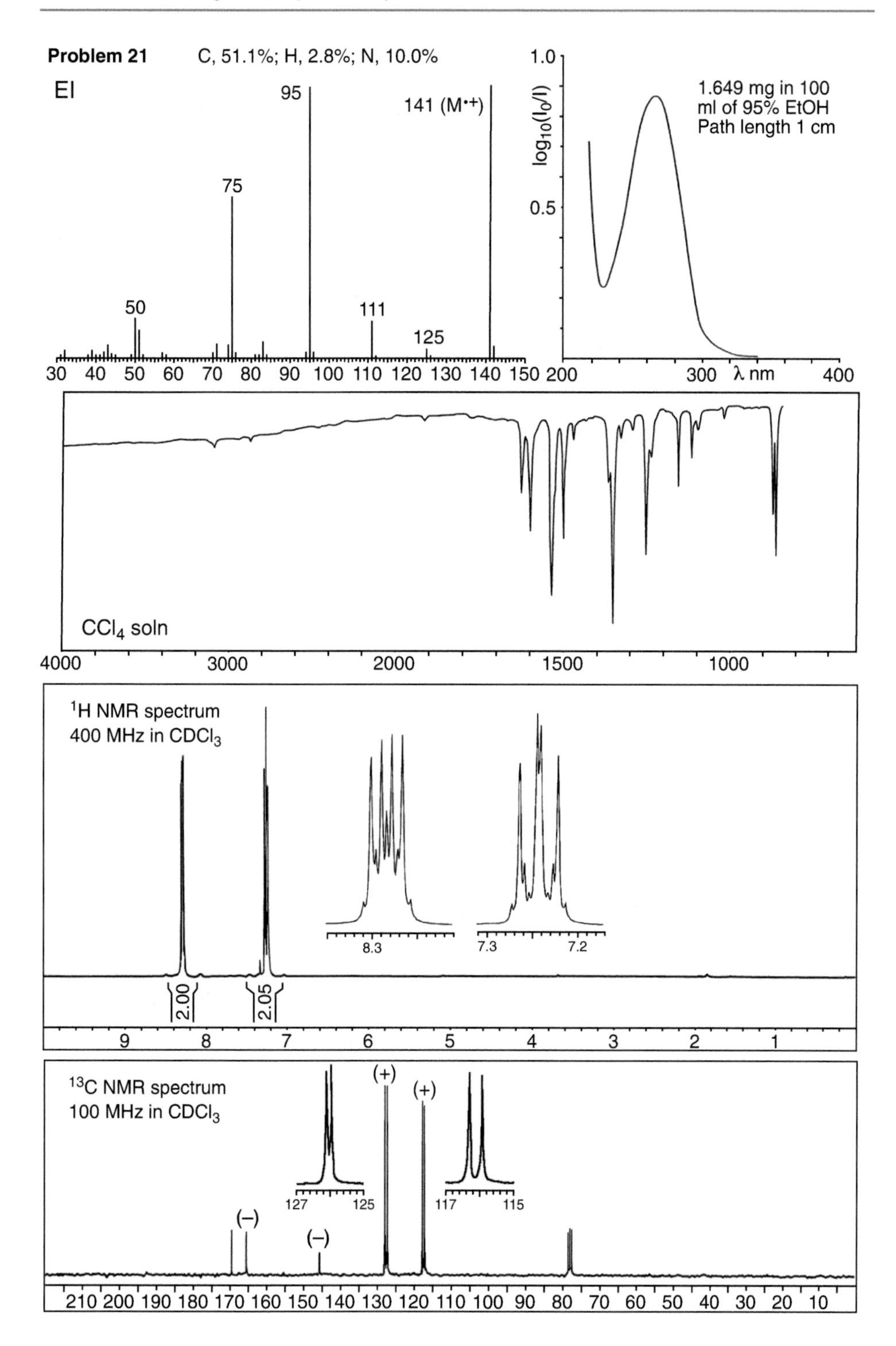
EI
95
141 (M•+)
75
50
111
125
30 40 50 60 70 80 90 100 110 120 130 140 150
1.0
0.5
log10(I0/I)
1.649 mg in 100
ml of 95% EtOH
Path length 1 cm
200 300 λ nm 400
CCl4 soln
4000 3000 2000 1500 1000
1H NMR spectrum
400 MHz in CDCl3
8.3
7.3 7.2
2.00
2.05
9 8 7 6 5 4 3 2 1
13C NMR spectrum
100 MHz in CDCl3
(+)
(+)
127 125
117 115
(–)
(–)
210 200 190 180 170 160 150 140 130 120 110 100 90 80 70 60 50 40 30 20 10

Problem 22 C, 64.3%; H, 8.8%

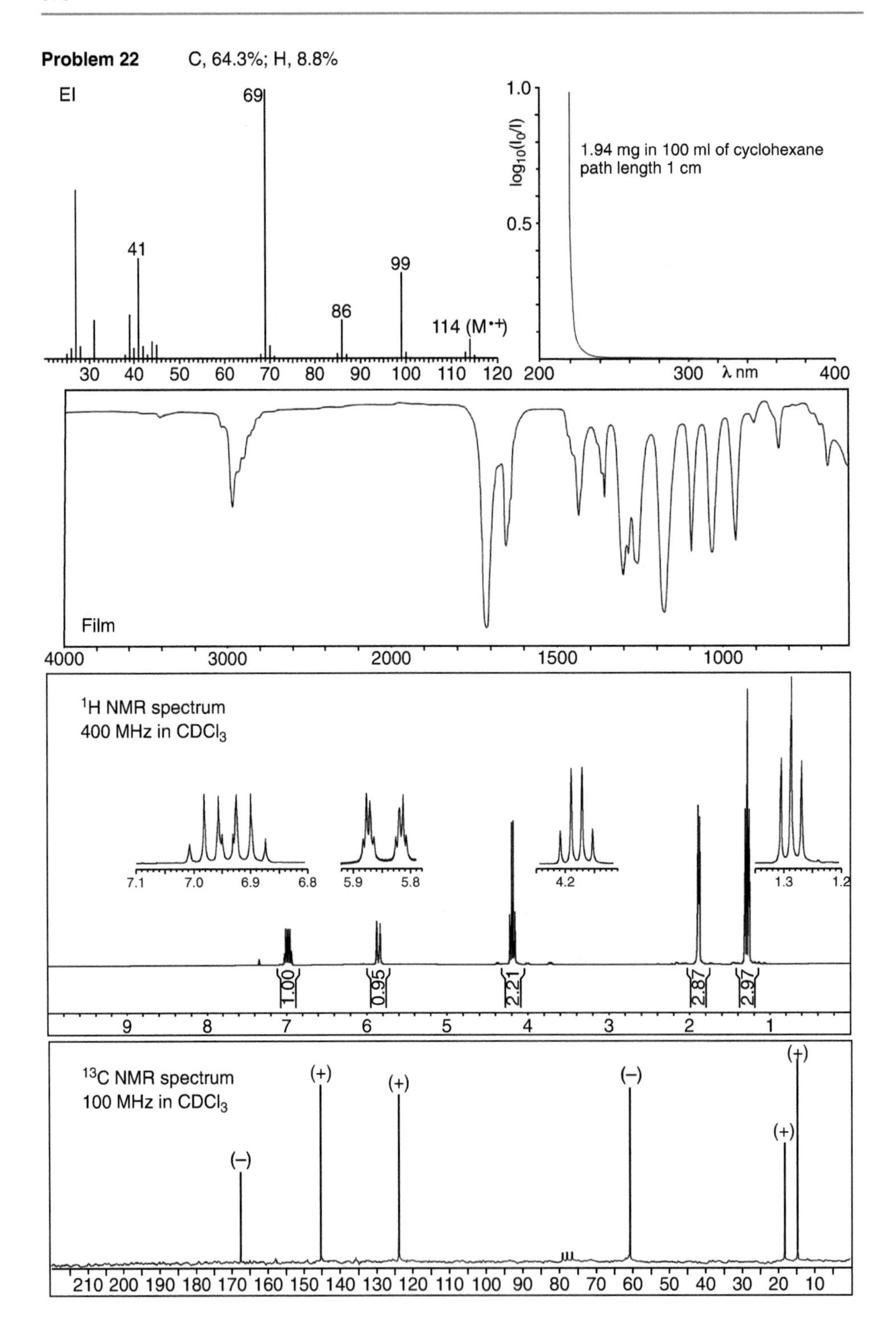
EI
69
41
99
86
114 (M•+)
30 40 50 60 70 80 90 100 110 120
1.0
0.5
$\log_{10}(I_0/I)$
1.94 mg in 100 ml of cyclohexane
path length 1 cm
200 300 λ nm 400
Film
4000 3000 2000 1500 1000
^{1}H NMR spectrum
400 MHz in $CDCl_3$
7.1 7.0 6.9 6.8
5.9 5.8
4.2
1.3 1.2
1.00
0.95
2.21
2.87
2.97
9 8 7 6 5 4 3 2 1
^{13}C NMR spectrum
100 MHz in $CDCl_3$
(–)
(+)
(+)
(–)
(+)
(+)
210 200 190 180 170 160 150 140 130 120 110 100 90 80 70 60 50 40 30 20 10

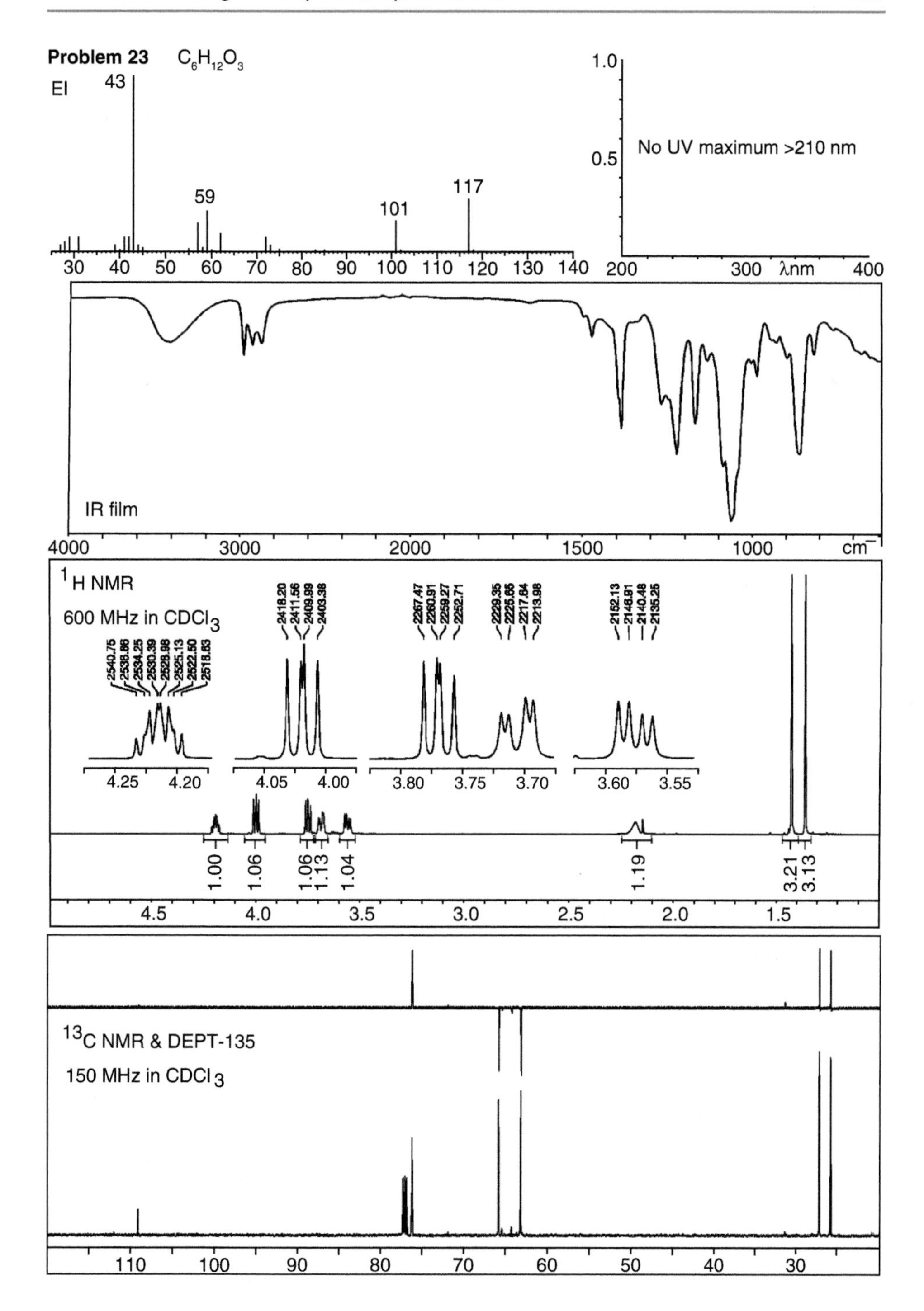

Problem 23 C6H12O3
EI
43
59
101
117
30 40 50 60 70 80 90 100 110 120 130 140
1.0
0.5
No UV maximum >210 nm
200 300 λnm 400
IR film
4000 3000 2000 1500 1000 cm⁻¹
1H NMR
600 MHz in CDCl3
2540.75 2536.86 2534.25 2530.39 2528.98 2525.13 2522.50 2518.63
2418.20 2411.56 2409.99 2403.38
2267.47 2260.91 2259.27 2252.71
2229.35 2225.65 2217.84 2213.98
2152.13 2148.81 2140.48 2135.25
4.25 4.20
4.05 4.00
3.80 3.75 3.70
3.60 3.55
1.00 1.06 1.06 1.13 1.04 1.19 3.21 3.13
4.5 4.0 3.5 3.0 2.5 2.0 1.5
13C NMR & DEPT-135
150 MHz in CDCl3
110 100 90 80 70 60 50 40 30

Problem 23 continued

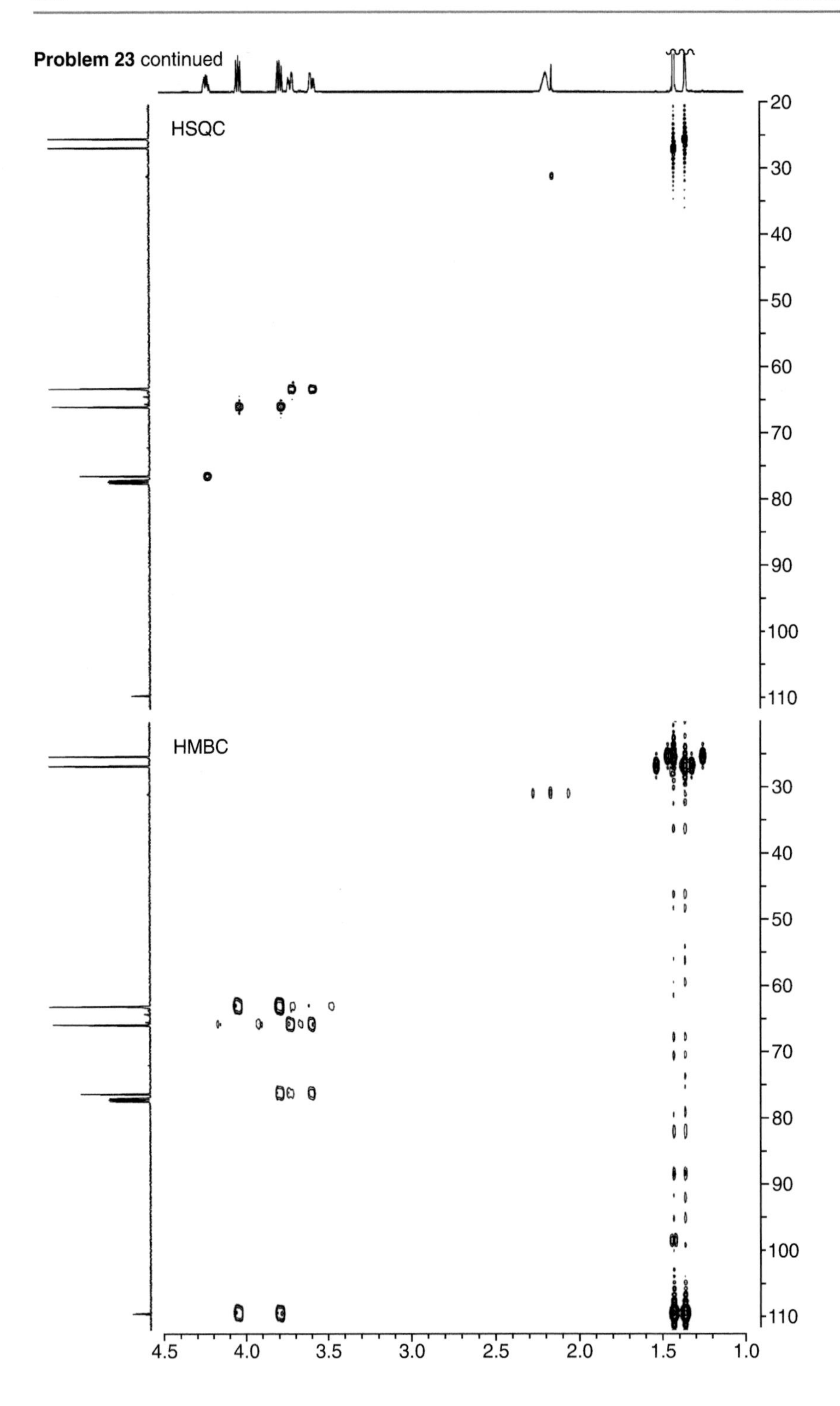

Problem 23 continued

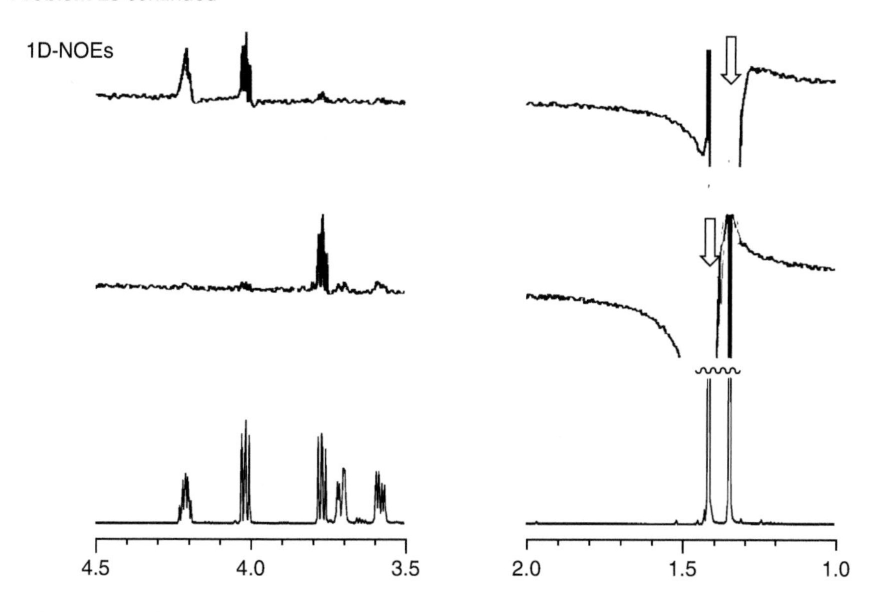

Problem 24 $C_{12}H_{16}O_4$

ν_{max} 3374, 1616, 1599 and 1456 cm^{-1} λ_{max} 220 nm (ε 5500).

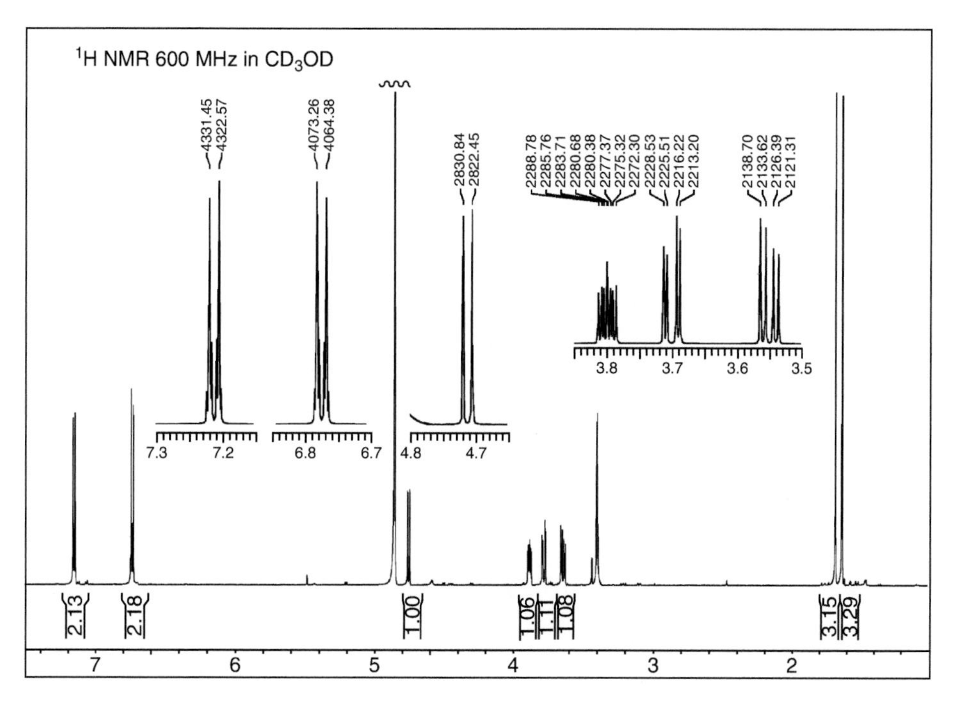

Problem 24 continued

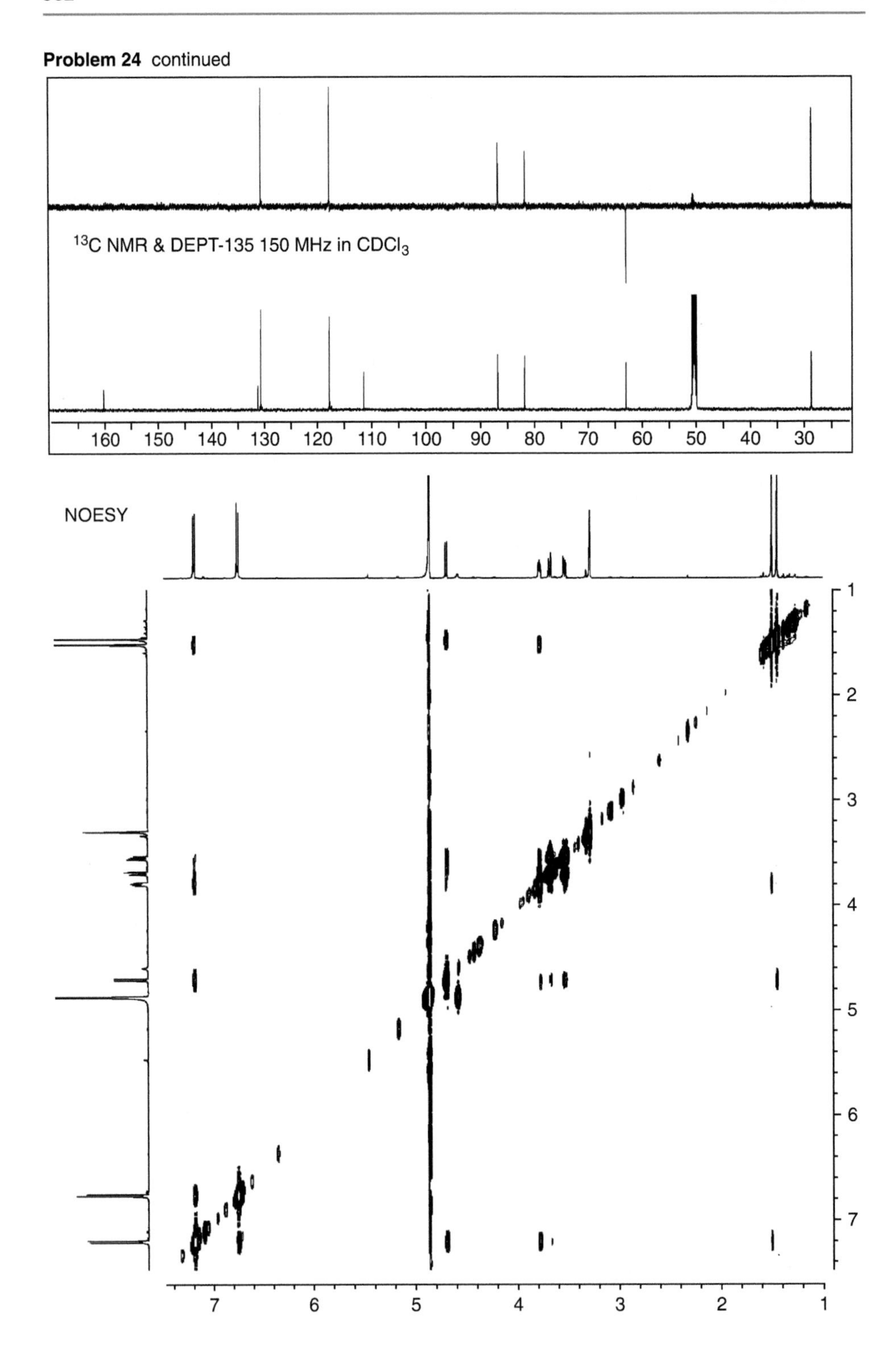

Problem 24 continued

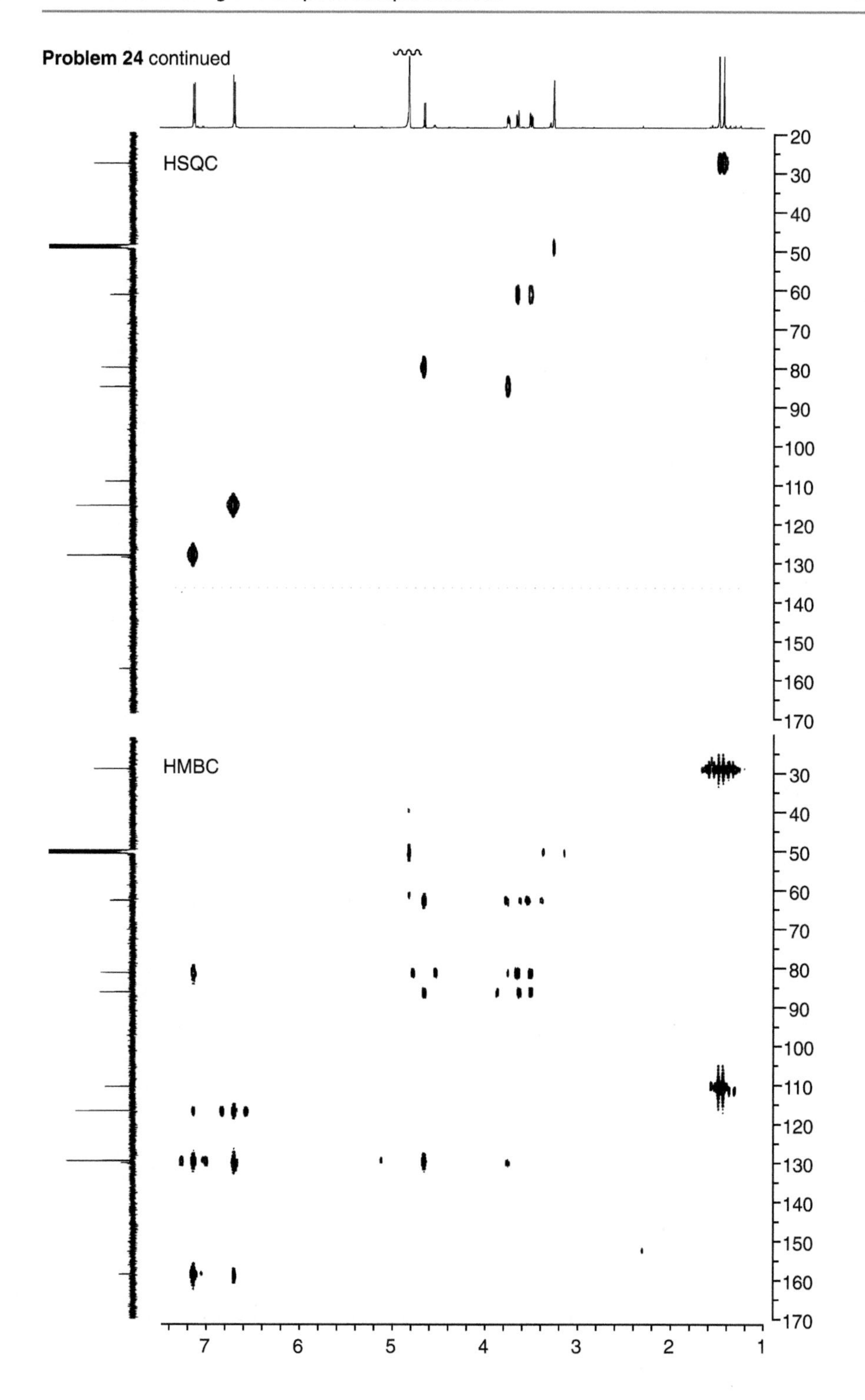

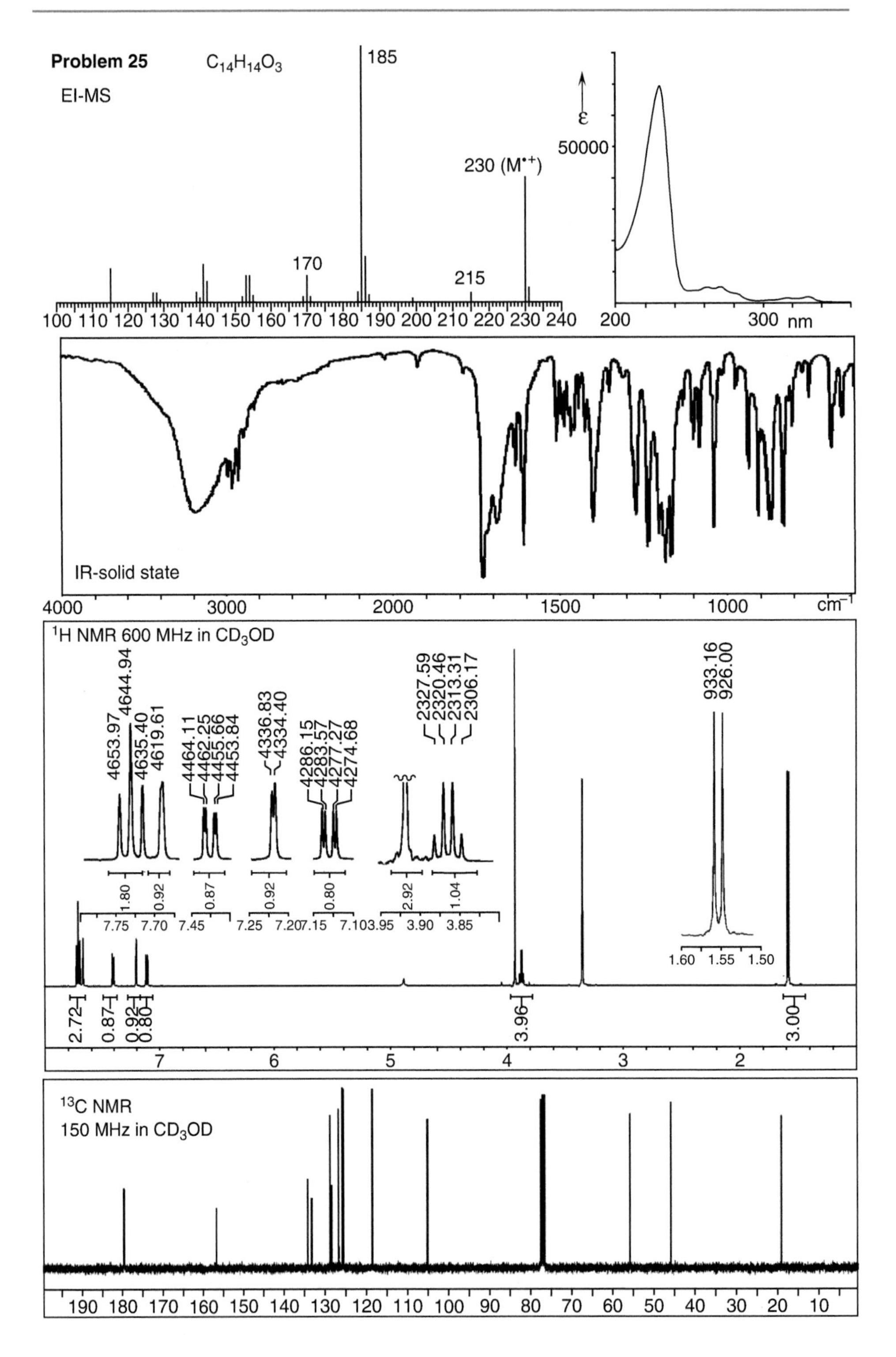

Problem 25
$C_{14}H_{14}O_3$
EI-MS
185
230 (M•+)
170
215
100 110 120 130 140 150 160 170 180 190 200 210 220 230 240
ε
50000
200
300 nm
IR-solid state
4000
3000
2000
1500
1000
cm−1
1H NMR 600 MHz in CD3OD
4653.97
4644.94
4635.40
4619.61
4464.11
4462.25
4455.66
4453.84
4336.83
4334.40
4286.15
4283.57
4277.27
4274.68
2327.59
2320.46
2313.31
2306.17
933.16
926.00
1.80
0.92
0.87
0.92
0.80
2.92
1.04
7.75 7.70 7.45
7.25 7.20 7.15 7.10 3.95 3.90 3.85
1.60 1.55 1.50
2.72
0.87
0.92
0.80
3.96
3.00
7
6
5
4
3
2
13C NMR
150 MHz in CD3OD
190 180 170 160 150 140 130 120 110 100 90 80 70 60 50 40 30 20 10

Problem 25 continued

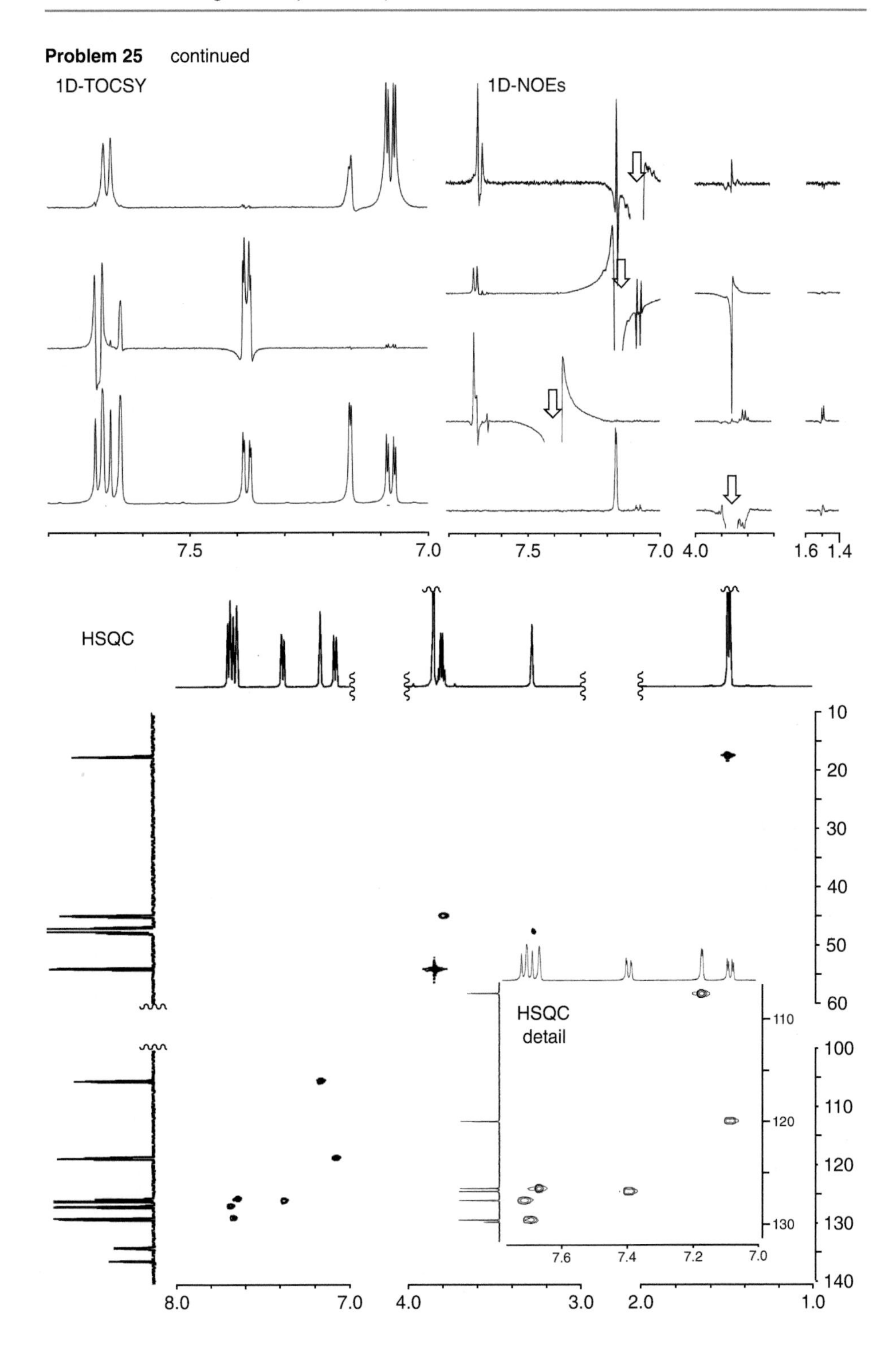

Problem 25 continued

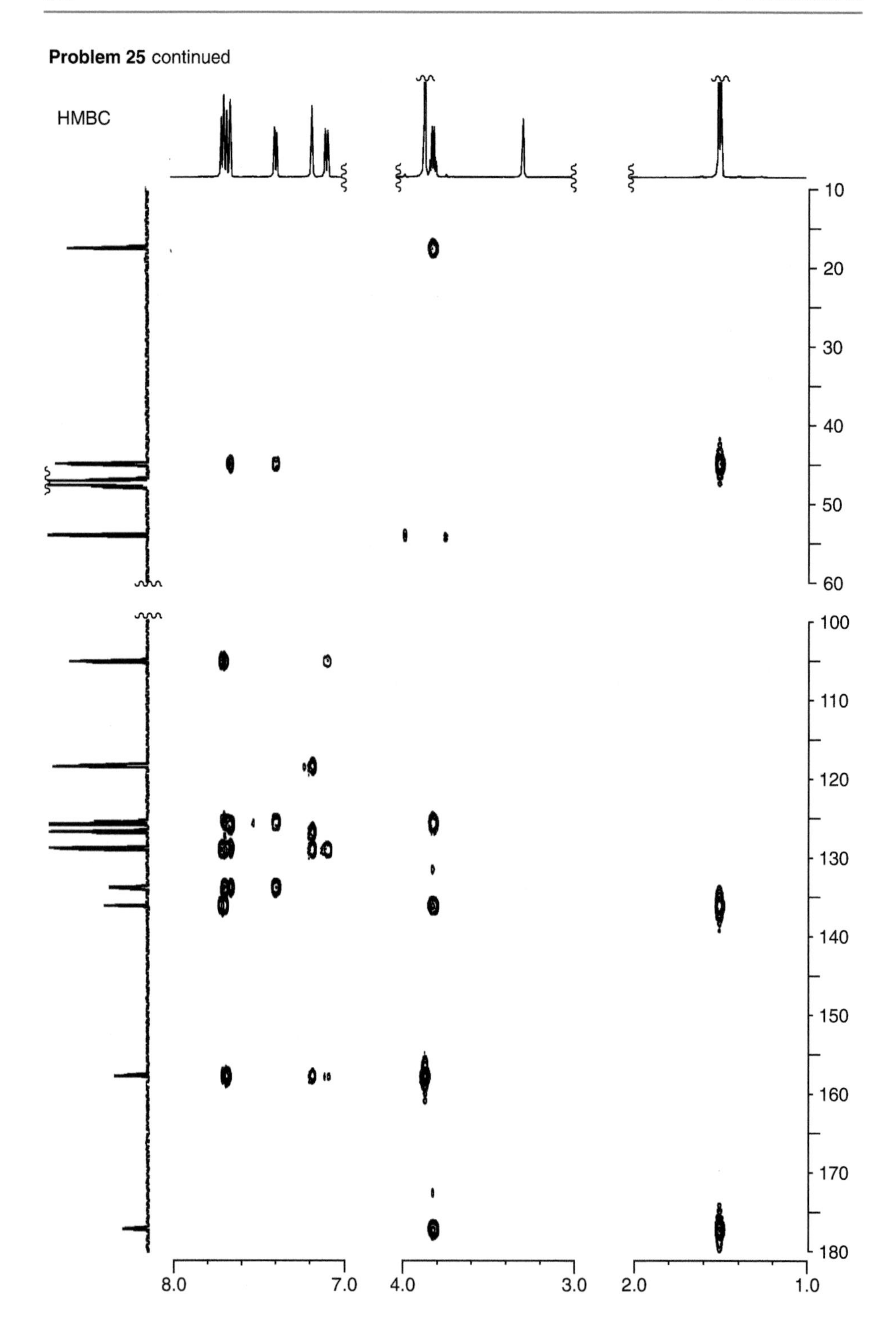

Problem 26 $C_{10}H_{15}NO_5$

ν_{max} 3330, 1740, 1715, 1670 and 1638 cm^{-1}

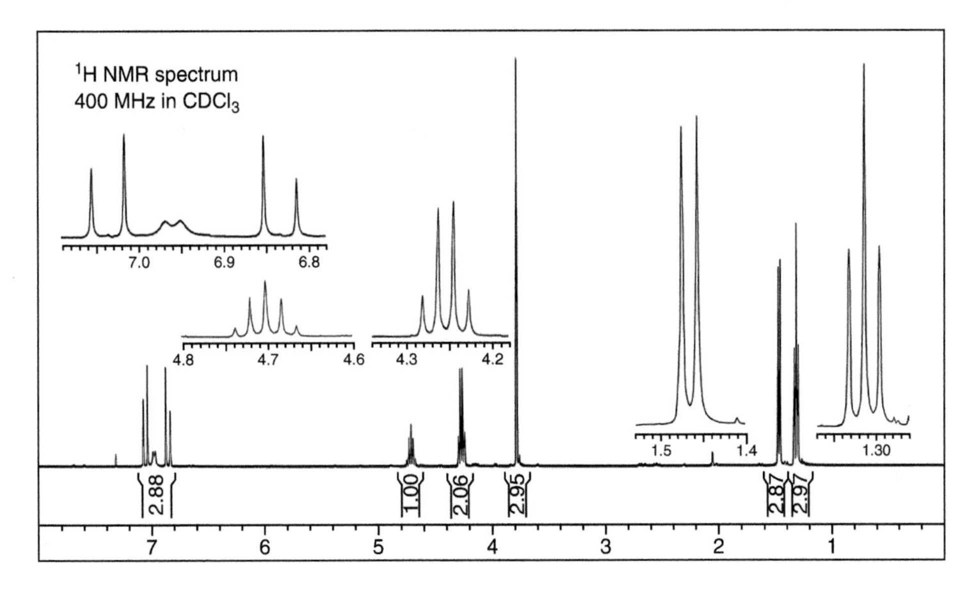

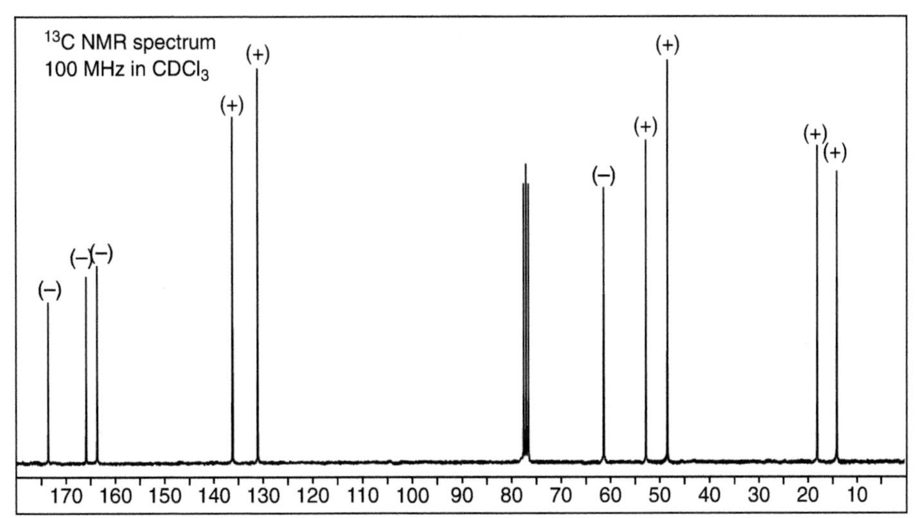

Problem 26 continued

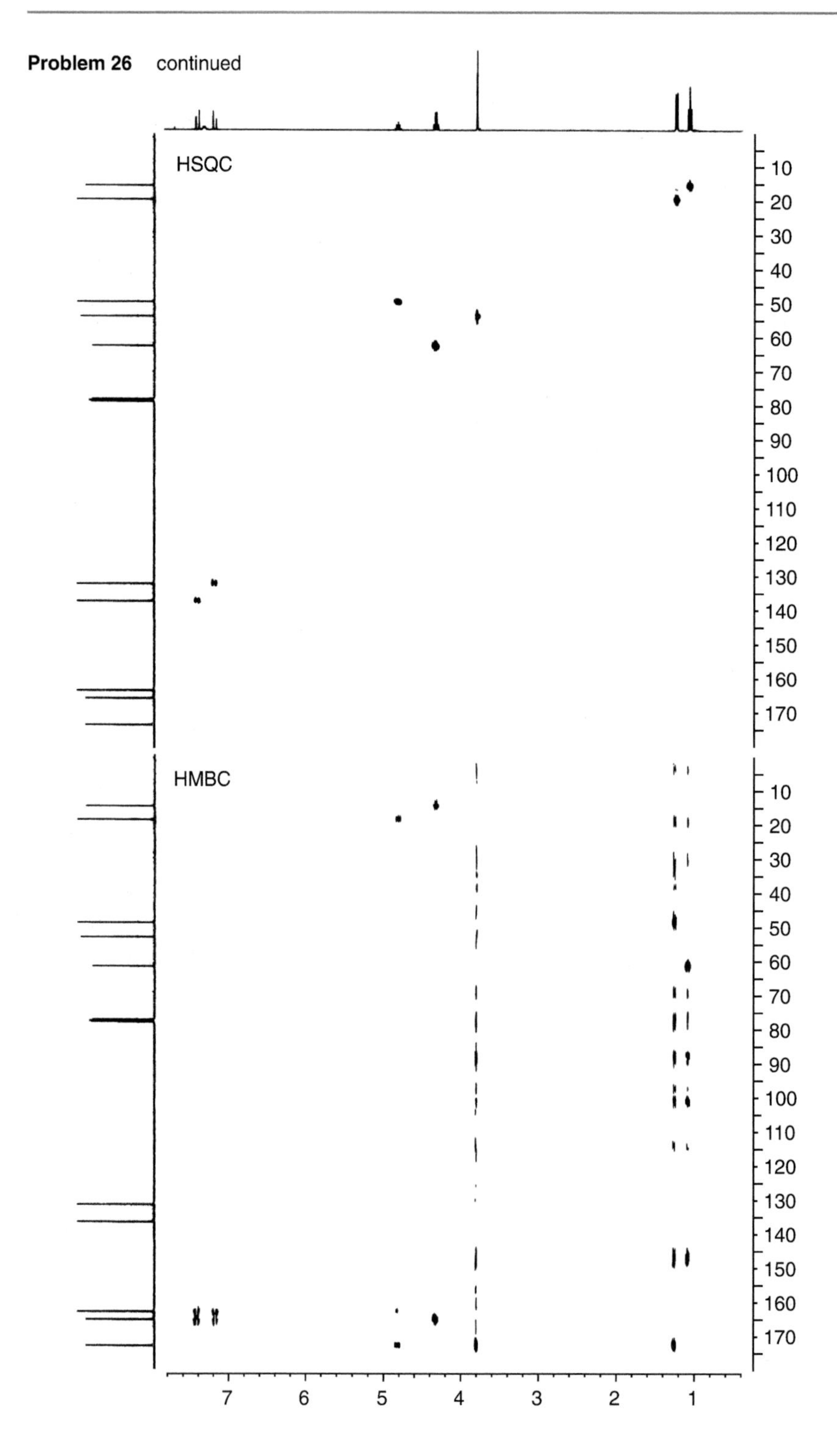

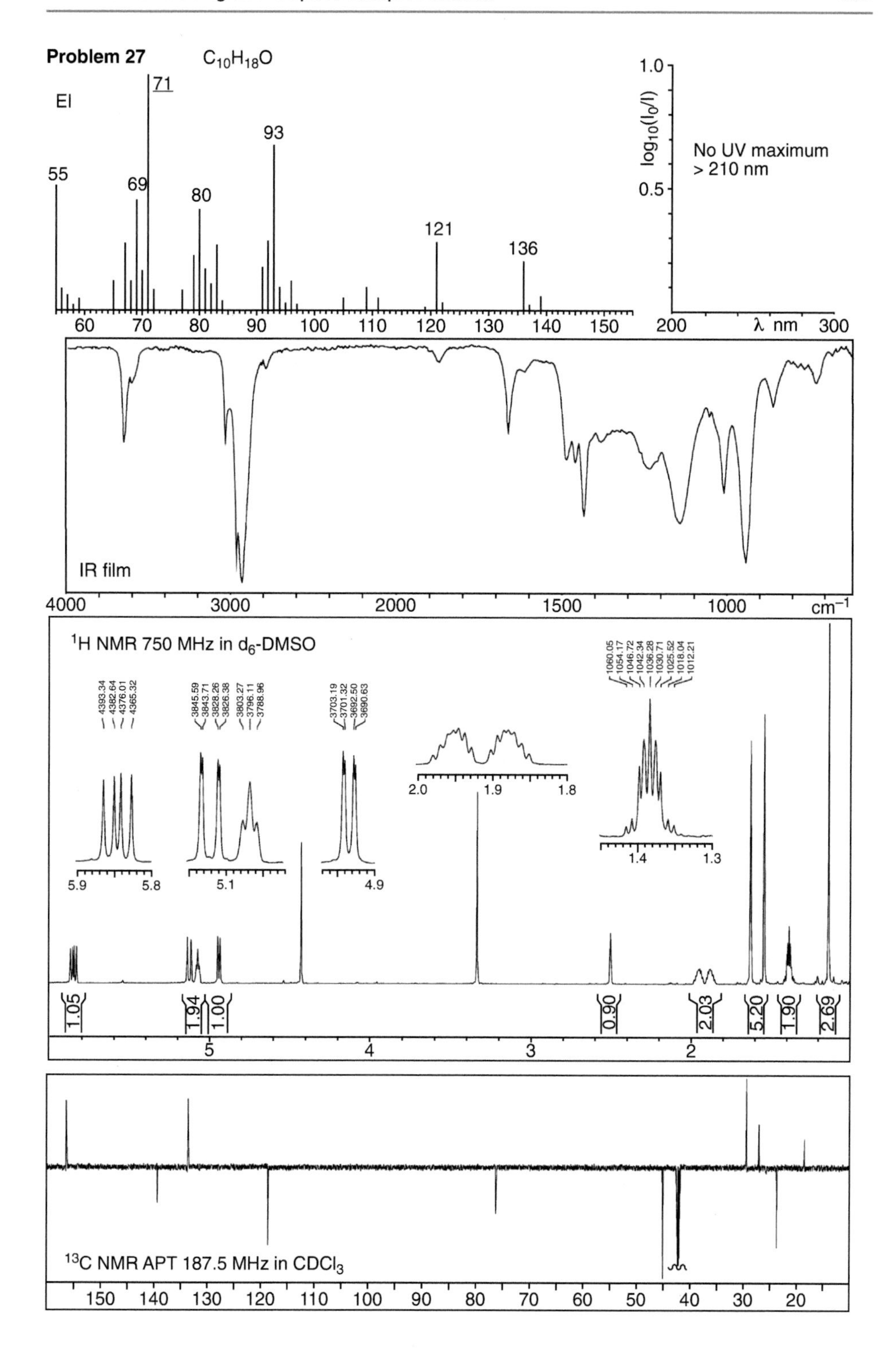

Problem 27
$C_{10}H_{18}O$
EI
55
69
71
80
93
121
136
No UV maximum > 210 nm
IR film
¹H NMR 750 MHz in d6-DMSO
4393.34
4382.64
4376.01
4365.32
3845.59
3843.71
3828.26
3826.38
3803.27
3796.11
3788.96
3703.19
3701.32
3692.50
3690.63
1060.05
1054.17
1046.72
1042.34
1036.28
1030.71
1025.52
1018.04
1012.21
1.05
1.94
1.00
0.90
2.03
5.20
1.90
2.69
¹³C NMR APT 187.5 MHz in CDCl3

Problem 27 continued

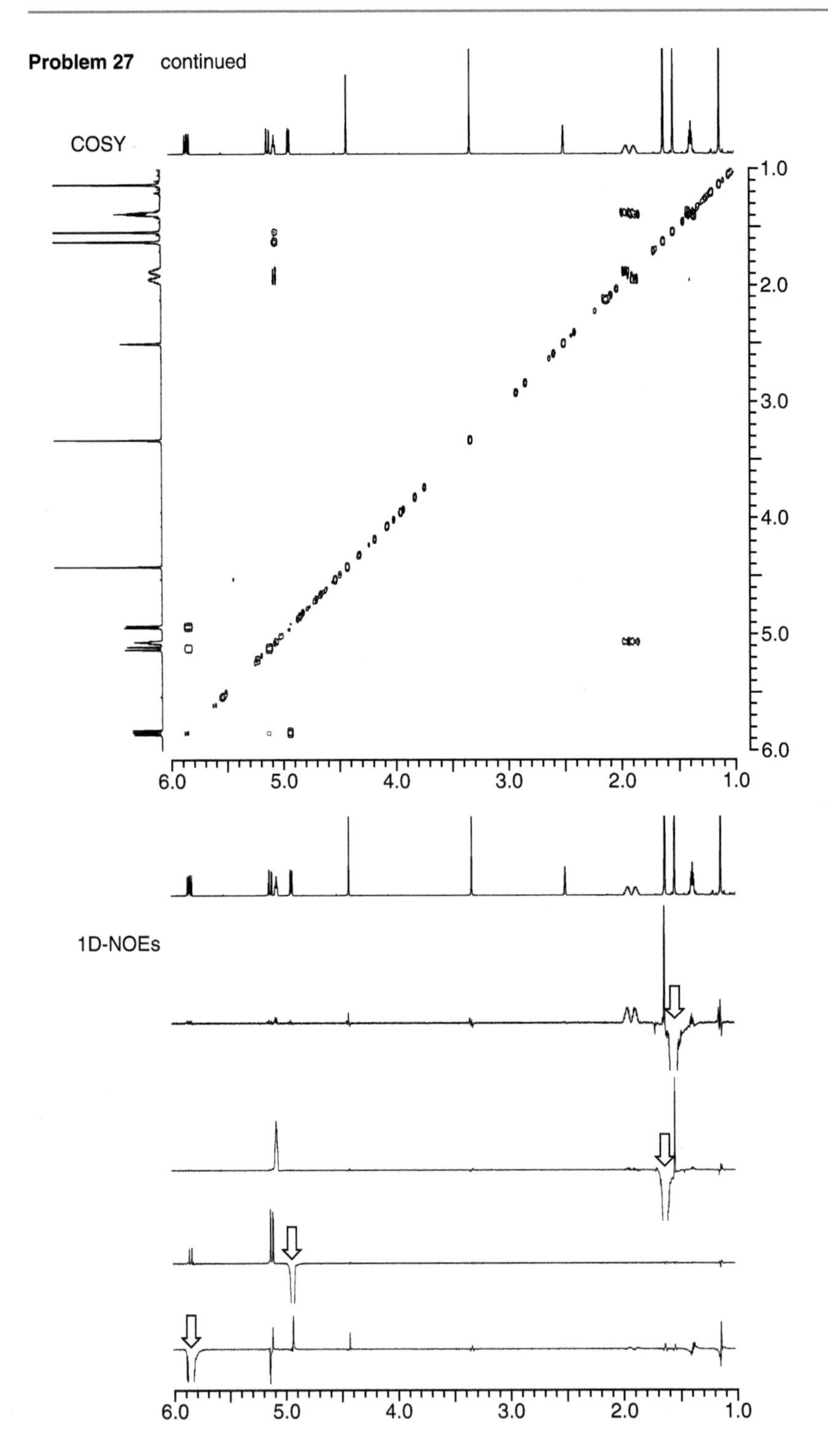

Problem 27 continued

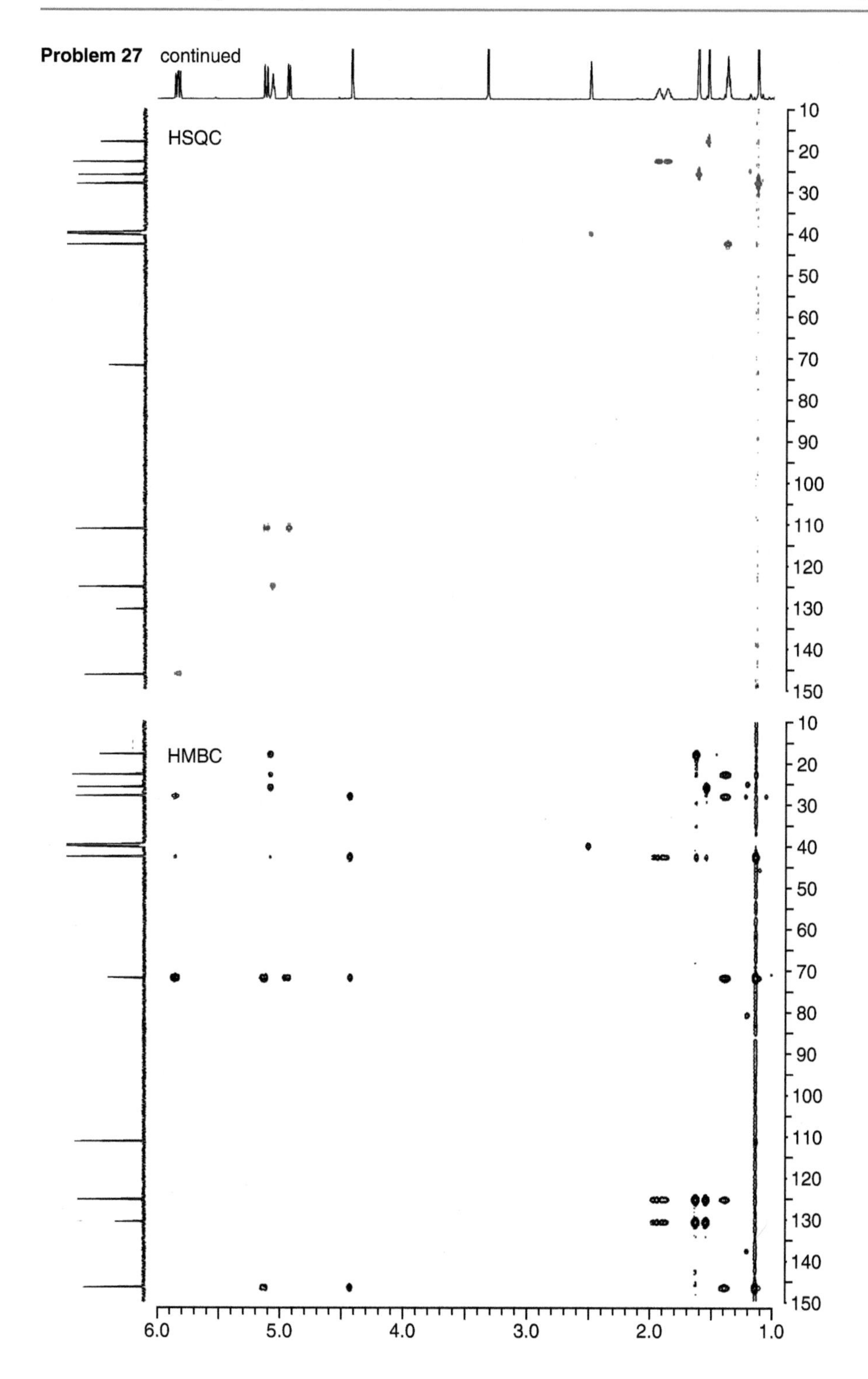

Problem 28 $C_{11}H_{14}O_3$

ν_{max} 3362, 1612 and 1596 cm^{-1}

^{1}H NMR spectrum 400 MHz in CDCl$_3$

2871.93
2863.69
2855.40
2618.29
2609.97
2585.79
2577.63
1985.65
1984.25
1981.49
1980.05
1712.87
1711.32
1706.60
1705.04
1700.31
1698.95
1694.55
1693.15
1688.42
1686.90
1682.19
1680.64
835.81
834.17
832.52
821.40
819.76
818.12
695.68
691.32
683.75
681.19
679.51
676.95
669.34
665.02
577.40
571.12

7.2 7.1 6.6 6.5 6.4 5.0 4.9 4.3 4.2 2.1 2.0 1.8 1.7 1.6

0.93 1.94 0.96 1.03 3.08 1.41 1.11 1.38 3.32

7 6 5 4 3 2

^{13}C NMR 100 MHz in CDCl$_3$

(–)(–) (+) (–) (+) (+) (+) (+) (+) (–) (+)

160 150 140 130 120 110 100 90 80 70 60 50 40 30 20

COSY

1.0 1.5 2.0 2.5 3.0 3.5 4.0 4.5 5.0 5.5 6.0 6.5 7.0 7.5

7.5 7.0 6.5 6.0 5.5 5.0 4.5 4.0 3.5 3.0 2.5 2.0 1.5 1.0

Problem 28 continued

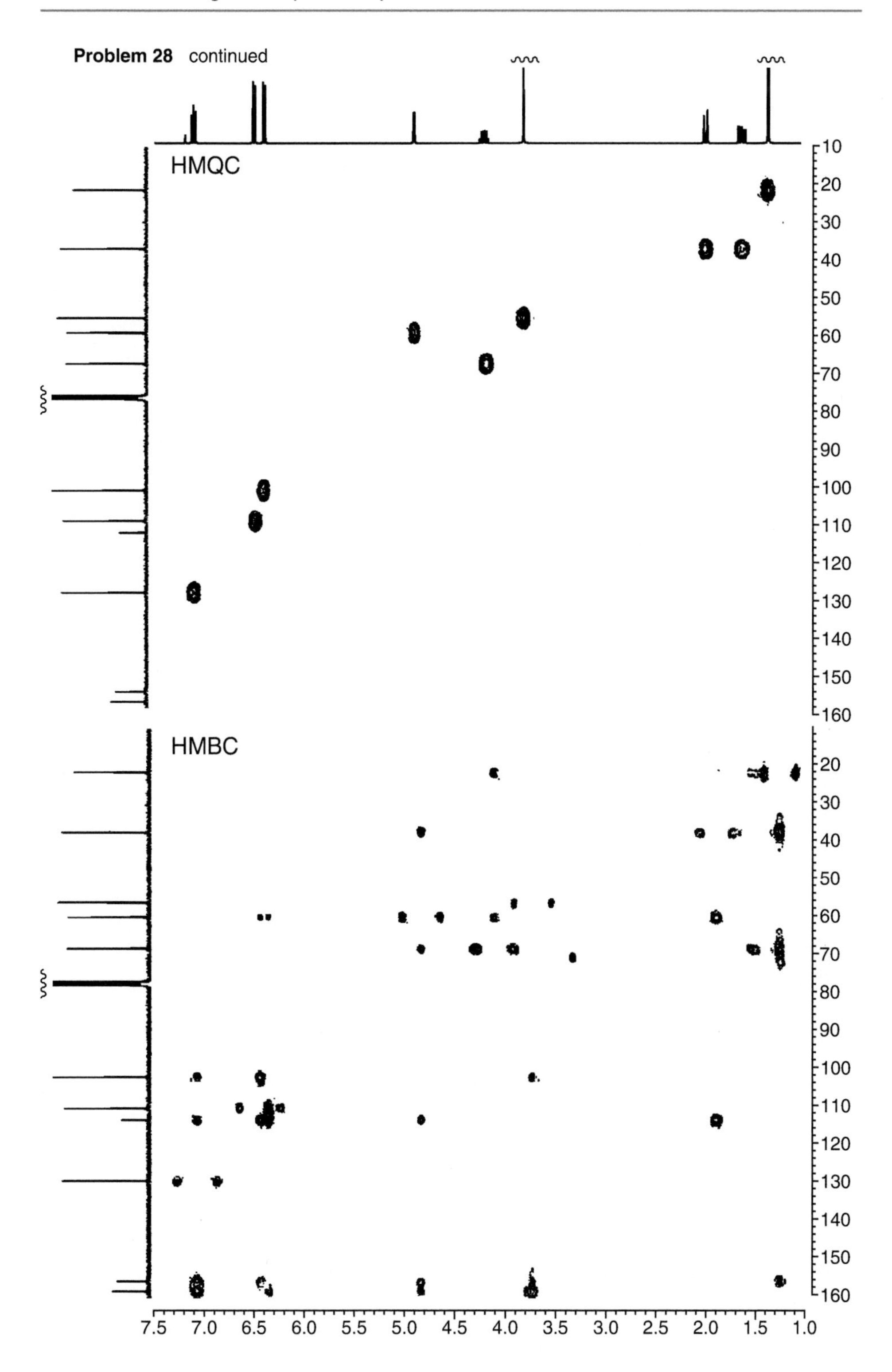

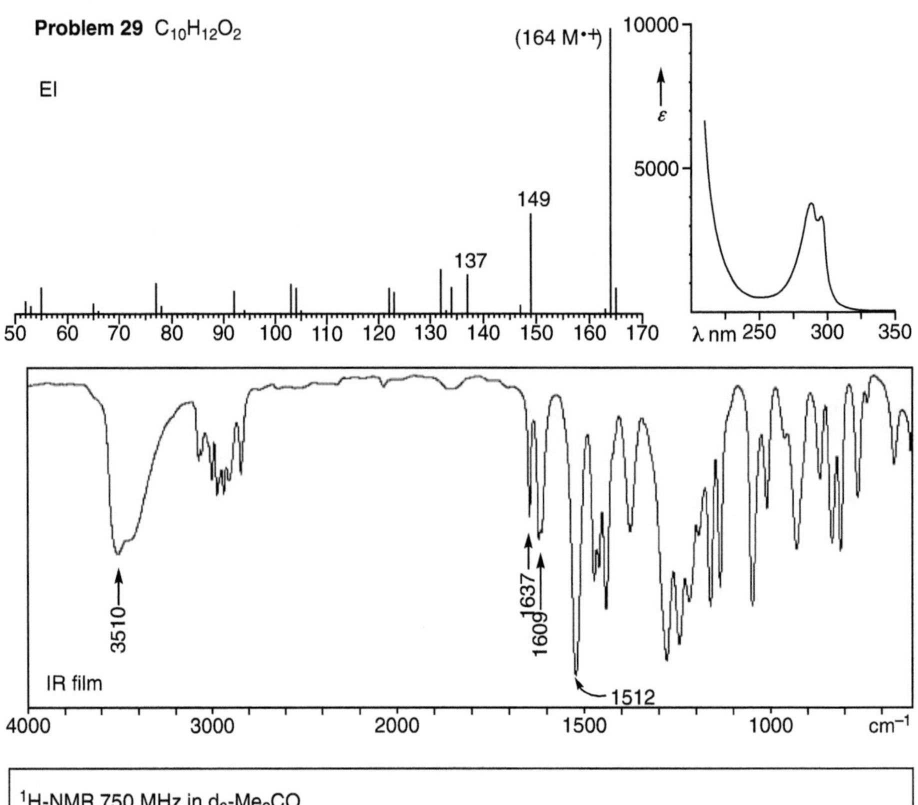
Problem 29 C10H12O2
EI
(164 M•+)
149
137
10000
ε
5000
λ nm
3510
1637
1609
1512
IR film
cm−1

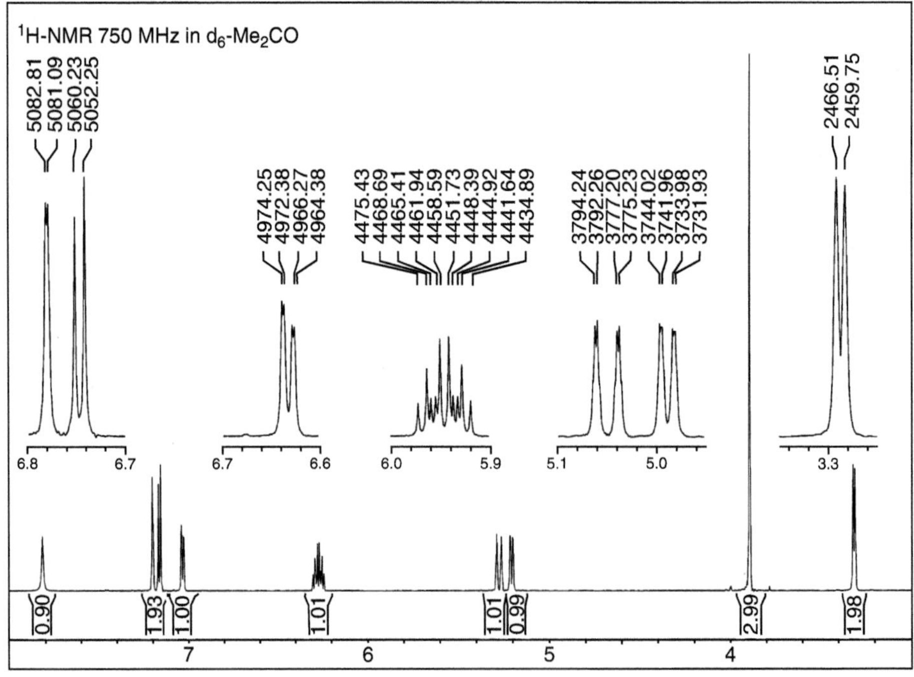
1H-NMR 750 MHz in d6-Me2CO
5082.81
5081.09
5060.23
5052.25
4974.25
4972.38
4966.27
4964.38
4475.43
4468.69
4465.41
4461.94
4458.59
4451.73
4448.39
4444.92
4441.64
4434.89
3794.24
3792.26
3777.20
3775.23
3744.02
3741.96
3733.98
3731.93
2466.51
2459.75
0.90
1.93
1.00
1.01
1.01
0.99
2.99
1.98

Problem 29 continued

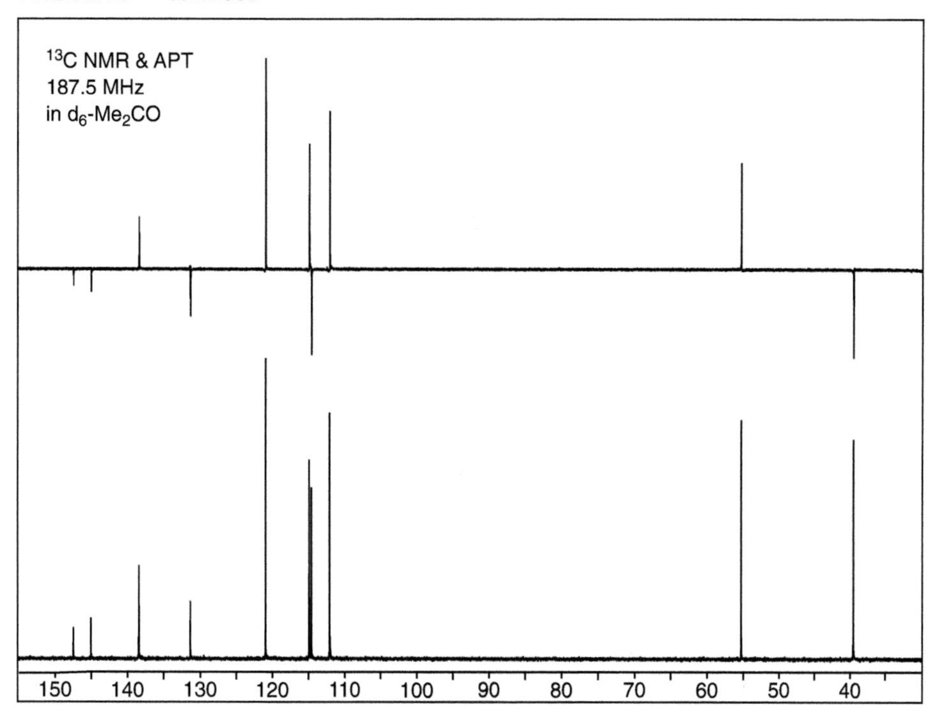

1D-NOEs

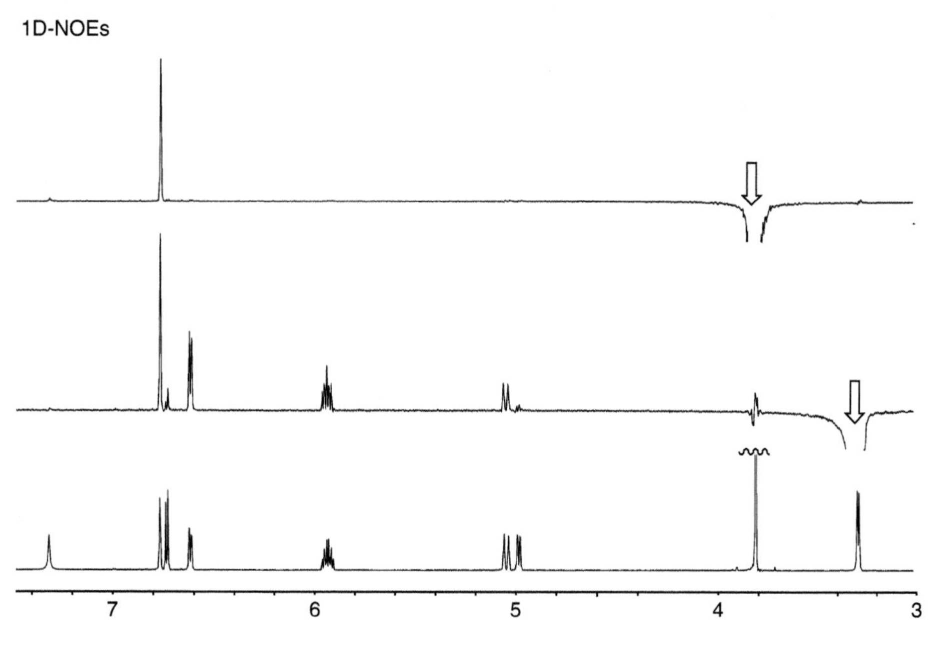

Problem 29 continued

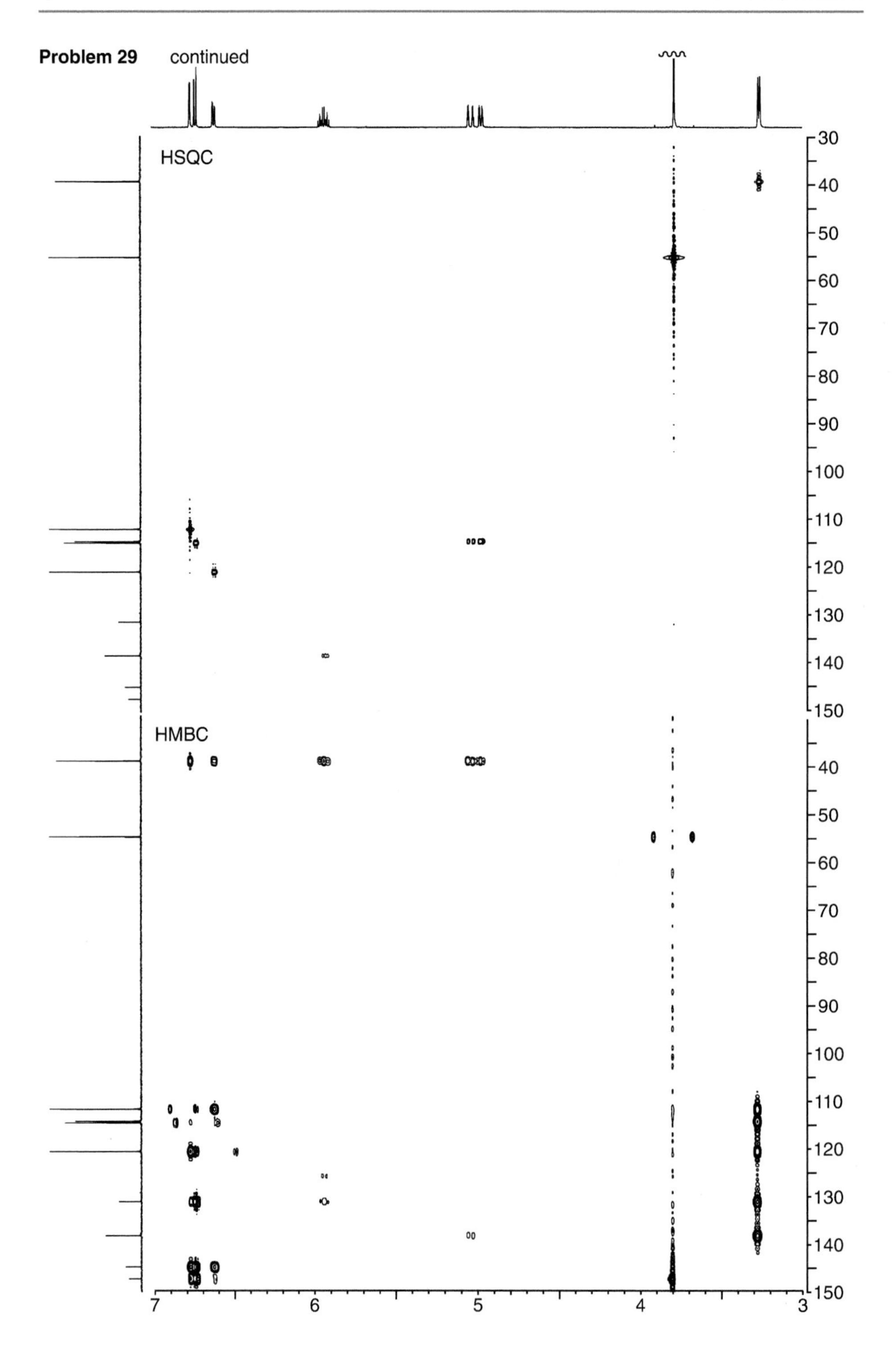

Problem 30 $C_{17}H_{21}NO$

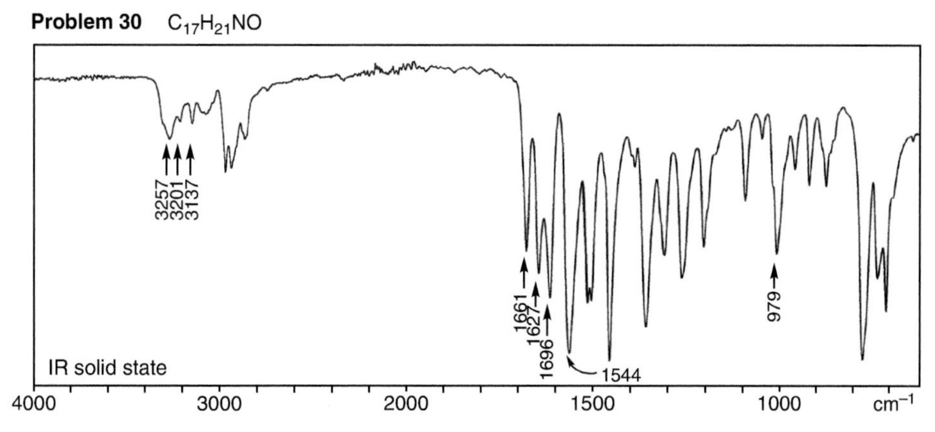

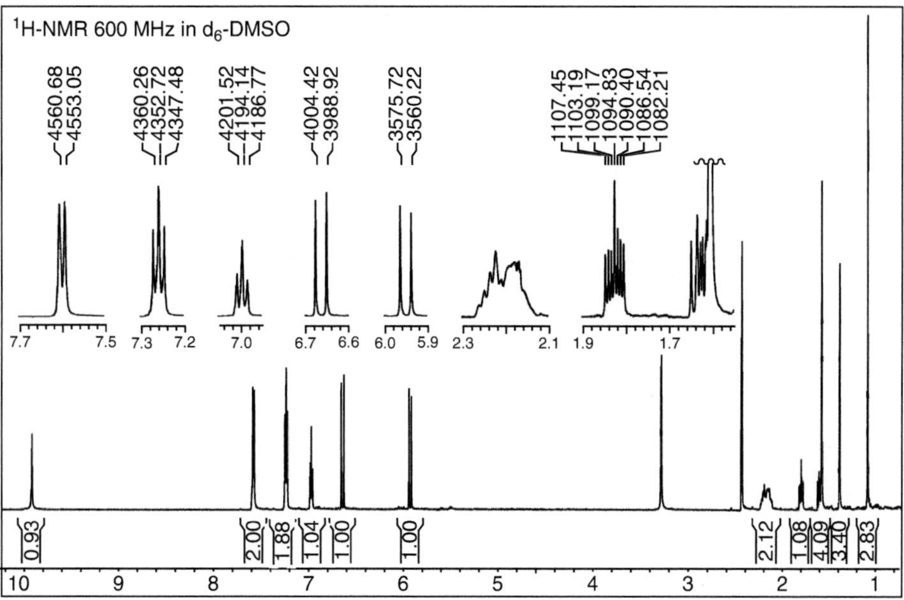

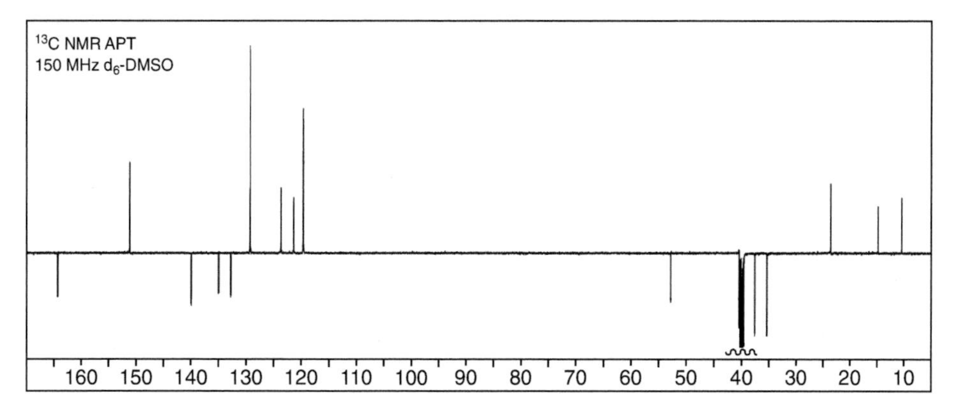

Problem 30 continued

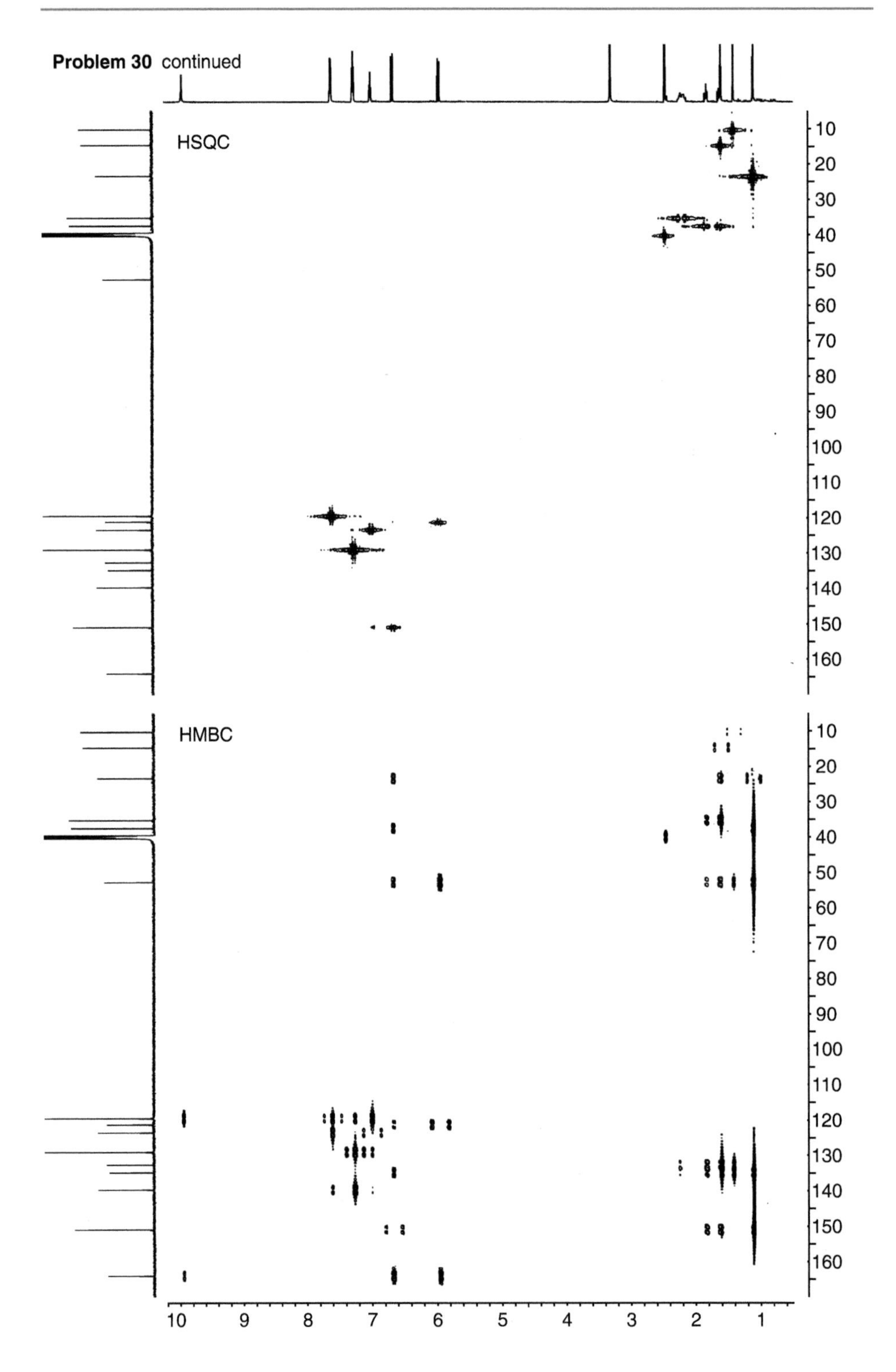

Problem 30 continued

1D-NOEs

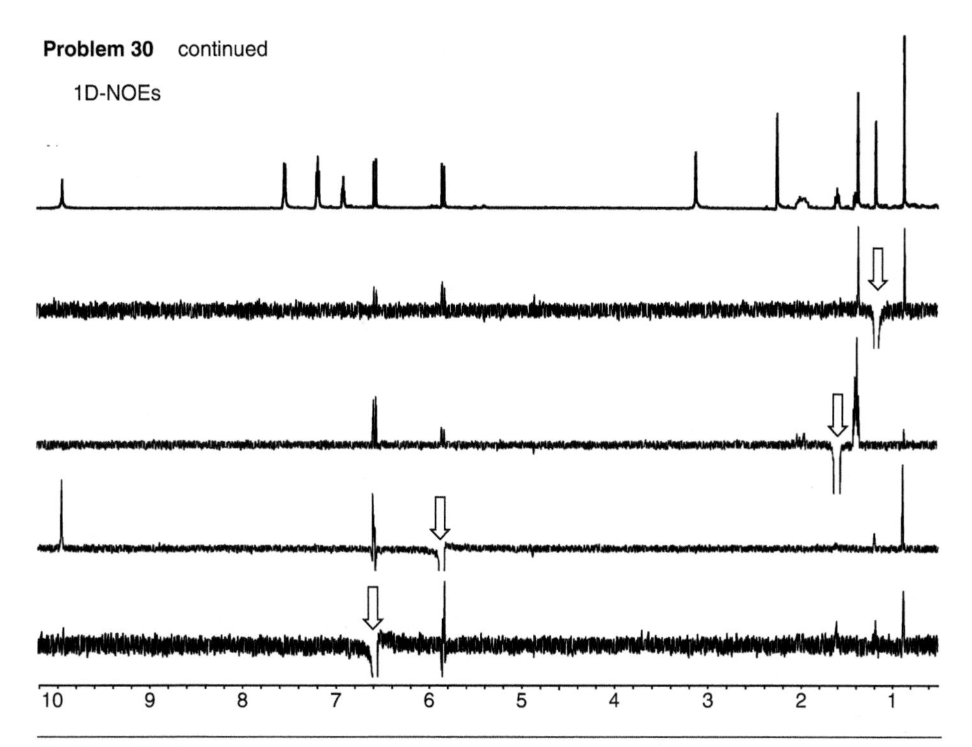

Problem 31 $C_{10}H_{16}O_3$

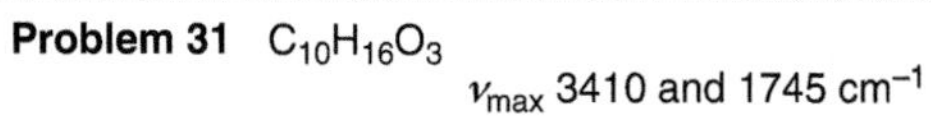

ν_{max} 3410 and 1745 cm^{-1}

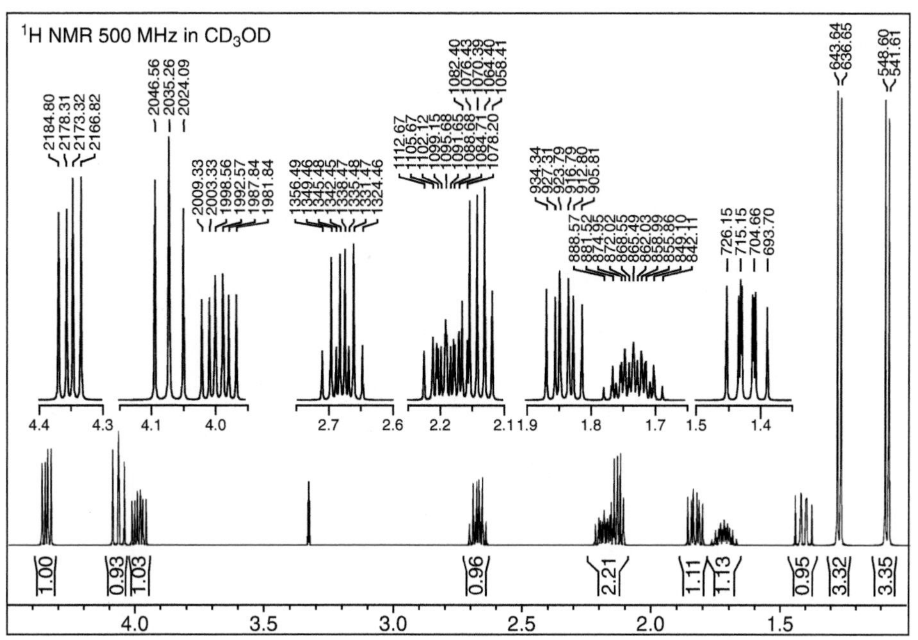

Problem 31 continued

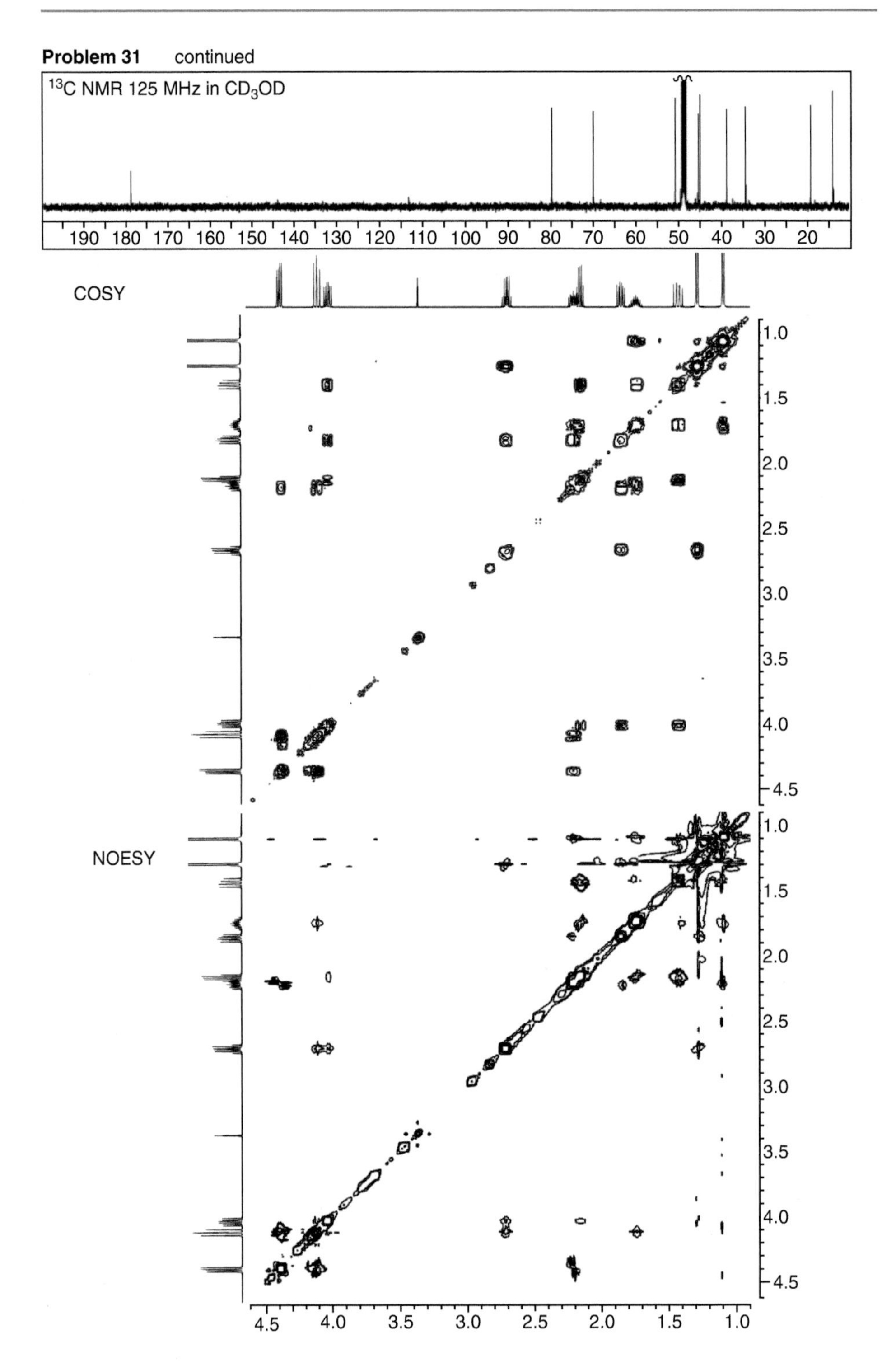

Problem 31 continued

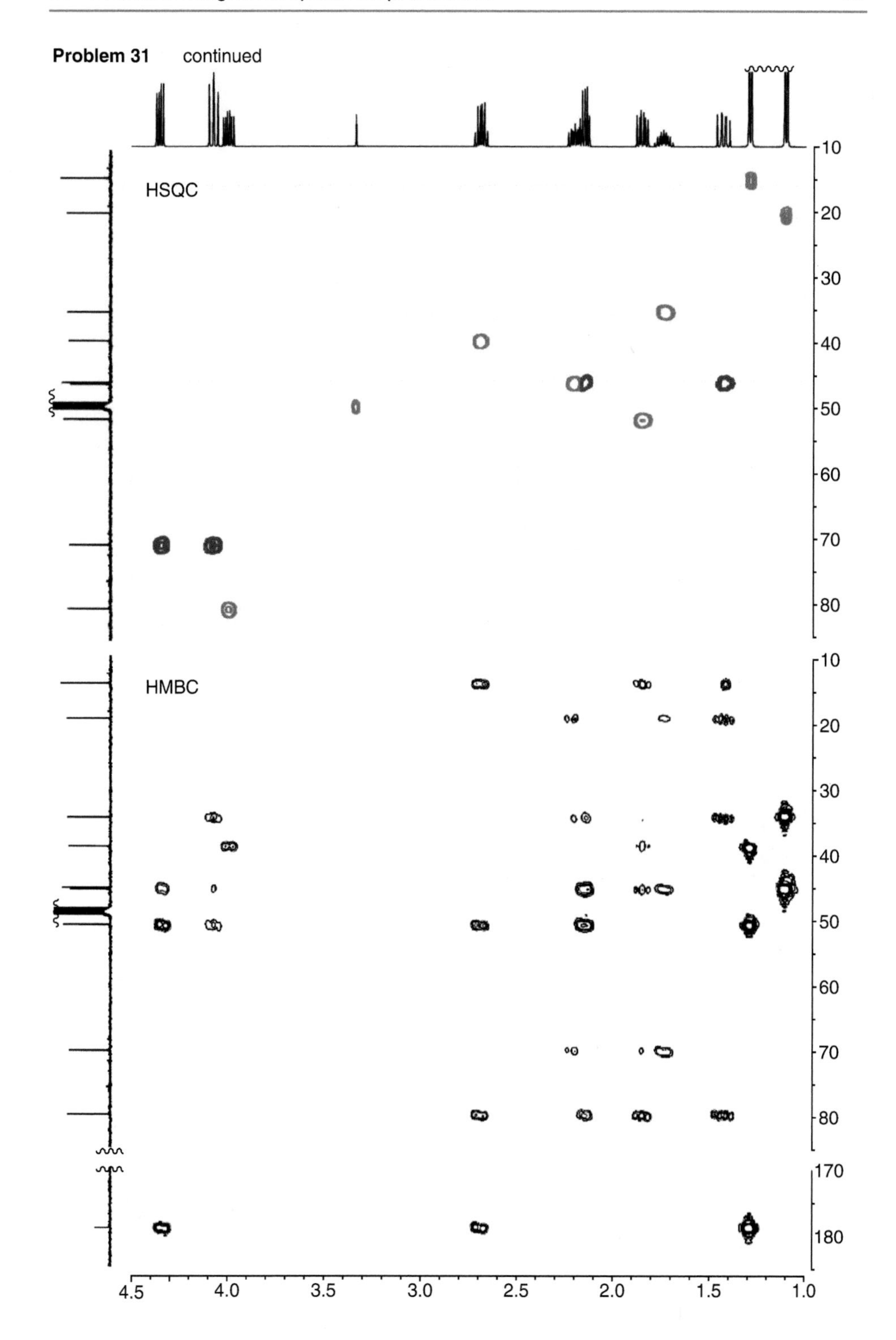

Problem 32 $C_7H_{10}O_5$

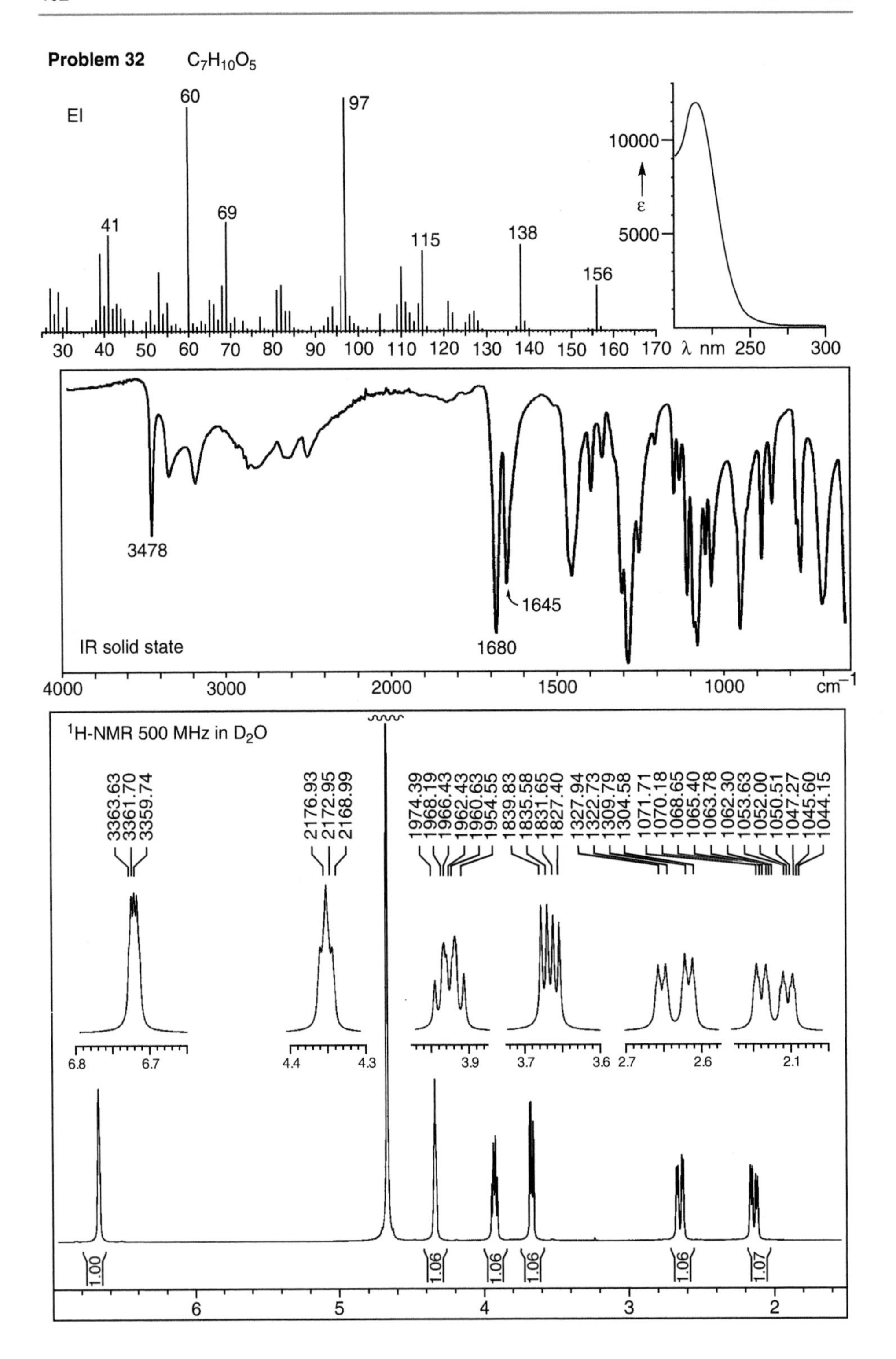

Problem 32 continued

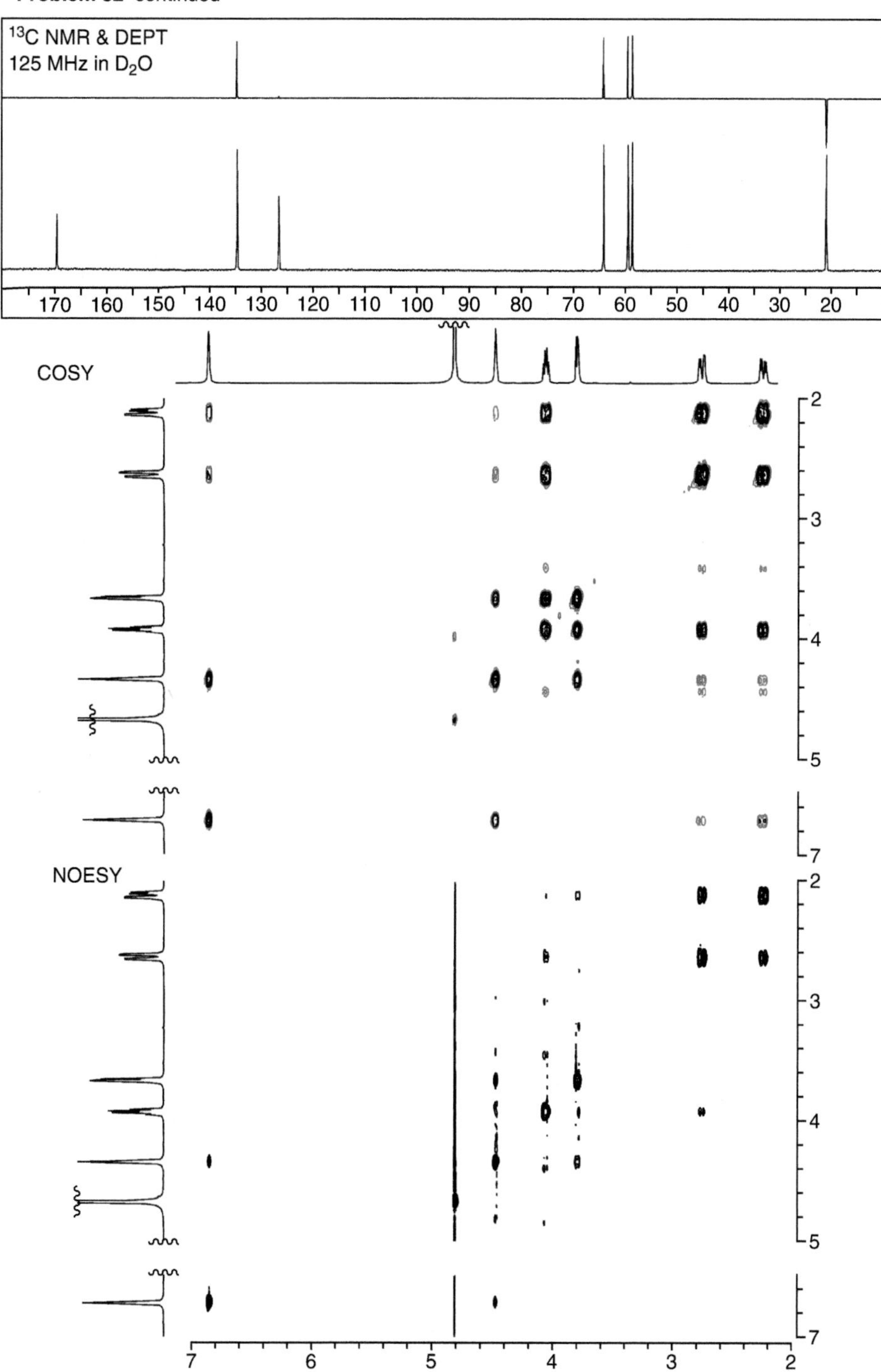

Problem 32 continued

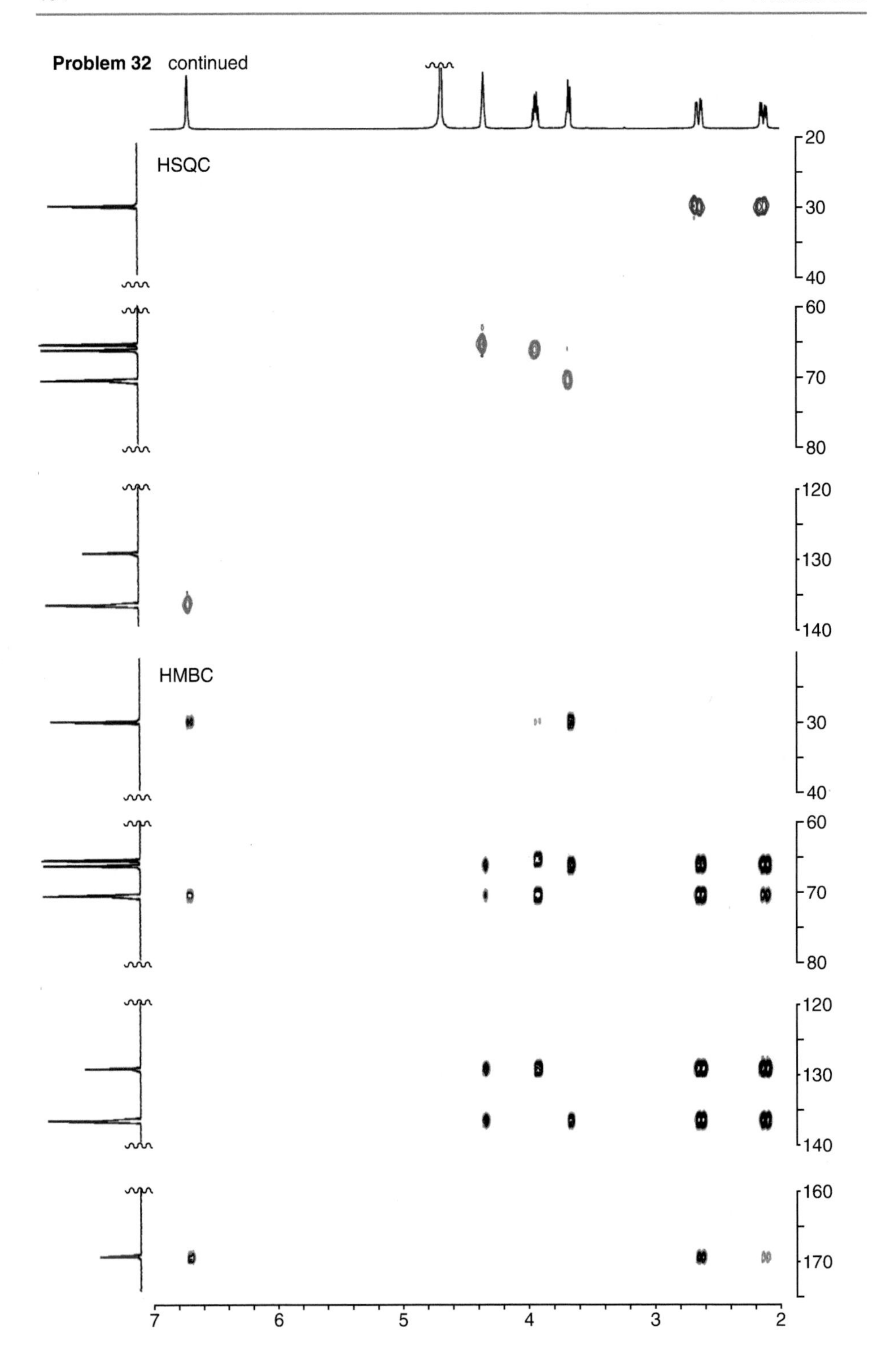

Problem 33 $C_{15}H_{18}O_3$

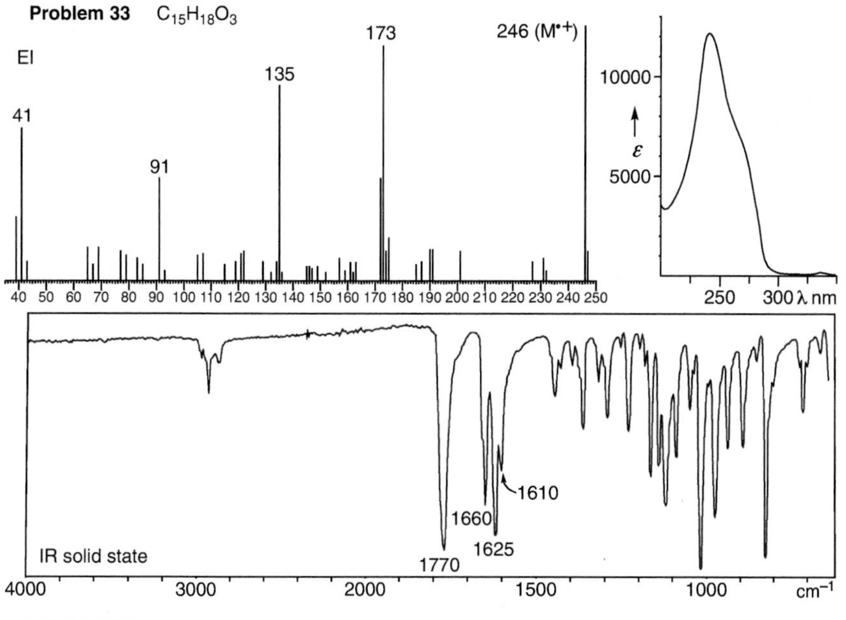

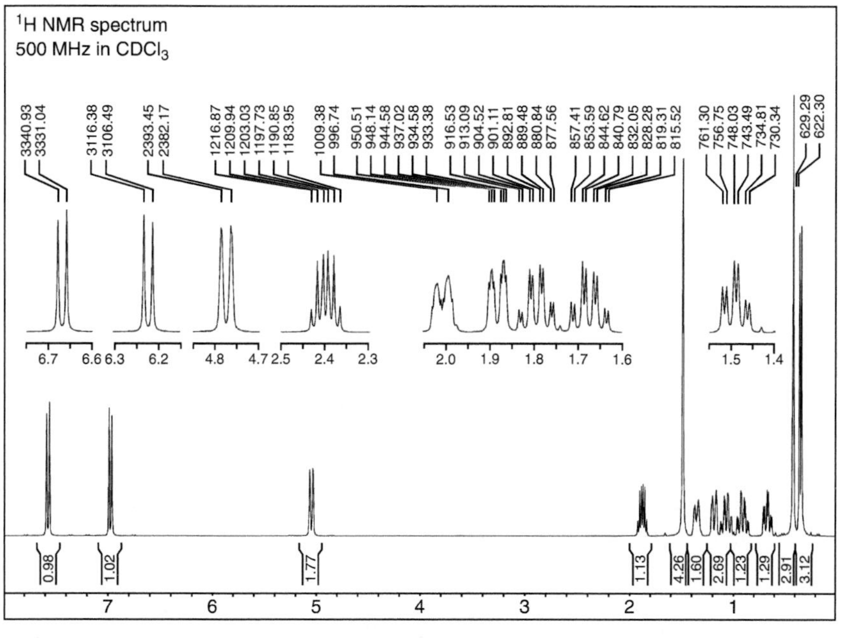

Problem 33 continued

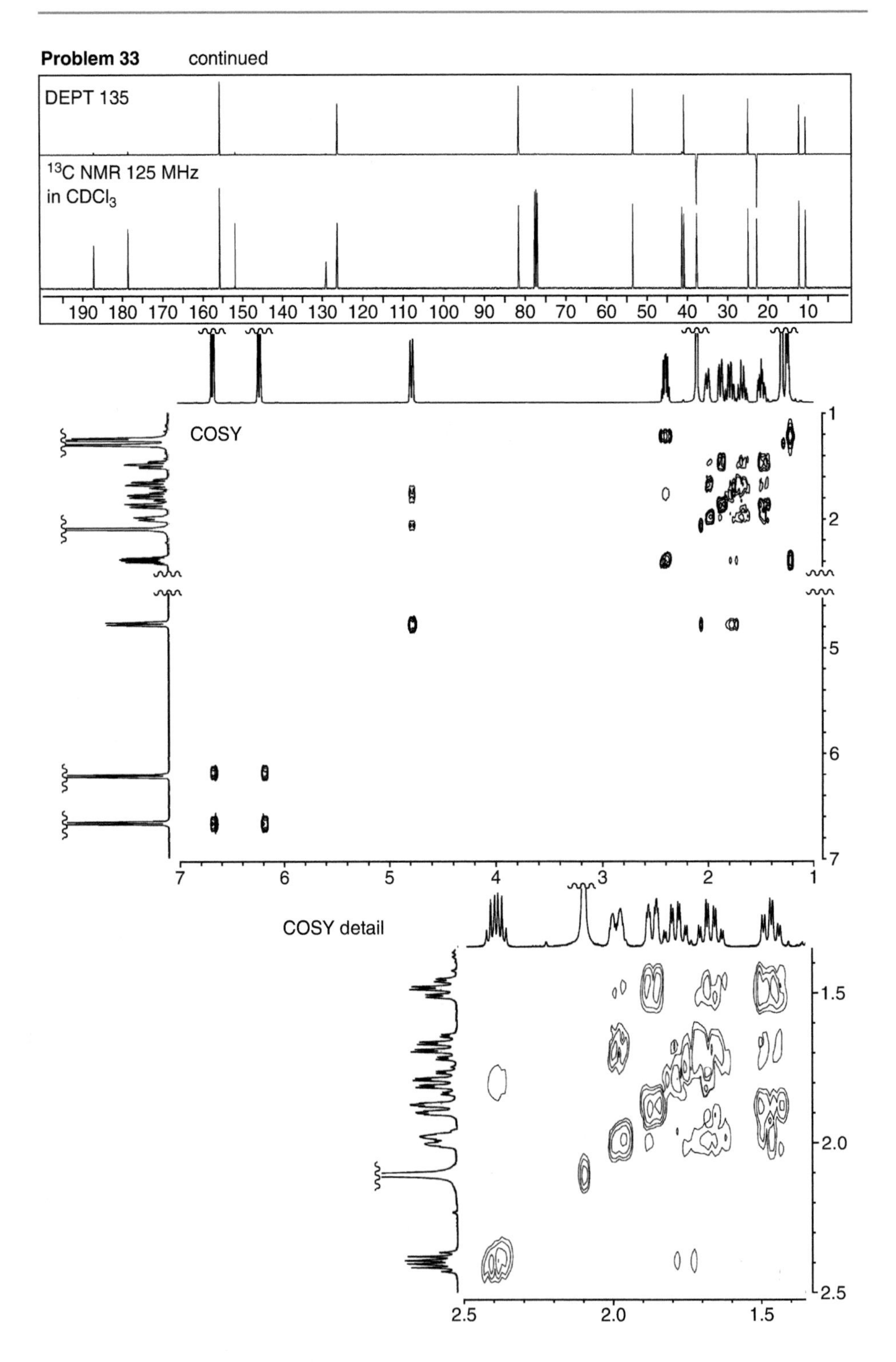

Problem 33 continued

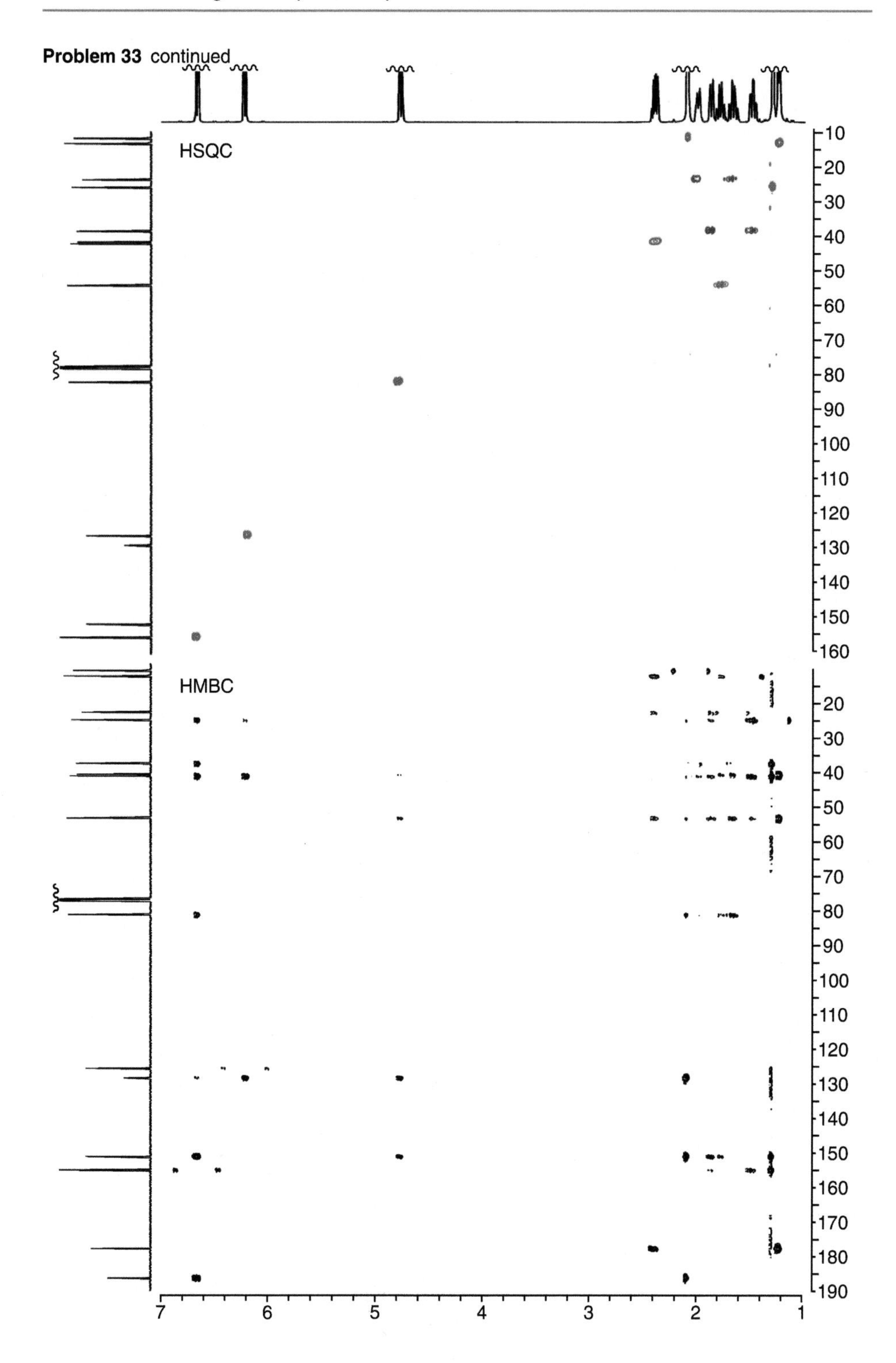

Problem 33 continued

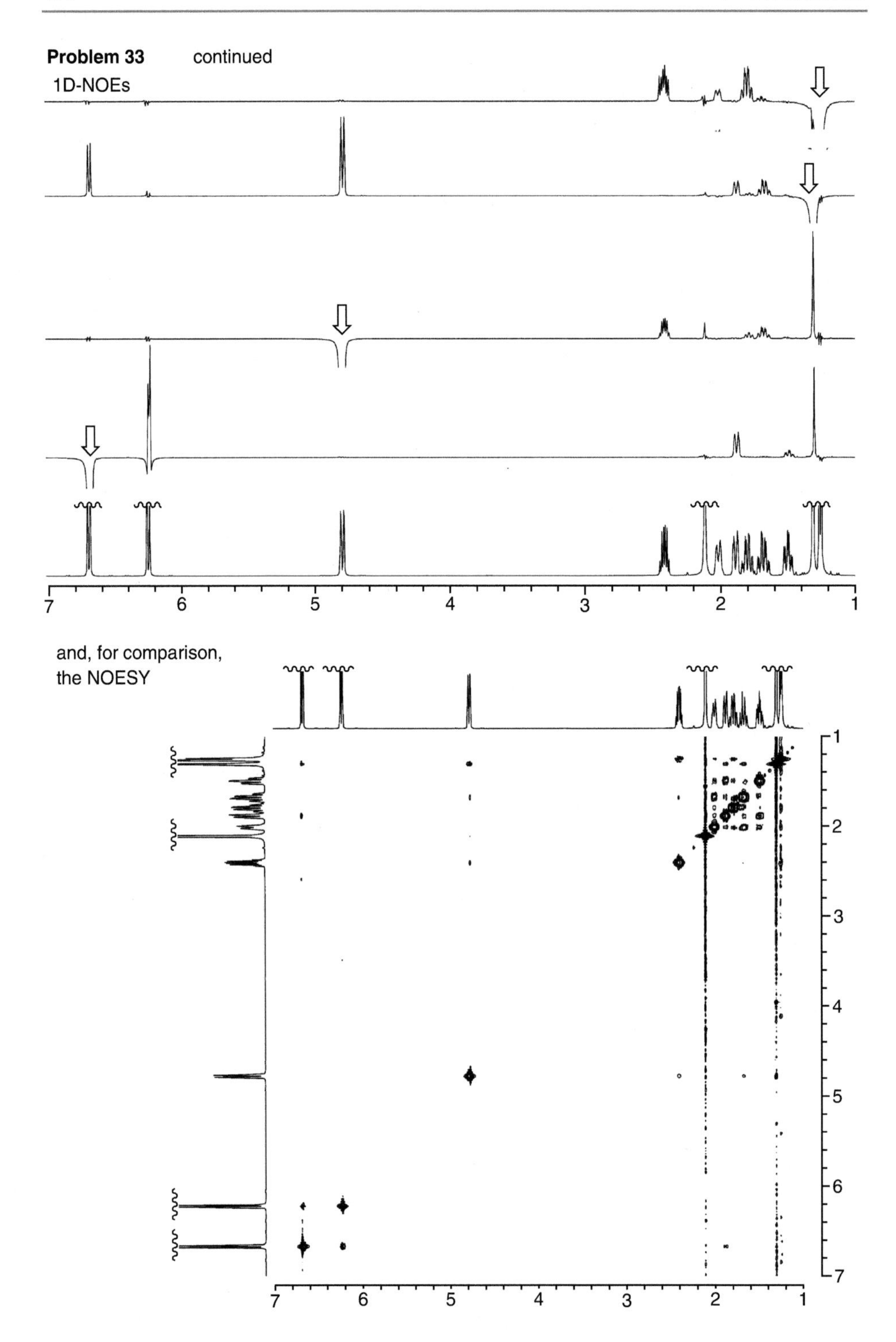

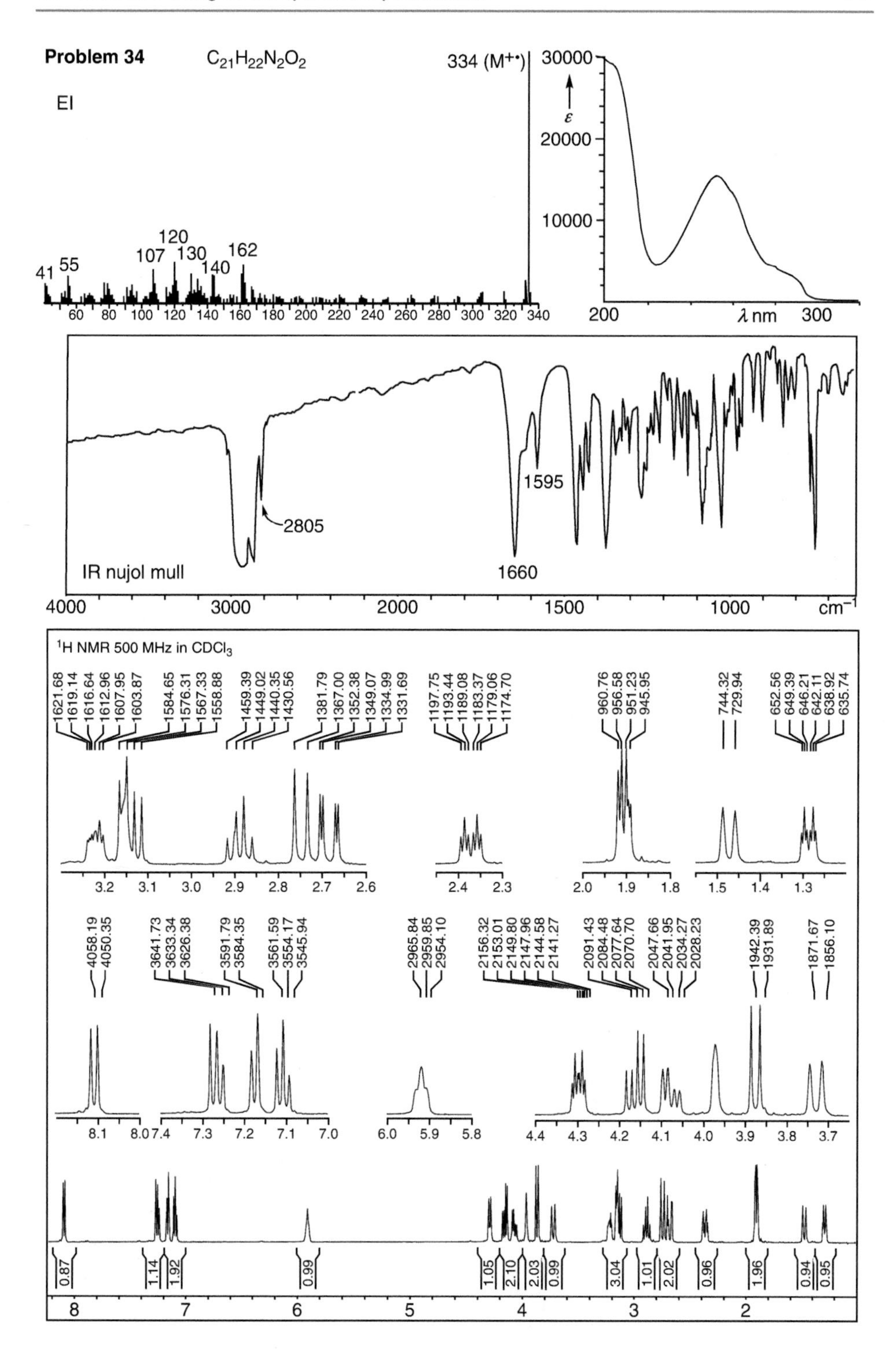

Problem 34
$C_{21}H_{22}N_2O_2$
EI
334 (M+•)
41 55 107 120 130 140 162
60 80 100 120 140 160 180 200 220 240 260 280 300 320 340
30000
ε
20000
10000
200
λ nm
300
IR nujol mull
2805
1595
1660
4000 3000 2000 1500 1000 cm−1
1H NMR 500 MHz in CDCl3
1621.68 1619.14 1616.64 1612.96 1607.95 1603.87 1584.65 1576.31 1567.33 1558.88
1459.39 1449.02 1440.35 1430.56
1381.79 1367.00 1352.38 1349.07 1334.99 1331.69
1197.75 1193.44 1189.08 1183.37 1179.06 1174.70
960.76 956.58 951.23 945.95
744.32 729.94
652.56 649.39 646.21 642.11 638.92 635.74
3.2 3.1 3.0 2.9 2.8 2.7 2.6
2.4 2.3
2.0 1.9 1.8
1.5 1.4 1.3
4058.19 4050.35
3641.73 3633.34 3626.38
3591.79 3584.35
3561.59 3554.17 3545.94
2965.84 2959.85 2954.10
2156.32 2153.01 2149.80 2147.96 2144.58 2141.27
2091.43 2084.48 2077.64 2070.70
2047.66 2041.95 2034.27 2028.23
1942.39 1931.89
1871.67 1856.10
8.1 8.0
7.4 7.3 7.2 7.1 7.0
6.0 5.9 5.8
4.4 4.3 4.2 4.1 4.0 3.9 3.8 3.7
0.87 1.14 1.92 0.99 1.05 2.10 2.03 0.99 3.04 1.01 2.02 0.96 1.96 0.94 0.95
8 7 6 5 4 3 2

Problem 34 continued

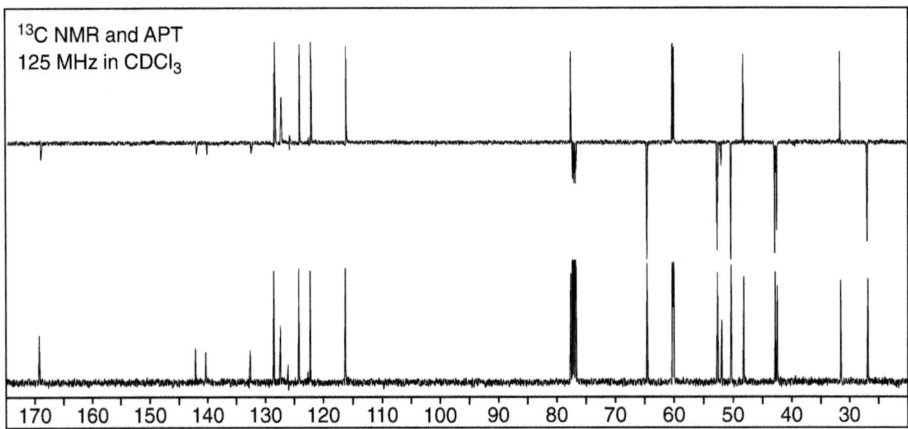

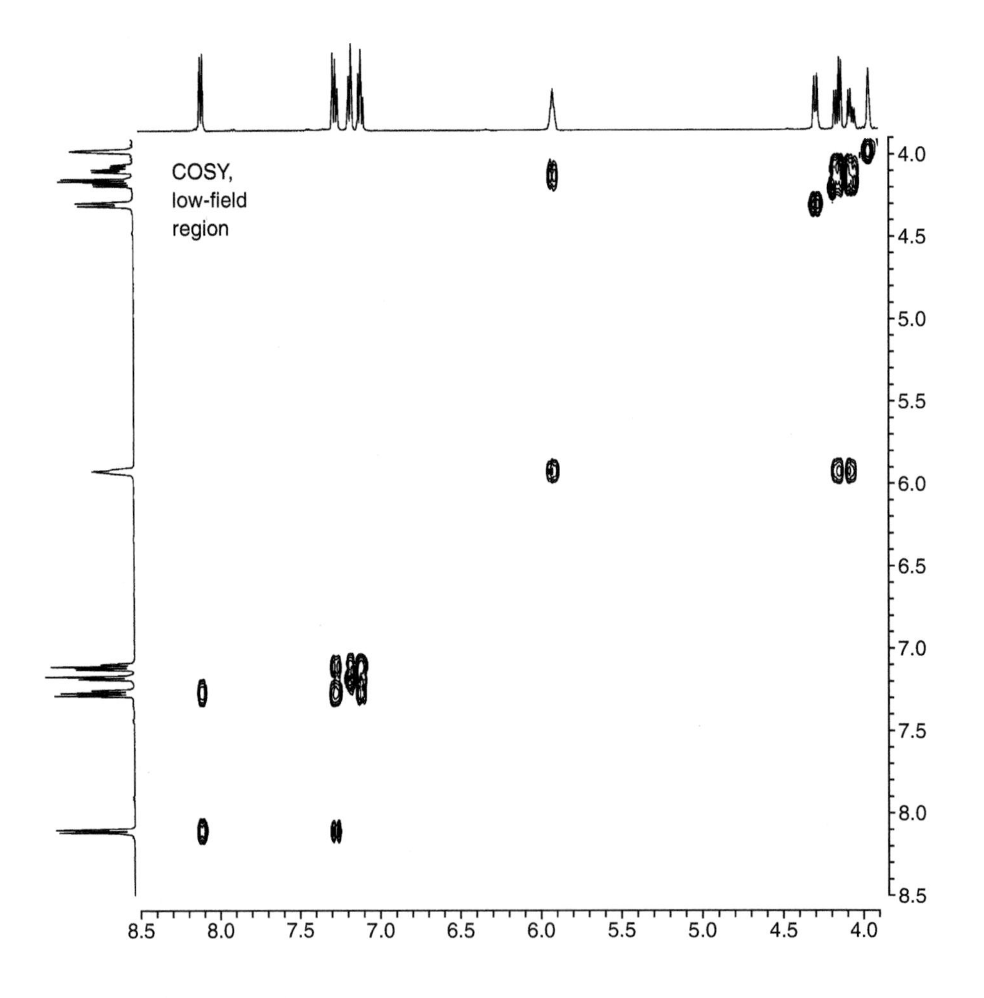

Problem 34 continued

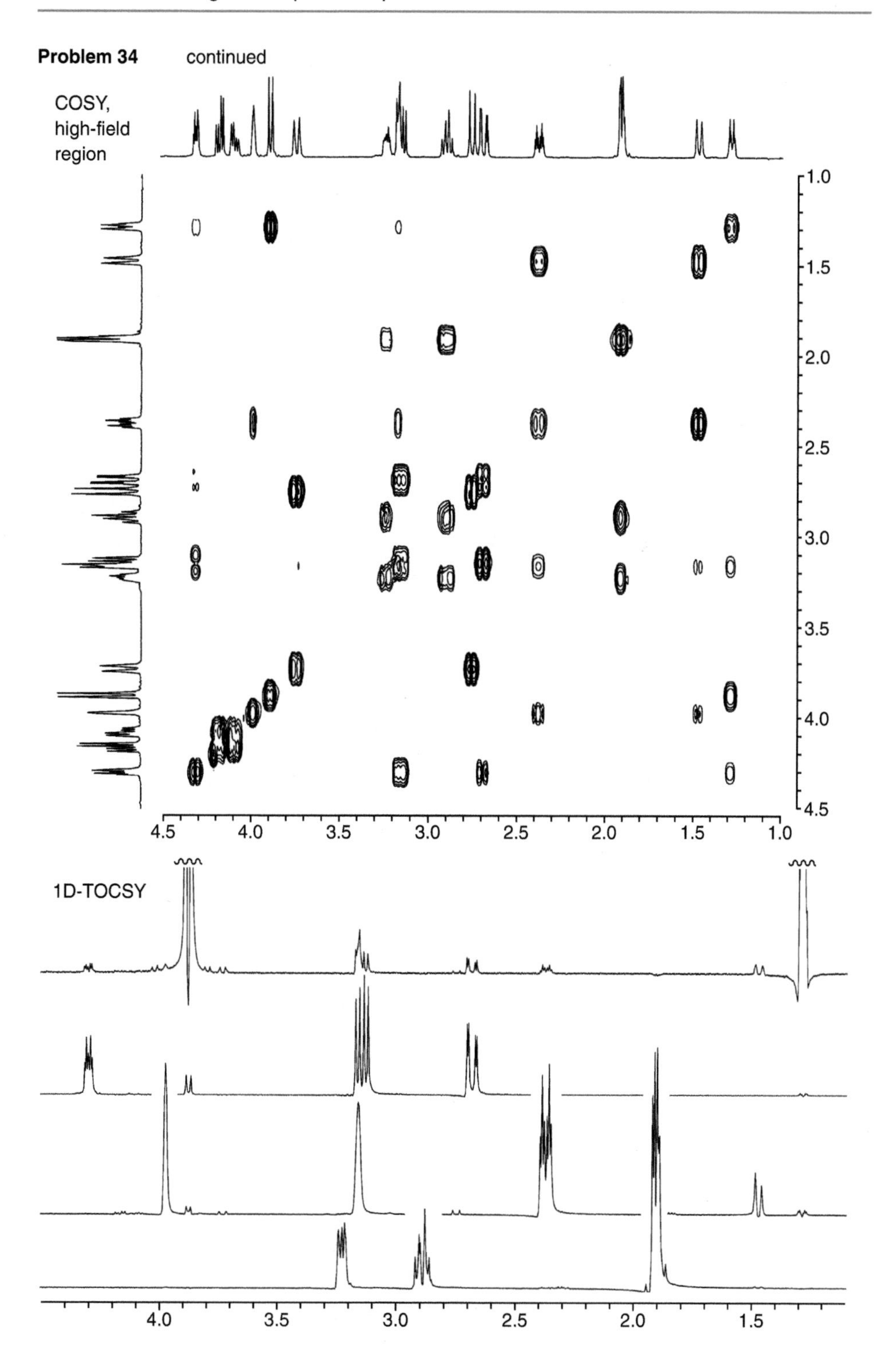

Problem 34 continued

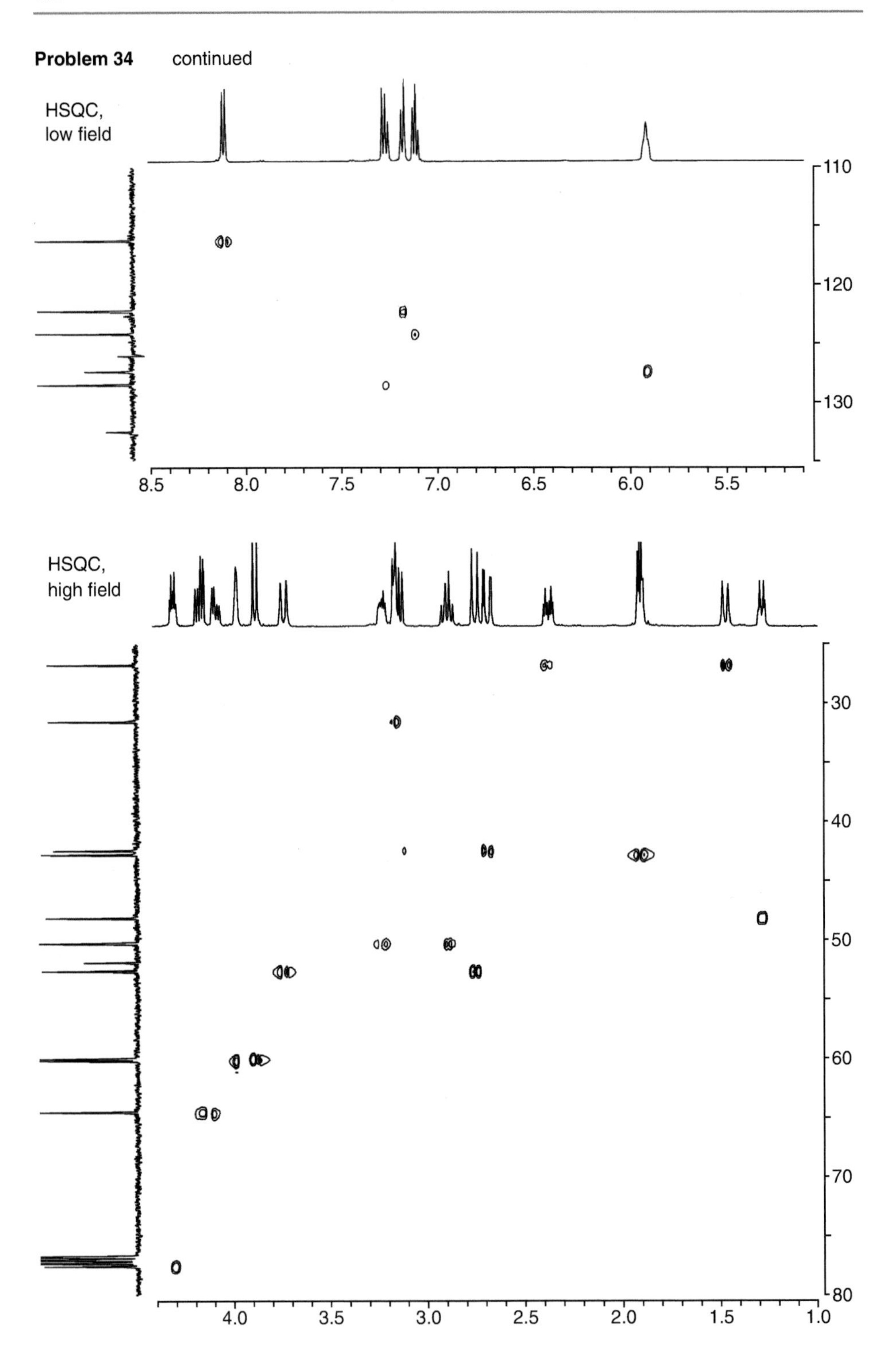

Problem 34 continued

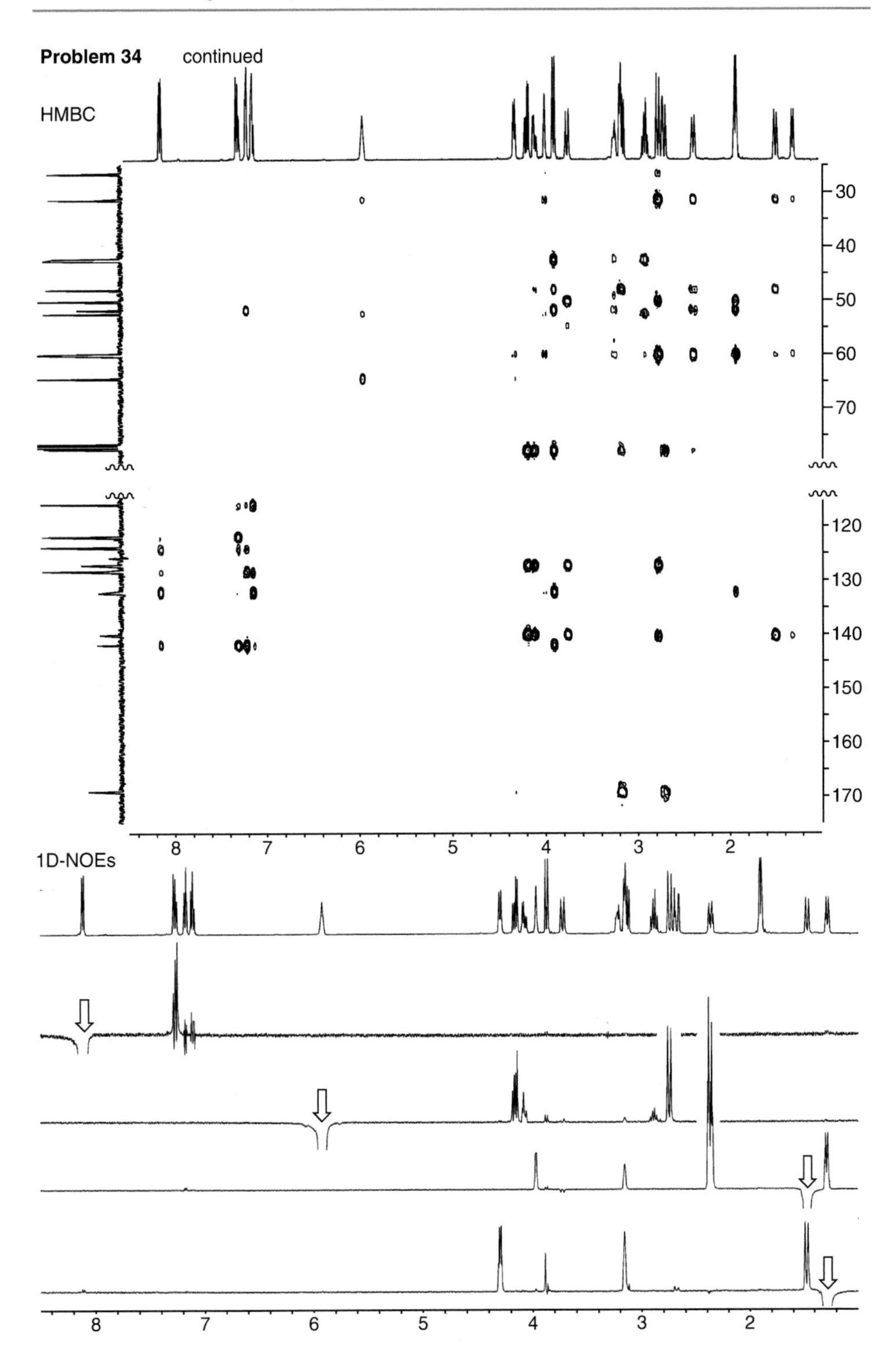

Problem 34 continued

1,1-ADEQUATE

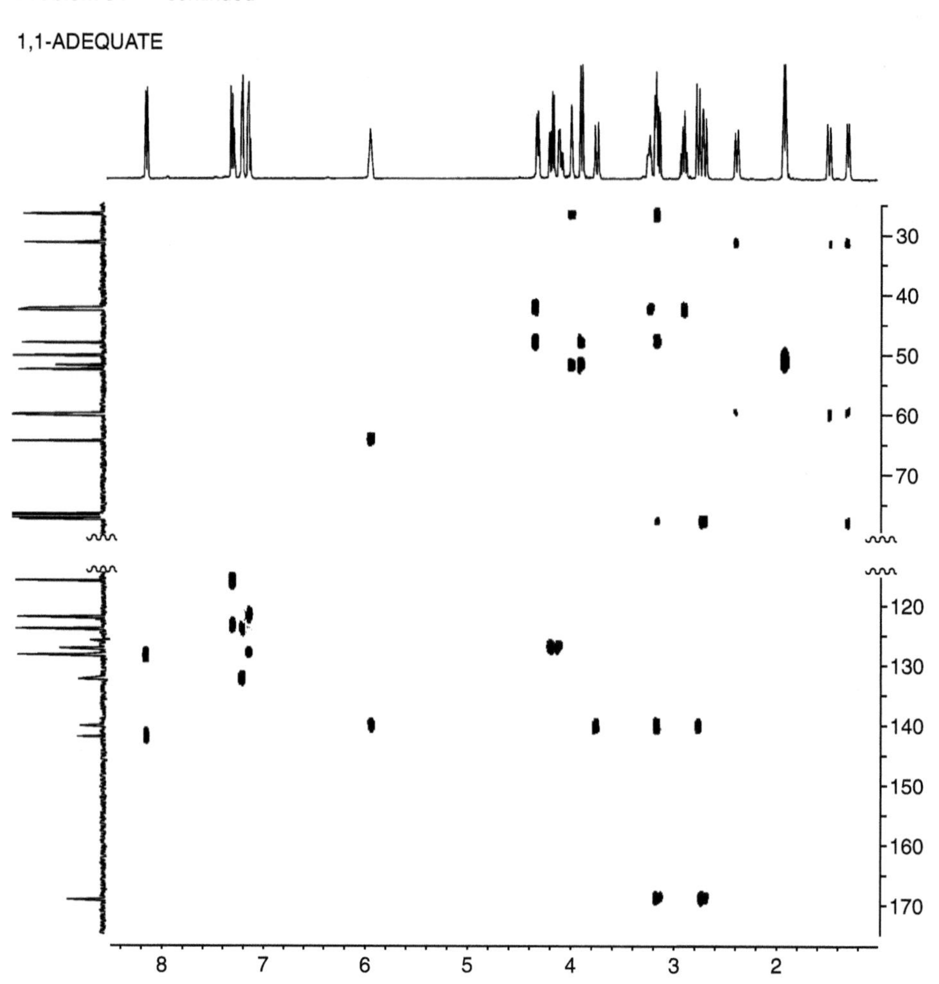

Index

I. Fleming, D. Williams, *Spectroscopic Methods in Organic Chemistry*,
https://doi.org/10.1007/978-3-030-18252-6

B

I

J

K

L